INTERNATIONAL CODE OF BOTANICAL NOMENCLATURE

(VIENNA CODE)

2006

Regnum Vegetabile

a series of publications for the use of plant taxonomists published under
the auspices of the International Association for Plant Taxonomy

Volume 146

ISSN 0080-0694

International Code
of
Botanical Nomenclature

(Vienna Code)

adopted by the Seventeenth International Botanical Congress
Vienna, Austria, July 2005

prepared and edited by

J. McNEILL, Chairman
F. R. BARRIE, H. M. BURDET, V. DEMOULIN,
D. L. HAWKSWORTH, K. MARHOLD, D. H. NICOLSON,
J. PRADO, P. C. SILVA, J. E. SKOG, J. H. WIERSEMA, Members
N. J. TURLAND, Secretary
of the Editorial Committee

2006

A.R.G. Gantner Verlag KG

ISBN 3-906166-48-1

Published for IAPT by

A.R.G. Gantner Verlag, Ruggell, Liechtenstein

Distributed by

KOELTZ SCIENTIFIC BOOKS
P.O.Box 1360, D 61453 Koenigstein / Germany
E-mail: koeltz@t-online.de
Internet: https://www.koeltz.com

CONTENTS

Contents

PREFACE

Unambiguous names for organisms are essential for effective scientific communication; names can only be unambiguous if there are internationally accepted rules governing their formation and use. The rules that govern scientific naming in botany (including phycology and mycology) are revised at Nomenclature Section meetings at successive International Botanical Congresses. The present edition of the *International code of botan_ical nomenclature* embodies the decisions of the XVII International Botanical Congress held in Vienna in 2005 and supersedes the *Saint Louis Code*, published six years ago subsequent to the XVI International Botanical Congress in St Louis, Missouri, U.S.A. It is written entirely in (British) English. The *St Louis Code* was translated into Chinese, French, Japanese, Portuguese, Russian, Slovak, and Spanish; it is therefore anticipated that the *Vienna Code,* too, will become available in several languages in due course.

One of the reasons invoked for the choice of Vienna as the site of the seventeenth Congress, was that the second International Botanical Congress had been held there exactly 100 years earlier. It was that Congress that accepted the first internationally developed rules governing the naming of plants, *Règles internationales de la Nomenclature botanique adoptées par le Congrès International de Botanique de Vienne 1905 / International rules of Botanical Nomenclature ... / Internationale Regeln der Botanischen Nomenclatur ...* – or simply the *Vienna Rules*, thus obviating confusion with this *Vienna Code*, and not requiring this Code to bear any qualifying numeral.

The *Vienna Code* does not differ substantially in overall presentation and arrangement from the *St Louis Code*, and the numbering of Articles remains the same, although there have been a few additions to, and modifications of, paragraphs, Recommendations, and Examples, often involving changes in their numbering. One small change has also been made in the numbering of the Appendices to make this more logical: the former App. IIIA, dealing with conserved names of genera is now simply App. III, and the former App. IIIB, with names of species, becomes App. IV. With App. IIA & IIB continuing to contain the two sorts of conserved family names, there is now a logical sequence for the lists of conserved names: II for families, III for genera, and IV for species. The subsequent Appendices increase in number accordingly, so that names rejected "utique" under Art. 56 form App. V, and suppressed works, App. VI. The *St*

Louis Code omitted the "Important Dates in the Code" that had appeared in the *Berlin* & *Tokyo Codes*; this has been restored in the *Vienna Code*, from a draft by D. L. Hawksworth.

In overall presentation the most notable feature, however, is the inclusion for the first time of a Glossary, which appears as Appendix VII. This was requested by the Vienna Congress and it was made clear that it should be an integral part of the Code with all the authority thereof. This has meant that the Glossary is very tightly linked to the wording of the Code, and only nomenclatural terms defined in the Code can be included. A few other terms in more general use and not defined in the Code (e.g. description, position, rank) but with distinctive application in the Code have, however, also been included; they are distinguished by the statement "not defined" followed by an explanation of the way in which, in the opinion of the Editorial Committee, they are applied in the Code. For the preparation of the Glossary, the Committee is particularly grateful to P. C. Silva, who initiated the project and who prepared the first draft for consideration by the Editorial Committee and who has worked over several subsequent ones, ensuring precision and consistency.

The text of the Code uses three different sizes of print, the Recommendations and Notes being set in smaller type than the Articles, and the Examples and footnotes in smaller type than the Recommendations and Notes. The type sizes reflect the distinction between the rules which are mandatory (Articles), complementary information or advice (Notes and Recommendations), and explanatory material (Examples and footnotes). A Note has binding effect but does not introduce any new provision or concept; rather, it explains something that may not at first be readily apparent but is covered explicitly or implicitly elsewhere in the *Code.* Some Examples, which were deliberately agreed by a Nomenclature Section, contain material which is not fully, or not explicitly, covered in the rules. Such "voted examples" are prefixed by an asterisk (*). If, by a change of the corresponding provision in a subsequent edition of the *Code,* a "voted example" becomes fully covered, the asterisk is removed.

As in the previous edition, scientific names under the jurisdiction of the *Code,* irrespective of rank, are consistently printed in *italic type.* The *Code* sets no binding standard in this respect, as typography is a matter of editorial style and tradition not of nomenclature. Nevertheless, editors and authors, in the interest of international uniformity, may wish to consider adhering to the practice exemplified by the *Code,* which has been well received in general and is followed in a number of botanical and mycological journals. To set off scientific plant names even better, the abandonment in the *Code* of italics for technical terms and other words in Latin, traditional but inconsistent in early editions, has been maintained.

Like its forerunners, the Editorial Committee has tried hard to achieve uniformity in bibliographic style and formal presentation – a sound educational exercise for its members, and a worthwhile goal because the *Code* is considered a model to follow by many of its users. The titles of books in bibliographic citations are abbreviated in conformity with *Taxonomic literature,* ed. 2, by Stafleu & Cowan (1976-1988; with Supplements 1-6 by Stafleu & Mennega, 1992-2000), or by analogy, but with capital initial letters. For journal titles, the abbreviations follow the *Botanico-periodicum-huntianum,* ed. 2 (2004).

Author citations of scientific names appearing in the *Code* are standardized in conformity with *Authors of plant names,* by Brummitt & Powell (1992), as mentioned in Rec. 46A Note 1; these are also adopted and updated by the International Plant Names Index, and may be accessed at http://www.ipni.org/index.html. One may note that the *Code* has no tradition of recording the ascription of names to pre-1753 authors by the validating author, although such "pre-ex" author citations are permitted (see Art. 46 Ex. 18).

Like its immediate predecessor, the Vienna Congress was conservative in nomenclatural matters in comparison with some earlier Congresses. Relatively few changes were accepted, but a small number of significant ones and many useful clarifications and improvements of the *Code,* both in wording and substance, were adopted. Here we only draw attention to changes of some note. A full report on the Section's decisions has been published elsewhere (McNeill & al. in Taxon 54: 1057-1064. 2006).

Perhaps the most important single decision incorporated into the *Vienna Code* was to deal with what many have recognized as a bomb waiting to explode, the publication status of theses submitted for a higher degree. In most, but certainly not all, countries, such theses have not traditionally been considered media for effective publication under the Code, and degree candidates have normally gone on to publish in journals or monograph series the taxonomic novelties and nomenclatural actions contained in their theses. However, as soon as theses ceased to be typewritten with carbon copies, or as soon as they were made available commercially by photo-reproduction, no provision existed in the Code to treat them as other than effectively published. Because of the fact that in some other countries, notably the Netherlands and some Scandinavian countries, theses, to be accepted, must be produced in substantial numbers and are intended as effectively published media, it has not hitherto been possible to resolve the issue. Nevertheless, despite the lack of any justification in the Code for treating most theses produced over the past 40 years as other than effectively published, the practice not to do so has persisted outside of a few countries. In consequence, the Section took the unusual step of accepting a retroactive change in the Code by deciding that no independent non-serial publication stated to be a thesis submitted for a higher degree on or after 1 January 1953 would be considered an effectively published work without a statement to that effect or other internal evidence. The Editorial Committee was instructed to provide examples of internal evidence that would best reflect current practice. The new Art. 30 Note 1 refers to the presence of an International Standard Book Number (ISBN) or a statement of the name of the printer, publisher, or distributor in the original printed version as such internal evidence.

Several proposals on criteria for valid publication of names were considered in Vienna. It was made explicit that names be composed only of letters of the Latin alphabet except as otherwise provided in the Code, and some clarification was accepted on what constitutes a description or diagnosis: statements on usage of plants, on cultural and cultivation features, and on geographical origin or geological age are not acceptable, nor is the mere mention of features but not their expression. Conceptually more significant, however, was the decision to make provision for binding decisions on whether or not a descriptive statement meets the requirement of Art. 32.1(d) for a "description or diagnosis" – the so-called "nomina subnuda" situation. This introduces into the Code an entirely new concept in botanical nomenclature, although one that is well-established in zoological nomenclature, namely rulings on interpretation of the Code itself. Since the Sydney Congress of 1981, there has been provision for rulings on whether or not two names or epithets are likely to be confused, and, of course, in the conservation and rejection of names, judgement must be made as to whether or not there will be "disadvantageous nomenclatural change", but these do not involve interpretation of the *Code* itself. The procedure established is the same as that for judgement on whether names or epithets are sufficiently alike to be confused (Art. 53.5) and the General Committee will probably need to establish mechanisms to ensure that proposed rulings coming from the different Permanent Committees are reasonably consistent in their interpretation of Art. 32.1(d).

Article 33, dealing with new combinations, although improved significantly at the previous Congress, was again the subject of clarification, principally in making a separation in the paragraphs of the Article between the situation before 1 January 1953 and the more precise requirements from that date onward. In addition it was made clearer that, prior to 1 January 1953, when the epithet of a previously and validly published name that applies to the same taxon is adopted, the "presumed new combination" is validly published if there is any indication at all of a basionym, however indirect, but if there is no such indication, the new combination is only validly published if it would otherwise be a validly published name. By contrast, it was accepted in Vienna that on or after 1 January 1953 a claimed new combination or avowed substitute, that lacks the full information required regarding the basionym or replaced synonym is not validly published even though the name would otherwise be validly published as the name of a new taxon. Although involving the somewhat cumbersome expression "a generic name with a basionym" it has been made explicit that most of the rules on combinations apply also to such generic names.

Three important sets of changes were accepted in Vienna applying to names in particular groups of organisms, fossil plants, pleomorphic fungi, and fungi that had previously been named under the ICZN, respectively.

That for fossil plants was a reversal of one component of the rules on morphotaxa introduced in the *St Louis Code*. At the St Louis Congress it was argued (and accepted) that all fossil taxa should be treated as morphotaxa. This has not, however, been considered appropriate by the majority of palaeobotanists and a distinction between a morphotaxon and a regular fossil taxon is now established. Whereas a morphotaxon comprises only the one part, life-history stage, or preservational state represented by the type of its name, any new fossil taxon that is described as including more than one part, life-history stage, or preservational state is not a morphotaxon. A corollary of this change is that Art. 11.7 of the *Tokyo Code* has had to be reinstated (as Art. 11.8 of the *Vienna Code*) because priority of a name of a taxon based on a non-fossil type competing with one for the same taxon based on a fossil type is no longer implicit. Opportunity has also been taken to make clear that later homonyms are illegitimate whether the type is fossil or non-fossil.

The *Code* has long provided for a dual nomenclature for fungi with a pleomorphic life history. Proposals to amend the article involved (Art. 59) in order to facilitate a single name for a fungal taxon for which the anamorph-teleomorph relationship is known were extensively debated amongst the mycologists present in Vienna who came to a consensus on one very significant change in Art. 59, through which, by using the epitype concept, a name, currently only applicable to an anamorph, may be applied in the future to the whole organism (the holomorph) – cf. Art. 59.7.

A very important change in the *Code,* as it affects certain groups of organisms now recognized as fungi, is the extension to fungi of the provision of the second sentence of Art. 45.4, previously applicable only to algae. This deals with the names of taxa originally assigned to a group not covered by the *ICBN*, but which are now considered to be either algae – or now also fungi. To be accepted as validly published under the *ICBN*, such names need only meet the requirements of the pertinent non-botanical *Code*. The particular situation that triggered the proposal was that of the *Microsporidia*, long considered protozoa and now recognized as fungi. In addition, species names in the genus *Pneumocystis* (*Archiascomycetes*), containing important human and other mammalian pathogens, none of which were validly published under the *St Louis Code* (usually because of the lack of a Latin diagnosis or description), are now also to be treated as validly published. The change may have negative effects on a few names in groups longer established as fungi such as slime moulds, labyrinthulids, and trichomycetes, at least on authorship, but the numbers and importance are considered small compared with the benefits for the microsporidians and the species of *Pneumosystis*.

In the *St Louis Code*, the previously rather ambiguous restrictions on illustrations as types of names published after the type method entered the *Code* were clarified by establishing that illustrations were permitted as types of names published before 1 January 1958, but were prohibited thereafter unless it were "impossible to preserve a specimen", a condition that many felt hard to define. Many at the Vienna Congress also felt that this "clarification" had had the effect of retroactively devalidating names published after 1957 with an illustration as type. The Congress agreed to move the date and decided that for names of microscopic algae and microfungi for which preservation of a type was technically difficult, the type might be an illustration, but that for all other organisms, names published on or after 1 January 2007 would require a specimen as type.

Stemming from the Report of the Special Committee on Suprageneric Names set up at the St Louis Congress, it was agreed that the starting date for valid publication of suprageneric names of spermatophytes, pterido-

phytes, and bryophytes (excluding those mosses already with a 1801 starting date) be 4 August 1789, the date of publication of Jussieu's *Genera
plantarum*. This restores the original basis of spermatophyte family names
in App. IIB, dating to the Montreal Congress of 1959, which had never
been included in any article of the *Code*, and which had had to be changed
in the *St Louis Code* as a result of the Tokyo Congress failing to support a
proposal similar to this one and the St Louis Congress deleting a pro-
tecting footnote. The Section also established that parenthetic author citation is not permitted at suprageneric ranks.

Full details of unavoidable changes made to Appendix IIB since the *St
Louis Code* were published in the Second Report of the Special Committee on Suprageneric Names (Turland & Watson in Taxon 54: 491-499.
2005). The amendment to Art. 18.2, new Note 1 and voted Ex. 5, accepted
at the Vienna Congress, have necessitated some additional changes since
that Committee's report and it is appropriate to detail these here. When, in
a work, taxa ranked as orders are subdivided into families, the names of
those taxa must be treated at the stated ranks and the orders cannot be
treated as having been published as families under Art. 18.2. The orders
and families in Berchtold & Presl's *O přirozenosti rostlin* (1820) were already treated at the stated ranks, although *Ambrosiaceae* and *Asteraceae,*
previously listed from Martinov's *Tekhno-botanicheskii Slovar* (1820)
have been updated because Berchtold & Presl published their book earlier
in 1820 (Jan-Apr) than Martinov (3 Aug) (A. Doweld, pers. comm.). In
Vines's *A student's text-book of botany* (1895) one order is subdivided into families, two of which, *Cymodoceaceae* and *Posidoniaceae,* have been
updated. Six names in Link's Handbuch, vols. 1 and 2 (1829) have been
updated because, in vol. 3, Link published two family names under the
order *Fungi,* which means that the names ranked as orders throughout the
work (Art. 35.5) must be treated as the names of orders, not as families as
has traditionally been done. The affected names are *Dodonaeaceae, Melianthaceae, Moraceae, Neuradaceae, Tetragoniaceae* and *Theophrastaceae*. In addition, *Cordiaceae,* which was updated to Link in the Special
Committee's report, remains as listed in App. IIB in the *St Louis Code*.
Moreover, four family names previously overlooked in Berchtold &
Presl's rare, later, multi-volume work of the same name (1823-1825) have
been updated: *Aquifoliaceae, Cornaceae, Potamogetonaceae* and *Punicaceae*.

The rules determining when a rank is denoted by a misplaced term (and
hence not validly published) were clarified and made more practical. This

introduced the concepts of "minimum invalidity" (Art. 33.10), by which only those names with rank-denoting terms that must be removed to provide the correct sequence would be considered not validly published, and of "informal usage" (Art. 33.11), for situations in which the same term was used for several different non-sequential ranks; such names are to be treated as validly published but unranked. It was established that having the ranks of both order and family in a work precluded application of Art. 18.2 (and similarly Art. 19.2 in the cases of suborder and subfamily), and that sequential use of the same rank did not preclude valid publication (Art. 33 Note 3).

One further date limit first appears in the *Vienna Code*. From 1 January 2007 a new combination, a new generic name with a basionym, or an avowed substitute is not validly published unless its basionym or replaced synonym is cited. Currently, although a full and direct reference to the place of publication must be given, the basionym or replaced synonym need only be indicated.

One portion of the *Code* that remains virtually unchanged after Vienna is that for which by far the largest number of amendment proposals (147) was submitted, namely orthography Of the 147 proposals, only five were accepted but 107 were referred to the Editorial Committee. After review of all these proposals by a subcommittee of the Editorial Committee (F. R. Barrie, D. H. Nicolson, and N. J. Turland, who gratefully acknowledge advice from R. Gereau, Missouri Botanical Garden), the changes incorporated into the *Vienna Code* are very few and none imposes a significant change in practice. The most notable is a clarification of the respective application of Rec. 60C.1 and 60C.2.

The *Code* establishes (Art. 12.1) that only if validly published does a name have any status; indeed, unless otherwise indicated, the word "name" in the *Code* means a name that has been validly published (Art. 6.3). For this reason recent editions of the *Code* have replaced "name" by "designation" when the requirements for valid publication have not been met, and the *Vienna Code* has taken this further by avoiding such contradictory expressions as a name being validated, or being invalid. Given the very different meaning of "valid" and "invalid" applied to names in zoological nomenclature (equivalent to the botanical "correct" and "incorrect"), it is convenient that neither "valid name" nor "invalid name" need be used in botanical nomenclature: either a name is validly published or else it is not a validly published name, i.e. not a name under the *Code*.

The *Vienna Code* was prepared according to the procedures outlined in Div. III, which have been operating with hardly any change since the Paris Congress of 1954. A total of 312 individual numbered amendment proposals were published in *Taxon* between February 2002 and November 2004. Their synopsis, with comments by the Rapporteurs, appeared in *Taxon* (54: 215-250) in February 2005 and served as the basis for the preliminary, non-binding mail vote by the members of the International Association for Plant Taxonomy (and some other persons), as specified in Division III of the *Code*. Tabulation of the mail vote was taken care of by the Nomenclature Section's Recorder, T. F. Stuessy and his assistants in Vienna. The results were made available to the members of the Nomenclature Section at the beginning of its meetings; they were also tabulated in the November 2005 issue of *Taxon* (54: 1057-1064), along with the action taken by Congress.

The Nomenclature Section met at the Uni-Campus, University of Vienna, Spitalgasse 2, Vienna, on 12-16 July 2005. With 198 registered members carrying 402 institutional votes in addition to their personal votes, the Vienna Section had a large attendance compared with many previous Congresses but was substantially smaller than that at St Louis (with 297 members carrying 494 institutional votes). The Section Officers, previously appointed in conformity with Division III of the *Code,* were D. H. Nicolson (President), T. F. Stuessy (Recorder), J. McNeill (Rapporteur-général), and N. J. Turland (Vice-Rapporteur). Each Nomenclature Section is entitled to define its own procedural rules within the limits set by the *Code,* but tradition is held sacred. As on previous occasions, at least a 60% assenting majority was required for any proposed change to the *Code* to be adopted. Proposals that received 75% or more "no" votes in the mail ballot were ruled as rejected unless raised anew from the floor. The proceedings of the nomenclature sessions are presently being edited, based on a tape transcript. They will be published later this year or early in 2007 in the serial *Englera*.

The Nomenclature Section also appointed the Editorial Committee for the *Vienna Code.* As is traditional, only persons present at the Section meetings were invited to serve on that Committee, which as the *Code* requires is chaired by the Rapporteur-général and as is logical includes the Vice-Rapporteur as its secretary. The Editorial Committee sadly lost one of its members, when Guanghua Zhu died on 2 November 2005; the other 12 members of the committee convened on 6 January 2006 at the Missouri Botanical Garden, St Louis, U.S.A., for a full week's hard work. The

Committee worked on the basis of a draft of the text of the main body of the *Code,* prepared by the Chairman to incorporate the changes decided by the Section, which was distributed by electronic mail in December 2005; and of a preliminary version of the proceedings of the Section meetings, as transcribed from tape and revised portion-wise by F. R. Barrie, D. L. Hawksworth, J. McNeill, D. H Nicolson, and N. J. Turland.

Each Editorial Committee has the task of addressing matters specifically referred to it, incorporating changes agreed by the Section, clarifying any ambiguous wording, ensuring consistency, and providing additional examples for inclusion. The terms of the Committee's mandate, as defined by the Section in Vienna at its constituent meeting, included the usual empowerment to alter the wording, the examples, or the location of Articles and Recommendations, in so far as the meaning was not affected; while retaining the present numbering in so far as possible.

The full Editorial Committee concentrated on the main body of the *Code,* including Appendix I (hybrids). A new electronic draft of these portions was completed prior to the end of its meeting, and provided to all Committee members for checking and for any further necessary clarification; as a result a revised draft was prepared and circulated in mid-May to all members for final proofreading. The contents of Appendices II-VI were revised and updated in a bilateral process involving the Chairman and a specialist for each of the groups concerned, normally a Committee member (V. Demoulin for the fungi, D. H. Nicolson for genera and species of vascular plants, P. C. Silva for the algae, J. E. Skog for fossil plants, N. J. Turland for family names of vascular plants), except for the bryophytes (G. Zijlstra, Utrecht, Secretary, Committee for Bryophyta, with assistance from P. Isoviita, Helsinki). The Secretaries of the other Permanent Committees for particular groups provided useful assistance to the responsible Editorial Committee member. The Subject index and the Index to scientific names were revised by J. Prado; the Index to the Appendices was updated by J. McNeill, who, with N. J. Turland, also cared for the final copy-editing; the time-consuming task of final formatting and production of camera-ready copy was carried out by N. J. Turland.

This is the proper place for us to thank all those who have contributed to the publication of the new *Code:* our fellow members of the Editorial Committee for their forbearance, helpfulness, and congeniality; all the persons, just named, who contributed in a special way and much beyond their normal commitment to particular editorial tasks; the botanists at large who volunteered advice and suggestions, including relevant new examples; the International Association for Plant Taxonomy and its Secretary, Tod Stuessy, for maintaining IAPT's traditional commitment to plant nomenclature by funding travel and some ancillary costs for the Editorial Committee meeting in St Louis; the Missouri Botanical Garden and its Director, Peter Raven, for providing accommodation free of charge and hospitality for that meeting; and the publisher, Sven Koeltz, for his helpfulness and the speed with which he once again guided the *Code* through the printing process.

In addition to those who have helped to make possible this new edition of the Code, botanical nomenclature depends on the scores of botanists who serve on the Permanent Nomenclature Committees that work continuously between Congresses, dealing principally with proposals for conservation or rejection of names, and also those who are members of Special Committees set up by the Nomenclature Section of the Congress to review and seek solutions to particular nomenclatural problems. Botanical nomenclature is remarkable for the large number of taxonomists who voluntarily work so effectively and so extensively to the immeasurable benefit of all those who use plant names. On their behalf we express our sincere thanks to all who participate in this work.

The International Code of Botanical Nomenclature is published under the ultimate authority of the International Botanical Congresses. Provisions for the modification of the Code are detailed in Division III (p. 117). The next International Botanical Congress will be held in Melbourne, Australia from 23-30 July 2011, with a Nomenclature Section meeting likely in the preceding week. Invitation for proposals to amend this Code and instructions on procedure and format will be published in Taxon during 2007.

Like other international codes of nomenclature the *ICBN* has no legal status and is dependent on the voluntary acceptance of its rules by authors, editors, and other users of plant names. We trust that this *Vienna Code* will make their work just that little easier.

Edinburgh & St Louis, 24 July 2006

John McNeill
Nicholas J. Turland

IMPORTANT DATES IN THE CODE

DATES UPON WHICH PARTICULAR PROVISIONS OF THE CODE BECOME EFFECTIVE

1 May	1753	Art. 7.7, 13.1(a), (c), (d), (e)
4 Aug	1789	Art. 13.1 (a), (c)
1 Jan	1801	Art. 13.1(b)
31 Dec	1801	Art. 13.1(d)
31 Dec	1820	Art. 13.1(f)
1 Jan	1821	Art. 13.1(d)
1 Jan	1848	Art, 13.1(e)
1 Jan	1886	Art. 13.1(e)
1 Jan	1890	Art. 35.4
1 Jan	1892	Art. 13.1(e)
1 Jan	1900	Art. 13.1(e)
1 Jan	1908	Art. 35.2, 42.3, 44.1
1 Jan	1912	Art. 20.2, 38.1
1 Jan	1935	Art. 36.1
1 Jan	1953	Art. 30.1, 30.3, 30.4, 30.5, 32.5, 33.2, 33.3, 33.4, 33.5, 33.7, 33.8, 34.2, 35.1, 35.3
1 Jan	1958	Art. 36.2, 37.1, 39.1
1 Jan	1973	Art. 30.3, 45.1
1 Jan	1990	Art. 9.20, 37.6, 37.7
1 Jan	1996	Art. 36.3
1 Jan	2001	Art. 7.11, 9.13, 9.21, 38.2
1 Jan	2007	Art. 33.4, 37.4, 59.4

ARTICLES INVOLVING DATES APPLICABLE TO THE MAIN TAXONOMIC GROUPS

All groups	Art. 9.20, 9.21, 20.2, 30.1, 30.3, 30.4, 32.5, 33.2, 33.3, 33.4, 33.5, 33.7, 33.8, 34.2, 35.1, 35.2, 35.3, 37.1, 37.4, 37.6, 37.7, 42.3, 44.1, 45.1
Algae	Art. 7.7, 13.1(e), 36.2, 39.1
Bryophytes	Art. 7.7, 13.1(b), (c)
Fossil Plants	Art. 7.7, 9.13, 13.1(f), 36.3, 38.1, 38.2
Fungi	Art. 13.1(d), 59.4
Vascular plants	Art. 13.1(a)

ARTICLES DEFINING THE DATES OF CERTAIN WORKS

Art. 13.1 (a-f), 13.5

INTERNATIONAL CODE OF BOTANICAL NOMENCLATURE

PREAMBLE

1. Botany requires a precise and simple system of nomenclature used by botanists in all countries, dealing on the one hand with the terms that denote the ranks of taxonomic groups or units, and on the other hand with the scientific names that are applied to the individual taxonomic groups of plants[1]. The purpose of giving a name to a taxonomic group is not to indicate its characters or history, but to supply a means of referring to it and to indicate its taxonomic rank. This *Code* aims at the provision of a stable method of naming taxonomic groups, avoiding and rejecting the use of names that may cause error or ambiguity or throw science into confusion. Next in importance is the avoidance of the useless creation of names. Other considerations, such as absolute grammatical correctness, regularity or euphony of names, more or less prevailing custom, regard for persons, etc., notwithstanding their undeniable importance, are relatively accessory.

2. The Principles form the basis of the system of botanical nomenclature.

3. The detailed Provisions are divided into Rules, set out in the Articles, and Recommendations. Examples (Ex.) are added to the rules and recommendations to illustrate them.

4. The object of the Rules is to put the nomenclature of the past into order and to provide for that of the future; names contrary to a rule cannot be maintained.

5. The Recommendations deal with subsidiary points, their object being to bring about greater uniformity and clarity, especially in future nomencla-

---

[1] In this *Code,* unless otherwise indicated, the word "plant" means any organism traditionally studied by botanists (see Pre. 7).

ture; names contrary to a recommendation cannot, on that account, be rejected, but they are not examples to be followed.

6. The provisions regulating the governance of this *Code* form its last division.

7. The rules and recommendations apply to all organisms traditionally treated as plants, whether fossil or non-fossil[1], e.g. blue-green algae *(Cyanobacteria)*[2]; fungi, including chytrids, oomycetes, and slime moulds; photosynthetic protists and taxonomically related non-photosynthetic groups. Provisions for the names of hybrids appear in App. I.

8. The *International code of nomenclature for cultivated plants* is prepared under the authority of the International Commission for the Nomenclature of Cultivated Plants and deals with the use and formation of names for special plant categories in agricultural, forestry, and horticultural nomenclature.

9. The only proper reasons for changing a name are either a more profound knowledge of the facts resulting from adequate taxonomic study or the necessity of giving up a nomenclature that is contrary to the rules.

10. In the absence of a relevant rule or where the consequences of rules are doubtful, established custom is followed.

11. This edition of the *Code* supersedes all previous editions.

---

[1] In this *Code,* the term "fossil" is applied to a taxon when its name is based on a fossil type and the term "non-fossil" is applied to a taxon when its name is based on a non-fossil type (see Art. 13.3).

[2] For the nomenclature of other prokaryotic groups, see the *International code of nomenclature of bacteria (Bacteriological Code).*

DIVISION I. PRINCIPLES

Principle I

Botanical nomenclature is independent of zoological and bacteriological nomenclature. The *Code* applies equally to names of taxonomic groups treated as plants whether or not these groups were originally so treated (see Pre. 7).

Principle II

The application of names of taxonomic groups is determined by means of nomenclatural types.

Principle III

The nomenclature of a taxonomic group is based upon priority of publication.

Principle IV

Each taxonomic group with a particular circumscription, position, and rank can bear only one correct name, the earliest that is in accordance with the Rules, except in specified cases.

Principle V

Scientific names of taxonomic groups are treated as Latin regardless of their derivation.

Principle VI

The Rules of nomenclature are retroactive unless expressly limited.

DIVISION II. RULES AND RECOMMENDATIONS

CHAPTER I. TAXA AND THEIR RANKS

Article 1

1.1. Taxonomic groups of any rank will, in this *Code,* be referred to as taxa (singular: taxon).

1.2. Fossil taxa (diatoms excepted) may be treated as morphotaxa. A morphotaxon is defined as a fossil taxon which, for nomenclatural purposes, comprises only the one part, life-history stage, or preservational state represented by the corresponding nomenclatural type.

Note 1. Any fossil taxon that is described as including more than one part, life-history stage, or preservational state is not a morphotaxon.

Ex.1. Alcicornopteris hallei J. Walton (in Ann. Bot, n.s., 13: 450. 1949) was described from fossil material that included a compression on the surface of a petrified nodule with anatomy permitting description of the rachides, sporangia, and spores of a pteridosperm. This species comprises two preservational stages, two life-history stages, and three parts of the plant and is therefore not a morphotaxon.

Ex.2. Protofagacea allonensis Herend. & al. (in Int. J. Pl. Sci. 56: 94. 1995) was described on the basis of dichasia of staminate flowers, with anthers containing pollen grains, fruits, and cupules. This species comprises more than one part and more than one life-history stage and is therefore not a morphotaxon.

1.3. As in the case of form-taxa for asexual forms (anamorphs) of certain pleomorphic fungi (Art. 59), the provisions of this *Code* authorize the publication and use of names of morphotaxa (Art. 11.7).

Article 2

2.1. Every individual plant is treated as belonging to an indefinite number of taxa of consecutively subordinate rank, among which the rank of species (species) is basic.

Article 3

3.1. The principal ranks of taxa in descending sequence are: kingdom (regnum), division or phylum (divisio, phylum), class (classis), order (ordo), family (familia), genus (genus), and species (species). Thus, each species is assignable to a genus, each genus to a family, etc.

Note 1. Species and subdivisions of genera must be assigned to genera, and infraspecific taxa must be assigned to species, because their names are combinations (Art. 21.1, 23.1, and 24.1), but this provision does not preclude the placement of taxa as incertae sedis with regard to ranks higher than genus.

Ex. 1. The genus *Haptanthus* Goldberg & C. Nelson (in Syst. Bot. 14: 16. 1989) was originally described without being assigned to a family.

Ex. 2. The family assignment of the fossil genus *Paradinandra* Schönenberger & E. M. Friis (in Amer. J. Bot. 88: 467. 2001) was given as "incertae sedis".

3.2. The principal ranks of nothotaxa (hybrid taxa) are nothogenus and nothospecies. These ranks are the same as genus and species. The prefix "notho" indicates the hybrid character (see App. I).

Article 4

4.1. The secondary ranks of taxa in descending sequence are tribe (tribus) between family and genus, section (sectio) and series (series) between genus and species, and variety (varietas) and form (forma) below species.

4.2. If a greater number of ranks of taxa is desired, the terms for these are made by adding the prefix "sub-" to the terms denoting the principal or secondary ranks. A plant may thus be assigned to taxa of the following ranks (in descending sequence): regnum, subregnum, divisio or phylum, subdivisio or subphylum, classis, subclassis, ordo, subordo, familia, subfamilia, tribus, subtribus, genus, subgenus, sectio, subsectio, series, subseries, species, subspecies, varietas, subvarietas, forma, subforma.

Note 1. Ranks formed by adding "sub-" to the principal ranks (Art. 3.1) may be formed and used whether or not any secondary ranks (Art. 4.1) are adopted.

4.3. Further ranks may also be intercalated or added, provided that confusion or error is not thereby introduced.

4.4. The subordinate ranks of nothotaxa are the same as the subordinate ranks of non-hybrid taxa, except that nothogenus is the highest rank permitted (see App. I).

Note 2. Throughout this *Code* the phrase "subdivision of a family" refers only to taxa of a rank between family and genus and "subdivision of a genus" refers only to taxa of a rank between genus and species.

Note 3. For the designation of certain categories of plants used in agriculture, forestry, and horticulture, see Art. 28 Notes 2-5.

Note 4. In classifying parasites, especially fungi, authors who do not give specific, subspecific, or varietal value to taxa characterized from a physiological standpoint but scarcely or not at all from a morphological standpoint may distinguish within the species special forms (formae speciales) characterized by their adaptation to different hosts, but the nomenclature of special forms is not governed by the provisions of this *Code*.

Article 5

5.1. The relative order of the ranks specified in Art. 3 and 4 must not be altered (see Art. 33.9 and 33.12).

Recommendation 5A

5A.1. For purposes of standardization, the following abbreviations are recommended: cl. (class), ord. (order), fam. (family), tr. (tribe), gen. (genus), sect. (section), ser. (series), sp. (species), var. (variety), f. (forma). The abbreviations for additional ranks created by the addition of the prefix sub-, or for nothotaxa with the prefix notho-, should be formed by adding the prefixes, e.g. subsp. (subspecies), nothosp. (nothospecies), but subg. (subgenus).

CHAPTER II. STATUS, TYPIFICATION, AND PRIORITY OF NAMES

SECTION 1. STATUS DEFINITIONS

Article 6

6.1. Effective publication is publication in accordance with Art. 29-31.

6.2. Valid publication of names is publication in accordance with Art. 32-45 or H.9 (see also Art. 61).

Note 1. For nomenclatural purposes, valid publication creates a name, and sometimes also an autonym (Art. 22.1 and 26.1), but does not itself imply any taxonomic circumscription beyond inclusion of the type of the name (Art. 7.1).

6.3. In this *Code,* unless otherwise indicated, the word "name" means a name that has been validly published, whether it is legitimate or illegitimate (see Art. 12).

Note 2. When the same name, based on the same type, has been published independently at different times by different authors, then only the earliest of these "isonyms" has nomenclatural status. The name is always to be cited from its original place of valid publication, and later isonyms may be disregarded.

Ex. 1. Baker (Summary New Ferns: 9. 1892) and Christensen (Index Filic.: 44. 1905) independently published the name *Alsophila kalbreyeri* as a substitute for *A. podophylla* Baker (1891) non Hook. (1857). As published by Christensen, *Alsophila kalbreyeri* is a later isonym of *A. kalbreyeri* Baker, without nomenclatural status (see also Art. 33 Ex. 19).

Ex. 2. In publishing "*Canarium pimela* Leenh. nom. nov.", Leenhouts (in Blumea 9: 406. 1959) reused the illegitimate *C. pimela* K. D. Koenig (1805), attributing it to himself and basing it on the same type. He thereby created a later isonym without nomenclatural status.

6.4. An illegitimate name is one that is designated as such in Art. 18.3, 19.5, or 52-54 (see also Art. 21 Note 1 and Art. 24 Note 2). A name which according to this *Code* was illegitimate when published cannot become legitimate later unless it is conserved or sanctioned.

Ex. 3. Anisothecium Mitt. (1869) when published included the previously designated type of *Dicranella* (Müll. Hal.) Schimp. (1856). When *Dicranella* was conserved with a different type, *Anisothecium* did not thereby become legitimate.

Ex. 4. Skeletonemopsis P. A. Sims (1995) was illegitimate when published because it included the original type of *Skeletonema* Grev. (1865). When *Skeletonema* was conserved with a different type, *Skeletonemopsis* nevertheless remained illegitimate and had to be conserved in order to be available for use.

6.5. A legitimate name is one that is in accordance with the rules, i.e. one that is not illegitimate as defined in Art. 6.4.

6.6. At the rank of family or below, the correct name of a taxon with a particular circumscription, position, and rank is the legitimate name which must be adopted for it under the rules (see Art. 11).

Ex. 5. The generic name *Vexillifera* Ducke (1922), based on the single species *V. micran-thera*, is legitimate. The same is true of the generic name *Dussia* Krug & Urb. ex Taub. (1892), based on the single species *D. martinicensis.* Both generic names are correct when the genera are thought to be separate. Harms (in Repert. Spec. Nov. Regni Veg. 19: 291. 1924), however, united *Vexillifera* and *Dussia* in a single genus; the latter name is the correct one for the genus with this particular circumscription. The legitimate name *Vexillifera* may therefore be correct or incorrect according to different taxonomic concepts.

6.7. The name of a taxon below the rank of genus, consisting of the name of a genus combined with one or two epithets, is termed a combination (see Art. 21, 23, and 24).

Ex. 6. Combinations: *Mouriri* subg. *Pericrene, Arytera* sect. *Mischarytera, Gentiana lutea, Gentiana tenella* var. *occidentalis, Equisetum palustre* var. *americanum, Equisetum palustre* f. *fluitans.*

6.8. Autonyms are such names as can be established automatically under Art. 22.3 and 26.3, whether or not they appear in print in the publication in which they are created (see Art. 32.8, Rec. 22B.1 and 26B.1).

SECTION 2. TYPIFICATION

Article 7

7.1. The application of names of taxa of the rank of family or below is determined by means of nomenclatural types (types of names of taxa). The application of names of taxa in the higher ranks is also determined by means of types when the names are ultimately based on generic names (see Art. 10.7).

7.2. A nomenclatural type (typus) is that element to which the name of a taxon is permanently attached, whether as the correct name or as a synonym. The nomenclatural type is not necessarily the most typical or representative element of a taxon.

7.3. A new name published as an avowed substitute (replacement name, nomen novum) for an older name is typified by the type of the older name (see Art. 33.4; but see Art. 33 Note 2).

Ex. 1. Myrcia lucida McVaugh (1969) was published as a nomen novum for *M. laevis* O. Berg (1862), an illegitimate homonym of *M. laevis* G. Don (1832). The type of *M. lucida* is therefore the type of *M. laevis* O. Berg (non G. Don), namely, *Spruce 3502* (BR).

7.4. A new name formed from a previously published legitimate name (stat. nov., comb. nov.) is, in all circumstances, typified by the type of the basionym, even though it may have been applied erroneously to a taxon now considered not to include that type (but see Art. 48.1 and 59.6).

Ex. 2. Pinus mertensiana Bong. was transferred to the genus *Tsuga* by Carrière, who, however, as is evident from his description, erroneously applied the new combination *T. mertensiana* to another species of *Tsuga,* namely *T. heterophylla* (Raf.) Sarg. The combination *Tsuga mertensiana* (Bong.) Carrière must not be applied to *T. heterophylla* but must be retained for *P. mertensiana* when that species is placed in *Tsuga;* the citation in parentheses (under Art. 49) of the name of the original author, Bongard, indicates the basionym, and hence the type, of the name.

Ex. 3. Delesseria gmelinii J. V. Lamour. (1813) is a legitimate replacement name for *Fucus palmetta* S. G. Gmel. (1768), the change of epithet being necessitated by the simultaneous publication of *D. palmetta* (Stackh.) J. V. Lamour (see Art. 11 Note 1). All intended combinations based on *D. gmelinii* (and not excluding the type of *F. palmetta;* see Art. 48.1) have the same type as *F. palmetta,* even though the material possessed by Lamouroux is now assigned to a different species, *D. bonnemaisonii* C. Agardh (1822).

7.5. A name that is illegitimate under Art. 52 is typified either by the type of the name that ought to have been adopted under the rules (automatic typification), or by a different type designated or definitely indicated by the author of the illegitimate name. However, if no type was designated or definitely indicated and the type of the earlier name was included (see Art. 52.2) in a subordinate taxon that did not include the evidently intended type of the illegitimate name, typification is not automatic. Automatic typification does not apply to names sanctioned under Art. 15.

Ex. 4. Bauhinia semla Wunderlin (1976) is illegitimate under Art. 52 (see Art. 52 Ex. 10), but its publication as a replacement name for *B. retusa* Roxb. (1832) non Poir. (1811) is definite indication of a different type (that of *B. retusa*) from that of the name (*B. roxburghiana* Voigt, 1845), which ought to have been adopted.

Ex. 5. Hewittia bicolor Wight & Arn. (1837), the type of *Hewittia* Wight & Arn., is illegitimate under Art. 52 because, in addition to the illegitimate intended basionym *Convolvulus bicolor* Vahl (1794) non Desr. (1792), the legitimate *C. bracteatus* Vahl (1794) was cited as a synonym. Wight & Arnott's adoption of the epithet *"bicolor"* is definite indication that the type of *H. bicolor,* and therefore the type of *Hewittia,* is the type of *C. bicolor,* and not that of *C. bracteatus* whose epithet ought to have been adopted.

Ex. 6. Gilia splendens, when validly published by Mason & Grant (in Madroño 9: 212. 1948), included, as "a long-tubed form of the species", *G. splendens* subsp. *grinnellii* based on *G. grinnellii* Brand (1907) and is thus superfluous and illegitimate. Although Mason & Grant, believing that *G. splendens* was already validly published, did not indicate its type, it is not automatically that of *G. grinnellii;* the specimen that has since been adopted as the conserved type could have been selected as lectotype.

7.6. The type of an autonym is the same as that of the name from which it is derived.

7.7. A name validly published by reference to a previously and effectively published description or diagnosis (Art. 32.1(d)) is to be typified by an element selected from the context of the validating description or diagnosis, unless the validating author has definitely designated a different type (but see Art. 10.2). However, the type of a name of a taxon assigned to a group with a nomenclatural starting-point later than 1 May 1753 (see Art. 13.1) is to be determined in accordance with the indication or descriptive and other matter accompanying its valid publication (see Art. 32-45).

Ex. 7. Since the name *Adenanthera bicolor* Moon (1824) is validated solely by reference to Rumphius (Herb. Amboin. 3: t. 112. 1743), the type of the name, in the absence of the specimen from which it was figured, is the illustration referred to. It is not the specimen, at Kew, collected by Moon and labelled *"Adenanthera bicolor",* since Moon did not definitely designate the latter as the type.

Ex. 8. Echium lycopsis L. (Fl. Angl.: 12. 1754) was published without a description or diagnosis but with reference to Ray (Syn. Meth. Stirp. Brit., ed. 3: 227. 1724), in which a *"Lycopsis"* species was discussed with no description or diagnosis but with citation of earlier references, including Bauhin (Pinax: 255. 1623). The accepted validating description of *E. lycopsis* is that of Bauhin, and the type must be chosen from the context of his work. Consequently the Sherard specimen in the Morison herbarium (OXF), selected by Klotz (in Wiss. Z. Martin-Luther-Univ. Halle-Wittenberg Math.-Naturwiss. Reihe 9: 375-376. 1960), although probably consulted by Ray, is not eligible as type. The first acceptable choice is that of the illustration, cited by both Ray and Bauhin, of *"Echii altera species"* in Dodonaeus (Stirp. Hist. Pempt.: 620. 1583), suggested by Gibbs (in Lagascalia 1: 60-61. 1971) and formally made by Stearn (in Ray Soc. Publ. 148, Introd.: 65. 1973).

7.8. Typification of names adopted in one of the works specified in Art. 13.1(d), and thereby sanctioned (Art. 15), may be effected in the light of anything associated with the name in that work.

7.9. The typification of names of morphotaxa of plant fossils (Art. 1.2), of fungal anamorphs (Art. 59), and of any other analogous taxa at or below the rank of genus does not differ from that indicated above.

Note 1. See also Art. 59 for details regarding typification of names in certain pleomorphic fungi.

7.10. For purposes of priority (Art. 9.17, 9.18, and 10.5), designation of a type is achieved only by effective publication (Art. 29-31).

7.11. For purposes of priority (Art. 9.17, 9.18, and 10.5), designation of a type is achieved only if the type is definitely accepted as such by the typifying author, if the type element is clearly indicated by direct citation including the term "type" (typus) or an equivalent, and, on or after 1 January 2001, if the typification statement includes the phrase "designated here" (hic designatus) or an equivalent.

Note 2. Art. 7.10 and 7.11 apply only to the designation of lectotypes (and their equivalents under Art. 10), neotypes, and epitypes; for the indication of a holotype see Art. 37.

Ex. 9. Chlorosarcina Gerneck (1907) originally comprised two species, *C. minor* and *C. elegans.* Vischer (1933) transferred the former to *Chlorosphaera* G. A. Klebs and retained the latter in *Chlorosarcina.* He did not, however, use the term "type" or an equivalent, so that his action does not constitute typification of *Chlorosarcina.* The first to designate a type, as "LT.", was Starr (in ING Card No. 16528, Nov 1962), who selected *Chlorosarcina elegans.*

*Ex. 10. The phrase "standard species" as used by Hitchcock & Green (in Anonymous, Nomencl. Prop. Brit. Botanists: 110-199. 1929) is now treated as equivalent to "type", and hence type designations in this work are acceptable.

Recommendation 7A

7A.1. It is strongly recommended that the material on which the name of a taxon is based, especially the holotype, be deposited in a public herbarium or other public collection with a policy of giving bona fide researchers access to deposited material, and that it be scrupulously conserved.

Article 8

8.1. The type (holotype, lectotype, or neotype) of a name of a species or infraspecific taxon is either a single specimen conserved in one herbarium or other collection or institution, or an illustration (but see also Art. 37.4 and 37.6 for names published on or after 1 January 1958).

8.2. For the purpose of typification a specimen is a gathering, or part of a gathering, of a single species or infraspecific taxon made at one time, disregarding admixtures (see Art. 9.12). It may consist of a single plant, parts

---

* Here and elsewhere in the Code, a prefixed asterisk denotes a "voted Example", accepted by a Congress in order to legislate nomenclatural practice when the corresponding Article of the *Code* is open to divergent interpretation or does not adequately cover the matter.

of one or several plants, or of multiple small plants. A specimen is usually mounted on a single herbarium sheet or in an equivalent preparation, such as a box, packet, jar or microscope slide.

Ex. 1. *"Echinocereus sanpedroensis"* (Raudonat & Rischer in Echinocereenfreund 8(4): 91-92. 1995) was based on a "holotype" consisting of a complete plant with roots, a detached branch, an entire flower, a flower cut in halves, and two fruits, which according to the label were taken from the same cultivated individual at different times and preserved, in alcohol, in a single jar. This material belongs to more than one gathering and cannot be accepted as a type. Raudonat & Rischer's name is not validly published under Art. 37.2.

8.3. A specimen may be mounted as more than one preparation, as long as the parts are clearly labelled as being part of that same specimen. Multiple preparations from a single gathering that are not clearly labelled as being part of a single specimen are duplicates[1], irrespective of whether the source was one plant or more than one (but see Art. 8.5).

Ex. 2. The holotype specimen of *Delissea eleeleensis* H. St. John, *Christensen 261* (BISH), is mounted as two preparations, a herbarium sheet (BISH No. 519675) bearing the annotation "fl. bottled" and an inflorescence preserved in alcohol in a jar labelled *"Cyanea, Christensen 261"*. The annotation indicates that the inflorescence is part of the holotype specimen and not a duplicate, nor is it part of the isotype specimen (BISH No. 519676), which is not labelled as including additional material preserved in a separate preparation.

Ex. 3. The holotype specimen of *Johannesteijsmannia magnifica* J. Dransf., *Dransfield 862* (K), consists of a leaf mounted on five herbarium sheets, an inflorescence and infructescence in a box, and liquid-preserved material in a bottle.

Ex. 4. The holotype of *Cephaëlis acanthacea* Steyerm., *Cuatrecasas 16752* (F), consists of a single specimen mounted on two herbarium sheets, labelled "sheet 1" and "sheet 2". Although the two sheets have separate herbarium accession numbers, F-1153741 and F-1153742, respectively, the cross-labelling indicates that they constitute a single specimen. A third sheet of *Cuatrecasas 16572,* F-1153740, is not cross-labelled and is therefore a duplicate.

Ex. 5. The holotype specimen of *Eugenia ceibensis* Standl., *Yuncker & al. 8309,* is mounted on a single herbarium sheet at F. A fragment was removed from the specimen subsequent to its designation as holotype and is now conserved in LL. The fragment is mounted on a herbarium sheet along with a photograph of the holotype and is labelled "fragment of type!". The fragment is no longer part of the holotype specimen because it is not permanently conserved in the same herbarium as the holotype. Such fragments have the status of a duplicate, i.e. an isotype.

8.4. Type specimens of names of taxa must be preserved permanently and may not be living plants or cultures. However, cultures of fungi and algae,

---

[1] Here and elsewhere in this *Code,* the word duplicate is given its usual meaning in herbarium curatorial practice. It is part of a single gathering of a single species or infraspecific taxon made by the same collector(s) at one time. The possibility of a mixed gathering must always be considered by an author choosing a lectotype, and corresponding caution used.

if preserved in a metabolically inactive state (e.g. by lyophilization or deep-freezing), are acceptable as types.

Ex. 6. The strain CBS 7351 is acceptable as the type of the name *Candida populi* Hagler & al. (in Int. J. Syst. Bacteriol. 39: 98. 1989) because it is permanently preserved in a metabolically inactive state by lyophilization (see also Rec. 8B.2).

8.5. The type, epitypes (Art. 9.7) excepted, of the name of a taxon of fossil plants of the rank of species or below is always a specimen (see Art. 9.13). One whole specimen is to be considered as the nomenclatural type (see Rec. 8A.3).

Recommendation 8A

8A.1. When a holotype, a lectotype, or a neotype is an illustration, the specimen or specimens upon which that illustration is based should be used to help determine the application of the name (see also Art. 9.13).

8A.2. When an illustration is designated as the type of a name under Art. 37.5, the collection data of the illustrated material should be given (see also Rec. 32D.2).

8A.3. If the type specimen of a name of a fossil plant is cut into pieces (sections of fossil wood, pieces of coalball plants, etc.), all parts originally used in establishing the diagnosis should be clearly marked.

8A.4. When a single specimen designated as type is mounted as multiple preparations, this should be stated in the protologue[1], and the preparations appropriately labelled.

Recommendation 8B

8B.1. Whenever practicable a living culture should be prepared from the holotype material of the name of a newly described taxon of fungi or algae and deposited in at least two institutional culture or genetic resource collections. (Such action does not obviate the requirement for a holotype specimen under Art. 8.4.)

8B.2. In cases where the type of a name is a culture permanently preserved in a metabolically inactive state (see Art. 8 Ex. 6), any living isolates obtained from that should be referred to as "ex-type" (ex typo), "ex-holotype" (ex holotypo), "ex-isotype" (ex isotypo), etc., in order to make it clear they are derived from the type but are not themselves the nomenclatural type.

Article 9

9.1. A holotype of a name of a species or infraspecific taxon is the one specimen or illustration (but see Art. 37.4) used by the author, or designated

---

[1] Protologue (from Greek πρῶτος, *protos,* first; λόγος, *logos,* discourse): everything associated with a name at its valid publication, i.e. description or diagnosis, illustrations, references, synonymy, geographical data, citation of specimens, discussion, and comments.

by the author as the nomenclatural type. As long as a holotype is extant, it fixes the application of the name concerned (but see Art. 9.13; see also Art. 10).

Note 1. Any designation made by the original author, if definitely expressed at the time of the original publication of the name of the taxon, is final (but see Art. 9.9 and 9.13). If the author used only one element, that one must be accepted as the holotype. If a new name is based on a previously published description or diagnosis of the taxon, the same considerations apply to material included by the earlier author (see Art. 7.7 and 7.8).

9.2. A lectotype is a specimen or illustration designated from the original material as the nomenclatural type, in conformity with Art. 9.9 and 9.10, if no holotype was indicated at the time of publication, or if it is missing, or if it is found to belong to more than one taxon (see also Art. 9.12).

Note 2. For the purposes of this *Code,* the original material comprises: *(a)* those specimens and illustrations (both unpublished and published either prior to or together with the protologue) upon which it can be shown that the description or diagnosis validating the name was based; *(b)* the holotype and those specimens which, even if not seen by the author of the description or diagnosis validating the name, were indicated as types (syntypes or paratypes) of the name at its valid publication; and *(c)* the isotypes or isosyntypes of the name irrespective of whether such specimens were seen by either the author of the validating description or diagnosis, or the author of the name (but see also Art. 7.7, second sentence, and 7.8).

9.3. An isotype is any duplicate of the holotype; it is always a specimen.

9.4. A syntype is any specimen cited in the protologue when there is no holotype, or any one of two or more specimens simultaneously designated as types (see also Art. 37 Note 1).

Ex. 1. In the protologue of *Laurentia frontidentata* E. Wimm. (see Art. 37 Ex. 2) a single gathering in two herbaria was designated as the type. There must exist, therefore, at least two specimens and these are syntypes.

9.5. A paratype is a specimen cited in the protologue that is neither the holotype nor an isotype, nor one of the syntypes if two or more specimens were simultaneously designated as types.

Ex. 2. The holotype of the name *Rheedia kappleri* Eyma (1932), which applies to a polygamous species, is a male specimen, *Kappler 593a* (U). The author designated a hermaphroditic specimen, *Forestry Service of Surinam B. W. 1618* (U), as a paratype.

Note 3. In most cases in which no holotype was designated there will also be no paratypes, since all the cited specimens will be syntypes. However, when an author designated two or more specimens as types (Art. 9.4), any remaining cited specimens are paratypes and not syntypes.

Ex. 3. In the protologue of *Eurya hebeclados* Y. Ling (1951) the author simultaneously designated two specimens as types, *Y. Ling 5014* as "typus, ♂" and *Y. Y. Tung 315* as "typus, ♀", which are therefore syntypes. Ling also cited the specimen *Y. Ling 5366* but without designating it as a type; it is therefore a paratype.

9.6. A neotype is a specimen or illustration selected to serve as nomenclatural type if no original material is extant, or as long as it is missing (see also Art. 9.14).

9.7. An epitype is a specimen or illustration selected to serve as an interpretative type when the holotype, lectotype, or previously designated neotype, or all original material associated with a validly published name, is demonstrably ambiguous and cannot be critically identified for purposes of the precise application of the name of a taxon (but see also Art. 59.7). When an epitype is designated, the holotype, lectotype, or neotype that the epitype supports must be explicitly cited (see Art. 9.18).

Ex. 4. The holotype of *Vitellaria paradoxa* C. F. Gaertn. (1807) is a seed of unknown provenance (P), clearly belonging to the species currently known as *Butyrospermum paradoxum* (C. F. Gaertn.) Hepper. However, the two subspecies recognized within that species can only be distinguished by characters of foliage or inflorescence. Hall & Hindle (in Taxon 44: 410. 1995) designated an epitype with foliage, *Mungo Park* (BM). It belongs to the western subspecies, now to be known as *B. paradoxum* subsp. *paradoxum*.

Ex. 5. Podlech (in Taxon 46: 465. 1997) designated Herb. Linnaeus No. 926.43 (LINN) as the lectotype of *Astragalus trimestris* L. (1753). He simultaneously designated an epitype (Egypt. Düben oberhalb Rosetta am linken Nilufer bei Schech Mantur, 9 May 1902, *Anonymous* (BM)), because the lectotype lacked fruits, "which show important diagnostic features for this species."

9.8. The use of a term defined in the *Code* (Art. 9.1-9.7) as denoting a type, in a sense other than that in which it is so defined, is treated as an error to be corrected (for example, the use of the term lectotype to denote what is in fact a neotype).

Note 4. Correction can be effected only if the requirements of Art. 7.11 are met.

Ex. 6. Borssum Waalkes (in Blumea 14: 198. 1966) cited Herb. Linnaeus No. 866.7 (LINN) as the holotype of *Sida retusa* L. (1763). The term is incorrectly used because illustrations in Plukenet (Phytographia: t. 9, f. 2. 1691) and Rumphius (Herb. Amboin. 6: t. 19. 1750) were cited by Linnaeus in the protologue of *S. retusa*. Since all three elements are original material (Art. 9 Note 2), Borssum Waalkes's use of holotype is an error to be corrected to lectotype.

9.9. If no holotype was indicated by the author of a name of a species or infraspecific taxon, or when the holotype has been lost or destroyed, or when the material designated as type is found to belong to more than one taxon, a lectotype or, if permissible (Art. 9.6), a neotype as a substitute for it may be designated (Art. 7.10 and 7.11).

9.10. In lectotype designation, an isotype must be chosen if such exists, or otherwise a syntype if such exists. If no isotype, syntype or isosyntype (duplicate of syntype) is extant, the lectotype must be chosen from among the paratypes if such exist. If no cited specimens exist, the lectotype must be chosen from among the uncited specimens and cited and uncited illustrations which comprise the remaining original material, if such exist.

9.11. If no original material is extant or as long as it is missing, a neotype may be selected. A lectotype always takes precedence over a neotype, except as provided by Art. 9.14.

9.12. When a type specimen (herbarium sheet or equivalent preparation) contains parts belonging to more than one taxon (see Art. 9.9), the name must remain attached to that part which corresponds most nearly with the original description or diagnosis.

Ex. 7. The type of the name *Tillandsia bryoides* Griseb. ex Baker (1878) is *Lorentz 128* (BM); this specimen, however, proved to be mixed. Smith (in Proc. Amer. Acad. Arts 70: 192. 1935) acted in accordance with Art. 9.12 in designating one part of Lorentz's specimen as the lectotype.

9.13. The holotype (or lectotype) of a name of a species or infraspecific taxon of fossil plants (Art. 8.5) is the specimen (or one of the specimens) on which the validating illustrations (Art. 38) are based. When, prior to 1 January 2001 (see Art. 38.2), in the protologue of a name of a new taxon of fossil plants of the rank of species or below, a type specimen is indicated (Art. 37.1) but not identified among the validating illustrations, a lectotype must be designated from among the specimens illustrated in the protologue. This choice is superseded if it can be demonstrated that the original type specimen corresponds to another validating illustration.

9.14. When a holotype or a previously designated lectotype has been lost or destroyed and it can be shown that all the other original material differs taxonomically from the destroyed type, a neotype may be selected to preserve the usage established by the previous typification (see also Art. 9.16).

9.15. A designation of a lectotype or neotype that later is found to refer to a single gathering but to more than one specimen must nevertheless be accepted (subject to Art. 9.17), but may be further narrowed to a single one of these specimens by way of a subsequent lectotypification or neotypification.

Ex. 8. Erigeron plantagineus Greene (1898) was described from material collected by R. M. Austin in California. Cronquist (in Brittonia 6: 173. 1947) wrote "Type: *Austin s.n.,* Modoc County, California (ND)", thereby designating the Austin material in ND as the [first-step] lectotype. Strother & Ferlatte (in Madroño 35: 85. 1988), noting that there were two specimens of this gathering at ND, designated one of them (ND-G No. 057228) as the [second-step] lectotype. In subsequent references, both lectotypification steps may be cited in sequence.

9.16. A neotype selected under Art. 9.14 may be superseded if it can be shown to differ taxonomically from the holotype or lectotype that it replaced.

9.17. The author who first designates a lectotype or a neotype must be followed, but that choice is superseded if *(a)* the holotype or, in the case of a neotype, any of the original material is rediscovered; the choice may also be superseded if one can show that *(b)* it is in serious conflict with the protologue and another element is available that is not in conflict with the protologue, or that *(c)* it is contrary to Art. 9.12.

9.18. The author who first designates an epitype must be followed; a different epitype may be designated only if the original epitype is lost or destroyed. A lectotype or neotype supported by an epitype may be superseded in accordance with Art. 9.17 or, in the case of a neotype, Art. 9.16. If it can be shown that an epitype and the type it supports differ taxonomically and that neither Art. 9.16 nor 9.17 applies, the name may be proposed for conservation with a conserved type (Art. 14.9; see also Art. 57).

Note 5. An epitype supports only the type to which it is linked by the typifying author. If the supported type is superseded, the epitype has no standing with respect to the replacement type.

9.19. Designation of an epitype is not effected unless the herbarium or institution in which the epitype is conserved is specified or, if the epitype is a published illustration, a full and direct bibliographic reference to it is provided.

9.20. On or after 1 January 1990, lectotypification or neotypification of a name of a species or infraspecific taxon by a specimen or unpublished illustration is not effected unless the herbarium or institution in which the type is conserved is specified.

9.21. On or after 1 January 2001, lectotypification or neotypification of a name of a species or infraspecific taxon is not effected unless indicated by use of the term "lectotypus" or "neotypus", its abbreviation, or its equivalent in a modern language (but see Art. 9.8).

Recommendation 9A

9A.1. Typification of names for which no holotype was designated should only be carried out with an understanding of the author's method of working; in particular it should be realized that some of the material used by the author in describing the taxon may not be in the author's own herbarium or may not even have survived, and conversely, that not all the material surviving in the author's herbarium was necessarily used in describing the taxon.

9A.2. Designation of a lectotype should be undertaken only in the light of an understanding of the group concerned. In choosing a lectotype, all aspects of the protologue should be considered as a basic guide. Mechanical methods, such as the automatic selection of the first element cited or of a specimen collected by the person after whom a species is named, should be avoided as unscientific and productive of possible future confusion and further changes.

9A.3. In choosing a lectotype, any indication of intent by the author of a name should be given preference unless such indication is contrary to the protologue. Such indications are manuscript notes, annotations on herbarium sheets, recognizable figures, and epithets such as *typicus, genuinus,* etc.

9A.4. When a single gathering is cited in the protologue, but a particular institution housing it is not designated, it should be assumed that the specimen housed in the institution where the author is known to have worked is the holotype, unless there is evidence that further material of the same gathering was used.

9A.5. When two or more heterogeneous elements were included in or cited with the original description or diagnosis, the lectotype should be so selected as to preserve current usage. In particular, if another author has already segregated one or more elements as other taxa, one of the remaining elements should be designated as the lectotype provided that this element is not in conflict with the original description or diagnosis (see Art. 9.17).

Recommendation 9B

9B.1. In selecting a neotype, particular care and critical knowledge should be exercised because the reviewer usually has no guide except personal judgement as to what best fits the protologue; if this selection proves to be faulty it will inevitably result in further change.

Article 10

10.1. The type of a name of a genus or of any subdivision of a genus is the type of a name of a species (except as provided by Art. 10.4). For purposes of designation or citation of a type, the species name alone suffices, i.e. it is considered as the full equivalent of its type.

Note 1. Terms such as "holotype", "syntype", and "lectotype", as presently defined in Art. 9, although not applicable, strictly speaking, to the types of names in ranks higher than species, are so used by analogy.

10.2. If in the protologue of the name of a genus or of any subdivision of a genus the holotype or lectotype of one or more previously or simultaneously published species name(s) is definitely included (see Art. 10.3), the type must be chosen (Art. 7.10 and 7.11) from among these types unless the type was indicated (Art. 22.6, 22.7, 37.1, and 37.3) or designated by the author of the name. If no type of a previously or simultaneously published species name was definitely included, a type must be otherwise chosen, but the choice is to be superseded if it can be demonstrated that the selected type is not conspecific with any of the material associated with the protologue.

Ex. 1. The genus *Anacyclus,* as originally circumscribed by Linnaeus (1753), comprised three validly named species. Cassini (in Cuvier, Dict. Sci. Nat. 34: 104. 1825) designated *Anthemis valentina* L. (1753) as type of *Anacyclus,* but this was not an original element of the genus. Green (in Anonymous, Nomencl. Prop. Brit. Botanists: 182. 1929) designated *Anacyclus valentinus* L. (1753), "the only one of the three original species still retained in the genus", as the "standard species" (see Art. 7 Ex. 10), and her choice must be followed (Art. 10.5). Humphries (in Bull. Brit. Mus. (Nat. Hist.), Bot. 7: 109. 1979) designated a specimen in the Clifford Herbarium (BM) as lectotype of *Anacyclus valentinus,* and that specimen thereby became the ultimate type of the generic name.

Ex. 2. Castanella Spruce ex Benth. & Hook. f. (1862) was described on the basis of a single specimen collected by Spruce and without mention of a species name. Swart (in ING Card No. 2143. 1957) was the first to designate a type (as "T."): *C. granatensis* Triana & Planch. (1862), based on *Linden 1360.* As long as the Spruce specimen is considered to be conspecific with Linden's material, Swart's type designation cannot be superseded, even though the Spruce specimen became the type of *Paullinia paullinioides* Radlk. (1896), because the latter is not a "previously or simultaneously published species name".

10.3. For the purposes of Art. 10.2, definite inclusion of the type of a name of a species is effected by citation of, or reference (direct or indirect) to, a validly published name, whether accepted or synonymized by the author, or by citation of the holotype or lectotype of a previously or simultaneously published name of a species.

Ex. 3. The protologue of *Elodes* Adans. (1763) included references to *"Elodes"* of Clusius (1601), *"Hypericum"* of Tournefort (1700), and *Hypericum aegypticum* L. (1753). The last is the only reference to a validly published name of a species, and neither of the other elements is the type of a name of a species. The type of *H. aegypticum* is therefore the type of *Elodes,* even though subsequent authors designated *H. elodes* L. (1759) as the type (see Robson in Bull. Brit. Mus. (Nat. Hist.), Bot. 5: 305, 336. 1977).

10.4. By and only by conservation (Art. 14.9), the type of a name of a genus may be a specimen or illustration, preferably used by the author in the preparation of the protologue, other than the type of a name of an included species.

Ex. 4. Physconia Poelt (1965) was originally conserved with the specimen "*'Lichen pulverulentus',* Germania, Lipsia in *Tilia,* 1767, *Schreber* (M)" as the type. That specimen is the type of *P. pulverulacea* Moberg (1979), which name is now cited in the type entry in App. III.

Note 2. If the element designated under Art. 10.4 is the type of a species name, that name may be cited as the type of the generic name. If the element is not the type of a species name, a parenthetical reference to the correct name of the type element may be added.

Ex. 5. Pseudolarix Gordon (1858) was conserved with a specimen from the Gordon herbarium as its conserved type. As this specimen is not the type of any species name, its accepted identity "[= *P. amabilis* (J. Nelson) Rehder ...]" has been added to the corresponding entry in App. III.

10.5. The author who first designates a type of a name of a genus or subdivision of a genus must be followed, but the choice may be superseded if *(a)* it can be shown that it is in serious conflict with the protologue and another element is available which is not in conflict with the protologue, or *(b)* that it was based on a largely mechanical method of selection.

Ex. 6. Fink (in Contr. U.S. Natl. Herb. 14(1): 2. 1910) specified that he was "stating the types of the genera according to the 'first species' rule". His type designations may therefore be superseded under Art. 10.5(b). For example, Fink had designated *Biatorina griffithii* (Ach.) A. Massal. as the type of *Biatorina* A. Massal.; but his choice was superseded when the next subsequent designation, by Santesson (in Symb. Bot. Upsal. 12(1): 428. 1952), stated a different type, *B. atropurpurea* (Schaer.) A. Massal.

*Ex. 7. Authors following the *American code of botanical nomenclature,* Canon 15 (in Bull. Torrey Bot. Club 34: 172. 1907), designated as the type "the first binomial species in order" eligible under certain provisions. This method of selection is to be considered as largely mechanical. Thus the first type designation for *Delphinium* L., by Britton (in Britton & Brown, Ill. Fl. N. U.S., ed. 2, 2: 93. 1913), who followed the *American code* and chose *D. consolida* L., has been superseded under Art. 10.5(b) by the designation of *D. peregrinum* L. by Green (in Anonymous, Nomencl. Prop. Brit. Botanists: 162. 1929). The unicarpellate *D. consolida* could not have been superseded as type by the tricarpellate *D. peregrinum* under Art. 10.5(a), however, because it is not in serious conflict with the generic protologue, which specifies "germina tria vel unum", the assignment of the genus to "Polyandria Trigynia" by Linnaeus notwithstanding.

10.6. The type of a name of a family or of any subdivision of a family is the same as that of the generic name on which it is based (see Art. 18.1). For purposes of designation or citation of a type, the generic name alone suffices. The type of a name of a family or subfamily not based on a generic name is the same as that of the corresponding alternative name (Art. 18.5 and 19.7).

10.7. The principle of typification does not apply to names of taxa above the rank of family, except for names that are automatically typified by being based on generic names (see Art. 16). The type of such a name is the same as that of the generic name on which it is based.

Note 3. For the typification of some names of subdivisions of genera see Art. 22.6 and 22.7.

Recommendation 10A

10A.1. When a combination in a rank of subdivision of a genus has been published under a generic name that has not yet been typified, the type of the generic name should be selected from the subdivision of the genus that was designated as nomenclaturally typical, if that is apparent.

SECTION 3. PRIORITY

Article 11

11.1. Each family or taxon of lower rank with a particular circumscription, position, and rank can bear only one correct name, special exceptions being made for 9 families and 1 subfamily for which alternative names are permitted (see Art. 18.5 and 19.7). However, the use of separate names for the form-taxa of fungi and for morphotaxa of fossil plants is allowed under Art. 1.3, 59.4, and 59.5.

11.2. In no case does a name have priority outside the rank in which it is published (but see Art. 53.4).

Ex. 1. Campanula sect. *Campanopsis* R. Br. (Prodr.: 561. 1810) when treated as a genus is called *Wahlenbergia* Roth (1821), a name conserved against the taxonomic (heterotypic) synonym *Cervicina* Delile (1813), and not *Campanopsis* (R. Br.) Kuntze (1891).

Ex. 2. Magnolia virginiana var. *foetida* L. (1753) when raised to specific rank is called *M. grandiflora* L. (1759), not *M. foetida* (L.) Sarg. (1889).

Ex. 3. Lythrum intermedium Ledeb. (1822) when treated as a variety of *L. salicaria* L. (1753) is called *L. salicaria* var. *glabrum* Ledeb. (Fl. Ross. 2: 127. 1843), not *L. salicaria* var. *intermedium* (Ledeb.) Koehne (in Bot. Jahrb. Syst. 1: 327. 1881).

Ex. 4. When the two varieties constituting *Hemerocallis lilioasphodelus* L. (1753), var. *flava* L. and var. *fulva* L., are considered to be distinct species, the one not including the lectotype of the species name is called *H. fulva* (L.) L. (1762), but the other one bears the name *H. lilioasphodelus* L., which in the rank of species has priority over *H. flava* (L.) L. (1762).

11.3. For any taxon from family to genus inclusive, the correct name is the earliest legitimate one with the same rank, except in cases of limitation of priority by conservation (see Art. 14) or where Art. 11.7, 15, 19.4, 56, 57, or 59 apply.

Ex. 5. When *Aesculus* L. (1753), *Pavia* Mill. (1754), *Macrothyrsus* Spach (1834) and *Calothyrsus* Spach (1834) are referred to a single genus, its name is *Aesculus* L.

11.4. For any taxon below the rank of genus, the correct name is the combination of the final epithet[1] of the earliest legitimate name of the taxon in the same rank, with the correct name of the genus or species to which it is assigned, except *(a)* in cases of limitation of priority under Art. 14, 15, 56, or 57, or *(b)* if the resulting combination could not be validly published under Art. 32.1(c) or would be illegitimate under Art. 53, or *(c)* if Art. 11.7, 22.1, 26.1, or 59 rules that a different combination is to be used.

Ex. 6. *Primula* sect. *Dionysiopsis* Pax (in Jahresber. Schles. Ges. Vaterländ. Kultur 87: 20. 1909) when transferred to *Dionysia* Fenzl becomes *D.* sect. *Dionysiopsis* (Pax) Melch. (in Mitt. Thüring. Bot. Vereins 50: 164-168. 1943); the substitute name *D.* sect. *Ariadna* Wendelbo (in Bot. Not. 112: 496. 1959) is illegitimate under Art. 52.1.

Ex. 7. *Antirrhinum spurium* L. (1753) when transferred to *Linaria* Mill. is called *L. spuria* (L.) Mill. (1768).

Ex. 8. When transferring *Serratula chamaepeuce* L. (1753) to *Ptilostemon* Cass., Cassini illegitimately (Art. 52.1) named the species *P. muticus* Cass. (1826). In that genus, the correct name is *P. chamaepeuce* (L.) Less. (1832).

Ex. 9. The correct name for *Rubus aculeatiflorus* var. *taitoensis* (Hayata) T. S. Liu & T. Y. Yang (in Annual Taiwan Prov. Mus. 12: 12. 1969) is *R. taitoensis* Hayata var. *taitoensis,* because *R. taitoensis* Hayata (1911) has priority over *R. aculeatiflorus* Hayata (1915).

Ex. 10. When transferring *Spartium biflorum* Desf. (1798) to *Cytisus* Desf., Spach correctly proposed the substitute name *C. fontanesii* Spach (1849) because of the previously and validly published *C. biflorus* L'Hér. (1791); the combination *C. biflorus* based on *S. biflorum* would be illegitimate under Art. 53.1.

Ex. 11. *Spergula stricta* Sw. (1799) when transferred to *Arenaria* L. is called *A. uliginosa* Schleich. ex Schltdl. (1808) because of the existence of the name *A. stricta* Michx. (1803), based on a different type; but on further transfer to the genus *Minuartia* L. the epithet *stricta* is again available and the species is called *M. stricta* (Sw.) Hiern (1899).

Ex. 12. *Arum dracunculus* L. (1753) when transferred to *Dracunculus* Mill. is named *D. vulgaris* Schott (1832), as use of the Linnaean epithet would result in a tautonym (Art. 23.4).

Ex. 13. *Cucubalus behen* L. (1753) when transferred to *Behen* Moench was legitimately renamed *B. vulgaris* Moench (1794) to avoid the tautonym *"B. behen"*. In *Silene* L., the epithet *behen* is unavailable because of the existence of *S. behen* L. (1753). Therefore, the substitute name *S. cucubalus* Wibel (1799) was proposed. This, however, is illegitimate (Art. 52.1) since the specific epithet *vulgaris* was available. In *Silene,* the correct name of the species is *S. vulgaris* (Moench) Garcke (1869).

Ex. 14. *Helianthemum italicum* var. *micranthum* Gren. & Godr. (Fl. France 1: 171. 1847) when transferred as a variety to *H. penicillatum* Thibaud ex Dunal retains its varietal epithet and is named *H. penicillatum* var. *micranthum* (Gren. & Godr.) Grosser (in Engler, Pflanzenr. 14: 115. 1903).

---

[1] Here and elsewhere in this *Code,* the phrase "final epithet" refers to the last epithet in sequence in any particular combination, whether in the rank of a subdivision of a genus, or of a species, or of an infraspecific taxon.

Ex. 15. The final epithet of the combination *Thymus praecox* subsp. *arcticus* (Durand) Jalas (in Veröff. Geobot. Inst. ETH Stiftung Rübel Zürich 43: 190. 1970), based on *T. serpyllum* var. *arcticus* Durand (Pl. Kaneanae Groenl. 196. 1856), was first used at the rank of subspecies in the combination *T. serpyllum* L. subsp. *arcticus* (Durand) Hyl. (in Uppsala Univ. Årsskr. 1945(7): 276. 1945). However, if *T. britannicus* Ronniger (1924) is included in this taxon, the correct name at subspecies rank is *T. praecox* subsp. *britannicus* (Ronniger) Holub (in Preslia 45: 359. 1973), for which the final epithet was first used at this rank in the combination *T. serpyllum* subsp. *britannicus* (Ronniger) P. Fourn. (Quatre Fl. France: 841. 1938, "S.-E. [Sous-Espèce] *Th. Britannicus*").

Note 1. The valid publication of a name at a rank lower than genus precludes any simultaneous homonymous combination (Art. 53), irrespective of the priority of other names with the same final epithet that may require transfer to the same genus or species.

Ex. 16. Tausch included two species in his new genus *Alkanna: A. tinctoria* Tausch (1824), a new species based on *"Anchusa tinctoria"* in the sense of Linnaeus (1762), and *A. matthioli* Tausch (1824), a nomen novum based on *Lithospermum tinctorium* L. (1753). Both names are legitimate and take priority from 1824.

Ex. 17. Raymond-Hamet transferred to the genus *Sedum* both *Cotyledon sedoides* DC. (1808) and *Sempervivum sedoides* Decne. (1844). He combined the epithet of the later name, *Sempervivum sedoides,* under *Sedum* as *S. sedoides* (Decne.) Raym.-Hamet (1929), and published a new name, *S. candollei* Raym.-Hamet (1929), for the earlier name. Both names are legitimate.

11.5. When, for any taxon of the rank of family or below, a choice is possible between legitimate names of equal priority in the corresponding rank, or between available final epithets of names of equal priority in the corresponding rank, the first such choice to be effectively published (Art. 29-31) establishes the priority of the chosen name, and of any legitimate combination with the same type and final epithet at that rank, over the other competing name(s) (but see Art. 11.6).

Note 2. A choice as provided for in Art. 11.5 is effected by adopting one of the competing names, or its final epithet in the required combination, and simultaneously rejecting or relegating to synonymy the other(s), or nomenclatural (homotypic) synonyms thereof.

Ex. 18. When *Dentaria* L. (1753) and *Cardamine* L. (1753) are united, the resulting genus is called *Cardamine* because that name was chosen by Crantz (Cl. Crucif. Emend.: 126. 1769), who first united them.

Ex. 19. When *Claudopus* Gillet (1876), *Eccilia* (Fr. : Fr.) P. Kumm. (1871), *Entoloma* (Fr. ex Rabenh.) P. Kumm. (1871), *Leptonia* (Fr. : Fr.) P. Kumm. (1871), and *Nolanea* (Fr. : Fr.) P. Kumm. (1871) are united, one of the generic names simultaneously published by Kummer must be used for the combined genus. Donk, who did so (in Bull. Jard. Bot. Buitenzorg, ser. 3, 18(1): 157. 1949), selected *Entoloma,* which is therefore treated as having priority over the other names.

Ex. 20. Brown (in Tuckey, Narr. Exped. Zaire: 484. 1818) was the first to unite *Waltheria americana* L. (1753) and *W. indica* L. (1753). He adopted the name *W. indica* for the combined species, and this name is accordingly treated as having priority over *W. americana*.

Ex. 21. Baillon (in Adansonia 3: 162. 1863), when uniting for the first time *Sclerocroton integerrimus* Hochst. (1845) and *S. reticulatus* Hochst. (1845), adopted the name *Stillingia integerrima* (Hochst.) Baill. for the combined taxon. Consequently *Sclerocroton integerrimus* is treated as having priority over *S. reticulatus* irrespective of the genus *(Sclerocroton, Stillingia, Excoecaria,* or *Sapium)* to which the species is assigned.

Ex. 22. Linnaeus (1753) simultaneously published the names *Verbesina alba* and *V. prostrata*. Later (1771), he published *Eclipta erecta*, an illegitimate name because *V. alba* was cited in synonymy, and *E. prostrata*, based on *V. prostrata*. The first author to unite these taxa was Roxburgh (Fl. Ind., ed. 1832, 3: 438. 1832), who adopted the name *E. prostrata* (L.) L. Therefore *V. prostrata* is treated as having priority over *V. alba*.

Ex. 23. *Donia speciosa* and *D. formosa*, which were simultaneously published by Don (1832), were illegitimately renamed *Clianthus oxleyi* and *C. dampieri*, respectively, by Lindley (1835). Brown (in Sturt, Narr. Exped. C. Australia 2: 71. 1849) united both in a single species, adopting the illegitimate name *C. dampieri* and citing *D. speciosa* and *C. oxleyi* as synonyms; his choice is not of the kind provided for by Art. 11.5. *Clianthus speciosus* (G. Don) Asch. & Graebn. (1909), published with *D. speciosa* and *C. dampieri* listed as synonyms, is an illegitimate later homonym of *C. speciosus* (Endl.) Steud. (1840); again, conditions for a choice under Art. 11.5 were not satisfied. Ford & Vickery (1950) published the legitimate combination *C. formosus* (G. Don) Ford & Vickery and cited *D. formosa* and *D. speciosa* as synonyms, but since the epithet of the latter was unavailable in *Clianthus* a choice was not possible and again Art. 11.5 does not apply. Thompson (1990) was the first to effect an acceptable choice when publishing the combination *Swainsona formosa* (G. Don) Joy Thomps. and indicating that *D. speciosa* was a synonym of it.

11.6. An autonym is treated as having priority over the name or names of the same date and rank that established it.

Note 3. When the final epithet of an autonym is used in a new combination under the requirements of Art. 11.6, the basionym of that combination is the name from which the autonym is derived, or its basionym if it has one.

Ex. 24. By describing *Synthyris* subg. *Plagiocarpus,* Pennell (in Proc. Acad. Nat. Sci. Philadelphia 85: 86. 1933) established the name *Synthyris* Benth. subg. *Synthyris* (although using the designation *"Eusynthyris"*), and when this group is included in *Veronica*, *V.* subg. *Synthyris* (Benth.) M. M. Mart. Ort. & al. (in Taxon 53: 440. 2004) has precedence over a combination in *Veronica* based on *S.* subg. *Plagiocarpus* Pennell.

Ex. 25. *Heracleum sibiricum* L. (1753) includes *H. sibiricum* subsp. *lecokii* (Godr. & Gren.) Nyman (Consp. Fl. Eur.: 290. 1879) and *H. sibiricum* subsp. *sibiricum* automatically established at the same time. When *H. sibiricum* is included in *H. sphondylium* L. (1753) as a subspecies, the correct name for the taxon is *H. sphondylium* subsp. *sibiricum* (L.) Simonk. (Enum. Fl. Transsilv.: 266. 1887), not subsp. *lecokii*, whether or not subsp. *lecokii* is treated as distinct.

Ex. 26. The publication of *Salix tristis* var. *microphylla* Andersson (Salices Bor.-Amer.: 21. 1858) created the autonym *S. tristis* Aiton (1789) var. *tristis*, dating from 1858. If *S. tristis*, including var. *microphylla*, is recognized as a variety of *S. humilis* Marshall (1785), the correct name is *S. humilis* var. *tristis* (Aiton) Griggs (in Proc. Ohio Acad. Sci. 4: 301. 1905). However, if both varieties of *S. tristis* are recognized as varieties of *S. humilis,* then the names *S. humilis* var. *tristis* and *S. humilis* var. *microphylla* (Andersson) Fernald (in Rhodora 48: 46. 1946) are both used.

Ex. 27. In the classification adopted by Rollins and Shaw, *Lesquerella lasiocarpa* (Hook. ex A. Gray) S. Watson (1888) is composed of two subspecies, subsp. *lasiocarpa* (which includes the type of the name of the species and is cited without an author) and subsp. *berlandieri* (A. Gray) Rollins & E. A. Shaw. The latter subspecies is composed of two varieties. In that classification the correct name of the variety which includes the type of subsp. *berlandieri* is *L. lasiocarpa* var. *berlandieri* (A. Gray) Payson (1922), not *L. lasiocarpa* var. *berlandieri* (cited without an author) or *L. lasiocarpa* var. *hispida* (S. Watson) Rollins & E. A. Shaw (1972), based on *Synthlipsis berlandieri* var. *hispida* S. Watson (1882), since publication of the latter name established the autonym *S. berlandieri* A. Gray var. *berlandieri* which, at varietal rank, is treated as having priority over var. *hispida*.

11.7. For purposes of priority, names of fossil morphotaxa compete only with names based on a fossil type representing the same part, life-history stage, or preservational state (see Art. 1.2).

Ex. 28. The generic name *Sigillaria* Brongn. (1822), established for bark fragments, may in part represent the same biological taxon as the "cone-genus" *Mazocarpon* M. J. Benson (1918), which represents permineralizations, or *Sigillariostrobus* (Schimp.) Geinitz (1873), which represents compressions. Certain species of all three genera, *Sigillaria, Mazocarpon,* and *Sigillariostrobus,* have been assigned to the family *Sigillariaceae*. All these generic names can be used concurrently in spite of the fact that they may, at least in part, apply to the same organism.

Ex. 29. The morphogeneric name *Tuberculodinium* D. Wall (1967) may be retained for a genus of fossil cysts even though cysts of the same kind are known to be part of the life cycle of an extant genus that bears an earlier name, *Pyrophacus* F. Stein (1883).

Ex 30. A common Jurassic leaf-compression fossil is referred to by different authors either as *Ginkgo huttonii* (Sternb.) Heer or *Ginkgoites huttonii* (Sternb.) M. Black. Both names are in accordance with the *Code,* and either name can correctly be used, depending on whether this Jurassic morphospecies is regarded as rightly assigned to the living (non-fossil) genus *Ginkgo* L. or whether it is more appropriate to assign it to the morphogenus *Ginkgoites* Seward (type, *G. obovata* (Nath.) Seward, a Triassic leaf compression).

11.8. Names of plants (diatoms excepted) based on a non-fossil type are treated as having priority over names of the same rank based on a fossil (or subfossil) type.

Ex. 31. If *Platycarya* Siebold & Zucc. (1843), a non-fossil genus, and *Petrophiloides* Bowerb. (1840), a fossil genus, are united, the name *Platycarya* is correct for the combined genus, although it is antedated by *Petrophiloides*.

Ex. 32. Boalch and Guy-Ohlson (in Taxon 41: 529-531. 1992) united the two prasinophyte genera *Pachysphaera* Ostenf. (1899) and *Tasmanites* E. J. Newton (1875). *Pachysphaera* is based on a non-fossil type and *Tasmanites* on a fossil type. Under the *Code* in effect in 1992, *Tasmanites* had priority and was therefore adopted. Under the current *Code,* in which the exemption in Art. 11.8 applies only to diatoms and not to algae in general, *Pachysphaera* is correct for the combined genus.

Ex. 33. The generic name *Metasequoia* Miki (1941) was based on the fossil type of *M. disticha* (Heer) Miki. After discovery of the non-fossil species *M. glyptostroboides* Hu & W. C. Cheng, conservation of *Metasequoia* Hu & W. C. Cheng (1948) as based on the non-fossil type was approved. Otherwise, any new generic name based on *M. glyptostroboides* would have had to be treated as having priority over *Metasequoia* Miki.

Note 4. The provisions of Art. 11 determine priority between different names applicable to the same taxon; they do not concern homonymy. In accordance with Art. 53, later homonyms are illegitimate whether the type is fossil or non-fossil.

Ex. 34. *Endolepis* Torr. (1861), based on a non-fossil type, is an illegitimate later homonym of, and does not have priority over, *Endolepis* Schleid. (1846), based on a fossil type.

Ex. 35. *Cornus paucinervis* Hance (1881), based on a non-fossil type, is an illegitimate later homonym and does not have priority over *C. paucinervis* Heer (Fl. Tert. Helv. 3: 289. 1859), based on a fossil type.

Ex. 36. *Ficus crassipes* F. M. Bailey (1889), *F. tiliifolia* Baker (1885), and *F. tremula* Warb. (1894), each based on a non-fossil type, were illegitimate later homonyms of, respectively, *F. crassipes* (Heer) Heer (1882), *F. tiliifolia* (A. Braun) Heer (1856), and *F. tremula* (Heer) Heer (1874), each based on a fossil type. The three names with non-fossil types have been conserved against their earlier homonyms in order to maintain their use.

11.9. For purposes of priority, names in Latin form given to hybrids are subject to the same rules as are those of non-hybrid taxa of equivalent rank.

Ex. 37. The name ×*Solidaster* H. R. Wehrh. (1932) antedates ×*Asterago* Everett (1937) for the hybrids between *Aster* L. and *Solidago* L.

Ex. 38. *Anemone* ×*hybrida* Paxton (1848) antedates *A.* ×*elegans* Decne. (1852), pro sp., as the binomial for the hybrids derived from *A. hupehensis* (Lemoine & É. Lemoine) Lemoine & É. Lemoine × *A. vitifolia* Buch.-Ham. ex DC.

Ex. 39. Camus (in Bull. Mus. Natl. Hist. Nat. 33: 538. 1927) published the name ×*Agroelymus* A. Camus for a nothogenus, without a Latin description or diagnosis, mentioning only the names of the parents involved (*Agropyron* Gaertn. and *Elymus* L.). Since this name was not validly published under the *Code* then in force, Rousseau (in Mém. Jard. Bot. Montréal 29: 10-11. 1952) published a Latin diagnosis. However, the date of valid publication of ×*Agroelymus* under this *Code* (Art. H.9) is 1927, not 1952, so it antedates the name ×*Elymopyrum* Cugnac (in Bull. Soc. Hist. Nat. Ardennes 33: 14. 1938).

11.10. The principle of priority does not apply above the rank of family (but see Rec. 16B).

Article 12

12.1. A name of a taxon has no status under this *Code* unless it is validly published (see Art. 32-45).

SECTION 4. LIMITATION OF THE PRINCIPLE OF PRIORITY

Article 13

13.1. Valid publication of names for plants of the different groups is treated as beginning at the following dates (for each group a work is mentioned which is treated as having been published on the date given for that group):

Non-fossil plants:

(a) SPERMATOPHYTA and PTERIDOPHYTA, 1 May 1753 (Linnaeus, *Species plantarum,* ed. 1), except suprageneric names, 4 August 1789 (Jussieu, *Genera plantarum*).

(b) MUSCI (the *Sphagnaceae* excepted), 1 January 1801 (Hedwig, *Species muscorum*).

(c) SPHAGNACEAE and HEPATICAE, 1 May 1753 (Linnaeus, *Species plantarum,* ed. 1), except suprageneric names, 4 August 1789 (Jussieu, *Genera plantarum*).

(d) FUNGI (including slime moulds and lichen-forming fungi), 1 May 1753 (Linnaeus, *Species plantarum,* ed. 1). Names in the *Uredinales, Ustilaginales,* and *Gasteromycetes* (s. l.) adopted by Persoon (*Synopsis methodica fungorum,* 31 December 1801) and names of other fungi (excluding slime moulds) adopted by Fries (*Systema mycologicum,* vol. 1 (1 January 1821) to 3, with additional *Index* (1832), and *Elenchus fungorum,* vol. 1-2), are sanctioned (see Art. 15). For nomenclatural purposes names given to lichens apply to their fungal component.

(e) ALGAE, 1 May 1753 (Linnaeus, *Species plantarum,* ed. 1). Exceptions:

NOSTOCACEAE HOMOCYSTEAE, 1 January 1892 (Gomont, "Monographie des Oscillariées", in Ann. Sci. Nat., Bot., ser. 7, 15: 263-368; 16: 91-264). The two parts of Gomont's "Monographie", which appeared in 1892 and 1893, respectively, are treated as having been published simultaneously on 1 January 1892.

NOSTOCACEAE HETEROCYSTEAE, 1 January 1886 (Bornet & Flahault, "Révision des Nostocacées hétérocystées", in Ann. Sci. Nat., Bot., ser. 7, 3: 323-381; 4: 343-373; 5: 51-129; 7: 177-262). The four parts of the "Révision", which appeared in 1886, 1886, 1887, and 1888, respectively, are treated as having been published simultaneously on 1 January 1886.

DESMIDIACEAE (s. l.), 1 January 1848 (Ralfs, *British* Desmidieae).

OEDOGONIACEAE, 1 January 1900 (Hirn, "Monographie und Icono-graphie der Oedogoniaceen", in Acta Soc. Sci. Fenn. 27(1)).

Fossil plants:

(f) ALL GROUPS, 31 December 1820 (Sternberg, *Flora der Vorwelt, Versuch* 1: 1-24, t. 1-13). Schlotheim's *Petrefactenkunde* (1820) is regarded as published before 31 December 1820.

13.2. The group to which a name is assigned for the purposes of this Article is determined by the accepted taxonomic position of the type of the name.

Ex. 1. The genus *Porella* and its single species, *P. pinnata,* were referred by Linnaeus (1753) to the *Musci;* since the type specimen of *P. pinnata* is now accepted as belonging to the *Hepaticae,* the names were validly published in 1753.

Ex. 2. The designated type of *Lycopodium* L. (1753) is *L. clavatum* L. (1753) and the type specimen of this is currently accepted as a pteridophyte. Accordingly, although the genus is listed by Linnaeus among the *Musci,* the generic name and the names of the pteridophyte species included by Linnaeus under it were validly published in 1753.

13.3. For nomenclatural purposes, a name is treated as pertaining to a non-fossil taxon unless its type is fossil in origin. Fossil material is distinguished from non-fossil material by stratigraphic relations at the site of original occurrence. In cases of doubtful stratigraphic relations, provisions for non-fossil taxa apply.

13.4. Generic names which appear in Linnaeus's *Species plantarum,* ed. 1 (1753) and ed. 2 (1762-1763), are associated with the first subsequent description given under those names in Linnaeus's *Genera plantarum,* ed. 5 (1754) and ed. 6 (1764). The spelling of the generic names included in *Species plantarum,* ed. 1, is not to be altered because a different spelling has been used in *Genera plantarum,* ed. 5.

13.5. The two volumes of Linnaeus's *Species plantarum,* ed. 1 (1753), which appeared in May and August, 1753, respectively, are treated as having been published simultaneously on 1 May 1753.

Ex. 3. The generic names *Thea* L. (Sp. Pl.: 515. 24 Mai 1753), and *Camellia* L. (Sp. Pl.: 698. 16 Aug 1753; Gen. Pl., ed. 5: 311. 1754), are treated as having been published simultaneously on 1 May 1753. Under Art. 11.5 the combined genus bears the name *Camellia,* since Sweet (Hort. Suburb. Lond.: 157. 1818), who was the first to unite the two genera, chose that name, and cited *Thea* as a synonym.

13.6. Names of anamorphs of fungi with a pleomorphic life cycle do not, irrespective of priority, affect the nomenclatural status of the names of the correlated holomorphs (see Art. 59.4).

Article 14

14.1. In order to avoid disadvantageous nomenclatural changes entailed by the strict application of the rules, and especially of the principle of priority in starting from the dates given in Art. 13, this *Code* provides, in App. II-IV, lists of names of families, genera, and species that are conserved (nomina conservanda) (see Rec. 50E). Conserved names are legitimate even though initially they may have been illegitimate.

14.2. Conservation aims at retention of those names which best serve stability of nomenclature.

14.3. The application of both conserved and rejected names is determined by nomenclatural types. The type of the specific name cited as the type of a conserved generic name may, if desirable, be conserved and listed in App. III.

14.4. A conserved name of a family or genus is conserved against all other names in the same rank based on the same type (nomenclatural, i.e. homotypic, synonyms, which are to be rejected) whether or not these are cited in the corresponding list as rejected names, and against those names based on different types (taxonomic, i.e. heterotypic, synonyms) that are listed as rejected[1]. A conserved name of a species is conserved against all names listed as rejected, and against all combinations based on the rejected names.

Note 1. The *Code* does not provide for conservation of a name against itself, i.e. against an "isonym" (Art. 6 Note 2), the same name with the same type but with a different place and date of valid publication and perhaps with a different authorship (but see Art. 14.9) than is given in the relevant entry in App. II, III, or IV.

Note 2. A species name listed as conserved or rejected in App. IV may have been published as the name of a new taxon, or as a combination based on an earlier name. Rejection of a name based on an earlier name does not in itself preclude the use of the earlier name since that name is not "a combination based on a rejected name" (Art. 14.4).

Ex. 1. Rejection of *Lycopersicon lycopersicum* (L.) H. Karst. in favour of *L. esculentum* Mill. does not preclude the use of the homotypic *Solanum lycopersicum* L.

14.5. When a conserved name competes with one or more names based on different types and against which it is not explicitly conserved, the earliest of the competing names is adopted in accordance with Art. 11, except for some

---

[1] The *International code of zoological nomenclature* and the *International code of nomenclature of bacteria* use the terms "objective synonym" and "subjective synonym" for nomenclatural and taxonomic synonym, respectively.

conserved family names (App. IIB), which are conserved against unlisted names.

Ex. 2. If *Weihea* Spreng. (1825) is united with *Cassipourea* Aubl. (1775), the combined genus will bear the prior name *Cassipourea,* although *Weihea* is conserved and *Cassipourea* is not.

Ex. 3. If *Mahonia* Nutt. (1818) is united with *Berberis* L. (1753), the combined genus will bear the prior name *Berberis,* although *Mahonia* is conserved and *Berberis* is not.

Ex. 4. Nasturtium R. Br. (1812) was conserved only against the homonym *Nasturtium* Mill. (1754) and the nomenclatural (homotypic) synonym *Cardaminum* Moench (1794); consequently if reunited with *Rorippa* Scop. (1760) it must bear the name *Rorippa.*

14.6. When a name of a taxon has been conserved against an earlier name based on a different type, the latter is to be restored, subject to Art. 11, if it is considered the name of a taxon at the same rank distinct from that of the nomen conservandum, except when the earlier rejected name is a homonym of the conserved name.

Ex. 5. The generic name *Luzuriaga* Ruiz & Pav. (1802) is conserved against the earlier names *Enargea* Banks ex Gaertn. (1788) and *Callixene* Comm. ex Juss. (1789). If, however, *Enargea* is considered to be a separate genus, the name *Enargea* is retained for it.

Ex. 6. To preserve the name *Roystonea regia* (Kunth) O. F. Cook (1900), its basionym *Oreodoxa regia* Kunth (1816) is conserved against *Palma elata* W. Bartram (1791). However, the latter remains available as the basionym of *R. elata* (W. Bartram) F. Harper (1946), if this name is applied to a species distinct from *R. regia.*

14.7. A rejected name, or a combination based on a rejected name, may not be restored for a taxon that includes the type of the corresponding conserved name.

Ex. 7. Enallagma Baill. (1888) is conserved against *Dendrosicus* Raf. (1838), but not against *Amphitecna* Miers (1868); if *Enallagma* and *Amphitecna* are united, the combined genus must bear the name *Amphitecna,* although the latter is not explicitly conserved against *Dendrosicus.*

14.8. The listed type of a conserved name may not be changed except by the procedure outlined in Art. 14.12.

Ex. 8. Bullock & Killick (in Taxon 6: 239. 1957) published a proposal that the listed type of *Plectranthus* L'Hér. be changed from *P. punctatus* (L. f.) L'Hér. to *P. fruticosus* L'Hér. This proposal was approved by the appropriate Committees and by an International Botanical Congress.

14.9. A name may be conserved with a different type from that designated by the author or determined by application of the *Code* (see also Art. 10.4). Such a name may be conserved either from its place of valid publication (even though the type may not then have been included in the named taxon) or from a later publication by an author who did include the type as conserved. In the latter case the original name and the name as conserved are treated as if they were homonyms (Art. 53), whether or not the name as conserved was accompanied by a description or diagnosis of the taxon named.

Ex. 9. Bromus sterilis L. (1753) has been conserved from its place of valid publication even though its conserved type, a specimen (*Hubbard 9045,* E) collected in 1932, was not originally included in Linnaeus's species.

Ex. 10. Protea L. (1753) did not include the conserved type of the generic name, *P. cynaroides* (L.) L. (1771), which in 1753 was placed in the genus *Leucadendron. Protea* was therefore conserved from the 1771 publication, and *Protea* L. (1771), although not designed to be a new generic name and still including the original type elements, is treated as if it were a validly published homonym of *Protea* L. (1753).

14.10. A conserved name, with any corresponding autonym, is conserved against all earlier homonyms. An earlier homonym of a conserved name is not made illegitimate by that conservation but is unavailable for use; if not otherwise illegitimate, it may serve as basionym of another name or combination based on the same type (see also Art. 55.3).

Ex. 11. The generic name *Smithia* Aiton (1789), conserved against *Damapana* Adans. (1763), is thereby conserved automatically against the earlier homonym *Smithia* Scop. (1777).

14.11. A name may be conserved in order to preserve a particular spelling or gender. A name so conserved is to be attributed without change of priority to the author who validly published it, not to an author who later introduced the conserved spelling or gender.

Ex. 12. The spelling *Rhodymenia,* used by Montagne (1839), has been conserved against the original spelling *Rhodomenia,* used by Greville (1830). The name is to be cited as *Rhodymenia* Grev. (1830).

Note 3. The date of conservation does not affect the priority (Art. 11) of a conserved name, which is determined only on the basis of the date of valid publication (Art. 32-45; but see Art. 14.9).

14.12. The lists of conserved names will remain permanently open for additions and changes. Any proposal of an additional name must be accompanied by a detailed statement of the cases both for and against its conservation. Such proposals must be submitted to the General Committee (see Div. III), which will refer them for examination to the committees for the various taxonomic groups.

14.13. Entries of conserved names may not be deleted.

14.14. When a proposal for the conservation of a name, or of its rejection under Art. 56, has been approved by the General Committee after study by the Committee for the taxonomic group concerned, retention (or rejection) of that name is authorized subject to the decision of a later International Botanical Congress.

Recommendation 14A

14A.1. When a proposal for the conservation of a name, or of its rejection under Art. 56, has been referred to the appropriate Committee for study, authors should follow existing usage of names as far as possible pending the General Committee's recommendation on the proposal.

Article 15

15.1. Names sanctioned under Art. 13.1(d) are treated as if conserved against earlier homonyms and competing synonyms. Such names, once sanctioned, remain sanctioned even if elsewhere in the sanctioning works the sanctioning author does not recognize them.

Ex. 1. Agaricus ericetorum Fr. was accepted by Fries in *Systema mycologicum* (1821), but later (1828) regarded by him as a synonym of *A. umbelliferus* L. and not included in his *Index* (1832) as an accepted name. Nevertheless *A. ericetorum* is a sanctioned name.

15.2. An earlier homonym of a sanctioned name is not made illegitimate by that sanctioning but is unavailable for use; if not otherwise illegitimate, it may serve as a basionym of another name or combination based on the same type (see also Art. 55.3).

Ex. 2. Patellaria Hoffm. (1789) is an earlier homonym of the sanctioned generic name *Patellaria* Fr. (1822) : Fr. Hoffmann's name is legitimate but unavailable for use. *Lecanidion* Endl. (1830), based on the same type as *Patellaria* Fr. : Fr., is illegitimate under Art. 52.1.

Ex. 3. Agaricus cervinus Schaeff. (1774) is an earlier homonym of the sanctioned *A. cervinus* Hoffm. (1789) : Fr.; Schaeffer's name is unavailable for use, but it is legitimate and may serve as basionym for combinations in other genera. In *Pluteus* Fr. the combination is cited as *P. cervinus* (Schaeff.) P. Kumm. and has priority over the taxonomic (heterotypic) synonym *P. atricapillus* (Batsch) Fayod, based on *A. atricapillus* Batsch (1786).

15.3. When, for a taxon from family to and including genus, two or more sanctioned names compete, Art. 11.3 governs the choice of the correct name (see also Art. 15.5).

15.4. When, for a taxon below the rank of genus, two or more sanctioned names and/or two or more names with the same final epithet and type as a sanctioned name compete, Art. 11.4 governs the choice of the correct name.

Note 1. The date of sanctioning does not affect the priority (Art. 11) of a sanctioned name, which is determined only on the basis of valid publication. In particular, when two or more homonyms are sanctioned only the earliest of them may be used, the later being illegitimate under Art. 53.2.

Ex. 4. Fries (Syst. Mycol. 1: 41. 1821) accepted *Agaricus flavovirens* Pers. (1793), treating *A. equestris* L. (1753) as a synonym. Later (Elench. Fung. 1: 6. 1828) he stated "Nomen prius et aptius arte restituendum" and accepted *A. equestris*. Both names are sanctioned, but when they are considered synonyms *A. equestris,* having priority, is to be used.

15.5. A name which neither is sanctioned nor has the same type and final epithet as a sanctioned name in the same rank may not be applied to a taxon which includes the type of a sanctioned name in that rank the final epithet of which is available for the required combination (see Art. 11.4(b)).

15.6. Conservation (Art. 14) and explicit rejection (Art. 56.1) override sanctioning.

CHAPTER III. NOMENCLATURE OF TAXA ACCORDING TO THEIR RANK

SECTION 1. NAMES OF TAXA ABOVE THE RANK OF FAMILY

Article 16

16.1. The name of a taxon above the rank of family is treated as a noun in the plural and is written with an initial capital letter. Such names may be either *(a)* automatically typified names, formed by replacing the termination *-aceae* in a legitimate name of an included family based on a generic name by the termination denoting their rank (preceded by the connecting vowel *-o-* if the termination begins with a consonant), as specified in Rec. 16A.1-3 and Art. 17.1; or *(b)* descriptive names, not so formed, which may be used unchanged at different ranks.

Ex. 1. Automatically typified names above the rank of family: *Magnoliophyta*, based on *Magnoliaceae; Gnetophytina*, based on *Gnetaceae; Pinopsida*, based on *Pinaceae; Marattiidae*, based on *Marattiaceae; Caryophyllidae* and *Caryophyllales*, based on *Caryophyllaceae; Fucales*, based on *Fucaceae; Bromeliineae*, based on *Bromeliaceae*.

Ex. 2. Descriptive names above the rank of family: *Anthophyta, Chlorophyta, Parietales; Ascomycota, Ascomycotina, Ascomycetes; Angiospermae, Centrospermae, Coniferae, Enantioblastae, Gymnospermae*.

16.2. For automatically typified names, the name of the subdivision or subphylum that includes the type of the adopted name of a division or phylum, the name of the subclass that includes the type of the adopted name of a class, and the name of the suborder that includes the type of the adopted name of an order are to be based on the same type as the corresponding higher-ranked name.

Ex. 3. Pteridophyta Bergen & B. M. Davis (1906) and *Pteridophytina* B. Boivin (1956); *Gnetopsida* Engl. (1898) and *Gnetidae* Cronquist & al. (1966); *Liliales* Perleb (1826) and *Liliineae* Rchb. (1841).

16.3. When an automatically typified name above the rank of family has been published with an improper Latin termination, not agreeing with those provided for in Rec. 16A.1-3 and Art. 17.1, the termination must be changed to conform with these standards, without change of the author citation or date of publication (see Art. 32.7). However, if such names are published with a non-Latin termination they are not validly published.

Ex. 4. *"Cactarieae"* (Dumortier, 1829, based on *Cactaceae*) and *"Coriales"* (Lindley, 1833, based on *Coriariaceae*), both published for taxa of the rank of order, are to be corrected to *Cactales* Dumort. (1829) and *Coriariales* Lindl. (1833), respectively.

Ex. 5. However, Acoroidées (Kirschleger, Fl. Alsace 2: 103. 1853 - Jul 1857), published for a taxon of the rank of order, is not to be accepted as *"Acorales* Kirschl.", as it has a French rather than a Latin termination. The name *Acorales* was later validly published by Reveal (in Phytologia 79: 72. 1996).

Note 1. The terms "divisio" and "phylum", and their equivalents in modern languages, are treated as referring to one and the same rank. When "divisio" and "phylum" are used simultaneously to denote different ranks, this is to be treated as informal usage of rank-denoting terms (see Art. 33.11).

16.4. Where one of the word elements *-clad-, -cocc-, -cyst-, -monad-, -myces-, -nemat-,* or *-phyton-,* being the genitive singular stem of the second part of a name of an included genus, has been omitted before the termination *-phyceae, -phycota* (algae), *-mycetes, -mycota* (fungi), *-opsida,* or *-phyta* (other groups of plants), the shortened class name or division or phylum name is regarded as based on the generic name in question if such derivation is obvious or is indicated at establishment of the group name. These word elements may also be omitted before the termination for sub-division or subphylum as appropriate in each case.

Ex. 6. The name *Raphidophyceae* Chadef. ex P. C. Silva (1980) was indicated by its author to be based on *Raphidomonas* F. Stein (1878). The name *Saccharomycetes* G. Winter (1881) is regarded as being based on *Saccharomyces* Meyen (1838). The name *Trimerophytina* H. P. Banks (1975) was indicated by its author to be based on *Trimerophyton* Hopping (1956).

Note 2. The principle of priority does not apply above the rank of family (Art. 11.10; but see Rec. 16B).

Recommendation 16A

16A.1. A name of a division or phylum should end in *-phyta* unless the taxon is a division or phylum of fungi, in which case its name should end in *-mycota*.

16A.2. A name of a subdivision or subphylum should end in *-phytina,* unless it is a subdivision or subphylum of fungi, in which case it should end in *-mycotina*.

16A.3. A name of a class or of a subclass should end as follows:

(a) In the algae: *-phyceae* (class) and *-phycidae* (subclass);

(b) In the fungi: *-mycetes* (class) and *-mycetidae* (subclass);

(c) In other groups of plants: *-opsida* (class) and *-idae,* but not *-viridae* (subclass).

Recommendation 16B

16B.1. In choosing among typified names for a taxon above the rank of family, authors should generally follow the principle of priority.

Article 17

17.1. Automatically typified names of orders or suborders are to end in *-ales* (but not *-virales*) and *-ineae,* respectively.

17.2. Names intended as names of orders, but published with their rank denoted by a term such as "cohors", "nixus", "alliance", or "Reihe" instead of "order", are treated as having been published as names of orders.

Recommendation 17A

17A.1. Authors should not publish new names for orders that include a family from the name of which an existing ordinal name is derived.

SECTION 2. NAMES OF FAMILIES AND SUBFAMILIES, TRIBES AND SUBTRIBES

Article 18

18.1. The name of a family is a plural adjective used as a noun; it is formed from the genitive singular of a name of an included genus by replacing the genitive singular inflection (Latin *-ae, -i, -us, -is;* transliterated Greek *-ou, -os, -es, -as,* or *-ous,* and its equivalent *-eos*) with the termination *-aceae* (but see Art. 18.5). For generic names of non-classical origin, when analogy with classical names is insufficient to determine the genitive singular, *-aceae* is added to the full word. Likewise, when formation from the genitive singular of a generic name results in a homonym, *-aceae* may be added to the nominative singular. For generic names with alternative genitives the one implicitly used by the original author must be maintained, except that the genitive of names ending in *-opsis* is, in accordance with botanical tradition, always *-opsidis.*

Ex. 1. Family names based on a generic name of classical origin: *Rosaceae* (from *Rosa, Rosae*), *Salicaceae* (from *Salix, Salicis*), *Plumbaginaceae* (from *Plumbago, Plumbaginis*), *Rhodophyllaceae* (from *Rhodophyllus, Rhodophylli*), *Rhodophyllidaceae* (from *Rhodophyllis, Rhodophyllidos*), *Sclerodermataceae* (from *Scleroderma, Sclerodermatos*), *Aextoxicaceae* (from *Aextoxicon, Aextoxicou*), *Potamogetonaceae* (from *Potamogeton, Potamogetonos*).

Ex. 2. Family names based on a generic name of non-classical origin: *Nelumbonaceae* (from *Nelumbo, Nelumbonis,* declined by analogy with *umbo, umbonis*), *Ginkgoaceae* (from *Ginkgo,* indeclinable).

18.2. Names intended as names of families, but published with their rank denoted by one of the terms "order" (ordo) or "natural order" (ordo naturalis) instead of "family", are treated as having been published as names of families (see also Art. 19.2), unless this treatment would result in a taxonomic sequence with a misplaced rank-denoting term.

Ex. 3. *Cyperaceae* Juss. (1789), *Lobeliaceae* Juss. (1813), and *Xylomataceae* Fr. (1820) were published as "ordo *Cyperoideae*", "ordo naturalis *Lobeliaceae*", and "ordo *Xylomaceae*", respectively.

Note 1. If the term "family" is simultaneously used to denote a rank different from "order" or "natural order", a name published for a taxon at the latter rank cannot be considered to have been published as the name of a family.

*Ex. 4. Names published at the rank of order ("řad") by Berchtold & Presl (*O přirozenosti rostlin ....* 1820) are not to be treated as having been published at the rank of family, since the term family ("čeled'") was sometimes used to denote a rank below the rank of order.

18.3. A name of a family based on an illegitimate generic name is illegitimate unless conserved.

Ex. 5. *Caryophyllaceae* Juss., nom. cons. (from *Caryophyllus* Mill. non L.); *Winteraceae* R. Br. ex Lindl., nom. cons. (from *Wintera* Murray, an illegitimate synonym of *Drimys* J. R. Forst. & G. Forst.).

18.4. When a name of a family has been published with an improper Latin termination, the termination must be changed to conform with the rule, without change of the author citation or date of publication (see Art. 32.7). However, if such a name is published with a non-Latin termination, it is not validly published.

Ex. 6. "*Coscinodisceae*" (Kützing 1844), published to designate a family, is to be accepted as *Coscinodiscaceae* Kütz. 1844 and not attributed to De Toni, who first used the correct spelling (in Notarisia 5: 915. 1890).

Ex. 7. "*Atherospermeae*" (Brown 1814), published to designate a family, is to be accepted as *Atherospermataceae* R. Br. and not attributed to Airy Shaw (in Willis, Dict. Fl. Pl., ed. 7: 104. 1966), who first used the correct spelling, or to Lindley (Veg. Kingd.: 300. 1846), who used the spelling "*Atherospermaceae*".

Ex. 8. However, Tricholomées (Roze in Bull. Soc. Bot. France 23: 49. 1876), published to designate a family, is not to be accepted as "*Tricholomataceae* Roze", as it has a French rather than a Latin termination. The name *Tricholomataceae* was finally validly published by Pouzar (1983; see App. IIA).

18.5. The following names, of long usage, are treated as validly published: *Compositae* (*Asteraceae;* type, *Aster* L.); *Cruciferae* (*Brassicaceae;* type,

Brassica L.); *Gramineae* (*Poaceae;* type, *Poa* L.); *Guttiferae* (*Clusiaceae;* type, *Clusia* L.); *Labiatae* (*Lamiaceae;* type, *Lamium* L.); *Leguminosae* (*Fabaceae;* type, *Faba* Mill. [= *Vicia* L.]); *Palmae* (*Arecaceae;* type, *Areca* L.); *Papilionaceae* (*Fabaceae;* type, *Faba* Mill.); *Umbelliferae* (*Apiaceae;* type, *Apium* L.). When the *Papilionaceae* are regarded as a family distinct from the remainder of the *Leguminosae,* the name *Papilionaceae* is conserved against *Leguminosae.*

18.6. The use, as alternatives, of the family names indicated in parentheses in Art. 18.5 is authorized.

Article 19

19.1. The name of a subfamily is a plural adjective used as a noun; it is formed in the same manner as the name of a family (Art. 18.1) but by using the termination *-oideae* instead of *-aceae.*

19.2. Names intended as names of subfamilies, but published with their rank denoted by the term "suborder" (subordo) instead of subfamily, are treated as having been published as names of subfamilies (see also Art. 18.2), unless this would result in a taxonomic sequence with a misplaced rank-denoting term.

Ex. 1. Cyrilloideae Torr. & A. Gray (Fl. N. Amer. 1: 256. 1838) and *Sphenocleoideae* Lindl. (Intr. Nat. Syst. Bot., ed. 2: 238. 1836) were published as "suborder *Cyrilleae*" and "Sub-Order ? *Sphenocleaceae*", respectively.

Note 1. If the term "subfamily" is simultaneously used to denote a rank different from "suborder", a name published for a taxon at the latter rank cannot be considered to have been published as the name of a subfamily.

19.3. A tribe is designated in a similar manner, with the termination *-eae,* and a subtribe similarly with the termination *-inae* (but not *-virinae*).

19.4. The name of any subdivision of a family that includes the type of the adopted, legitimate name of the family to which it is assigned is to be based on the generic name equivalent to that type (Art. 10.6; but see Art. 19.7).

Ex. 2. The type of the family name *Rosaceae* Juss. is *Rosa* L. and hence the subfamily and tribe which include *Rosa* are to be called *Rosoideae* Endl. and *Roseae* DC.

Ex. 3. The type of the family name *Gramineae* Juss. (nom. alt., *Poaceae* Barnhart – see Art. 18.5) is *Poa* L. and hence the subfamily, tribe and subtribe which include *Poa* are to be called *Pooideae* Asch., *Poëae* R. Br. and *Poinae* Dumort.

Note 2. This provision applies only to the names of those subordinate taxa that include the type of the adopted name of the family (but see Rec. 19A.2).

Ex. 4. The subfamily including the type of the family name *Ericaceae* Juss. (*Erica* L.), irrespective of priority, is to be called *Ericoideae* Endl., and the tribe including this type is called *Ericeae* D. Don. However, the correct name of the tribe including both *Rhododendron* L., the type of the subfamily name *Rhododendroideae* Endl., and *Rhodora* L. is *Rhodoreae* D. Don (1834) not *Rhododendreae* Brongn. (1843).

19.5. A name of a subdivision of a family based on an illegitimate generic name that is not the base of a conserved family name is illegitimate.

Ex. 5. The name *Caryophylloideae* Arn. (1832), based on *Caryophyllaceae* Juss., nom. cons., is legitimate although it is ultimately based on the illegitimate *Caryophyllus* Mill. non L.

19.6. When a name of a taxon assigned to one of the above categories has been published with an improper Latin termination, such as *-eae* for a subfamily or *-oideae* for a tribe, the termination must be changed to accord with the rule, without change of the author citation or date of publication (see Art. 32.7). However, if such names are published with a non-Latin termination they are not validly published.

Ex. 6. "*Climacieae*" (Grout, Moss Fl. N. Amer. 3: 4. 1928), published to designate a subfamily, is to be changed to *Climacioideae* Grout (1928).

Ex. 7. However, Melantheen (Kittel in Richard, Nouv. Elém. Bot., ed. 3, Germ. Transl.: 727. 1840), published to designate a tribe, is not to be accepted as "*Melanthieae* Kitt.", as it has a German rather than a Latin termination. The name *Melanthieae* was later validly published by Grisebach (Spic. Fl. Rumel. 2: 377. 1846).

19.7. When the *Papilionaceae* are included in the family *Leguminosae* (nom. alt., *Fabaceae;* see Art. 18.5) as a subfamily, the name *Papilionoid-eae* may be used as an alternative to *Faboideae.*

Recommendation 19A

19A.1. When a family is changed to the rank of a subdivision of a family, or the inverse change occurs, and no legitimate name is available in the new rank, the name should be retained, and only its termination *(-aceae, -oideae, -eae, -inae)* altered.

Ex. 1. The subtribe *Drypetinae* Griseb. (1859) when raised to the rank of tribe was named *Drypeteae* Hurus. (1954); the subtribe *Antidesmatinae* Müll. Arg. (1865) when raised to the rank of subfamily was named *Antidesmatoideae* Hurus. (1954).

19A.2. When a subdivision of a family is changed to another such rank, and no legitimate name is available in the new rank, its name should be based on the same generic name as the name in the former rank.

Ex. 2. Among the tribes of the family *Ericaceae* are *Pyroleae* D. Don, *Monotropeae* D. Don, and *Vaccinieae* D. Don, none of which includes the type of the family name (*Erica* L.). The later names *Pyroloideae* A. Gray, *Monotropoideae* A. Gray, and *Vaccinioideae* Endl. are based on the same generic names.

SECTION 3. NAMES OF GENERA AND SUBDIVISIONS OF GENERA

Article 20

20.1. The name of a genus is a noun in the nominative singular, or a word treated as such, and is written with an initial capital letter (see Art. 60.2). It may be taken from any source whatever, and may even be composed in an absolutely arbitrary manner, but it must not end in *-virus*.

Ex. 1. Rosa, Convolvulus, Hedysarum, Bartramia, Liquidambar, Gloriosa, Impatiens, Rhododendron, Manihot, Ifloga (an anagram of *Filago*).

20.2. The name of a genus may not coincide with a Latin technical term in use in morphology at the time of publication unless it was published before 1 January 1912 and accompanied by a specific name published in accordance with the binary system of Linnaeus.

Ex. 2. "Radicula" (Hill, 1756) coincides with the Latin technical term "radicula" (radicle) and was not accompanied by a specific name in accordance with the binary system of Linnaeus. The name *Radicula* is correctly attributed to Moench (1794), who first combined it with specific epithets.

Ex. 3. Tuber F. H. Wigg. : Fr., when published in 1780, was accompanied by a binary specific name (*Tuber gulosorum* F. H. Wigg.) and is therefore validly published, even though it coincides with a Latin technical term.

Ex. 4. The intended generic names *"Lanceolatus"* (Plumstead, 1952) and *"Lobata"* (Chapman, 1952) coincide with Latin technical terms and are therefore not validly published.

Ex. 5. Cleistogenes Keng (1934) coincides with "cleistogenes", the English plural of a technical term in use at the time of publication. Keng's name is validly published, however, because the technical term is not Latin. *Kengia* Packer (1960), published as a replacement name for *Cleistogenes,* is illegitimate under Art. 52.1.

Ex. 6. Words such as *"radix", "caulis", "folium", "spina",* etc., cannot now be validly published as generic names.

20.3. The name of a genus may not consist of two words, unless these words are joined by a hyphen.

Ex. 7. "Uva ursi", as originally published by Miller (1754), consisted of two separate words unconnected by a hyphen, and is therefore not validly published (Art. 32.1(c)); the name is correctly attributed to Duhamel (1755) as *Uva-ursi* (hyphenated when published).

Ex. 8. However, names such as *Quisqualis* L. (formed by combining two words into one when originally published), *Neves-armondia* K. Schum., *Sebastiano-schaueria* Nees, and *Solms-laubachia* Muschl. ex Diels (all hyphenated when originally published) are validly published.

Note 1. The names of intergeneric hybrids are formed according to the provisions of Art. H.6.

20.4. The following are not to be regarded as generic names:

(a) Words not intended as names.

Ex. 9. The designation *"Anonymos"* was applied by Walter (Fl. Carol.: 2, 4, 9, etc. 1788) to 28 different genera to indicate that they were without names.

Ex. 10. *"Schaenoides"* and *"Scirpoides"*, as used by Rottbøll (Descr. Pl. Rar.: 14, 27. 1772) to indicate unnamed genera resembling *Schoenus* and *Scirpus* which he stated (on p. 7) that he intended to name later, are token words and not generic names. These unnamed genera were later legitimately named *Kyllinga* Rottb. and *Fuirena* Rottb.

(b) Unitary designations of species.

Note 2. Examples such as *"Leptostachys"* and *"Anthopogon"*, listed in pre-Tokyo editions of the *Code,* were from publications now listed in App. VI.

Recommendation 20A

20A.1. Authors forming generic names should comply with the following advice:

(a) To use Latin terminations insofar as possible.

(b) To avoid names not readily adaptable to the Latin language.

(c) Not to make names which are very long or difficult to pronounce in Latin.

(d) Not to make names by combining words from different languages.

(e) To indicate, if possible, by the formation or ending of the name the affinities or analogies of the genus.

(f) To avoid adjectives used as nouns.

(g) Not to use a name similar to or derived from the epithet in the name of one of the species of the genus.

(h) Not to dedicate genera to persons quite unconnected with botany or at least with natural science.

(i) To give a feminine form to all personal generic names, whether they commemorate a man or a woman (see Rec. 60B).

(j) Not to form generic names by combining parts of two existing generic names, because such names are likely to be confused with nothogeneric names (see Art. H.6).

Ex. 1. *Hordelymus* (Jess.) Harz is based on *Hordeum* [unranked] *Hordelymus* Jess. The epithet was formed by combining parts of the generic names *Hordeum* L. and *Elymus* L. (see also Art. H.3 Ex. 2).

Article 21

21.1. The name of a subdivision of a genus is a combination of a generic name and a subdivisional epithet. A connecting term (subgenus, sectio, series, etc.) is used to denote the rank.

Note 1. Names of subdivisions of the same genus, even if they differ in rank, are treated as homonyms if they have the same epithet but are based on different types (Art. 53.4), the connecting term not being part of the name.

21.2. The epithet is either of the same form as a generic name, or a noun in the genitive plural, or a plural adjective agreeing in gender with the generic name, but not a noun in the genitive singular. It is written with an initial capital letter (see Art. 32.7 and 60.2).

21.3. The epithet in the name of a subdivision of a genus is not to be formed from the name of the genus to which it belongs by adding the prefix *Eu-* (see also Art. 22.2).

Ex. 1. Costus subg. *Metacostus; Ricinocarpos* sect. *Anomodiscus; Valeriana* sect. *Valerianopsis; Euphorbia* sect. *Tithymalus; Pleione* subg. *Scopulorum; Euphorbia* subsect. *Tenellae; Sapium* subsect. *Patentinervia; Arenaria* ser. *Anomalae;* but not *Carex* sect. *"Eucarex".*

21.4. The use of a binary combination instead of a subdivisional epithet is not admissible. Contrary to Art. 32.1(c), names so constructed are validly published but are to be altered to the proper form without change of author citation or date of publication.

Ex. 2. Sphagnum "b. *Sph. rigida*" (Lindberg in Öfvers. Förh. Kongl. Svenska Vetensk.-Akad. 19: 135. 1862) and *S.* sect. *"Sphagna rigida"* (Limpricht, Laubm. Deutschl. 1: 116. 1885) are to be cited as *Sphagnum* [unranked] *Rigida* Lindb. and *S.* sect. *Rigida* (Lindb.) Limpr., respectively.

Note 2. The names of hybrids with the rank of a subdivision of a genus are formed according to the provisions of Art. H.7.

Recommendation 21A

21A.1. When it is desired to indicate the name of a subdivision of the genus to which a particular species belongs in connection with the generic name and specific epithet, the subdivisional epithet should be placed in parentheses between the two; when desirable, the subdivisional rank may also be indicated.

Ex. 1. Astragalus (Cycloglottis) contortuplicatus; A. (Phaca) umbellatus; Loranthus (sect. *Ischnanthus) gabonensis.*

Recommendation 21B

21B.1. Recommendations made for forming the name of a genus (Rec. 20A) apply equally to an epithet of a subdivision of a genus, unless Rec. 21B.2-4 recommend otherwise.

21B.2. The epithet in the name of a subgenus or section is preferably a noun, that in the name of a subsection or lower subdivision of a genus preferably a plural adjective.

21B.3. Authors, when proposing new epithets for names of subdivisions of genera, should avoid those in the form of a noun when other co-ordinate subdivisions of the same genus have them in the form of a plural adjective, and vice-versa. They should also avoid, when proposing an epithet for a name of a subdivision of a genus, one already used for a subdivision of a closely related genus, or one which is identical with the name of such a genus.

21B.4. When a section or a subgenus is raised to the rank of genus, or the inverse change occurs, the original name or epithet should be retained unless the resulting name would be contrary to this *Code*.

Article 22

22.1. The name of any subdivision of a genus that includes the type of the adopted, legitimate name of the genus to which it is assigned is to repeat that generic name unaltered as its epithet, not followed by an author citation (see Art. 46). Such names are termed autonyms (Art. 6.8; see also Art. 7.6).

Ex. 1. The subgenus which includes the type of the name *Rhododendron* L. is to be named *Rhododendron* L. subg. *Rhododendron.*

Note 1. This provision applies only to the names of those subordinate taxa that include the type of the adopted name of the genus (but see Rec. 22A).

22.2. A name of a subdivision of a genus that includes the type (i.e. the original type or all elements eligible as type or the previously designated type) of the adopted, legitimate name of the genus is not validly published unless its epithet repeats the generic name unaltered. For the purposes of this provision, explicit indication that the nomenclaturally typical element is included is considered as equivalent to inclusion of the type, whether or not it has been previously designated (see also Art. 21.3).

Ex. 2. "*Dodecatheon* sect. *Etubulosa*" (Knuth in Engler, Pflanzenr. 22: 234. 1905) was not validly published since it was proposed for a section that included *D. meadia* L., the original type of the generic name *Dodecatheon* L.

Ex. 3. *Cactus* [unranked] *Melocactus* L. (Gen. Pl., ed. 5: 210. 1754) was proposed for one of four unranked (Art. 35.3), named subdivisions of the genus *Cactus,* comprising *C. melocactus* L. (its type under Art. 22.6) and *C. mammillaris* L. It is validly published, even though *C. mammillaris* was subsequently designated as the type of *Cactus* L. (by Coulter in Contr. U. S. Natl. Herb. 3: 95. 1894).

22.3. The first instance of valid publication of a name of a subdivision of a genus under a legitimate generic name automatically establishes the corresponding autonym (see also Art. 11.6 and 32.8).

Ex. 4. The subgenus of *Malpighia* L. that includes the lectotype of the generic name (*M. glabra* L.) is called *M.* subg. *Malpighia,* not *M.* subg. *Homoiostylis* Nied.; and the section of *Malpighia* that includes the lectotype of the generic name is called *M.* sect. *Malpighia,* not *M.* sect. *Apyrae* DC.

Ex. 5. However, the correct name of the section of the genus *Rhododendron* L. that includes *R. luteum* Sweet, the type of *R.* subg. *Anthodendron* (Rchb.) Rehder, is *R.* sect. *Pentanthera* G. Don, the oldest legitimate name for the section, and not *R.* sect. *Anthodendron.*

22.4. The epithet in the name of a subdivision of a genus may not repeat unchanged the correct name of the genus, unless the two names have the same type.

22.5. The epithet in the name of a subdivision of a genus may not repeat the generic name unaltered if the latter is illegitimate.

22.6. When the epithet in the name of a subdivision of a genus is identical with or derived from the epithet of one of its constituent species, the type of the name of the subdivision of the genus is the same as that of the species name, unless the original author of the subdivisional name designated another type.

Ex. 6. The type of *Euphorbia* subg. *Esula* Pers. is *E. esula* L.; the designation of *E. peplus* L. as lectotype by Croizat (in Revista Sudamer. Bot. 6: 13. 1939) has no standing.

Ex. 7. The type of *Plantago* sect. *Oliganthos* (Greek for few-flowered) Barnéoud (Monogr. Plantag.: 17. 1845) is necessarily *P. pauciflora* (Latin for few-flowered) Lam.; the later lectotype designation of *P. pauciflora* by Rahn (in Nordic J. Bot. 4: 609. 1984) was superfluous.

22.7. When the epithet in the name of a subdivision of a genus is identical with or derived from the epithet in a specific name that is a later homonym, its type is the type of that later homonym, the correct name of which necessarily has a different epithet.

Recommendation 22A

22A.1. A section including the type of the correct name of a subgenus, but not including the type of the correct name of the genus, should, where there is no obstacle under the rules, be given a name with the same epithet and type as the subgeneric name.

22A.2. A subgenus not including the type of the correct name of the genus should, where there is no obstacle under the rules, be given a name with the same epithet and type as the correct name of one of its subordinate sections.

Ex. 1. Instead of using a new epithet at the subgeneric level, Brizicky raised *Rhamnus* sect. *Pseudofrangula* Grubov to the rank of subgenus as *R.* subg. *Pseudofrangula* (Grubov) Brizicky. The type of both names is the same, *R. alnifolia* L'Hér.

Recommendation 22B

22B.1. When publishing a name of a subdivision of a genus that will also establish an autonym, the author should mention this autonym in the publication.

SECTION 4. NAMES OF SPECIES

Article 23

23.1. The name of a species is a binary combination consisting of the name of the genus followed by a single specific epithet in the form of an adjective, a noun in the genitive, or a word in apposition, or several words, but not a phrase name of one or more descriptive nouns and associated adjectives in the ablative (see Art. 23.6(a)), nor certain other irregularly formed designations (see Art. 23.6(c)). If an epithet consists of two or more words, these are to be united or hyphenated. An epithet not so joined when originally published is not to be rejected but, when used, is to be united or hyphenated, as specified in Art. 60.9.

23.2. The epithet in the name of a species may be taken from any source whatever, and may even be composed arbitrarily (but see Art. 60.1).

Ex. 1. Cornus sanguinea, Dianthus monspessulanus, Papaver rhoeas, Uromyces fabae, Fumaria gussonei, Geranium robertianum, Embelia sarasiniorum, Atropa bella-donna, Impatiens noli-tangere, Adiantum capillus-veneris, Spondias mombin (an indeclinable epithet).

23.3. Symbols forming part of specific epithets proposed by Linnaeus do not prevent valid publication of the relevant names but must be transcribed.

Ex. 2. Scandix pecten ♀ L. is to be transcribed as *Scandix pecten-veneris; Veronica anagallis* ▽ L. is to be transcribed as *Veronica anagallis-aquatica.*

23.4. The specific epithet, with or without the addition of a transcribed symbol, may not exactly repeat the generic name (such repetition would result in a tautonym).

Ex. 3. "Linaria linaria" and *"Nasturtium nasturtium-aquaticum"* are contrary to this rule and cannot be validly published.

Ex. 4. Linum radiola L. (1753) when transferred to *Radiola* Hill may not be named *"Radiola radiola",* as was done by Karsten (1882), since that combination cannot be validly published (see Art. 32.1(c)). The next oldest name, *L. multiflorum* Lam. (1779), is illegitimate, being a superfluous name for *L. radiola.* Under *Radiola,* the species has been given the legitimate name *R. linoides* Roth (1788).

23.5. The specific epithet, when adjectival in form and not used as a noun, agrees grammatically with the generic name; when it is a noun in apposition or a genitive noun, it retains its own gender and termination irrespective of the gender of the generic name. Epithets not conforming to this rule are to be corrected (see Art. 32.7). In particular, the usage of the word element *-cola* as an adjective is a correctable error.

Ex. 5. Adjectival epithets: *Helleborus niger* L., *Brassica nigra* (L.) W. D. J. Koch, *Verbascum nigrum* L.; *Rumex cantabricus* Rech. f., *Daboecia cantabrica* (Huds.) K. Koch (≡ *Vaccinium cantabricum* Huds.); *Vinca major* L., *Tropaeolum majus* L.; *Bromus mollis* L., *Geranium molle* L.; *Peridermium balsameum* Peck, derived from the epithet of *Abies balsamea* (L.) Mill., treated as an adjective.

Ex. 6. Names with a noun for an epithet: *Convolvulus cantabrica* L., *Gentiana pneumonanthe* L., *Lythrum salicaria* L., *Schinus molle* L., all with epithets featuring pre-Linnaean generic names. *Gloeosporium balsameae* Davis, derived from the epithet of *Abies balsamea* (L.) Mill., treated as a noun.

Ex. 7. Correctable errors: The epithet of *Polygonum segetum* Kunth (1817) is a genitive plural noun (of the corn fields); the combination *Persicaria "segeta"*, proposed by Small, is a correctable error for *Persicaria segetum* (Kunth) Small (1903). – In *Masdevallia echidna* Rchb. f. (1855), the epithet corresponds to the generic name of an animal; upon transfer to *Porroglossum* Schltr., the combination *P. "echidnum"* was proposed by Garay, which is a correctable error for *P. echidna* (Rchb. f.) Garay (1953).

Ex. 8. *Rubus "amnicolus"* is a correctable error for *R. amnicola* Blanch. (1906).

23.6. The following designations are not to be regarded as specific names:

(a) Descriptive designations consisting of a generic name followed by a phrase name (Linnaean "nomen specificum legitimum") of one or more descriptive nouns and associated adjectives in the ablative.

Ex. 9. *Smilax "caule inermi"* (Aublet, Hist. Pl. Guiane 2, Tabl.: 27. 1775) is an abbreviated descriptive reference to an imperfectly known species which is not given a binomial in the text but referred to merely by a phrase name cited from Burman.

(b) Other designations of species consisting of a generic name followed by one or more words not intended as a specific epithet.

Ex. 10. *Viola "qualis"* (Krocker, Fl. Siles. 2: 512, 517. 1790); *Urtica "dubia?"* (Forsskål, Fl. Aegypt.-Arab.: cxxi. 1775), the word "dubia?" (doubtful) being repeatedly used in Forsskål's work for species which could not be reliably identified.

Ex. 11. *Atriplex "nova"* (Winterl, Index Hort. Bot. Univ. Hung.: fol. A [8] recto et verso. 1788), the word "nova" (new) being here used in connection with four different species of *Atriplex*. However, in *Artemisia nova* A. Nelson (in Bull. Torrey Bot. Club 27: 274. 1900), *nova* was intended as a specific epithet, the species having been newly distinguished from others.

Ex. 12. Cornus "gharaf" (Forsskål, Fl. Aegypt.-Arab.: xci, xcvi. 1775) is an interim designation not intended as a species name. An interim designation in Forsskål's work is an original designation (for an accepted taxon and thus not a "provisional name" as defined in Art. 34.1(b)) with an epithet-like vernacular which is not used as an epithet in the "Centuriae" part of the work. *Elcaja "roka"* (Forsskål, Fl. Aegypt.-Arab.: xcv. 1775) is another example of such an interim designation; in other parts of the work (p. c, cxvi, 127) this species is not named.

Ex. 13. In *Agaricus "octogesimus nonus"* and *Boletus "vicesimus sextus"* (Schaeffer, Fung. Bavar. Palat. Nasc. 1: t. 100. 1762; 2: t. 137. 1763), the generic names are followed by ordinal adjectives used for enumeration. The corresponding species were given validly published names, *A. cinereus* Schaeff. and *B. ungulatus* Schaeff., in the final volume of the same work (1774).

Ex. 14. Honckeny (1782; see Art. 46 Ex. 38) used species designations such as, in *Agrostis, "A. Reygeri I.", "A. Reyg. II.", "A. Reyg. III."* (all referring to species described but not named in Reyger, Tent. Fl. Gedan.: 36-37. 1763), and also *"A. alpina. II"* for a newly described species following after *A. alpina* Scop. These are informal designations used for enumeration, not validly published binomials; they may not be expanded into, e.g., *"Agrostis reygeri-prima"*.

(c) Designations of species consisting of a generic name followed by two or more adjectival words in the nominative case.

Ex. 15. Salvia "africana coerulea" (Linnaeus, Sp. Pl.: 26. 1753) and *Gnaphalium "fruticosum flavum"* (Forsskål, Fl. Aegypt.-Arab.: cxix. 1775) are generic names followed by two adjectival words in the nominative case. They are not to be regarded as species names.

Ex. 16. However, *Rhamnus "vitis idaea"* Burm. f. (Fl. Ind.: 61. 1768) is to be regarded as a species name, since the generic name is followed by a noun and an adjective, both in the nominative case; these words are to be hyphenated *(R. vitis-idaea)* under the provisions of Art. 23.1 and 60.9. In *Anthyllis "Barba jovis"* L. (Sp. Pl.: 720. 1753) the generic name is followed by nouns in the nominative and in the genitive case, respectively, and they are to be hyphenated *(A. barba-jovis)*. Likewise, *Hyacinthus "non scriptus"* L. (Sp. Pl.: 316. 1753), where the generic name is followed by a negative particle and a past participle used as an adjective, is corrected to *H. non-scriptus,* and *Impatiens "noli tangere"* L. (Sp. Pl.: 938. 1753), where the generic name is followed by two verbs, is corrected to *I. noli-tangere*.

Ex. 17. Similarly, in *Narcissus "Pseudo Narcissus"* L. (Sp. Pl.: 289. 1753) the generic name is followed by an independent prefix and a noun in the nominative case, and the name is to be corrected to *N. pseudonarcissus* under the provisions of Art. 23.1 and 60.9.

(d) Formulae designating hybrids (see Art. H.10.3).

23.7. Phrase names used by Linnaeus as specific epithets ("nomina trivialia") are to be corrected in accordance with later usage by Linnaeus himself.

Ex. 18. Apocynum "fol. [foliis] androsaemi" L. is to be cited as *A. androsaemifolium* L. (Sp. Pl.: 213. 1753 [corr. L., Syst. Nat., ed. 10, 2: 946. 1759]); and *Mussaenda "fr. [fructu] frondoso"* L., as *M. frondosa* L. (Sp. Pl.: 177. 1753 [corr. L., Syst. Nat., ed. 10, 2: 931. 1759]).

23.8. Where the status of a designation of a species is uncertain under Art. 23.6, established custom is to be followed (Pre. 10).

*Ex. 19. Polypodium "F. mas", P. "F. femina", and *P. "F. fragile"* (Linnaeus, Sp. Pl.: 1090-1091. 1753) are, in accordance with established custom, to be treated as *P. filix-mas* L., *P. filix-femina* L., and *P. fragile* L., respectively. Likewise, *Cambogia "G. gutta"* is to be treated as *C. gummi-gutta* L. (Gen. Pl.: [522]. 1754). The intercalations *"Trich." [Tricho-manes]* and *"M." [Melilotus]* in the names of Linnaean species of *Asplenium* and *Trifolium,* respectively, are to be deleted, so that names in the form *Asplenium "Trich. dentatum"* and *Trifolium "M. indica",* for example, are treated as *A. dentatum* L. and *T. indicum* L. (Sp. Pl.: 765, 1080. 1753).*

Recommendation 23A

23A.1. Names of persons and also of countries and localities used in specific epithets should take the form of nouns in the genitive *(clusii, porsildiorum, saharae)* or of adjectives *(clusianus, dahuricus)* (see also Art. 60, Rec. 60C and 60D).

23A.2. The use of the genitive and the adjectival form of the same word to designate two different species of the same genus should be avoided (e.g. *Lysimachia hemsleyana* Oliv. and *L. hemsleyi* Franch.).

23A.3. In forming specific epithets, authors should comply also with the following suggestions:

(a) To use Latin terminations insofar as possible.

(b) To avoid epithets which are very long and difficult to pronounce in Latin.

(c) Not to make epithets by combining words from different languages.

(d) To avoid those formed of two or more hyphenated words.

(e) To avoid those which have the same meaning as the generic name (pleonasm).

(f) To avoid those which express a character common to all or nearly all the species of a genus.

(g) To avoid in the same genus those which are very much alike, especially those which differ only in their last letters or in the arrangement of two letters.

(h) To avoid those which have been used before in any closely allied genus.

(i) Not to adopt epithets from unpublished names found in correspondence, travellers' notes, herbarium labels, or similar sources, attributing them to their authors, unless these authors have approved publication (see Rec. 34A).

(j) To avoid using the names of little-known or very restricted localities unless the species is quite local.

SECTION 5. NAMES OF TAXA BELOW THE RANK OF SPECIES (INFRASPECIFIC TAXA)

Article 24

24.1. The name of an infraspecific taxon is a combination of the name of a species and an infraspecific epithet. A connecting term is used to denote the rank.

Ex. 1. Saxifraga aizoon subf. *surculosa* Engl. & Irmsch. This taxon may also be referred to as *Saxifraga aizoon* var. *aizoon* subvar. *brevifolia* f. *multicaulis* subf. *surculosa* Engl. & Irmsch.; in this way a full classification of the subforma within the species is given, not only its name.

24.2. Infraspecific epithets are formed like specific epithets and, when adjectival in form and not used as nouns, they agree grammatically with the generic name (see Art. 32.7).

Ex. 2. Solanum melongena var. *insanum* Prain (Bengal Pl.: 746. 1903, *'insana'*).

24.3. Infraspecific names with final epithets such as *typicus, originalis, originarius, genuinus, verus,* and *veridicus,* purporting to indicate the taxon containing the type of the name of the next higher taxon, are not validly published unless they are autonyms (Art. 26).

Ex. 3. Lobelia spicata "var. *originalis*" (McVaugh in Rhodora 38: 308. 1936) was not validly published (see Art. 26 Ex. 1), whereas the autonyms *Galium verum* L. subsp. *verum* and *G. verum* var. *verum* are validly published.

Ex. 4. Aloe perfoliata var. *vera* L. (Sp. Pl.: 320. 1753) is validly published because it does not purport to contain the type of *A. perfoliata* L. (1753).

24.4. The use of a binary combination instead of an infraspecific epithet is not admissible. Contrary to Art. 32.1(c), names so constructed are validly published but are to be altered to the proper form without change of the author citation or date of publication.

Ex. 5. Salvia grandiflora subsp. *"S. willeana"* (Holmboe in Bergens Mus. Skr., ser. 2, 1(2): 157. 1914) is to be cited as *S. grandiflora* subsp. *willeana* Holmboe.

Ex. 6. Phyllerpa prolifera var. *"Ph. firma"* (Kützing, Sp. Alg.: 495. 1849) is to be altered to *P. prolifera* var. *firma* Kütz.

Note 1. Infraspecific taxa within different species may bear names with the same final epithet; those within one species may bear names with the same final epithet as the names of other species (but see Rec. 24B.1).

Ex. 7. Rosa glutinosa var. *leioclada* H. Christ (in Boissier, Fl. Orient. Suppl.: 222. 1888) and *Rosa jundzillii* f. *leioclada* Borbás (in Math. Term. Közlem. 16: 376, 383. 1880) are both permissible, as is *Viola tricolor* var. *hirta* Ging. (in Candolle, Prodr. 1: 304. 1824), in spite of the previous existence of a species named *Viola hirta* L.

Note 2. Names of infraspecific taxa within the same species, even if they differ in rank, are treated as homonyms if they have the same final epithet but are based on different types (Art. 53.4), the connecting term not being part of the name.

Recommendation 24A

24A.1. Recommendations made for forming specific epithets (Rec. 23A) apply equally for infraspecific epithets.

Recommendation 24B

24B.1. Authors proposing new infraspecific names should avoid final epithets previously used as specific epithets in the same genus.

24B.2. When an infraspecific taxon is raised to the rank of species, or the inverse change occurs, the final epithet of its name should be retained unless the resulting combination would be contrary to this *Code*.

Article 25

25.1. For nomenclatural purposes, a species or any taxon below the rank of species is regarded as the sum of its subordinate taxa, if any. In fungi, a holomorph also includes its correlated form-taxa (see Art. 59).

Ex. 1. When *Montia parvifolia* (DC.) Greene is treated as comprising two subspecies, one must write *M. parvifolia* subsp. *parvifolia* for that part of the species that includes the nomenclatural type and excludes the type of the name of the other subspecies, *M. parvifolia* subsp. *flagellaris* (Bong.) Ferris. The name *M. parvifolia* applies to the species in its entirety.

Article 26

26.1. The name of any infraspecific taxon that includes the type of the adopted, legitimate name of the species to which it is assigned is to repeat the specific epithet unaltered as its final epithet, not followed by an author citation (see Art. 46). Such names are termed autonyms (Art. 6.8; see also Art. 7.6).

Ex. 1. The variety which includes the type of the name *Lobelia spicata* Lam. is to be named *Lobelia spicata* Lam. var. *spicata* (see also Art. 24 Ex. 3).

Note 1. This provision applies only to the names of those subordinate taxa that include the type of the adopted name of the species (but see Rec. 26A).

26.2. A name of an infraspecific taxon that includes the type (i.e. the holotype or all syntypes or the previously designated type) of the adopted, legitimate name of the species to which it is assigned is not validly published unless its final epithet repeats the specific epithet unaltered. For the purpose of this provision, explicit indication that the nomenclaturally typical element of the species is included is considered as equivalent to inclusion of the type, whether or not it has been previously designated (see also Art. 24.3).

Ex. 2. The intended combination "*Vulpia myuros* subsp. *pseudomyuros* (Soy.-Will.) Maire & Weiller" was not validly published in Maire (Fl. Afrique N. 3: 177. 1955) because it included "*F. myuros* L., Sp. 1, p. 74 (1753) sensu stricto" in synonymy, *Festuca myuros* L. being the basionym of *Vulpia myuros* (L.) C. C. Gmel.

Ex. 3. Linnaeus (Sp. Pl.: 3. 1753) recognized two named varieties under *Salicornia europaea*. Since *S. europaea* has no holotype and no syntypes are cited, both varietal names are validly published irrespective of the facts that the lectotype of *S. europaea,* designated by Jafri and Rateeb (in Jafri & El-Gadi, Fl. Libya 58: 57. 1979), can be attributed to *S. europaea* var. *herbacea* L. (1753) and that the latter name was subsequently lectotypified by Piirainen (in Ann. Bot. Fenn. 28: 82. 1991) by the same specimen as the species name.

Ex. 4. Linnaeus (Sp. Pl.: 779-781. 1753) recognized 13 named varieties under *Medicago polymorpha*. Since *M. polymorpha* L. has neither a holotype nor syntypes, all varietal names are validly published, and indeed the lectotype subsequently designated (by Heyn in Bull. Res. Council Israel, Sect. D, Bot., 7: 163. 1959) is not part of the original material for any of the varietal names of 1753.

26.3. The first instance of valid publication of a name of an infraspecific taxon under a legitimate species name automatically establishes the corresponding autonym (see also Art. 32.8 and 11.6).

Ex. 5. The publication of the name *Lycopodium inundatum* var. *bigelovii* Tuck. (in Amer. J. Sci. Arts 45: 47. 1843) automatically established the name of another variety, *L. inundatum* L. var. *inundatum,* the type of which is that of the name *L. inundatum* L.

Ex. 6. *Utricularia stellaris* L. f. (1782) includes *U. stellaris* var. *coromandeliana* A. DC. (Prodr. 8: 3. 1844) and *U. stellaris* L. f. var. *stellaris* (1844) automatically established at the same time. When *U. stellaris* is included in *U. inflexa* Forssk. (1775) as a variety, the correct name of that variety, under Art. 11.6, is *U. inflexa* var. *stellaris* (L. f.) P. Taylor (1961).

Ex. 7. Pangalo (in Trudy Prikl. Bot. 23: 258. 1930) when describing *Cucurbita mixta* Pangalo distinguished two varieties, *C. mixta* var. *cyanoperizona* Pangalo and var. *stenosperma* Pangalo, together encompassing the entire circumscription of the species. Since neither a holotype nor any syntypes were indicated for *C. mixta,* both varietal names were validly published (see Art. 26.2). Merrick & Bates (in Baileya 23: 96, 101. 1989), in the absence of known type material, neotypified *C. mixta* by an element that can be attributed to *C. mixta* var. *stenosperma*. As long as their choice of neotype is followed, the correct name for that variety is *C. mixta* var. *mixta,* not *C. mixta* var. *stenosperma.* When it is treated as a variety of *C. argyrosperma* Huber (1867), as was done by Merrick & Bates, its correct name under Art. 11.6 is not *C. argyrosperma* var. *stenosperma* (Pangalo) Merrick & D. M. Bates; a combination based on *C. mixta* is required.

Recommendation 26A

26A.1. A variety including the type of the correct name of a subspecies, but not including the type of the correct name of the species, should, where there is no obstacle under the rules, be given a name with the same final epithet and type as the subspecies name.

26A.2. A subspecies not including the type of the correct name of the species should, where there is no obstacle under the rules, be given a name with the same final epithet and type as a name of one of its subordinate varieties.

26A.3. A taxon of rank lower than variety which includes the type of the correct name of a subspecies or variety, but not the type of the correct name of the species, should, where there is no obstacle under the rules, be given a name with the same final epithet and type as the name of the subspecies or variety. On the other hand, a subspecies or variety which does not include the type of the correct name of the species should not be given a name with the same final epithet as a name of one of its subordinate taxa below the rank of variety.

Ex. 1. Fernald treated *Stachys palustris* subsp. *pilosa* (Nutt.) Epling (in Repert. Spec. Nov. Regni Veg. Beih. 8: 63. 1934) as composed of five varieties, for one of which (that including the type of *S. palustris* subsp. *pilosa*) he made the combination *S. palustris* var. *pilosa* (Nutt.) Fernald (in Rhodora 45: 474. 1943), there being no legitimate varietal name available.

Ex. 2. There being no legitimate name available at the rank of subspecies, Bonaparte made the combination *Pteridium aquilinum* subsp. *caudatum* (L.) Bonap. (Notes Ptérid. 1: 62. 1915), using the same final epithet that Sadebeck had used earlier in the combination *P. aquilinum* var. *caudatum* (L.) Sadeb. (in Jahrb. Hamburg. Wiss. Anst. Beih. 14(3): 5. 1897), both combinations being based on *Pteris caudata* L. Each name is legitimate, and both can be used, as by Tryon (in Rhodora 43: 52-54. 1941), who treated *P. aquilinum* var. *caudatum* as one of four varieties under subsp. *caudatum* (see Art. 34.2).

Recommendation 26B

26B.1. When publishing a name of an infraspecific taxon that will also establish an autonym, the author should mention this autonym in the publication.

Article 27

27.1. The final epithet in the name of an infraspecific taxon may not repeat unchanged the epithet of the correct name of the species to which the taxon is assigned unless the two names have the same type.

27.2. The final epithet in the name of an infraspecific taxon may not repeat unchanged the epithet of the species name if that species name is illegitimate.

Ex. 1. When Honda (in Bot. Mag. (Tokyo) 41: 385. 1927) published *Agropyron japonicum* var. *hackelianum* Honda under the illegitimate *A. japonicum* Honda (1927), which is a later homonym of *A. japonicum* (Miq.) P. Candargy (1901), he did not validly publish an autonym *"A. japonicum* var. *japonicum"* (see also Art. 55 Ex. 2).

SECTION 6. NAMES OF PLANTS IN CULTIVATION

Article 28

28.1. Plants brought from the wild into cultivation retain the names that are applied to the same taxa growing in nature.

Note 1. Hybrids, including those arising in cultivation, may receive names as provided in App. I (see also Art. 11.9, 40, and 50).

Note 2. Additional, independent designations for special categories of plants used in agriculture, forestry, and horticulture (and arising either in nature or cultivation) are dealt with in the *International code of nomenclature for cultivated plants,* where the term "cultivar" is defined and regulations are provided for the formation and use of cultivar epithets.

Note 3. Nothing precludes the use, for cultivated plants, of names published in accordance with the requirements of the botanical *Code.*

Note 4. Epithets in names published in conformity with the botanical *Code* may be used as cultivar epithets under the rules of the *International code of nomenclature for cultivated plants,* when cultivar is considered to be the appropriate status for the groups concerned.

Ex. 1. Mahonia japonica DC. (1821) may be treated as a cultivar, which is then designated as *Mahonia* 'Japonica'; *Taxus baccata* var. *variegata* Weston (1770), when treated as a cultivar, is designated as *Taxus baccata* 'Variegata'.

Note 5. The *International code of nomenclature for cultivated plants* provides for the establishment of cultivar epithets differing markedly from epithets in Latin form.

Ex. 2. ×*Disophyllum* 'Frühlingsreigen'; *Eriobotrya japonica* 'Golden Ziad' and *E. japonica* 'Maamora Golden Yellow'; *Phlox drummondii* 'Sternenzauber'; *Quercus frainetto* 'Hungarian Crown'.

Ex. 3. Juniperus ×*pfitzeriana* 'Wilhelm Pfitzer' (P. A. Schmidt 1998) was established for a tetraploid cultivar presumed to result from the original cross between *J. chinensis* L. and *J. sabina* L.

CHAPTER IV. EFFECTIVE AND VALID PUBLICATION

SECTION 1. CONDITIONS AND DATES OF EFFECTIVE PUBLICATION

Article 29

29.1. Publication is effected, under this *Code,* only by distribution of printed matter (through sale, exchange, or gift) to the general public or at least to botanical institutions with libraries accessible to botanists generally. It is not effected by communication of new names at a public meeting, by the placing of names in collections or gardens open to the public, by the issue of microfilm made from manuscripts, typescripts or other unpublished material, or solely by distribution electronically or through any electronic medium.

Ex. 1. Cusson announced his establishment of the genus *Physospermum* in a memoir read at the Société des Sciences de Montpellier in 1770, and later in 1782 or 1783 at the Société de Médecine de Paris, but its effective publication dates from 1787 (in Hist. Soc. Roy. Méd. 5(1): 279).

Recommendation 29A

29A.1. Publication of nomenclatural novelties in periodicals (see Rec. 30A.2) that distribute an electronic version as well as a printed version, should only be in those with the following features:

(a) The printed and electronic versions are identical in content and pagination;

(b) The electronic version is in a platform-independent and printable format;

(c) The electronic version is publicly available via the World Wide Web or its successors;

(d) The presence of nomenclatural novelties is prominently indicated in the work (see Rec. 30A.2).

Article 30

30.1. Publication by indelible autograph before 1 January 1953 is effective. Indelible autograph produced at a later date is not effectively published.

Ex. 1. Salvia oxyodon Webb & Heldr. was effectively published in an indelible autograph catalogue placed on sale (Webb & Heldreich, *Catalogus plantarum hispanicarum ... ab A. Blanco lectarum,* Paris, Jul 1850, folio).

Ex. 2. The *Journal of the International Conifer Preservation Society,* vol. 5[1]. 1997 ("1998"), consists of duplicated sheets of typewritten text with handwritten additions and corrections in several places. The handwritten portions, being indelible autograph published after 1 January 1953, are not effectively published. Intended new combinations (*"Abies koreana* var. *yuanbaoshanensis",* p. 53) for which the basionym reference is handwritten are not validly published. The entirely handwritten account of a new taxon (p. 61: name, Latin description, statement of type) is treated as unpublished (see also Rec. 34A.1).

30.2. For the purpose of this Article, indelible autograph is handwritten material reproduced by some mechanical or graphic process (such as lithography, offset, or metallic etching).

Ex. 3. Léveillé, *Flore du Kouy Tchéou* (1914-1915), is a work lithographed from a handwritten text.

30.3. Publication on or after 1 January 1953 in trade catalogues or non-scientific newspapers, and on or after 1 January 1973 in seed-exchange lists, does not constitute effective publication.

30.4. The distribution on or after 1 January 1953 of printed matter accompanying exsiccatae does not constitute effective publication.

Note 1. If the printed matter is also distributed independently of the exsiccata, it is effectively published.

Ex. 4. The printed labels of Fuckel's *Fungi rhenani exsiccati* (1863-1874) are effectively published even though not independently issued. The labels antedate Fuckel's subsequent accounts (e.g. in Jahrb. Nassauischen Vereins Naturk. 23-24. 1870).

Ex. 5. Vězda's *Lichenes selecti exsiccati* (1967-) were issued with printed labels that were also distributed independently as printed fascicles; the latter are effectively published and new names appearing in Vězda's exsiccata are to be cited from the fascicles.

30.5. Publication on or after 1 January 1953 of an independent non-serial work stated to be a thesis submitted to a university or other institute of education for the purpose of obtaining a degree is not effectively published unless it includes an explicit statement (referring to the requirements of the *Code* for effective publication) or other internal evidence that it is regarded as an effective publication by its author or publisher.

Note 2. The presence of an International Standard Book Number (ISBN) or a statement of the name of the printer, publisher, or distributor in the original printed version is regarded as internal evidence that the work was intended to be effectively published.

Ex. 6. "*Meclatis* in *Clematis;* yellow flowering *Clematis* species – Systematic studies in *Clematis* L. *(Ranunculaceae),* inclusive of cultonomic aspects" a "Proefschrft ter verkrijging van de graad van doctor ... van Wageningen Universiteit" by Brandenburg, was effectively published on 8 June 2000, because it bore the ISBN 90-5808-237-7.

Ex. 7. The thesis "Comparative investigations on the life-histories and reproduction of some species in the siphoneous green algal genera *Bryopsis* and *Derbesia*" by Rietema, submitted to Rijksuniversiteit te Groningen in 1975, is stated to have been printed ("Druck") by Verenigde Reproduktie Bedrijven, Groningen and is therefore effectively published.

Ex. 8. The dissertation "Die Gattung *Mycena* s.l." by Rexer, submitted to the Eberhard-Karls-Universität Tübingen, was effectively published in 1994 because it bore the statement "Druck: Zeeb-Druck, Tübingen 7 (Hagelloch)", referring to a commercial printer. The generic name *Roridomyces* Rexer, typified by *Agaricus roridus* Scop., and combinations in *Mycena* are therefore validly published. The generic name *Roridella* E. Horak (Röhrlinge und Blätterpilze in Europa: 509. 2005), also published with *A. roridus* Scop. as type, is illegitimate (Art. 52.1).

Ex. 9. The thesis by Demoulin, "Le genre *Lycoperdon* en Europe et en Amérique du Nord", defended in 1971, does not contain internal evidence that it is regarded as effectively published. Even if photocopies of it can be found in some libraries, new species of *Lycoperdon*, e.g. "*L. americanum*", "*L. cokeri*", and "*L. estonicum*", introduced there, were validly published in the effectively published "Espèces nouvelles ou méconnues du genre *Lycoperdon* (Gastéromycètes)" (Demoulin in Lejeunia, n.s., 62: 1–28. 1972).

Ex. 10. The dissertation "*Nasa* and the conquest of South America – Systematic Rearrangements in *Loasaceae* Juss." submitted in June 1997 to the Ludwig-Maximilians-Universität München by Weigend is not effectively published as it does not include an ISBN, the name of any printer or publisher or distributor, or any statement that it was intended to be effectively published under the *Code*, even though 40 copies were distributed, all the other formalities for the publication of new taxa were met, and statements were made implying effective publication but not mentioning the *Code*, such as that although "the majority of names will be published elsewhere ... for some ... groups new names are here provided". The names intended to be published in the thesis were validly published in Taxon 55: 463-468. 2006.

Ex. 11. *Montanoa imbricata* V. A. Funk was validly published in "The systematics of *Montanoa (Asteraceae, Heliantheae)*" (Funk in Mem. New York Bot. Gard. 36: 116. 1982), not in Funk's dissertation "The Systematics of *Montanoa* Cerv. *(Asteraceae)*" submitted to the Ohio State University in 1980, nor in facsimile copies of the dissertation printed from microfiche and distributed, on demand, by University Microfilms, Ann Arbor, beginning in 1980.

Recommendation 30A

30A.1. It is strongly recommended that authors avoid publishing new names and descriptions or diagnoses of new taxa (nomenclatural novelties) in ephemeral printed matter of any kind, in particular printed matter that is multiplied in restricted and uncertain numbers, in which the permanence of the text may be limited, for which effective publication in terms of number of copies is not obvious, or that is unlikely to reach the general public. Authors should also avoid publishing new names and descriptions or diagnoses in popular periodicals, in abstracting journals, or on correction slips.

Ex. 1. Kartesz provided an unpaged, printed insert titled "Nomenclatural innovations" to accompany the electronic version (1.0) of the *Synthesis of the North American flora* produced on compact disk (CD-ROM; distribution through an electronic medium in terms of Art. 29.1). This insert, which is effectively published under Art. 29-30, is the place of valid publication of 41 new combinations, which also appear on the disk, in an item authored by Kartesz: "A synonymized checklist and atlas with biological attributes for the vascular flora of the United States, Canada, and Greenland" (e.g. *Dichanthelium hirstii* (Swallen) Kartesz in Kartesz & Meacham, Synth. N. Amer. Fl., Nomencl. Innov.: [1]. Aug 1999). Kartesz's procedure is not to be recommended, as the insert is unlikely to be permanently stored and catalogued in botanical libraries and so reach the general public.

30A.2. To aid availability through time and place, authors publishing nomenclatural novelties should give preference to periodicals that regularly publish taxonomic articles, or else printed copies of a publication (even if also distributed electronically) should be deposited in at least ten, but preferably more, botanical or other generally accessible libraries throughout the world including a name-indexing centre appropriate to the taxonomic group.

30A.3. Authors and editors are encouraged to mention nomenclatural novelties in the summary or abstract, or list them in an index in the publication.

Article 31

31.1. The date of effective publication is the date on which the printed matter became available as defined in Art. 29 and 30. In the absence of proof establishing some other date, the one appearing in the printed matter must be accepted as correct.

Ex. 1. Individual parts of Willdenow's *Species plantarum* were published as follows: 1(1), Jun 1797; 1(2), Jul 1798; 2(1), Mar 1799; 2(2), Dec 1799; 3(1), 1800; 3(2), Nov 1802; 3(3), Apr-Dec 1803; 4(1), 1805; 4(2), 1806; these dates are presently accepted as the dates of effective publication (see Stafleu & Cowan in Regnum Veg. 116: 303. 1988).

Ex. 2. T. M. Fries first published *Lichenes arctoi* in 1860 as an independently paginated preprint, which predates the identical version published in a journal (Nova Acta Reg. Soc. Sci. Upsal., ser. 3, 3: 103-398. 1861).

Ex. 3. Diatom Research, vol. 2, no. 2, bears a title-page date of Dec 1987, but the authors of a paper included in a later issue (vol. 3, p. 265) stated that the date of publication was 18 Feb 1988, which therefore should be taken as the date of all nomenclatural novelties in that issue of the journal.

Note 1. Effective publication requires distribution of printed matter, which establishes the date of effective publication, even if a name is published in a periodical with parallel printed and electronic versions.

Ex. 4. The paper in which the name *Ceratocystis omanensis* Al-Subhi & al. is described was available online in final form on Science Direct on 7 November 2005, and distributed in a printed version in *Mycological Research* 110(2): 237-245 on 7 March 2006, The date of effective publication of the name for the purposes of this Article is 7 March 2006 and not 7 November 2005.

31.2. When separates from periodicals or other works placed on sale are issued in advance, the date on the separate is accepted as the date of effective publication unless there is evidence that it is erroneous.

Ex. 5. The names of the *Selaginella* species published by Hieronymus (in Hedwigia 51: 241-272) were effectively published on 15 October 1911, since the volume in which the paper appeared, though dated 1912, states (p. ii) that the separate appeared on that date.

Recommendation 31A

31A.1. The date on which the publisher or publisher's agent delivers printed matter to one of the usual carriers for distribution to the public should be accepted as its date of effective publication.

SECTION 2. CONDITIONS AND DATES OF VALID PUBLICATION OF NAMES

Article 32

32.1. In order to be validly published, a name of a taxon (autonyms excepted) must: *(a)* be effectively published (see Art. 29-31) on or after the starting-point date of the respective group (Art. 13.1); *(b)* be composed only of letters of the Latin alphabet, except as provided in Art. 23.3 and Art. 60.4, 60.6, 60.9, and 60.10; *(c)* have a form which complies with the provisions of Art. 16-27 (but see Art. 21.4 and 24.4), and Art. H.6 and H.7; *(d)* be accompanied by a description or diagnosis or by a reference to a previously and effectively published description or diagnosis (except as provided in Art. 42.3, 44.1, and H.9); and *(e)* comply with the special provisions of Art. 33-45 (see also Art. 61).

32.2. A diagnosis of a taxon is a statement of that which in the opinion of its author distinguishes the taxon from other taxa.

Ex. 1. *"Egeria"* (Néraud in Gaudichaud, Voy. Uranie, Bot.: 25, 28. 1826), published without a description or a diagnosis or a reference to a former one, was not validly published.

Ex. 2. "*Loranthus macrosolen* Steud." originally appeared without a description or diagnosis on the printed labels issued about the year 1843 with Sect. II, No. 529, 1288, of Schimper's herbarium specimens of Abyssinian plants; the name was not validly published, however, until Richard (Tent. Fl. Abyss. 1: 340. 1847) supplied a description.

*Ex. 3. In Don, *Sweet's Hortus britannicus,* ed. 3 (1839), for each listed species the flower colour, the duration of the plant, and a translation into English of the specific epithet are given in tabular form. In many genera the flower colour and duration may be identical for all species and clearly their mention is not intended as a validating description or diagnosis.

New names appearing in that work are therefore not validly published, except in some cases where reference is made to earlier descriptions or diagnoses or to validly published basionyms.

Ex. 4. "*Crepis praemorsa* subsp. *tatrensis*" (Dvořák & Dadáková in Biológia (Bratislava) 32: 755. 1977) appeared with "a subsp. *praemorsa* karyotypo achaeniorumque longitudine praecipue differt". This statement specifies the features by which the two taxa differ but not how these features differ and so it does not satisfy the requirement of Art. 32.1(d) for a "description or diagnosis".

Ex. 5. The generic name *Epilichen* Clem. (Gen. Fungi 174. 1909) is validly published with the two-word diagnosis "Karschia lichenicola", referring to the ability of the included species formerly included in *Karschia* to grow on lichens. This statement, in the opinion of Clements, distinguished the genus from others although provision of such a diagnosis would not be considered good practice today.

32.3. The requirements of Art. 32.1(d) are not met by statements describing properties such as purely aesthetic features, economic, medicinal or culinary usage, cultural significance, cultivation techniques, geographical origin, or geological age.

Ex. 6. "*Musa basjoo*" (Siebold in Verh. Bat. Genootsch. Kunsten 12: 18. 1830) appeared with "Ex insulis Luikiu introducta, vix asperitati hiemis resistens. Ex foliis linteum, praesertim in insulis Luikiu ac quibusdam insulis provinciae Satzuma conficitur. Est haud dubie linteum, quod Philippinis incolis audit Nippis". This statement gives information about the economic use (linen is made from the leaves), horticultural attribute (scarcely survives the winter), and on its origin (introduced from the Ryukyu Islands), but since there is no descriptive information given for the "leaves", the only descriptive feature mentioned, it does not satisfy the requirement of Art. 32.1(d) for a "description or diagnosis". *Musa basjoo* Siebold & Zucc. ex Iinuma was later validly published in Iinuma, Sintei Somoku Dzusetsu [Illustrated Flora of Japan], ed. 2, 3: pl. 1. 1874 with floral details and an extensive description in Japanese on the page facing the plate.

32.4. When it is doubtful whether a descriptive statement satisfies the requirement of Art. 32.1(d) for a "description or diagnosis", a request for a decision may be submitted to the General Committee (see Div. III), which will refer it for examination to the committee for the appropriate taxonomic group. A recommendation whether or not to treat the name concerned as validly published may then be put forward to an International Botanical Congress, and if ratified will become a binding decision.

32.5. For the purpose of valid publication of a name, reference to a previously and effectively published description or diagnosis may be direct or indirect (Art. 32.6). For names published on or after 1 January 1953 it must, however, be full and direct as specified in Art. 33.4.

32.6. An indirect reference is a clear (if cryptic) indication, by an author citation or in some other way, that a previously and effectively published description or diagnosis applies.

Ex. 7. *"Kratzmannia"* (Opiz in Berchtold & Opiz, Oekon.-Techn. Fl. Böhm. 1: 398. 1836) was published with a diagnosis but was not definitely accepted by the author and therefore was not validly published. *Kratzmannia* Opiz (Seznam: 56. 1852), lacking description or diagnosis, is however definitely accepted, and its citation as *"Kratzmannia* O." constitutes indirect reference to the diagnosis published in 1836.

Ex. 8. Opiz published the name of the genus *Hemisphace* (Benth.) Opiz (1852) without a description or diagnosis, but as he wrote *"Hemisphace* Benth." he indirectly referred to the previously effectively published description by Bentham (Labiat. Gen. Spec.: 193. 1833) of *Salvia* sect. *Hemisphace.*

Ex. 9. The new combination *Cymbopogon martini* (Roxb.) Will. Watson (1882) is validly published through the cryptic notation "309", which, as explained at the top of the same page, is the running-number of the species (*Andropogon martini* Roxb.) in Steudel (Syn. Pl. Glumac. 1: 388. 1854). Although the reference to the basionym *Andropogon martini* is indirect, it is unambiguous (but see Art. 45 Ex. 1; see also Rec. 60C.2).

Ex. 10. Miller (1768), in the preface to *The gardeners dictionary,* ed. 8, stated that he had "now applied Linnaeus's method entirely except in such particulars ...", of which he gave examples. In the main text, he often referred to Linnaean genera under his own generic headings, e.g., to *Cactus* L. [pro parte] under *Opuntia* Mill. Therefore an implicit reference to a Linnaean binomial may be assumed when this is appropriate, and Miller's binomials are then accepted as new combinations (e.g., *O. ficus-indica* (L.) Mill., based on *C. ficus-indica* L.) or *nomina nova* (e.g., *O. vulgaris* Mill., based on *C. opuntia* L.: both names have the reference to *"Opuntia vulgo herbariorum"* of Bauhin & Cherler in common).

Ex. 11. Although no authors are cited for the names in Kummer's *Führer in die Pilzkunde* (1871) statements therein allow implicit reference to earlier authors such as Fries (see Art. 33 Ex. 7 and Pennycook in *Mycotaxon* 84: 163-219, 2002)

32.7. Names or epithets published with an improper Latin termination but otherwise in accordance with this *Code* are regarded as validly published; they are to be changed to accord with Art. 16-19, 21, 23, and 24, without change of the author citation or date of publication (see also Art. 60.11).

32.8. Autonyms (Art. 6.8) are accepted as validly published names, dating from the publication in which they were established (see Art. 22.3 and 26.3), whether or not they appear in print in that publication.

32.9. Names in specified ranks included in publications listed as suppressed works (opera utique oppressa; App. VI) are not validly published. Proposals for the addition of publications to App. VI must be submitted to the General Committee (see Div. III), which will refer them for examination to the committees for the various taxonomic groups (see Rec. 32F; see also Art. 14.14).

32.10. When a proposal for the suppression of a publication has been approved by the General Committee after study by the committees for the taxonomic groups concerned, suppression of that publication is authorized subject to the decision of a later International Botanical Congress.

Note 1. For valid publication of names of plant taxa that were originally not treated as plants, see Art. 45.4.

Recommendation 32A

32A.1. A name should not be validated solely by a reference to a description or diagnosis published before 1753.

Recommendation 32B

32B.1. The description or diagnosis of any new taxon should mention the points in which the taxon differs from its allies.

Recommendation 32C

32C.1. When naming a new taxon, authors should not adopt a name that has been previously but not validly published for a different taxon.

Recommendation 32D

32D.1. In describing or diagnosing new taxa, authors should, when possible, supply figures with details of structure as an aid to identification.

32D.2. In the explanation of the figures, authors should indicate the specimen(s) on which they are based (see also Rec. 8A.2).

32D.3. Authors should indicate clearly and precisely the scale of the figures which they publish.

Recommendation 32E

32E.1. Descriptions or diagnoses of parasitic plants should always be followed by indication of the hosts, especially those of parasitic fungi. The hosts should be designated by their scientific names and not solely by names in modern languages, the applications of which are often doubtful.

Recommendation 32F

32F.1. When a proposal for the suppression of a publication under Art. 32.9 has been referred to the appropriate committees for study, authors should follow existing usage of names as far as possible pending the General Committee's recommendation on the proposal.

Article 33

33.1. A combination (autonyms excepted) is not validly published unless the author definitely associates the final epithet with the name of the genus or species, or with its abbreviation.

Ex. 1. Combinations validly published: In Linnaeus's *Species plantarum* the placing of the epithet in the margin opposite the name of the genus clearly associates the epithet with the name of the genus. The same result is attained in Miller's *Gardeners dictionary,* ed. 8, by the inclusion of the epithet in parentheses immediately after the name of the genus, in Steudel's *Nomenclator botanicus* by the arrangement of the epithets in a list headed by the name of the genus, and in general by any typographical device which associates an epithet with a particular generic or specific name.

Ex. 2. Combinations not validly published: Rafinesque's statement under *Blephilia* that "Le type de ce genre est la *Monarda ciliata* Linn." (in J. Phys. Chim. Hist. Nat. Arts 89: 98. 1819) does not constitute valid publication of the combination *B. ciliata,* since Rafinesque did not definitely associate the epithet *ciliata* with the generic name *Blephilia.* Similarly, the combination *Eulophus peucedanoides* is not to be attributed to Bentham & Hooker (Gen. Pl. 1: 885. 1867) on the basis of their listing of "*Cnidium peucedanoides,* H. B. et K." under *Eulophus.*

Ex. 3. *Erioderma polycarpum* subsp. *verruculosum* Vain. (Etude Lich. Brésil 1: 202. 1890) is validly published since Vainio clearly linked the subspecific epithet to the specific epithet by an asterisk.

Ex. 4. Tuckerman (in Proc. Amer. Acad. Arts 12: 168, 1877) described "*Erioderma velligerum* subsp. nov.", but did not associate the subspecific epithet with the epithet of any species name. His statement that his new subspecies was "very near: *E. chilense*", from which he provided distinguishing features, does not effect valid publication of his intended subspecies name.

33.2. Before 1 January 1953 an indirect reference to a basionym or replaced synonym is sufficient for valid publication of a new combination, a new generic name with a basionym, or a nomen novum. Thus, errors in the citation of the basionym or replaced synonym, or in author citation (Art. 46), do not affect valid publication of such names.

Ex. 5. The name "*Persicaria runcinata* (Hamilt.)" was included in a list of names by Masamune (in Bot. Mag. (Tokyo) 51: 234. 1937) with no further information. The name *Polygonum runcinatum* was validly published by Don (Prodr. Fl. Nepal.: 73. 1825) and ascribed there to "Hamilton mss". The mention by Masamune of "Hamilt." is regarded as an indirect reference through Buchanan-Hamilton to the name published by Don, and the combination *Persicaria runcinata* (Buch.-Ham. ex D. Don) Masam. must be accepted as validly published.

Ex. 6. The new binomials in Miller's *The gardeners dictionary,* ed. 8 (1768) that adopt epithets used by Linnaeus are regarded as new combinations, e.g. *Opuntia ficus-indica* (L.) Mill., based on *Cactus ficus-indica* L. (see Art. 32 Ex. 10).

Ex. 7. In Kummer's *Führer in die Pilzkunde* (1871) the statement that the author intended to adopt at generic rank the subdivisions of *Agaricus* then in use, which at the time were those of Fries, and the general arrangement of the work, which faithfully follows that of Fries, provide indirect reference to Fries's earlier names of "tribes". Therefore, names such as *Hypholoma* (Fr. : Fr.) P. Kumm. and *H. fasciculare* (Huds. : Fr.) are accepted as being based on the corresponding Friesian names (here: *A.* "tribus" *Hypholoma* Fr. : Fr. and *A. fascicularis* Huds. : Fr.) although Kummer did not explicitly refer to Fries.

33.3. Before 1 January 1953, if, for a presumed new combination, no reference to a basionym is given but the epithet of a previously and validly published name that applies to the same taxon is adopted and that name is neither cited nor indicated in any way, the new combination is validly published as such if, and only if, it would otherwise be a validly published name. This provision also applies to a new generic name presumed to be based on the epithet of an earlier validly published name of a subdivision of a genus.

Ex. 8. Scaevola taccada was validly published by Roxburgh (1814) by reference to an illustration in Rheede (Hort. Malab. 4: t. 59. 1683) that appears to be its sole basis. As the name applies to the species previously described as *Lobelia taccada* Gaertn. (1788), it is treated as a new combination, *S. taccada* (Gaertn.) Roxb., not as the name of a new species, even though *L. taccada* is neither cited nor indicated in any way in Roxburgh's protologue.

Ex. 9. Brachiolejeunea was published by Stephani & Spruce (in Hedwigia 28: 167. 1889) for a taxon that had previously been described as *Lejeunea* subg. *Brachiolejeunea* Spruce (in Trans. & Proc. Bot. Soc. Edinburgh 15: 75, 129. 1884) but without any reference to Spruce's earlier publication. Because Stephani & Spruce provided a description of *B. plagiochiloides* that under Art. 42 is a descriptio generico-specifica of a monotypic genus the name would be validly published as a new genus. It is, however, to be treated as a new generic name based on Spruce's subgeneric name, even though *L.* subg. *Brachiolejeunea* is neither cited nor indicated in any way in the protologue of Stephani & Spruce.

Ex. 10. When Sampaio published "*Psorama murale* Samp." (in Sampaio & Crespo in Bol. Real Soc. Esp. Hist. Nat. 27: 142. 1927), he adopted the epithet of *Lichen murale* Schreb. (1771), a name applied to the same taxon, without indicating that name directly or indirectly. He cited *Lecanora saxicola* Ach. in synonymy. *Psorama murale* is to be treated as a new combination based on *Lichen murale* because otherwise it would be a validly published but illegitimate replacement name for *Lecanora saxicola*.

33.4. On or after 1 January 1953, a new combination, a new generic name with a basionym, or an avowed substitute (replacement name, nomen novum) based on a previously and validly published name is not validly published unless its basionym (name-bringing or epithet-bringing synonym) or the replaced synonym (when a new name is proposed) is clearly indicated and a full and direct reference given to its author and place of valid publication, with page or plate reference and date (but see Art. 33.5 and 33.7). On or after 1 January 2007, a new combination, a new generic name with a basionym, or an avowed substitute is not validly published unless its basionym or replaced synonym is cited.

Ex. 11. In transferring *Ectocarpus mucronatus* D. A. Saunders to *Giffordia*, Kjeldsen & Phinney (in Madroño 22: 90. 27 Apr 1973) cited the basionym and its author but without reference to its place of valid publication. They later (in Madroño 22: 154. 2 Jul 1973) validly published the binomial *G. mucronata* (D. A. Saunders) Kjeldsen & H. K. Phinney by giving a full and direct reference to the place of valid publication of the basionym.

Note 1. For the purpose of this *Code,* a page reference (for publications with a consecutive pagination) is a reference to the page or pages on which the basionym or replaced synonym was validly published or on which the protologue is printed, but not to the pagination of the whole publication unless it is coextensive with that of the protologue.

Ex. 12. When proposing *"Cylindrocladium infestans",* Peerally (in Mycotaxon 40: 337. 1991) cited the basionym as *"Cylindrocladiella infestans* Boesew., Can. J. Bot. 60: 2288-2294. 1982". As this refers to the pagination of Boesewinkel's entire paper, not of the protologue of the intended basionym alone, the combination was not validly published by Peerally.

Ex. 13. The new combination *Conophytum marginatum* subsp. *littlewoodii* (L. Bolus) S. A. Hammer (Dumpling & His Wife: New Views Gen. Conophytum: 181. 2002), being made prior to 1 January 2007, was validly published even though Hammer did not cite the basionym *(Conophytum littlewoodii)* but only indicated it by citing its bibliographic reference.

33.5. For names published on or after 1 January 1953, errors in the citation of the basionym or replaced synonym, including incorrect author citation (Art. 46), but not omissions (Art. 33.4), do not preclude valid publication of a new combination, new generic name with a basionym, or nomen novum.

Ex. 14. Aronia arbutifolia var. *nigra* (Willd.) F. Seym. (Fl. New England: 308. 1969) was published as a new combination "Based on *Mespilus arbutifolia* L. var. *nigra* Willd., in Sp. Pl. 2: 1013. 1800." Willdenow treated these plants in the genus *Pyrus,* not *Mespilus,* and publication was in 1799, not 1800; these errors are treated as bibliographic errors of citation and do not prevent valid publication of the new combination.

Ex. 15. The new combination *Agropyron desertorum* var. *pilosiusculum* (Melderis) H. L. Yang (in Kuo, Fl. Reipubl. Popularis Sin. 9(3): 113. 1987) was unknowingly but validly published by Yang, who wrote *"Agropyron desertorum* ... var. *pilosiusculum* Meld. in Norlindh, Fl. Mong. Steppe. 1: 121. 1949", which constitutes a full and direct reference to the basionym, *A. desertorum* f. *pilosiusculum* Melderis, despite the error in citing the rank-denoting term.

33.6. Mere reference to the *Index kewensis,* the *Index of fungi,* or any work other than that in which the name was validly published does not constitute a full and direct reference to the original publication of a name (but see Art. 33.7).

Ex. 16. Ciferri (in Mycopathol. Mycol. Appl. 7: 86-89. 1954), in proposing 142 new combinations in *Meliola,* omitted references to places of publication of basionyms, stating that they could be found in Petrak's lists or in the *Index of fungi;* none of these combinations was validly published. Similarly, Grummann (Cat. Lich. Germ.: 18. 1963) introduced a new combination in the form *Lecanora campestris* f. *"pseudistera* (Nyl.) Grumm. c.n. – *L. p.* Nyl., Z 5: 521", in which "Z 5" referred to Zahlbruckner (Cat. Lich. Univ. 5: 521. 1928), who gave the full citation of the basionym, *Lecanora pseudistera* Nyl.; Grummann's combination was not validly published.

Note 2. The publication of a name for a taxon previously known under a misapplied name must be valid under Art. 32-45. This procedure is not the same as publishing an avowed substitute (replacement name, nomen novum) for a validly published but illegitimate name (Art. 58.1), the type of which is necessarily the same as that of the name which it replaced (Art. 7.3).

Ex. 17. Sadleria hillebrandii Rob. (1913) was introduced as a "nom. nov." for "*Sadleria pallida* Hilleb. Fl. Haw. Is. 582. 1888. Not Hook. & Arn. Bot. Beech. 75. 1832." Since the requirements of Art. 32-45 were satisfied (for valid publication, prior to 1935, simple reference to a previous description or diagnosis in any language was sufficient), the name is validly published. It is, however, to be considered the name of a new species, validated by Hillebrand's description of the taxon to which he misapplied the name *S. pallida* Hook. & Arn., and not a nomen novum as stated by Robinson; hence, Art. 7.3 does not apply.

Ex. 18. Juncus bufonius "var. *occidentalis*" (Hermann in U.S. Forest Serv., Techn. Rep. RM-18: 14. 1975) was published as a "nom. et stat. nov." for *J. sphaerocarpus* "auct. Am., non Nees". Since there is no Latin diagnosis, designation of type, or reference to any previous publication providing these requirements, the name is not validly published.

33.7. On or after 1 January 1953, in any of the following cases, a full and direct reference to a work other than that in which the basionym or replaced synonym was validly published is treated as an error to be corrected, not affecting the valid publication of a new combination, a new generic name with a basionym, or nomen novum:

(a) when the name cited as the basionym or replaced synonym was validly published earlier than in the cited publication, but in that cited publication, in which all conditions for valid publication are again fulfilled, there is no reference to the actual place of valid publication;

(b) when the failure to cite the place of valid publication of the basionym or replaced synonym is explained by the later nomenclatural starting-point for the group concerned, and in particular by the backward shift of the starting date for some fungi;

(c) when an intended new combination or new generic name with a basionym would otherwise be validly published as a (legitimate or illegitimate) nomen novum; or

(d) when an intended new combination, new generic name with a basionym, or nomen novum would otherwise be the validly published name of a new taxon.

Ex. 19. (a) The combination *Trichipteris kalbreyeri* was proposed by Tryon (1970) with a full and direct reference to "*Alsophila Kalbreyeri* C. Chr. Ind. Fil. 44. 1905". This, however, is not the place of valid publication of the intended basionym, which had previously been published, with the same type, by Baker (1891; see Art. 6 Ex. 1). As Christensen provided no reference to Baker's earlier publication, Tryon's error of citation does not affect the valid publication of his new combination, which is to be cited as *T. kalbreyeri* (Baker) R. M. Tryon.

Ex. 20. (a) The intended new combination *"Machaerina iridifolia"* was proposed by Koyama (in Bot. Mag. (Tokyo) 69: 64. 1956) with a full and direct reference to "*Cladium iridifolium* Baker, Flor. Maurit. 424 (1877)". However, *C. iridifolium* had been proposed by Baker as a new combination based on *Scirpus iridifolius* Bory (1804). As Baker provided an explicit reference to Bory, Art. 33.7(a) does not apply and the combination under *Machaerina* was not validly published by Koyama.

Ex. 21. (b) The combination *Lasiobelonium corticale* was proposed by Raitviir (1980) with a full and direct reference to *Peziza corticalis* in Fries (Syst. Mycol. 2: 96. 1822). This, however, is not the place of valid publication of the basionym, which, under the *Code* operating in 1980, was in Mérat (Nouv. Fl. Env. Paris, ed. 2, 1: 22. 1821), and under the current *Code* is in Persoon (Observ. Mycol. 1: 28. 1796). Raitviir's error of citation, being partly explained by the backward shift of the starting date for ascomycetes and partly by the absence of a reference to Mérat in Fries's work, does not negate valid publication of the new combination, which is to be cited as *L. corticale* (Pers. : Fr.) Raitv.

Ex. 22. (c) The intended new combination *Mirabilis laevis* subsp. *glutinosa* was proposed by Murray (in Kalmia 13: 32. 1983) with a full and direct reference to "*Mirabilis glutinosa* A. Nels., Proc. Biol. Soc. Wash. 17: 92 (1904)" as "basionymum". This, however, cannot be a basionym because it is an illegitimate later homonym of *M. glutinosa* Kuntze (1898); it is also the replaced synonym of *Hesperonia glutinosa* Standl. (1909). Under Art. 33.7(c) Murray validly published a new combination based on *H. glutinosa* because otherwise he would have published a nomen novum for *M. glutinosa*. The name is therefore to be cited as *M. laevis* subsp. *glutinosa* (Standl.) A. E. Murray.

Ex. 23. (d) The nomen novum *Agropyron kengii* was proposed by Tzvelev (1968) with a full and direct reference to "*Roegneria hirsuta* Keng, Fl. ill. sin., Gram. (1959) 407". This, however, is not the place of valid publication of the intended replaced synonym, which was subsequently validy published by Keng (1963). As Tzvelev also provided a Latin description and indicated a single gathering as the type, the nomen novum was validly published as such because it would otherwise have been the validly published name of a new taxon.

33.8. On or after 1 January 1953, if an author claims to be publishing a new combination, new generic name with a basionym, or avowed substitute, but fails to provide the full information required under Art. 33.4, as qualified by Art. 33.5 and 33.7, the name is not validly published even though the author may have at the same time provided other information that would have resulted in valid publication as the name of a new taxon.

33.9. A name given to a taxon of which the rank is at the same time, contrary to Art. 5, denoted by a misplaced term is not validly published. Such misplacements include forms divided into varieties, species containing genera, and genera containing families or tribes.

33.10. Only those names published with the rank-denoting terms that must be removed so as to achieve a proper sequence are to be regarded as not validly published. In cases where terms are switched, e.g. family-order, and a proper sequence can be achieved by removing either or both of the rank-denoting terms, names at neither rank are validly published unless one is a secondary rank (Art. 4.1) and one is a principal rank (Art. 3.1), e.g. family-genus-tribe, in which case only names published at the secondary rank are not validly published.

Ex. 24. "Sectio *Orontiaceae*" was not validly published by Brown (Prodr.: 337. 1810) since he misapplied the term "sectio" to a rank higher than genus.

Ex. 25. "Tribus *Involuta*" and "tribus *Brevipedunculata*" (Huth in Bot. Jahrb. Syst. 20: 365, 368. 1895) are not validly published names, since Huth misapplied the term "tribus" to a rank lower than section, within the genus *Delphinium*.

Note 3. Sequential use of the same rank-denoting term in a taxonomic sequence does not represent misplaced-rank denoting terms.

Ex. 26. Danser (in Recueil Trav. Bot. Néerl. 18: 125–210. 1921) published ten new names of subspecies in a treatment of *Polygonum* in which he recognized subspecies (indicated by Roman numerals) within subspecies (indicated by Arabic numerals). These do not represent misplaced rank-denoting terms, so Art. 33.9 does not apply and the new names are validly published.

33.11. Situations where the same rank-denoting term is used at more than one non-successive position in the taxonomic sequence represent informal usage of rank-denoting terms. Names published with such rank-denoting terms are treated as unranked (see Art. 35.1 and 35.3).

Ex. 27. Names published with the term "series" by Bentham & Hooker (Gen. Pl. 1-3. 1862-1883) are treated as unranked because this term was used at seven different hierarchical positions in the taxonomic sequence. Therefore, the sequence in *Rhynchospora* (3: 1058-1060. 1883) of genus-"series"-section does not contain a misplaced rank-denoting term.

33.12. An exception to Art. 33.9 is made for names of the subdivisions of genera termed tribes (tribus) in Fries's *Systema mycologicum,* which are treated as validly published names of subdivisions of genera.

Ex. 28. *Agaricus* "tribus" *Pholiota* Fr. (Syst. Mycol. 1: 240. 1821), sanctioned in the same work, is the validly published basionym of the generic name *Pholiota* (Fr. : Fr.) P. Kumm. (1871) (see Art. 33 Ex. 7).

Recommendation 33A

33A.1. The full and direct reference to the place of publication of the basionym or replaced synonym should immediately follow a proposed new combination or nomen novum. It should not be provided by mere cross-reference to a bibliography at the end of the publication or to other parts of the same publication, e.g. by use of the abbreviations "loc. cit." or "op. cit."

Article 34

34.1. A name is not validly published *(a)* when it is not accepted by the author in the original publication; *(b)* when it is merely proposed in anticipation of the future acceptance of the taxon concerned, or of a particular circumscription, position, or rank of the taxon (so-called provisional name), except as provided for in Art. 59; *(c)* when it is merely cited as a synonym; *(d)* by the mere mention of the subordinate taxa included in the taxon concerned. Art. 34.1(a) does not apply to names published with a question mark or other indication of taxonomic doubt, yet accepted by their author.

Ex. 1. (a) *"Sebertia"*, proposed by Pierre (ms.) for a monotypic genus, was not validly published by Baillon (in Bull. Mens. Soc. Linn. Paris 2: 945. 1891) because he did not accept the genus. Although he gave a description of it, he referred its only species *"Sebertia acuminata* Pierre (ms.)" to the genus *Sersalisia* R. Br. as *S. ? acuminata,* which he thereby validly published under the provision of Art. 34.1, last sentence. The name *Sebertia* was validly published by Engler (1897).

Ex. 2. (a) The designations listed in the lefthand column of the Linnaean thesis *Herbarium amboinense* defended by Stickman (1754) were not names accepted by Linnaeus upon publication and are not validly published.

Ex. 3. (a) *Coralloides gorgonina* Bory was validly published in a paper by Flörke (in Mag. Neusten Entdeck. Gesammten Naturk. Ges. Naturf. Freunde Berlin 3: 125. 1809), even though Flörke did not accept it as a new species. At Bory's request, Flörke included Bory's diagnosis (and name) making Bory the author of the name under Art. 46.2. The acceptance or otherwise of the name by Flörke is not, therefore, relevant for valid publication.

Ex. 4. (a) (b) The designation *"Conophyton"*, suggested by Haworth (Rev. Pl. Succ.: 82. 1821) for *Mesembryanthemum* sect. *Minima* Haw. (Rev. Pl. Succ.: 81. 1821) in the words "If this section proves to be a genus, the name of *Conophyton* would be apt", was not a validly published generic name since Haworth did not adopt it or accept the genus. The name was validly published as *Conophytum* N. E. Br. (1922).

Ex. 5. (b) *"Pteridospermaexylon"* and *"P. theresiae"* were published by Greguss (in Földt. Közl. 82: 171. 1952) for a genus and species of fossil wood. As Greguss explicitly stated "Vorläufig benenne ich es mit den Namen ..." [provisionally I designate it by the names ...], these are provisional names and as such are not validly published.

Ex. 6. (b) The designation *"Sterocaulon subdenudatum"* proposed by Havaas (in Bergens Mus. Årbok. 12: 13, 20. 1954) is not validly published in spite of it being presented as a new species with a Latin diagnosis, since on both pages it was indicated to be "ad int."

Ex. 7. (c) *"Ornithogalum undulatum* hort. Bouch." was not validly published by Kunth (Enum. Pl. 4: 348. 1843) when he cited it as a synonym under *Myogalum boucheanum* Kunth; the combination under *Ornithogalum* L. was validly published later: *O. boucheanum* (Kunth) Asch. (1866).

Ex. 8. (d) The family designation *"Rhaptopetalaceae"* was not validly published by Pierre (in Bull. Mens. Soc. Linn. Paris 2: 1296. Mai 1897), who merely mentioned the constituent genera, *Brazzeia* Baill., *"Scytopetalum"*, and *Rhaptopetalum* Oliv., but gave no description or diagnosis; the family bears the name *Scytopetalaceae* Engl. (Oct 1897), accompanied by a description.

Ex. 9. (d) The generic designation *"Ibidium"* was not validly published by Salisbury (in Trans. Hort. Soc. London 1: 291. 1812), who merely mentioned four included species but supplied no generic description or diagnosis.

Ex. 10. (final sentence) *Aponogetonaceae* Planch. (in Bot. Mag.: ad. t. 4894. 1856) was validly published by reference to the description of "Aponogétacées" (Planchon in Ann. Sci.
Nat., Bot., sér. 3, 1: 119. 1844) even though Hooker indicated taxonomic doubt when he
wrote (pp. [4-5]) "M. Planchon ... suggests that *Aponogeton* should form a suborder of *Al-
ismaceae,* or probably a new order, *Aponogetaceae*".

34.2. When, on or after 1 January 1953, two or more different names based
on the same type are proposed simultaneously for the same taxon by the
same author (so-called alternative names), none of them is validly published. This rule does not apply in those cases where the same combination
is simultaneously used at different ranks, either for infraspecific taxa within
a species or for subdivisions of a genus within a genus (see Rec. 22A.1-2
and 26A.1-3).

Ex. 11. The species of *Brosimum* Sw. described by Ducke (in Arch. Jard. Bot. Rio de
Janeiro 3: 23-29. 1922) were published with alternative names under *Piratinera* Aubl. added
in a footnote (pp. 23-24). The publication of both sets of names, being effected before 1
January 1953, is valid.

Ex. 12. *"Euphorbia jaroslavii"* (Poljakov in Bot. Mater. Gerb. Bot. Inst. Komarova Akad.
Nauk SSSR 15: 155. 1953) was published with an alternative designation, *"Tithymalus
jaroslavii".* Neither was validly published. However, one name, *Euphorbia yaroslavii* (with
a different transliteration of the initial letter), was validly published by Poljakov (1961), who
effectively published it with a reference to the earlier publication and simultaneously
rejected assignment to *Tithymalus*.

Ex. 13. Description of *"Malvastrum bicuspidatum* subsp. *tumidum* S. R. Hill var. *tumidum,*
subsp. et var. nov." (in Brittonia 32: 474. 1980) simultaneously validated both *M. bicuspidatum* subsp. *tumidum* S. R. Hill and *M. bicuspidatum* var. *tumidum* S. R. Hill.

Ex. 14. Hitchcock (in Univ. Wash. Publ. Biol. 17(1): 507-508. 1969) used the name *Bromus
inermis* subsp. *pumpellianus* (Scribn.) Wagnon and provided a full and direct reference to its
basionym, *B. pumpellianus* Scribn. Within that subspecies, he recognized varieties, one of
which he named *B. inermis* var. *pumpellianus* (without author citation but clearly based on
the same basionym and type). In so doing, he met the requirements for valid publication of
B. inermis var. *pumpellianus* (Scribn.) C. L. Hitchc.

Note 1. The name of a fungal holomorph and that of a correlated anamorph (see
Art. 59), even if proposed simultaneously, are not alternative names in the sense of
Art. 34.2, and both are validly published. They have different types, and the circumscription of the holomorph is considered to include the anamorph, but not vice
versa.

Ex. 15. *Lasiosphaeria elinorae* Linder (1929), the name of a fungal holomorph, and the
simultaneously published name of a correlated anamorph, *Helicosporium elinorae* Linder,
are both validly published, and both can be used under Art. 59.5.

Recommendation 34A

34A.1. Authors should avoid mentioning in their publications previously unpublished names that they do not accept, especially if the persons responsible for these

unpublished names have not formally authorized their publication (see Rec. 23A.3(i)).

Article 35

35.1. A new name or combination published on or after 1 January 1953 without a clear indication of the rank of the taxon concerned is not validly published.

35.2. For suprageneric names published on or after 1 January 1908, the use of one of the terminations specified in Rec. 16A.1-3, Art. 17.1, 18.1, 19.1, and 19.3 is accepted as an indication of the corresponding rank, unless this *(a)* would conflict with the explicitly designated rank of the taxon (which takes precedence), *(b)* would result in a rank sequence contrary to Art. 5 (in which case Art. 33.9 applies), or *(c)* would result in a rank sequence in which the same rank-denoting term occurs at more than one hierarchical position.

Ex. 1. Jussieu (in Mém. Mus. Hist. Nat. 12: 497. 1827) proposed *Zanthoxyleae* without specifying the rank. Although he employed the present termination for tribe *(-eae),* that name, being published prior to 1908, is unranked. *Zanthoxyleae* Dumort. (Anal. Fam. Pl.: 45. 1829), however, is a tribal name, as Dumortier specified its rank.

Ex. 2. Nakai (Chosakuronbun Mokuroku [Ord. Fam. Trib. Nov.], 1943) validly published the names *Parnassiales, Lophiolaceae, Ranzanioideae,* and *Urospatheae.* He indicated the respective ranks of order, family, subfamily, and tribe, by virtue of their terminations, even though he did not mention these ranks explicitly.

35.3. A new name or combination published before 1 January 1953 without a clear indication of its rank is validly published provided that all other requirements for valid publication are fulfilled; it is, however, inoperative in questions of priority except for homonymy (see Art. 53.4). If it is a new name, it may serve as a basionym for subsequent combinations or a replaced synonym for nomina nova in definite ranks.

Ex. 3. The groups *"Soldanellae", "Sepincoli", "Occidentales",* etc., were published without any indication of rank under *Convolvulus* L. by House (in Muhlenbergia 4: 50. 1908). The names *C.* [unranked] *Soldanellae,* etc., are validly published but they are not in any definite rank and have no status in questions of priority except for purposes of homonymy.

Ex. 4. In *Carex* L., the epithet *Scirpinae* was used in the name of a subdivision of a genus of no stated rank by Tuckerman (Enum. Meth. Caric.: 8. 1843); this taxon was assigned sectional rank by Kükenthal (in Engler, Pflanzenr. 38: 81. 1909) and its name may be cited as *Carex* sect. *Scirpinae* (Tuck.) Kük. (*C.* [unranked] *Scirpinae* Tuck.).

Ex. 5. Loesener published "*Geranium andicola* var. vel forma *longipedicellatum*" (Bull. Herb. Boissier, ser. 2, 3(2): 93. 1903) without a clear indication of infraspecific rank. The name is correctly cited as "*G. andicola* [unranked] *longipedicellatum* Loes." The epithet was used in a subsequent combination, *G. longipedicellatum* (Loes.) R. Knuth (1912).

35.4. If in one whole publication (Art. 35.5), prior to 1 January 1890, only one infraspecific rank is admitted, it is considered to be that of variety unless this would be contrary to the author's statements in the same publication.

35.5. In questions of indication of rank, all publications appearing under the same title and by the same author, such as different parts of a flora issued at different times (but not different editions of the same work), must be considered as a whole, and any statement made therein designating the rank of taxa included in the work must be considered as if it had been published together with the first instalment.

Ex. 6. In Link's *Handbuch* (1829-1833) the rank-denoting term "O." (ordo) was used in all three volumes. These names of orders cannot be considered as having been published as names of families (Art. 18.2) since the term family was used for *Agaricaceae* and *Tremellaceae* under the order *Fungi* in vol. 3 (pp. 272, 337; see Art. 18 Note 1). This applies to all three volumes of the *Handbuch,* even though vol. 3 was published later (Jul - 29 Sep 1833) than vols. 1 and 2 (4-11 Jul 1829).

Article 36

36.1. On or after 1 January 1935 a name of a new taxon (algal and all fossil taxa excepted) must, in order to be validly published, be accompanied by a Latin description or diagnosis or by a reference to a previously and effectively published Latin description or diagnosis (but see Art. H.9).

Ex. 1. Arabis "Sekt. *Brassicoturritis* O. E. Schulz" and "Sekt. *Brassicarabis* O. E. Schulz" (in Engler & Prantl, Nat. Pflanzenfam., ed. 2, 17b: 543-544. 1936), published with German but no Latin descriptions or diagnoses, are not validly published names.

Ex. 2. "Schiedea gregoriana" (Degener, Fl. Hawaiiensis, fam. 119. 9 Apr 1936) was accompanied by an English but no Latin description and is accordingly not a validly published name. *Schiedea kealiae* Caum & Hosaka (in Occas. Pap. Bernice Pauahi Bishop Mus. 11(23): 3. 10 Apr 1936), the type of which is part of the material used by Degener, is provided with a Latin description and is validly published.

Ex. 3. Alyssum flahaultianum Emb., first published without a Latin description or diagnosis (in Bull. Soc. Hist. Nat. Maroc 15: 199. 1936), was validly published posthumously when a Latin translation of Emberger's original French description was provided (in Willdenowia 15: 62-63. 1985).

36.2. In order to be validly published, a name of a new taxon of non-fossil algae published on or after 1 January 1958 must be accompanied by a Latin description or diagnosis or by a reference to a previously and effectively published Latin description or diagnosis.

Ex. 4. Although *Neoptilota* Kylin (Gatt. Rhodophyc.: 392. 1956) was accompanied by only a German description, it is a validly published name since it applies to an alga and was published before 1958.

36.3. In order to be validly published, a name of a new taxon of fossil plants published on or after 1 January 1996 must be accompanied by a Latin or English description or diagnosis or by a reference to a previously and effectively published Latin or English description or diagnosis.

Recommendation 36A

36A.1. Authors publishing names of new taxa of non-fossil plants should give or cite a full description in Latin in addition to the diagnosis.

Article 37

37.1. Publication on or after 1 January 1958 of the name of a new taxon of the rank of genus or below is valid only when the type of the name is indicated (see Art. 7-10; but see Art. H.9 Note 1 for the names of certain hybrids).

37.2. For the name of a new species or infraspecific taxon, indication of the type as required by Art. 37.1 can be achieved by reference to an entire gathering, or part thereof, even if it consists of two or more specimens as defined in Art. 8 (see also Art. 37.7).

Ex. 1. When Cheng described *"Gnetum cleistostachyum"* (in Acta Phytotax. Sin. 13(4): 89. 1975) the name was not validly published because two gatherings were designated as types: *K. H. Tsai 142* (as "♀ Typus") and *X. Jiang 127* (as "♂ Typus").

Note 1. When the type is indicated by reference to a gathering that consists of more than one specimen, those specimens are syntypes (see Art. 9.4).

Ex. 2. The protologue of *Laurentia frontidentata* E. Wimm. (in Engler, Pflanzenr. 108: 855. 1968) includes the type statement *"E. Esterhuysen No. 17070!* Typus – Pret., Bol." The name is validly published because a single gathering is cited, despite the mention of duplicate specimens (syntypes) in two different herbaria.

37.3. For the name of a new genus or subdivision of a genus, reference (direct or indirect) to one species name only, or the citation of the holotype or lectotype of one previously or simultaneously published species name only, even if that element is not explicitly designated as type, is acceptable as indication of the type (see also Art. 22.6; but see Art. 37.6). Similarly, for the name of a new species or infraspecific taxon, mention of a single specimen or gathering (Art. 37.2) or illustration (when permitted by Art. 37.4 or 37.5), even if that element is not explicitly designated as type, is acceptable as indication of the type (but see Art. 37.6).

Ex. 3. *"Baloghia pininsularis"* was published by Guillaumin (in Mém. Mus. Natl. Hist. Nat., B, Bot. 8: 260. 1962) with two cited gatherings: *Baumann 13813* and *Baumann 13823*. As the author failed to designate one of them as the type, he did not validly publish the name.

Valid publication was effected when McPherson & Tirel (in Fl. Nouv.-Caléd. 14: 58. 1987) wrote "Lectotype (désigné ici): *Baumann-Bodenheim 13823* (P!; iso-, Z)" while providing a full and direct reference to Guillaumin's Latin description (Art. 45.1; see Art. 46 Ex. 9); McPherson & Tirel's use of "lectotype" is correctable to "holotype" under Art. 9.8.

Note 2. Mere citation of a locality does not constitute mention of a single specimen or gathering. Concrete reference to some detail relating to the actual type, such as the collector's name or collecting number or date, is required.

Note 3. Cultures of fungi and algae preserved in a metabolically inactive state are acceptable as types (Art. 8.4; see also Rec. 8B.1).

37.4. For the purpose of this Article, the type of a name of a new species or infraspecific taxon (fossils excepted: see Art. 8.5) may be an illustration prior to 1 January 2007, on or after which date, the type must be a specimen (except as provided in Art. 37.5).

37.5. For the purpose of this Article, the type of a name of a new species or infraspecific taxon of microscopic algae or microfungi (fossils excepted: see Art. 8.5) may be an effectively published illustration if there are technical difficulties of preservation or if it is impossible to preserve a specimen that would show the features attributed to the taxon by the author of the name.

37.6. For the name of a new taxon of the rank of genus or below published on or after 1 January 1990, indication of the type must include one of the words "typus" or "holotypus", or its abbreviation, or its equivalent in a modern language (see also Rec. 37A).

37.7. For the name of a new species or infraspecific taxon published on or after 1 January 1990 of which the type is a specimen or unpublished illustration, the single herbarium or collection or institution in which the type is conserved must be specified.

Ex. 4. In the protologue of *Setaria excurrens* var. *leviflora* Keng ex S. L. Chen (in Bull. Nanjing Bot. Gard. 1988-1989: 3. 1990) the gathering *Guangxi Team 4088* was indicated as "模式" (Chinese for "type") and the herbarium where the type is conserved was specified as "中国科学院植研究所读" (Botanical Research Institute, Chinese Academy of Sciences, i.e. PE).

Note 4. Specification of the herbarium or collection or institution may be made in an abbreviated form, e.g. as given in *Index herbariorum, part I,* or in the *World directory of collections of cultures of microorganisms.*

Ex. 5. When 't Hart described *"Sedum eriocarpum* subsp. *spathulifolium"* (in Ot Sist. Bot. Dergisi 2(2): 7. 1995) the name was not validly published because no herbarium or collection or institution in which the holotype specimen was conserved was specified. Valid publication was effected when 't Hart (in Strid & Tan, Fl. Hellen. 2: 325. 2002) wrote "Type ...

't Hart HRT-27104 ... (U)" while providing a full and direct reference to his previously published Latin diagnosis (Art. 45.1).

Recommendation 37A

37A.1. The indication of the nomenclatural type should immediately follow the description or diagnosis and should include the Latin word "typus" or "holotypus".

Article 38

38.1. In order to be validly published, a name of a new taxon of fossil plants of specific or lower rank published on or after 1 January 1912 must be accompanied by an illustration or figure showing the essential characters, in addition to the description or diagnosis, or by a reference to a previously and effectively published illustration or figure.

38.2. For the name of a new species or infraspecific taxon of fossil plants published on or after 1 January 2001, at least one of the validating illustrations must be identified as representing the type specimen (see also Art. 9.13).

Article 39

39.1. In order to be validly published, a name of a new taxon of non-fossil algae of specific or lower rank published on or after 1 January 1958 must be accompanied by an illustration or figure showing the distinctive morphological features, in addition to the Latin description or diagnosis, or by a reference to a previously and effectively published illustration or figure.

Recommendation 39A

39A.1. The illustration or figure required by Art. 39 should be prepared from actual specimens, preferably including the holotype.

Article 40

40.1. In order to be validly published, names of hybrids of specific or lower rank with Latin epithets must comply with the same rules as names of non-hybrid taxa of the same rank.

Ex. 1. "Nepeta ×faassenii" (Bergmans, Vaste Pl. Rotsheesters, ed. 2: 544. 1939, with a description in Dutch; Lawrence in Gentes Herb. 8: 64. 1949, with a diagnosis in English) is not validly published, not being accompanied by or associated with a Latin description or diagnosis. The name *Nepeta ×faassenii* Bergmans ex Stearn (1950) is validly published, being accompanied by a Latin description.

Ex. 2. "*Rheum ×cultorum*" (Thorsrud & Reisaeter, Norske Plantenavn: 95. 1948), being there a nomen nudum, is not validly published.

Ex. 3. "*Fumaria ×salmonii*" (Druce, List Brit. Pl.: 4. 1908) is not validly published, as only the presumed parentage *F. densiflora × F. officinalis* is stated.

Note 1. For names of hybrids of the rank of genus or subdivision of a genus, see Art. H.9.

Article 41

41.1. In order to be validly published, a name of a family or subdivision of a family must be accompanied *(a)* by a description or diagnosis of the taxon, or *(b)* by a reference (direct or indirect) to a previously and effectively published description or diagnosis of a family or subdivision of a family.

Ex. 1. "*Pseudoditrichaceae* fam. nov." (Steere & Iwatsuki in Canad. J. Bot. 52: 701. 1974) was not a validly published name of a family as there was no Latin description or diagnosis nor reference to either, but only mention of the single included genus and species (see Art. 34.1(d)), "*Pseudoditrichum mirabile* gen. et sp. nov.", both validly published under Art. 42 by a single Latin diagnosis.

Ex. 2. Presl did not validly publish "*Cuscuteae*" (in Presl & Presl, Delic. Prag.: 87. 1822) as the name of a family (see "Praemonenda", pp. [3-4]) by direct reference to the previously and effectively published description of "*Cuscuteae*" (Berchtold & Presl, Přir. Rostlin: 247. 1820) because the latter is the name of an order (see Art. 18 *Ex. 5).

41.2. In order to be validly published, a name of a genus or subdivision of a genus must be accompanied *(a)* by a description or diagnosis of the taxon (but see Art. 42), or *(b)* by a reference (direct or indirect) to a previously and effectively published description or diagnosis of a genus or subdivision of a genus.

Ex. 3. Validly published generic names: *Carphalea* Juss., accompanied by a generic description; *Thuspeinanta* T. Durand, replacing the name of the previously described genus *Tapeinanthus* Boiss. ex Benth. (non Herb.); *Aspalathoides* (DC.) K. Koch, based on the name of a previously described section, *Anthyllis* sect. *Aspalathoides* DC.; *Scirpoides* Ség. (Pl. Veron. Suppl.: 73. 1754), accepted there but without a generic description or diagnosis, validly published by indirect reference (through the title of the book and a general statement in the preface) to the generic diagnosis and further direct references in Séguier (Pl. Veron. 1: 117. 1745).

Note 1. An exception to Art. 41.2 is made for the generic names first published by Linnaeus in *Species plantarum,* ed. 1 (1753) and ed. 2 (1762-1763), which are treated as having been validly published on those dates (see Art. 13.4).

Note 2. In certain circumstances, an illustration with analysis is accepted as equivalent to a generic description or diagnosis (see Art. 42.3).

41.3. In order to be validly published, a name of a species or infraspecific taxon must be accompanied *(a)* by a description or diagnosis of the taxon (but see Art. 42 and 44), or *(b)* by a reference to a previously and effectively published description or diagnosis of a species or infraspecific taxon. A name of a species may also be validly published *(c)*, under certain circumstances, by reference to a genus the name of which was previously and validly published simultaneously with its description or diagnosis. A reference as mentioned under (c) is acceptable only if neither the author of the name of the genus nor the author of the name of the species indicates that more than one species belongs to the genus in question.

Ex. 4. *Trilepisium* Thouars (1806) was validated by a generic description but without mention of a name of a species. *T. madagascariense* DC. (1825) was subsequently proposed without a description or diagnosis of the species. Neither author gave any indication that there was more than one species in the genus. Candolle's specific name is therefore validly published.

Article 42

42.1. The names of a genus and a species may be validly published simultaneously by provision of a single description (descriptio generico-specifica) or diagnosis, even though this may have been intended as only generic or specific, if all of the following conditions obtain: *(a)* the genus is at that time monotypic; *(b)* no other names (at any rank) have previously been validly published based on the same type; and *(c)* the names of the genus and species otherwise fulfil the requirements for valid publication. Reference to an earlier description or diagnosis is not acceptable in place of a descriptio generico-specifica.

42.2. For the purpose of Art. 42, a monotypic genus is one for which a single binomial is validly published, even though the author may indicate that other species are attributable to the genus.

Ex. 1. Nylander (1879) described the new species *"Anema nummulariellum"* in a new genus *"Anema"* without providing a generic description or diagnosis. Since at the same time he also transferred *Omphalaria nummularia* Durieu & Mont. to *"Anema"*, none of his names was validly published. They were later validated by Forsell (1885).

Ex. 2. The names *Kedarnatha* P. K. Mukh. & Constance (1986) and *K. sanctuarii* P. K. Mukh. & Constance, the latter designating the single, new species of the new genus, are both validly published although a Latin description was provided only under the generic name.

Ex. 3. *Piptolepis phillyreoides* Benth. (1840) was a new species assigned to the monotypic new genus *Piptolepis* published with a combined generic and specific description, and both names are validly published.

Ex. 4. In publishing *"Phaelypea"* without a generic description or diagnosis, P. Browne (Civ. Nat. Hist. Jamaica: 269. 1756) included and described a single species, but he gave the species a phrase-name not a validly published binomial. Art. 42 does not therefore apply and *"Phaelypea"* is not a validly published name.

42.3. Prior to 1 January 1908 an illustration with analysis, or for non-vascular plants a single figure showing details aiding identification, is acceptable, for the purpose of this Article, in place of a written description or diagnosis.

42.4. For the purpose of Art. 42, an analysis is a figure or group of figures, commonly separate from the main illustration of the plant (though usually on the same page or plate), showing details aiding identification, with or without a separate caption.

Ex. 5. The generic name *Philgamia* Baill. (1894) was validly published, as it appeared on a plate with analysis of the only included species, *P. hibbertioides* Baill., and was published before 1 January 1908.

Article 43

43.1. A name of a taxon below the rank of genus is not validly published unless the name of the genus or species to which it is assigned is validly published at the same time or was validly published previously.

Ex. 1. Binary designations for six species of *"Suaeda"*, including *"S. baccata"* and *"S. vera"*, were published with descriptions and diagnoses by Forsskål (Fl. Aegypt.-Arab.: 69-71. 1775), but he provided no description or diagnosis for the genus: these were not therefore validly published names.

Ex. 2. Müller (in Flora 63: 286. 1880) published the new genus *"Phlyctidia"* with the species *"P. hampeana* n. sp.", *"P. boliviensis"* (= *Phlyctis boliviensis* Nyl.), *"P. sorediiformis"* (= *Phlyctis sorediiformis* Kremp.), *"P. brasiliensis"* (= *Phlyctis brasiliensis* Nyl.), and *"P. andensis"* (= *Phlyctis andensis* Nyl.). These were not, however, validly published specific names in this place, because the intended generic name *"Phlyctidia"* was not validly published; Müller gave no generic description or diagnosis but only a description and a diagnosis of the new species *"P. hampeana"*. This description and diagnosis did not validate the generic name as a descriptio generico-specifica under Art. 42 since the new genus was not monotypic. Valid publication of the name *Phlyctidia* was by Müller (1895), who provided a short generic diagnosis and explicitly included only two species, the names of which, *P. ludoviciensis* Müll. Arg. and *P. boliviensis* (Nyl.) Müll. Arg., were also validly published in 1895.

Note 1. This Article applies also when specific and other epithets are published under words not to be regarded as generic names (see Art. 20.4).

Ex. 3. The binary designation *"Anonymos aquatica"* (Walter, Fl. Carol.: 230. 1788) is not a validly published name. The correct name for the species concerned is *Planera aquatica* J. F. Gmel. (1791), and the date of the name, for purposes of priority, is 1791. The name must not be cited as *"P. aquatica* (Walter) J. F. Gmel."

Ex. 4. Despite the existence of the generic name *Scirpoides* Ség. (1754), the binary designation *"S. paradoxus"* (Rottbøll, Descr. Pl. Rar.: 27. 1772) is not validly published since

"Scirpoides" in Rottbøll's context was a word not intended as a generic name. The first validly published name for this species is *Fuirena umbellata* Rottb. (1773).

Article 44

44.1. The name of a species or of an infraspecific taxon published before 1 January 1908 may be validly published even if only accompanied by an illustration with analysis (as defined in Art. 42.4).

Ex. 1. Panax nossibiensis Drake (1896) was validly published on a plate with analysis.

44.2. Single figures of non-vascular plants showing details aiding identification are considered as illustrations with analysis (see also Art. 42.4).

Ex. 2. Eunotia gibbosa Grunow (1881), a name of a diatom, was validly published by provision of a figure of a single valve.

Article 45

45.1. The date of a name is that of its valid publication. When the various conditions for valid publication are not simultaneously fulfilled, the date is that on which the last is fulfilled. However, the name must always be explicitly accepted in the place of its validation. A name published on or after 1 January 1973 for which the various conditions for valid publication are not simultaneously fulfilled is not validly published unless a full and direct reference (Art. 33.4) is given to the places where these requirements were previously fulfilled (but see Art. 33.6).

Ex. 1. "Clypeola minor" first appeared in the Linnaean thesis *Flora monspeliensis* (1756), in a list of names preceded by numerals but without an explanation of the meaning of these numerals and without any other descriptive matter; when the thesis was reprinted in vol. 4 of the *Amoenitates academicae* (1759), a statement was added explaining that the numbers referred to earlier descriptions published in Magnol's *Botanicon monspeliense.* However, *"Clypeola minor"* was absent from the reprint, being no longer accepted by Linnaeus, and was not therefore validly published.

Ex. 2. When proposing *"Graphis meridionalis"* as a new species, Nakanishi (in J. Sci. Hiroshima Univ., Ser. B(2), 11: 75. 1966) provided a Latin description but failed to designate a holotype. *Graphis meridionalis* M. Nakan. was validly published when Nakanishi (in J. Sci. Hiroshima Univ., Ser. B(2), 11: 265. 1967) designated the holotype of the name and provided a full and direct reference to his previous publication.

45.2. A correction of the original spelling of a name (see Art. 32.7 and 60) does not affect its date of valid publication.

Ex. 3. The correction of the erroneous spelling of *Gluta "benghas"* (Linnaeus, Mant. Pl.: 293. 1771) to *G. renghas* L. does not affect the date of publication of the name even though the correction dates only from 1883 (Engler in Candolle & Candolle, Monogr. Phan. 4: 225).

45.3. For purposes of priority only legitimate names are taken into consideration (see Art. 11, 52-54). However, validly published earlier homonyms, whether legitimate or not, shall cause rejection of their later homonyms, unless the latter are conserved or sanctioned (but see Art. 15 Note 1).

45.4. If a taxon originally assigned to a group not covered by this *Code* is treated as belonging to a group of plants other than algae or fungi, the authorship and date of any of its names are determined by the first publication that satisfies the requirements for valid publication under this *Code*. If the taxon is treated as belonging to the algae or fungi, any of its names need satisfy only the requirements of the pertinent non-botanical *Code* for status equivalent to valid publication under the present *Code* (but see Art. 54, regarding homonymy). However, a name generated in zoological nomenclature in accordance with the Principle of Coordination is not considered validly published under the botanical *Code* unless it appears in print and is applied to an accepted taxon.

Ex. 4. Amphiprora Ehrenb. (1843), an available[1] name for a genus of animals, was first treated as belonging to the algae by Kützing (1844). *Amphiprora* has priority in botanical nomenclature from 1843, not 1844.

Ex. 5. Petalodinium Cachon & Cachon-Enj. (in Protistologia 5: 16. 1969) is available under the *International code of zoological nomenclature* as the name of a genus of dinoflagellates. When the taxon is treated as belonging to the algae, its name retains its original authorship and date even though the original publication lacked a Latin description or diagnosis.

Ex. 6. Labyrinthodyction Valkanov (in Progr. Protozool. 3: 373. 1969), available under the *International code of zoological nomenclature* as the name of a genus of rhizopods, is considered to have been validly published in 1969 if the taxon is treated as belonging to the fungi even though the original publication lacked a Latin description or diagnosis.

Ex. 7. Protodiniferaceae Kof. & Swezy (in Mem. Univ. Calif. 5: 111. 1921, *"Protodiniferidae"*), available under the *International code of zoological nomenclature,* is validly published as a name of a family of algae with its original authorship and date but with the original termination changed in accordance with Art. 18.4 and 32.7.

Ex. 8. Pneumocystis P. Delanoë & Delanoë (in Comp. Rend. Acad. Hebd. Séances Acad. Sci. 155: 660. 1912) was published for a "protozoan" genus with a description expressing doubt as to its generic status, "Si celui-ci doit constituer un genre nouveau, nous proposons de lui donner le nom de *Pneumocystis Carinii*". Under Art. 34.1(b) *Pneumocystis* would not be validly published, but Art. 11.5.1 of the *International code of zoological nomenclature* allows for such qualified publication at that time and therefore *Pneumocystis* is an available name under the *ICZN* and, as provided by Art. 45.4, validly published under this *Code*.

Ex. 9. Pneumocystis jirovecii Frenkel (in Natl. Cancer Inst. Monogr. 43: 16. 1976, *'jiroveci'*), treated as a protozoan, was published with only an English description and without desig-

---

[1] The word "available" in the *International code of zoological nomenclature* is equivalent to "validly published" in the present *Code*.

nation of a type, but these conditions are no obstacle to availability under Art. 72.3 and Rec. 13B of the *International code of zoological nomenclature.* Therefore, when considered the name of a fungus, *P. jirovecii,* with modified termination (Art. 60.11), is accepted as validly published under Art. 45.4. Subsequent publication of a Latin diagnosis by Frenkel (J. Eukaryot. Microbiol. 46 Suppl.: 91S. 1999), who treated the species as a fungus, was necessary under the edition of the *ICBN* in operation at that time, but is no longer so; hence, under this *Code, P. jirovecii* has priority from 1976, not 1999.

Ex. 10. Fibrillanosema crangonycis Galbreath & al. (in Int. J. Parasitol. 34: 241-242. 2004), was described as belonging to the *Microsporidia,* which until recently were considered to constitute a protozoan phylum. Its name is available under the *International code of zoological nomenclature* and is considered to be validly published when treated as a fungus although it lacks a Latin description or diagnosis.

Recommendation 45A

45A.1. A new name should be followed by a direct citation indicating its novel status, including the word "novus" (-a, -um) or its abbreviation, e.g. genus novum (gen. nov.), species nova (sp. nov.), combinatio nova (comb. nov.), nomen novum (nom. nov.), or status novus (stat. nov.).

Recommendation 45B

45B.1. Authors should indicate precisely the dates of publication of their works. In a work appearing in parts the last-published sheet of the volume should indicate the precise dates on which the different fascicles or parts of the volume were published as well as the number of pages and plates in each.

Recommendation 45C

45C.1. On separately printed and issued copies of works published in a periodical, the name of the periodical, the number of its volume or parts, the original pagination, and the date (year, month, and day) should be indicated.

SECTION 3. AUTHOR CITATIONS

Article 46

46.1. In publications, particularly those dealing with taxonomy and nomenclature, it may be desirable, even when no bibliographic reference to the protologue is made, to cite the author(s) of the name concerned (see Art. 6 Note 2; see also Art. 22.1 and 26.1). In so doing, the following rules are to be followed.

Ex. 1. Rosaceae Juss., *Rosa* L., *Rosa gallica* L., *Rosa gallica* var. *eriostyla* R. Keller, *Rosa gallica* L. var. *gallica.*

46.2. A name of a new taxon must be attributed to the author or authors to whom both the name and the validating description or diagnosis were ascribed, even when authorship of the publication is different. A new combination or a nomen novum must be attributed to the author or authors to whom it was ascribed when, in the publication in which it appears, it is explicitly stated that they contributed in some way to that publication. Art. 46.4 notwithstanding, authorship of a new name or combination must always be accepted as ascribed, even when it differs from authorship of the publication, when at least one author is common to both.

Ex. 2. The name *Viburnum ternatum* was published in Sargent (Trees & Shrubs 2: 37. 1907). It was ascribed to "Rehd.", and the whole account of the species was signed "Alfred Rehder" at the end of the article. The name is therefore cited as *V. ternatum* Rehder.

Ex. 3. In a paper by Hilliard & Burtt (1986) names of new species of *Schoenoxiphium,* including *S. altum,* were ascribed to Kukkonen, preceded by a statement "The following diagnostic descriptions of new species have been supplied by Dr. I. Kukkonen in order to make the names available for use". The name is therefore cited as *S. altum* Kukkonen.

Ex. 4. In Torrey & Gray (1838) the names *Calyptridium* and *C. monandrum* were ascribed to "Nutt. mss.", and the descriptions were enclosed in double quotes indicating that Nuttall wrote them, as acknowledged in the preface. The names are therefore cited as *Calyptridium* Nutt. and *C. monandrum* Nutt.

Ex. 5. The name *Brachystelma* was published by Sims (1822) along with one new species listed as "*Brachystelma tuberosa.* Brown Mscr."; in addition, at the end of the generic description, Sims added "Brown, Mscr.", indicating that Brown wrote it. Because the generic and specific names were validly published simultaneously (Art. 42), the direct association of Brown's name with the specific name and the generic description establishes the correct citation of the generic name as *Brachystelma* R. Br.

Ex. 6. When publishing *Eucryphiaceae* (1848) the otherwise unnamed author "W.", in a review of Gay's *Flora chilena* (1845-1854), wrote "wird die Gattung *Eucryphia* als Typus einer neuen Familie, der *Eucryphiaceae*", thus ascribing both the name and its validating description to Gay (Fl. Chil. 1: 348. 1846), who used the name "Eucrifiaceas", which was not validly published under Art. 18.4. The name is therefore cited as *Eucryphiaceae* Gay.

Ex. 7. Green (1985) ascribed the new combination *Neotysonia phyllostegia* to Paul G. Wilson and elsewhere in the same publication acknowledged his assistance. The name is therefore cited as *N. phyllostegia* (F. Muell.) Paul G. Wilson.

Ex. 8. The authorship of *Steyerbromelia discolor* L. B. Sm. & H. Rob. (1984) is accepted as originally ascribed, although the new species was described in a paper authored by Smith alone. The same applies to the new combination *Sophora tomentosa* subsp. *occidentalis* (L.) Brummitt (in Kirkia 5: 265. 1966), thus ascribed, published in a paper authored jointly by Brummitt & Gillett.

Ex. 9. The appropriate author citation for *Baloghia pininsularis* (see Art. 37 Ex. 3) is Guillaumin, and not McPherson & Tirel, because both the name and validating description were ascribed to Guillaumin in the protologue.

Note 1. When authorship of a name differs from authorship of the publication in which it was validly published, both are sometimes cited, connected by the word "in". In such a case, "in" and what follows are part of a bibliographic citation and are better omitted unless the place of publication is being cited.

Ex. 10. The original description of the new species *Verrucaria aethiobola* Wahlenb. (in Acharius, Methodus, Suppl.: 17. 1803) is ascribed by Acharius to "Wahlenb. Msc.", and the name itself is ascribed to "Wahlenb." (not in the text of the Supplement but in the index to the Methodus, p. 392). The name is therefore appropriately cited as *V. aethiobola* Wahlenb., better not as *V. aethiobola* "Wahlenb. in Acharius" (unless followed by a bibliographic citation of the place of publication), and certainly not as *V. aethiobola* "Wahlenb. ex Ach."

Ex. 11. The name *Drymaria arenarioides* was published in Roemer & Schultes (Syst. Veg. 5: 406. 1819), with the name ascribed to "Humb. et Bonpl." and the description ascribed to "Reliqu. Willd. MS." Because of this ascription, and because vol. 5 of this work is authored by Schultes alone, the name is to be cited as *D. arenarioides* Humb. & Bonpl. ex Schult., not as *D. arenarioides* Willd. or *D. arenarioides* Willd. ex Roem. & Schult. or *D. arenarioides* Humb. & Bonpl. ex Willd.

Ex. 12. When publishing *Strasburgeriaceae* (1908) Solereder wrote of *Strasburgeria* Baill. "welche neuerdings von Van Tieghem als Typus einer eigenen Familie *(Strasburgeriaceae)* angesehen wird" thus ascribing both the family name and its validating description to Tieghem (in J. Bot. (Morot) 17: 204. 1903), who used the name "Strasburgériacées", which was not validly published under Art. 18.4. The name is therefore cited as *Strasburgeriaceae* Tiegh., or *Strasburgeriaceae* Tiegh. in Solereder when followed by a bibliographic citation, but not *Strasburgeriaceae* Tiegh. ex Soler.

Ex. 13. When publishing *Elaeocarpaceae* (1816) Candolle wrote "*Elaeocarpeae.* Juss., Ann. Mus. 11, p. 233" thus ascribing both the name and its validating diagnosis to Jussieu (in Ann. Mus. Natl. Hist. Nat. 11: 233. 1808), who provided a diagnosis separating an un-named family comprising *Elaeocarpus* L. from *Tiliaceae*. The family name is therefore cited as *Elaeocarpaceae* Juss., or *Elaeocarpaceae* Juss. in Candolle when followed by a bibliographic citation, but not *Elaeocarpaceae* Juss. ex DC.

46.3. For the purposes of this Article, ascription is the direct association of the name of a person or persons with a new name or description or diagnosis of a taxon. An author citation appearing in a list of synonyms does not constitute ascription, nor does reference to a basionym or a replaced synonym (regardless of bibliographic accuracy) or reference to a homonym, or a formal error.

Ex. 14. The name *Asperococcus pusillus* was published in Hooker (Brit. Fl., ed. 4, 2(1): 277. 1833), with the name and diagnosis ascribed simultaneously in a paragraph ending with "Carm. MSS." followed by a description ascribed similarly to Carmichael. Direct association of Carmichael with both the name and the diagnosis is thus inferred and the name must be cited as *A. pusillus* Carmich. However, the paragraph containing the name *A. castaneus* and its diagnosis, published by Hooker on the same page of the same work, ends with "*Scytosiphon castaneus,* Carm. MSS." Because Carmichael is directly associated with "*S. castaneus*" and not *A. castaneus,* the name of this species is correctly cited as *A. castaneus* Hook. even though the following description is ascribed to Carmichael.

Ex. 15. *Lichen debilis* Sm. (1812) was not ascribed to Turner and Borrer by Smith's citing "*Calicium debile* Turn. and Borr. Mss." as a synonym.

Ex. 16. *Malpighia emarginata* DC. (1824) was not ascribed to Moçino & Sessé by Candolle's writing "*M. emarginata* (fl. mex. ic. ined.)". However, *Sicyos triqueter* Moç. & Sessé ex Ser. (1830) was ascribed to these authors by Seringe's writing "*S. triqueter* (Moç. & Sessé, fl. mex. mss.)".

Ex. 17. When Opiz (1852) wrote "*Hemisphace* Bentham" he did not ascribe the generic name to Bentham but provided an indirect reference to the basionym, *Salvia* sect. *Hemisphace* Benth. (see Art. 32 Ex. 8).

Ex. 18. When Brotherus (1907) published "*Dichelodontium nitidulum* Hooker & Wilson" he provided an indirect reference to the basionym, *Leucodon nitidulus* Hook. f. & Wilson, and did not ascribe the new combination to Hooker and Wilson. He did, however, ascribe to them the simultaneously published name of his new genus, *Dichelodontium.*

Ex. 19. When She & Watson (in Wu & al., Fl. China 14: 72. 2005) wrote "*Bupleurum hamiltonii* var. *paucefulcrans* C. Y. Wu ex R. H. Shan & Yin Li, Acta Phytotax. Sin. 12: 291. 1974" they did not ascribe the new combination to any of those authors but provided a full and direct reference to the basionym, *B. tenue* var. *paucefulcrans* C. Y. Wu ex R. H. Shan & Yin Li.

Ex. 20. When Sirodot (1872) wrote "*Lemanea* Bory" he in fact published a later homonym (see Art. 48 Ex. 1). His reference to Bory's earlier homonym is not therefore ascription of the later homonym, *Lemanea* Sirodot, to Bory.

Ex. 21. When Piper (in Proc. Biol. Soc. Wash. 28: 42. 1915) wrote "*Andropogon sorghum drummondii* (Nees) Hackel" for one of eleven "wild subspecies" of *A. sorghum* (L.) Brot., this was not an ascription to Hackel, but is treated as a formal error, since Hackel (in Candolle & Candolle, Monogr. Phan. 6: 507. 1889) had actually published this as *A. sorghum* var. *drummondii* (Nees) Hack. Furthermore, because the basionym was published by Steudel (1854) as "*A. drummondii* Nees (mpt. sub *Sorghum*)" this reference to the unpublished name "*Sorghum drummondii* Nees" is also not ascription (see Note 2), therefore the correct author citation for Hackel's taxon is *A. sorghum* var. *drummondii* (Steud.) Hack. and for Piper's taxon *A. sorghum* subsp. *drummondii* (Steud.) Piper.

Note 2. When the epithet of a validly published name is taken up from and attributed to the author of a different binary designation that has not been validly published, only the author of the validly published name is to be cited.

Ex. 22. "*Catha edulis*" was published, but not validly so, by Forsskål (Fl. Aegypt.-Arab.: cvii, 63. 1775). The epithet was taken up by Vahl (Symb. Bot. 1: 21. 1790), who validly published the name *Celastrus edulis* citing "*Catha edulis* Forssk." in synonymy. The name *Celastrus edulis* must be attributed to Vahl alone, not to "Forssk. ex Vahl". The name *Catha edulis* was first validly published by Endlicher (Enchir. Bot.: 575. 1841), whose combination is to be cited as *Catha edulis* (Vahl) Endl.

46.4. A name of a new taxon must be attributed to the author or authors of the publication in which it appears when only the name but not the validating description or diagnosis was ascribed to a different author or to different authors. A new combination or a nomen novum must be attributed to the author or authors of the publication in which it appears, although it was ascribed to a different author or to different authors, when no separate statement was made that they contributed in some way to that publication. However, in both cases authorship as ascribed, followed by "ex", may be inserted before the name(s) of the publishing author(s).

Ex. 23. Seemann (1865) published *Gossypium tomentosum* "Nutt. mss.", followed by a validating description not ascribed to Nuttall; the name may be cited as *G. tomentosum* Nutt. ex Seem. or *G. tomentosum* Seem.

Ex. 24. Rudolphi published *Pinaceae* (1830) as "*Pineae*. Spreng.", followed by a validating diagnosis not ascribed to Sprengel; the name may be cited as *Pinaceae* Spreng. ex F. Rudolphi or *Pinaceae* F. Rudolphi.

Ex. 25. The name *Lithocarpus polystachyus* published by Rehder (1919) was based on *Quercus polystachya* A. DC. (1864), ascribed by Candolle to "Wall.! list n. 2789" but formerly a nomen nudum; Rehder's combination may be cited as *L. polystachyus* (Wall. ex A. DC.) Rehder or *L. polystachyus* (A. DC.) Rehder.

Ex. 26. *Lilium tianschanicum* was described by Grubov (1977) as a new species and its name was ascribed to Ivanova; since there is no indication that Ivanova provided the validating description, the name may be cited as *L. tianschanicum* N. A. Ivanova ex Grubov or *L. tianschanicum* Grubov.

Ex. 27. In a paper by Boufford, Tsi and Wang (1990) the name *Rubus fanjingshanensis* was ascribed to Lu with no indication that Lu provided the description; the name should be attributed to Boufford & al. or to L. T. Lu ex Boufford & al.

Ex. 28. Green (1985) ascribed the new combination *Tersonia cyathiflora* to "(Fenzl) A. S. George"; since Green nowhere mentioned that George had contributed in any way, the combining author must be cited as A. S. George ex J. W. Green or just J. W. Green.

Ex. 29. However, R. Brown is accepted as the author of the treatments of genera and species appearing under his name in Aiton's *Hortus kewensis,* ed. 2 (1810-1813), even when new names or the descriptions validating them are not explicitly ascribed to him. In a postscript to that work (5: 532. 1813), Aiton wrote: "Much new matter has been added by [Robert Brown] ... the greater part of his able improvements are distinguished by the signature *Brown mss.*" The latter phrase is therefore a statement of authorship not merely an ascription. For example, the combination *Oncidium triquetrum,* based by indirect reference on *Epidendrum triquetrum* Sw. (1788), is to be cited as *O. triquetrum* (Sw.) R. Br. (1813) and not attributed to "R. Br. ex Aiton", or to Aiton alone, because in the generic heading Brown is credited with authorship of the treatment of *Oncidium.*

46.5. For the purposes of this Article, the authorship of a publication is the authorship of that part of a publication in which a name appears regardless of the authorship or editorship of the publication as a whole.

Ex. 30. *Pittosporum buxifolium* was described as a new species, with its name ascribed to Feng, in Wu & Li, *Flora yunnanica,* vol. 3 (1983). The account of *Pittosporaceae* in that flora was authored by Yin, while the whole volume was edited by Wu & Li. The author of the publication (including the validating diagnosis) was Yin. The name may therefore be cited as either *P. buxifolium* K. M. Feng ex W. Q. Yin or just *P. buxifolium* W. Q. Yin, but not *P. buxifolium* K. M. Feng ex C. Y. Wu & H. W. Li, nor *P. buxifolium* C. Y. Wu & H. W. Li.

Ex. 31. *Vicia amurensis* f. *sanneensis* was described as a new form, with its name ascribed to Jiang & Fu in Ma & al., *Flora intramongolica,* ed. 2, vol. 3 (1989). The account of *Vicia* in that flora was authored by Jiang, while the whole volume was jointly edited by Ma & al. The author of the publication is Jiang, who is common to the authorship ascribed to the name, which must therefore be cited as *V. amurensis* f. *sanneensis* Y. C. Jiang & S. M. Fu and not *V. amurensis* f. *sanneensis* Y. C. Jiang & S. M. Fu ex Ma & al.

Ex. 32. *Centaurea funkii* var. *xeranthemoides* "Lge. ined." was described in *Prodromus florae hispanicae* (2: 154. 1865). On the title page of each volume Willkomm & Lange are given as authors ("auctoribus ..."). However, the different family treatments are by one or the other and Fam. 63 *Compositae* has a footnote "Auctore Willkomm". The full citation is therefore *C. funkii* var. *xeranthemoides* Lange ex Willk. [in Willkomm & Lange, ...].

46.6. The citation of an author who published the name before the starting-point of the group concerned may be indicated by the use of the word "ex". For groups with a starting-point later than 1753, when a pre-starting-point name was changed in rank or taxonomic position by the first author who validly published it, the name of the pre-starting-point author may be added in parentheses, followed by "ex".

Ex. 33. Linnaeus (1754) ascribed the name *Lupinus* to the pre-starting-point author Tournefort; the name may be cited as *Lupinus* Tourn. ex L. (1753) or *Lupinus* L. (see Art. 13.4).

Ex. 34. *Lyngbya glutinosa* C. Agardh (Syst. Alg.: 73. 1824) was taken up by Gomont in the publication which marks the starting-point of the *"Nostocaceae homocysteae"* (in Ann. Sci. Nat., Bot., ser. 7, 15: 339. 1892) as *Hydrocoleum glutinosum.* This may be cited as *H. glutinosum* (C. Agardh) ex Gomont.

46.7. In determining the correct author citation, only internal evidence in the publication (as defined in Art. 35.5) where the name was validly published is to be accepted, including ascription of the name, statements in the introduction, title, or acknowledgements, and typographical or stylistic distinctions in the text.

Ex. 35. Although the descriptions in Aiton's *Hortus kewensis* (1789) are generally considered to have been written by Solander or Dryander, the names of new taxa published there must be attributed to Aiton, the stated author of the work, except where a name and description were both ascribed in that work to somebody else.

Ex. 36. The name *Andreaea angustata* was published in a work of Limpricht (1885) with the ascription "nov. sp. Lindb. in litt. ad Breidler 1884", but there is no internal evidence that Lindberg had supplied the validating description. Authorship is therefore to be cited as "Limpr." or "Lindb. ex Limpr."

Note 3. External evidence may be used to determine authorship of new names and combinations included in a publication or article for which there is no internal evidence of authorship.

Ex. 37. No authorship appears anywhere in the work known as "Cat. Pl. Upper Louisiana. 1813", a catalogue of plants available from the Fraser Brothers Nursery. Based on external evidence (cf. Stafleu & Cowan in Regnum Veg. 105: 785. 1981), authorship of the document, and of new names such as *Oenothera macrocarpa* that are published in it, are attributed to Thomas Nuttall.

Ex. 38. The book that appeared under the title *Vollständiges systematisches Verzeichniß aller Gewächse Teutschlandes* ... (Leipzig 1782) bears no explicit authorship but is attributed to "einem Mitgliede der Gesellschaft Naturforschender Freunde". External evidence may be used to determine that G. A. Honckeny is the author of the work and of new names that appear in it (e.g. *Poa vallesiana* Honck., *Phleum hirsutum* Honck.; but see Art. 23 Ex. 14), as done by Pritzel (Thes. Lit. Bot.: 123. 1847).

Note 4. Authors publishing new names and wishing to establish that other persons' names followed by "ex" may precede theirs in authorship citation may adopt the "ex" citation in the protologue.

Ex. 39. In validating the name *Nothotsuga,* Page (1989) cited it as "*Nothotsuga* H.-H. Hu ex C. N. Page", noting that in 1951 Hu had published it as a nomen nudum; the name may be attributed to Hu ex C. N. Page or just C. N. Page.

Ex. 40. Atwood (1981) ascribed the name of a new species, *Maxillaria mombachoënsis,* to "Heller ex Atwood", with a note stating that it was originally named by Heller, then deceased; the name may be attributed to A. H. Heller ex J. T. Atwood or just J. T. Atwood.

Recommendation 46A

46A.1. For the purpose of author citation, prefixes indicating ennoblement (see Rec. 60C.5(d-e)) should be suppressed unless they are an inseparable part of the name.

Ex. 1. Lam. for J. B. P. A. Monet Chevalier de Lamarck, but De Wild. for E. De Wildeman.

46A.2. When a name in an author citation is abbreviated, the abbreviation should be long enough to be distinctive, and should normally end with a consonant that, in the full name, precedes a vowel. The first letters should be given without any omission, but one of the last characteristic consonants of the name may be added when this is customary.

Ex. 2. L. for Linnaeus; Fr. for Fries; Juss. for Jussieu; Rich. for Richard; Bertol. for Bertoloni, to distinguish it from Bertero; Michx. for Michaux, to distinguish it from Micheli.

46A.3. Given names or accessory designations serving to distinguish two botanists of the same name should be abridged in the same way.

Ex. 3. R. Br. for Robert Brown; A. Juss. for Adrien de Jussieu; Burm. f. for Burman filius; J. F. Gmel. for Johann Friedrich Gmelin, J. G. Gmel. for Johann Georg Gmelin, C. C. Gmel. for Carl Christian Gmelin, S. G. Gmel. for Samuel Gottlieb Gmelin; Müll. Arg. for Jean Müller argoviensis (of Aargau).

46A.4. When it is a well-established custom to abridge a name in another manner, it is advisable to conform to custom.

Ex. 4. DC. for Augustin-Pyramus de Candolle; St.-Hil. for Saint-Hilaire.

Note 1. Brummitt & Powell's *Authors of plant names* (1992) provides unambiguous standard abbreviations, in conformity with the present Recommendation, for a large number of authors of plant names, and these abbreviations, updated as necessary from the *International Plant Names Index* (www.ipni.org) and *Index Fungorum* (www.indexfungorum.org), have been used for author citations throughout the present *Code*.

Recommendation 46B

46B.1. In citing the author of the scientific name of a taxon, the romanization of the author's name given in the original publication should normally be accepted. Where an author failed to give a romanization, or where an author has at different times used different romanizations, then the romanization known to be preferred by the author or that most frequently adopted by the author should be accepted. In the absence of such information the author's name should be romanized in accordance with an internationally available standard.

46B.2. Authors of scientific names whose personal names are not written in Roman letters should romanize their names, preferably (but not necessarily) in accordance with an internationally available standard and, as a matter of typographical convenience, without diacritical signs. Once authors have selected the romanization of their personal names, they should use it consistently thereafter. Whenever possible, authors should not permit editors or publishers to change the romanization of their personal names.

Recommendation 46C

46C.1. After a name published jointly by two authors, both authors should be cited, linked by the word "et" or by an ampersand (&).

Ex. 1. Didymopanax gleasonii Britton et Wilson (or Britton & Wilson).

46C.2. After a name published jointly by more than two authors, the citation should be restricted to the first author followed by "et al." or "& al.", except in the original publication.

Ex. 2. Lapeirousia erythrantha var. *welwitschii* (Baker) Geerinck, Lisowski, Malaisse & Symoens (in Bull. Soc. Roy. Bot. Belgique 105: 336. 1972) should be cited as *L. erythrantha* var. *welwitschii* (Baker) Geerinck & al.

Recommendation 46D

46D.1. Authors should cite themselves by name after each new name they publish rather than refer to themselves by expressions such as "nobis" (nob.) or "mihi" (m.).

Article 47

47.1. An alteration of the diagnostic characters or of the circumscription of a taxon without the exclusion of the type does not warrant a change of the author citation of the name of the taxon.

Ex. 1. When the original material of *Arabis beckwithii* S. Watson (1887) is attributed to two different species, as by Munz (1932), that species not including the lectotype must bear a different name (*A. shockleyi* Munz) but the other one is still named *A. beckwithii* S. Watson.

Ex. 2. Myosotis as revised by Brown differs from the genus as originally circumscribed by Linnaeus, but the generic name remains *Myosotis* L. since the type of the name is still included in the genus (it may be cited as *Myosotis* L. emend. R. Br.: see Rec. 47A).

Ex. 3. The variously defined species that includes the types of *Centaurea jacea* L. (1753), *C. amara* L. (1763) and a variable number of other species names is still called *C. jacea* L. (or L. emend. Coss. & Germ., L. emend. Vis., or L. emend. Godr., as the case may be: see Rec. 47A).

Recommendation 47A

47A.1. When an alteration as mentioned in Art. 47 has been considerable, the nature of the change may be indicated by adding such words, abbreviated where suitable, as "emendavit" (emend.) followed by the name of the author responsible for the change, "mutatis characteribus" (mut. char.), "pro parte" (p. p.), "excluso genere" or "exclusis generibus" (excl. gen.), "exclusa specie" or "exclusis speciebus" (excl. sp.), "exclusa varietate" or "exclusis varietatibus" (excl. var.), "sensu amplo" (s. ampl.), "sensu lato" (s. l.), "sensu stricto" (s. str.), etc.

Ex. 1. Phyllanthus L. emend. Müll. Arg.; *Globularia cordifolia* L. excl. var. (emend. Lam.).

Article 48

48.1. When an author adopts an existing name but definitely excludes its original type, a later homonym that must be attributed solely to that author is considered to have been published. Similarly, when an author who adopts a name refers to an apparent basionym but explicitly excludes its type, a new name is considered to have been published that must be attributed solely to that author. Exclusion can be effected by simultaneous explicit inclusion of the type in a different taxon by the same author (see also Art. 59.6).

Ex. 1. Sirodot (1872) placed the type of *Lemanea* Bory (1808) in *Sacheria* Sirodot (1872); hence *Lemanea,* as treated by Sirodot (1872), is to be cited as *Lemanea* Sirodot non Bory and not as *Lemanea* Bory emend. Sirodot.

Ex. 2. The name *Amorphophallus campanulatus* Decne. (1834) was apparently based on the illegitimate *Arum campanulatum* Roxb. (1819). However, the type of the latter was explicitly excluded by Decaisne, and his name is therefore a legitimate name of a new species, to be attributed solely to him.

Note 1. Misapplication of a new combination to a different taxon, but without explicit exclusion of the type of the basionym, is dealt with under Art. 7.4.

Note 2. Retention of a name in a sense that excludes its original type, or its type designated under Art. 7-10, can be effected only by conservation (see Art. 14.9).

Article 49

49.1. When a genus or a taxon of lower rank is altered in rank but retains its name or the final epithet in its name, the author of the earlier, name- or epithet-bringing legitimate name (the author of the basionym) must be cited in parentheses, followed by the name of the author who effected the alteration (the author of the new name). The same provision holds when a taxon of lower rank than genus is transferred to another genus or species, with or without alteration of rank.

Ex. 1. Medicago polymorpha var. *orbicularis* L. (1753) when raised to the rank of species becomes *M. orbicularis* (L.) Bartal. (1776).

Ex. 2. Anthyllis sect. *Aspalathoides* DC. (Prodr. 2: 169. 1825) raised to generic rank, retaining the epithet *Aspalathoides* as its name, is cited as *Aspalathoides* (DC.) K. Koch (1853).

Ex. 3. Cineraria sect. *Eriopappus* Dumort. (Fl. Belg.: 65. 1827) when transferred to *Tephroseris* (Rchb.) Rchb. is cited as *T.* sect. *Eriopappus* (Dumort.) Holub (in Folia Geobot. Phytotax. 8: 173. 1973).

Ex. 4. Cistus aegyptiacus L. (1753) when transferred to *Helianthemum* Mill. is cited as *H. aegyptiacum* (L.) Mill. (1768).

Ex. 5. Fumaria bulbosa var. *solida* L. (1753) was elevated to specific rank as *F. solida* (L.) Mill. (1771). The name of this species when transferred to *Corydalis* DC. is to be cited as *C. solida* (L.) Clairv. (1811), not *C. solida* (Mill.) Clairv.

Ex. 6. However, *Pulsatilla montana* var. *serbica* W. Zimm. (in Feddes Repert. Spec. Nov. Regni Veg. 61: 95. 1958), originally placed under *P. montana* subsp. *australis* (Heuff.) Zämelis, retains the same author citation when placed under *P. montana* subsp. *dacica* Rummelsp. (see Art. 24.1) and is not cited as var. *serbica* "(W. Zimm.) Rummelsp." (in Feddes Repert. 71: 29. 1965).

Ex. 7. Salix subsect. *Myrtilloides* C. K. Schneid. (Ill. Handb. Laubholzk. 1: 63. 1904), originally placed under *S.* sect. *Argenteae* W. D. J. Koch, retains the same author citation when placed under *S.* sect. *Glaucae* Pax and is not cited as *S.* subsect. *Myrtilloides* "(C. K. Schneid.) Dorn" (in Canad. J. Bot. 54: 2777. 1976).

49.2. Parenthetical authors are not to be cited for suprageneric names.

Ex. 8. Even though *Illiciaceae* A. C. Sm. (1947) was validly published by reference to *Illicieae* DC. (1824) it is not to be cited as *Illiciaceae* "(DC.) A. C. Sm."

Note 1. Art. 46.6 provides for the use of parenthetical author citations preceding the word "ex", after some names in groups with a starting-point later than 1753.

Article 50

50.1. When a taxon at the rank of species or below is transferred from the non-hybrid category to the hybrid category of the same rank (Art. H.10.2), or vice versa, the author citation remains unchanged but may be followed by an indication in parentheses of the original category.

Ex. 1. Stachys ambigua Sm. (1809) was published as the name of a species. If regarded as applying to a hybrid, it may be cited as *Stachys ×ambigua* Sm. (pro sp.).

Ex. 2. The binary name *Salix ×glaucops* Andersson (1868) was published as the name of a hybrid. Later, Rydberg (in Bull. New York Bot. Gard. 1: 270. 1899) considered the taxon to be a species. If this view is accepted, the name may be cited as *Salix glaucops* Andersson (pro hybr.).

SECTION 4. GENERAL RECOMMENDATIONS ON CITATION

Recommendation 50A

50A.1. In the citation as a synonym of a name not validly published, the words "as synonym" or "pro syn." should be added.

Recommendation 50B

50B.1. In the citation of a nomen nudum, its status should be indicated by adding the words "nomen nudum" or "nom. nud."

Ex. 1. "Carex bebbii" (Olney, Car. Bor.-Am. 2: 12. 1871), published without a description or diagnosis, should be cited as *Carex bebbii* Olney, nomen nudum (or nom. nud.).

Recommendation 50C

50C.1. The citation of a later homonym should be followed by the name of the author of the earlier homonym preceded by the word "non", preferably with the date of publication added. In some instances it will be advisable to cite also any other homonyms, preceded by the word "nec".

Ex. 1. Ulmus racemosa Thomas in Amer. J. Sci. Arts 19: 170. 1831, non Borkh. 1800.

Ex. 2. Lindera Thunb., Nov. Gen. Pl.: 64. 1783, non Adans. 1763.

Ex. 3. Bartlingia Brongn. in Ann. Sci. Nat. (Paris) 10: 373. 1827, non Rchb. 1824 nec F. Muell. 1882.

Recommendation 50D

50D.1. Misidentifications should not be included in synonymies but added after them. A misapplied name should be indicated by the words "auct. non" followed by the name of the original author and the bibliographic reference of the misiden-tification.

Ex. 1. Ficus stortophylla Warb. in Ann. Mus. Congo Belge, Bot., ser. 4, 1: 32. 1904. *F. irumuënsis* De Wild., Pl. Bequaert. 1: 341. 1922. *"F. exasperata"* auct. non Vahl: De Wildeman & Durand in Ann. Mus. Congo Belge, Bot., ser. 2, 1: 54. 1899; De Wildeman, Miss. Em. Laurent: 26. 1905; Durand & Durand, Syll. Fl. Congol.: 505. 1909.

Recommendation 50E

50E.1. If a name of a family, genus, or species is accepted as a nomen conservandum (see Art. 14 and App. II-IV) the abbreviation "nom. cons." or, in the case of a conserved spelling, "orth. cons." should be added in a formal citation.

Ex. 1. Protea L., Mant. Pl.: 187. 1771, nom. cons., non L. 1753.

Ex. 2. Combretum Loefl. 1758, nom. cons. [= *Grislea* L. 1753].

Ex. 3. Glechoma L. 1753, orth. cons., *'Glecoma'*.

50E.2. If a name has been rejected and has been placed on the list of nomina utique rejicienda (see Art. 56 and App. V) the abbreviation "nom. rej." should be added in a formal citation.

Ex. 4. Betula alba L. 1753, nom. rej.

Note 1. Rec. 50E.2 also applies to any combination based on a nomen utique rejiciendum (see Art. 56.1).

Ex. 5. Dryobalanops sumatrensis (J. F. Gmel.) Kosterm. in Blumea 33: 346. 1988, nom. rej.

50E.3. If a name has been adopted by Fries or Persoon, and thereby sanctioned (see Art. 13.1(d) and 7.8), ": Fr." or ": Pers." should be added in a formal citation. The same convention should be used for the basionym of the sanctioned name, if it has one, and for all combinations based on either the sanctioned name or its basionym.

Ex. 6. Boletus piperatus Bull. (Herb. France: t. 451, f. 2. 1790) was accepted in Fries (Syst. Mycol. 1: 388. 1821) and was thereby sanctioned. It should thus be cited as *B. piperatus* Bull. : Fr., and a subsequent combination based on it, as *Chalciporus piperatus* (Bull. : Fr.) Bataille.

Ex. 7. Agaricus sarcocephalus Fr. 1815 : Fr. was sanctioned as *Agaricus compactus* [unranked] *sarcocephalus* (Fr. : Fr.) Fr. 1821; *Psathyrella sarcocephala* (Fr. : Fr.) Singer is a subsequent combination based on it.

Recommendation 50F

50F.1. If a name is cited with alterations from the form as originally published, it is desirable that in full citations the exact original form should be added, preferably between single or double quotation marks.

Ex. 1. Pyrus calleryana Decne. *(P. mairei* H. Lév. in Repert. Spec. Nov. Regni Veg. 12: 189. 1913, *'Pirus'*).

Ex. 2. Zanthoxylum cribrosum Spreng., Syst. Veg. 1: 946. 1825, *"Xanthoxylon"*. *(Z. caribaeum* var. *floridanum* (Nutt.) A. Gray in Proc. Amer. Acad. Arts 23: 225. 1888, *"Xanthoxylum"*).

Ex. 3. Spathiphyllum solomonense Nicolson in Amer. J. Bot. 54: 496. 1967, *'solomonensis'*.

CHAPTER V. REJECTION OF NAMES

Article 51

51.1. A legitimate name must not be rejected merely because it, or its epithet, is inappropriate or disagreeable, or because another is preferable or better known (but see Art. 56.1), or because it has lost its original meaning, or (in pleomorphic fungi with names governed by Art. 59) because the morph represented by its type is not in accordance with that of the type of the generic name.

Ex. 1. The following changes are contrary to the rule: *Staphylea* to *Staphylis, Tamus* to *Thamnos, Thamnus,* or *Tamnus, Mentha* to *Minthe, Tillaea* to *Tillia, Vincetoxicum* to *Alexitoxicum;* and *Orobanche rapum* to *O. sarothamnophyta, O. columbariae* to *O. columbarihaerens, O. artemisiae* to *O. artemisiepiphyta.*

Ex. 2. Ardisia quinquegona Blume (1825) is not to be changed to *A. pentagona* A. DC. (1834), although the specific epithet *quinquegona* is a hybrid word (Latin and Greek) (contrary to Rec. 23A.3(c)).

Ex. 3. The name *Scilla peruviana* L. (1753) is not to be rejected merely because the species does not grow in Peru.

Ex. 4. The name *Petrosimonia oppositifolia* (Pall.) Litv. (1911), based on *Polycnemum oppositifolium* Pall. (1771), is not to be rejected merely because the species has leaves only partly opposite, and partly alternate, although there is another closely related species, *Petrosimonia brachiata* (Pall.) Bunge, having all its leaves opposite.

Ex. 5. Richardia L. (1753) is not to be changed to *Richardsonia,* as was done by Kunth (1818), although the name was originally dedicated to the British botanist, Richardson.

Ex. 6. The name *Sphaeria tiliae* Pers. (Syn. Meth. Fung.: 84. 1801) is not to be rejected because the holotype represents an anamorphic fungus, whereas the type of *Sphaeria* Haller 1768, that of *S. fragiformis* Pers., is a teleomorphic fungus. The epithet may therefore be used in the combination *Rabenhorstia tiliae* (Pers.) Fr. (Summ. Veg. Scand.: 410. 1849) for the anamorph of *Hercospora tiliae* Tul. & C. Tul. (Sel. Fung. Carp. 2: 154. 1863).

Article 52

52.1. A name, unless conserved (Art. 14) or sanctioned (Art. 15), is illegitimate and is to be rejected if it was nomenclaturally superfluous when published, i.e. if the taxon to which it was applied, as circumscribed by its author, definitely included the type (as qualified in Art. 52.2) of a name which ought to have been adopted, or of which the epithet ought to have been adopted, under the rules (but see Art. 52.3).

52.2. For the purpose of Art. 52.1, definite inclusion of the type of a name is effected by citation *(a)* of the holotype under Art. 9.1 or the original type under Art. 10 or all syntypes under Art. 9.4 or all elements eligible as types under Art. 10.2; or *(b)* of the previously designated type under Art. 9.9-11 or 10.2; or *(c)* of the previously conserved type under Art. 14.9; or *(d)* of the illustrations of these. It is also effected *(e)* by citation of the name itself or any name homotypic at that time, unless the type is at the same time excluded either explicitly or by implication.

Ex. 1. The generic name *Cainito* Adans. (1763) is illegitimate because it was a superfluous name for *Chrysophyllum* L. (1753), which Adanson cited as a synonym.

Ex. 2. *Chrysophyllum sericeum* Salisb. (1796) is illegitimate, being a superfluous name for *C. cainito* L. (1753), which Salisbury cited as a synonym.

Ex. 3. On the other hand, *Salix myrsinifolia* Salisb. (1796) is legitimate, being explicitly based upon *S. myrsinites* of Hoffmann (Hist. Salic. Ill.: 71. 1787), a misapplication of the name *S. myrsinites* L. (1753), which Salisbury excluded by implication as he did not cite Linnaeus as he did under each of the other 14 species of *Salix* in his 1796 publication.

Ex. 4. *Picea excelsa* Link (1841) is illegitimate because it is based on *Pinus excelsa* Lam. (1778), a superfluous name for *Pinus abies* L. (1753). Under *Picea* the correct name is *Picea abies* (L.) H. Karst. (1881).

Ex. 5. On the other hand, *Cucubalus latifolius* Mill. and *C. angustifolius* Mill. are not illegitimate names, although Miller's species are now united with the species previously named *C. behen* L. (1753): *C. latifolius* and *C. angustifolius* as circumscribed by Miller (1768) did not include the type of *C. behen* L., which name he adopted for another species.

Ex. 6. Explicit exclusion of type: When publishing the name *Galium tricornutum,* Dandy (in Watsonia 4: 47. 1957) cited *G. tricorne* Stokes (1787) pro parte as a synonym, but explicitly excluded the type of the latter name.

Ex. 7. Exclusion of type by implication: *Tmesipteris elongata* P. A. Dang. (in Botaniste 2: 213. 1891) was published as a new species but *Psilotum truncatum* R. Br. was cited as a synonym. However, on the following page, *T. truncata* (R. Br.) Desv. is recognized as a different species and two pages later both are distinguished in a key, thus showing that the meaning of the cited synonym was either "*P. truncatum* R. Br. pro parte" or "*P. truncatum* auct. non R. Br."

Ex. 8. Exclusion of type by implication: *Solanum torvum* Sw. (Prodr.: 47. 1788) was published with a new diagnosis but *S. indicum* L. (1753) was cited as a synonym. In accordance with the practice in his *Prodromus,* Swartz indicated where the species was to be inserted in the latest edition [ed. 14, by Murray] of Linnaeus's *Systema vegetabilium. Solanum torvum* was to be inserted between species 26 *(S. insanum)* and 27 *(S. ferox),* the number of *S. indicum* being 32. *Solanum torvum* is thus a legitimate name.

Ex. 9. Under *Persicaria maculosa* Gray (1821), the name *Polygonum persicaria* L. (1753) was cited as the replaced synonym, and hence the type of *Polygonum persicaria* was definitely included. However, *Persicaria mitis* Delarbre (1806), as the earlier legitimate replacement name for *Polygonum persicaria,* is necessarily homotypic; hence, *Persicaria maculosa* when published was an illegitimate superfluous name for *Persicaria mitis* and its continued use has been made possible only by conservation.

Ex. 10. Under *Bauhinia semla* Wunderlin (1976), the name *B. retusa* Roxb. (1832), non Poir. (1811), was cited as the replaced synonym while *B. emarginata* Roxb. ex G. Don (1832), non Mill. (1768) nec Jack (1822), was also cited in synonymy, and hence the types of the two synonyms were definitely included. However, *B. roxburghiana* Voigt (1845), which was published as a replacement name for *B. emarginata,* is necessarily homotypic with it and should have been adopted by Wunderlin. Therefore, *B. semla* is an illegitimate superfluous name typified by the type of its replaced synonym, *B. retusa* (see Art. 7 Ex. 4).

Ex. 11. *Erythroxylum suave* O. E. Schulz (1907) is illegitimate because Schulz cited "*Erythroxylum brevipes* DC. var. *spinescens* (A. Rich.) Griseb." (1866) in synonymy. This citation constitutes inclusion of the type of *E. spinescens* A. Rich. (1841).

Note 1. The inclusion, with an expression of doubt, of an element in a new taxon, e.g. the citation of a name with a question mark, does not make the name of the new taxon nomenclaturally superfluous.

Ex. 12. The protologue of *Blandfordia grandiflora* R. Br. (1810) includes, in synonymy, "*Aletris punicea*. Labill. nov. holl. 1. p. 85. t. 111 ?", indicating that the new species might be the same as *Aletris punicea* Labill. (1805). *Blandfordia grandiflora* is nevertheless a legitimate name.

Note 2. The inclusion, in a new taxon, of an element that was subsequently designated as the type of a name which, so typified, ought to have been adopted, or of which the epithet ought to have been adopted, does not in itself make the name of the new taxon illegitimate.

Ex. 13. *Leccinum* Gray (1821) does not include all potential types (in fact, none) of *Boletus* L. (1753) and thus is not illegitimate, even though it included, as *L. edule* (Bull. : Fr.) Gray, the subsequently conserved type of *Boletus, B. edulis* Bull. : Fr.

52.3. A name that was nomenclaturally superfluous when published is not illegitimate on account of its superfluity if it is based on a name-bringing or epithet-bringing synonym (basionym), or if it is based on the stem of a legitimate generic name. When published it is incorrect, but it may become correct later.

Ex. 14. *Chloris radiata* (L.) Sw. (1788) was nomenclaturally superfluous when published, since Swartz cited *Andropogon fasciculatus* L. (1753) as a synonym. However, it is not illegitimate since it was based on the legitimate *Agrostis radiata* L. (1759). *Chloris radiata* is the correct name in the genus *Chloris* for *Agrostis radiata* when *Andropogon fasciculatus* is treated as a different species, as was done by Hackel (in Candolle & Candolle, Monogr. Phan. 6: 177. 1889).

Ex. 15. The generic name *Hordelymus* (Jess.) Harz (1885) was nomenclaturally superfluous when published because its type, *Elymus europaeus* L., is also the type of *Cuviera* Koeler (1802). However, it is not illegitimate since it was based on the legitimate *Hordeum* [unranked] *Hordelymus* Jess. (Deutschl. Gräser: 202. 1863). *Cuviera* Koeler has since been rejected in favour of its later homonym *Cuviera* DC., and *Hordelymus* can now be used as the correct name for a segregate genus containing *Elymus europaeus* L.

Ex. 16. Carpinaceae Vest (Anleit. Stud. Bot.: 265, 280. 1818) was nomenclaturally super-fluous when published because of the inclusion of *Salix* L., the type of *Salicaceae* Mirb. (1815). However, it is not illegitimate because it is based on the stem of a legitimate generic name, *Carpinus* L.

Note 3. In no case does a statement of parentage accompanying the publication of a name for a hybrid make the name illegitimate (see Art. H.5).

Ex. 17. The name *Polypodium* ×*shivasiae* Rothm. (1962) was proposed for hybrids between *P. australe* Fée and *P. vulgare* subsp. *prionodes* (Asch.) Rothm., while at the same time the author accepted *P.* ×*font-queri* Rothm. (1936) for hybrids between *P. australe* and *P. vul-gare* L. subsp. *vulgare*. Under Art. H.4.1, *P.* ×*shivasiae* is a synonym of *P.* ×*font-queri;* nevertheless, it is not an illegitimate name.

Article 53

53.1. A name of a family, genus or species, unless conserved (Art. 14) or sanctioned (Art. 15), is illegitimate if it is a later homonym, that is, if it is spelled exactly like a name based on a different type that was previously and validly published for a taxon of the same rank (see also Art. 6 Note 2, and Art. 53.2 and 53.4).

Ex. 1. The name *Tapeinanthus* Boiss. ex Benth. (1848), given to a genus of *Labiatae,* is a later homonym of *Tapeinanthus* Herb. (1837), a name previously and validly published for a genus of *Amaryllidaceae. Tapeinanthus* Boiss. ex Benth. is therefore unavailable for use. It was renamed *Thuspeinanta* T. Durand (1888).

Ex. 2. The name *Torreya* Arn. (1838) is a nomen conservandum and is therefore available for use in spite of the existence of the earlier homonym *Torreya* Raf. (1818).

Ex. 3. Astragalus rhizanthus Boiss. (1843) is a later homonym of the validly published name *A. rhizanthus* Royle (1835) and is therefore unavailable for use. Boissier renamed it *A. cariensis* Boiss. (1849).

Note 1. A later homonym is unavailable for use even if the earlier homonym is il-legitimate or is otherwise generally treated as a synonym.

Ex. 4. Zingiber truncatum S. Q. Tong (1987) is illegitimate, being a later homonym of *Z. truncatum* Stokes (1812), even though the latter name is itself illegitimate under Art. 52.1 because in its protologue the name *Amomum zedoaria* Christm. (1779) was cited in syno-nymy. It was renamed *Z. neotruncatum* T. L. Wu & al. (2000).

Ex. 5. The name *Amblyanthera* Müll. Arg. (1860) is a later homonym of the validly pub-lished *Amblyanthera* Blume (1849) and is therefore unavailable for use, although *Ambly-anthera* Blume is now considered to be a synonym of *Osbeckia* L. (1753).

53.2. A sanctioned name is illegitimate if it is a later homonym of another sanctioned name (see also Art. 15 Note 1).

53.3. When two or more generic or specific names based on different types are so similar that they are likely to be confused (because they are applied to related taxa or for any other reason) they are to be treated as homonyms (see also Art. 61.5). If established practice has been to treat two similar names as homonyms, this practice is to be continued if it is in the interests of nomenclatural stability.

*Ex. 6. Names treated as homonyms: *Asterostemma* Decne. (1838) and *Astrostemma* Benth. (1880); *Pleuropetalum* Hook. f. (1846) and *Pleuripetalum* T. Durand (1888); *Eschweilera* DC. (1828) and *Eschweileria* Boerl. (1887); *Skytanthus* Meyen (1834) and *Scytanthus* Hook. (1844).

*Ex. 7. The three generic names *Bradlea* Adans. (1763), *Bradleja* Banks ex Gaertn. (1790), and *Braddleya* Vell. (1827), all commemorating Richard Bradley, are treated as homonyms because only one can be used without serious risk of confusion.

*Ex. 8. The names *Acanthoica* Lohmann (1902) and *Acanthoeca* W. N. Ellis (1930), both designating flagellates, are sufficiently alike to be considered homonyms (Taxon 22: 313. 1973).

*Ex. 9. Epithets so similar that they are likely to be confused if combined under the same generic or specific name: *chinensis* and *sinensis; ceylanica* and *zeylanica; napaulensis, nepalensis,* and *nipalensis; polyanthemos* and *polyanthemus; macrostachys* and *macrostachyus; heteropus* and *heteropodus; poikilantha* and *poikilanthes; pteroides* and *pteroideus; trinervis* and *trinervius; macrocarpon* and *macrocarpum; trachycaulum* and *trachycaulon.*

*Ex. 10. Names not likely to be confused: *Rubia* L. (1753) and *Rubus* L. (1753); *Monochaetum* (DC.) Naudin (1845) and *Monochaete* Döll (1875); *Peponia* Grev. (1863) and *Peponium* Engl. (1897); *Iris* L. (1753) and *Iria* (Pers.) Hedw. (1806); *Desmostachys* Miers (1852) and *Desmostachya* (Stapf) Stapf (1898); *Symphyostemon* Miers (1841) and *Symphostemon* Hiern (1900); *Gerrardina* Oliv. (1870) and *Gerardiina* Engl. (1897); *Urvillea* Kunth (1821) and *Durvillaea* Bory (1826); *Peltophorus* Desv. (1810; *Gramineae*) and *Peltophorum* (Vogel) Benth. (1840; *Leguminosae*); *Senecio napaeifolius* (DC.) Sch. Bip. (1845, *'napeaefolius';* see Art. 60 Ex. 18) and *S. napifolius* MacOwan (1890; the epithets being derived, respectively, from *Napaea* and *Brassica napus*); *Lysimachia hemsleyana* Oliv. (1891) and *L. hemsleyi* Franch. (1895) (see, however, Rec. 23A.2); *Euphorbia peplis* L. (1753) and *E. peplus* L. (1753).

Ex. 11. Names conserved against earlier names treated as homonyms (see App. III): *Lyngbya* Gomont (vs *Lyngbyea* Sommerf.); *Columellia* Ruiz & Pav. (vs *Columella* Lour.), both commemorating Columella, the Roman writer on agriculture; *Cephalotus* Labill. (vs *Cephalotos* Adans.); *Simarouba* Aubl. (vs *Simaruba* Boehm.).

Ex. 12. The name *Gilmania* Coville (1936) was published as a substitute name for *Phyllogonum* Coville (1893) because the author considered the latter to be a later homonym of *Phyllogonium* Bridel (1827). Treating them as homonyms has become accepted, e.g. in *Index Nominum Genericorum,* and the name *Gilmania* has been accepted as legitimate ever since. Therefore the names *Phyllogonum* and *Phyllogonium* are to continue to be treated as homonyms.

53.4. The names of two subdivisions of the same genus, or of two infraspecific taxa within the same species, even if they are of different rank, are treated as homonyms, the later of which is illegitimate, if they have the same or a confusingly similar final epithet and are not based on the same type.

Ex. 13. The names *Andropogon sorghum* subsp. *halepensis* (L.) Hack. and *A. sorghum* var. *halepensis* (L.) Hack. (in Candolle & Candolle, Monogr. Phan. 6: 502. 1889) are legitimate since both have the same type; repetition of the final epithet is in accord with Rec. 26A.1.

Ex. 14. *Anagallis arvensis* var. *caerulea* (L.) Gouan (Fl. Monsp.: 30. 1765), based on *A. caerulea* L. (1759), makes illegitimate the name *A. arvensis* subsp. *caerulea* Hartm. (Sv. Norsk Exc.-Fl.: 32. 1846), based on the later homonym *A. caerulea* Schreb. (1771).

Ex. 15. *Scenedesmus armatus* var. *brevicaudatus* (Hortob.) Pankow (in Arch. Protistenk. 132: 153. 1986), based on *S. carinatus* var. *brevicaudatus* Hortob. (in Acta Bot. Acad. Sci. Hung. 26: 318. 1981), is a later homonym of *S. armatus* f. *brevicaudatus* L. S. Péterfi (in Stud. Cercet. Biol. (Bucharest), Ser. Biol. Veg. 15: 25. 1963) even though the two names apply to taxa of different infraspecific rank. *Scenedesmus armatus* var. *brevicaudatus* (L. S. Péterfi) E. H. Hegew. (in Arch. Hydrobiol. Suppl. 60: 393. 1982), however, is not a later homonym since it is based on the same type as *S. armatus* f. *brevicaudatus* L. S. Péterfi.

Note 2. The same final epithet may be used in the names of subdivisions of different genera, and of infraspecific taxa within different species.

Ex. 16. *Verbascum* sect. *Aulacosperma* Murb. (Monogr. Verbascum: 34, 593. 1933) is permissible, although there is an earlier *Celsia* sect. *Aulacospermae* Murb. (Monogr. Celsia: 34, 56. 1926). This, however, is not an example to be followed, since it is contrary to Rec. 21B.2.

53.5. When it is doubtful whether names or their epithets are sufficiently alike to be confused, a request for a decision may be submitted to the General Committee (see Div. III), which will refer it for examination to the committee(s) for the appropriate taxonomic group(s). A recommendation, whether or not to treat the names concerned as homonyms, may then be put forward to an International Botanical Congress, and, if ratified, will become a binding decision.

Ex. 17. Names ruled as likely to be confused, and therefore to be treated as homonyms: *Ficus gomelleira* Kunth (1847) and *F. gameleira* Standl. (1937) (Taxon 42: 111. 1993); *Solanum saltiense* S. Moore (1895) and *S. saltense* (Bitter) C. V. Morton (1944) (Taxon 42: 434. 1993); *Balardia* Cambess. (1829; *Caryophyllaceae*) and *Ballardia* Montrouz. (1860; *Myrtaceae*) (Taxon 42: 434. 1993).

Ex. 18. Names ruled as not likely to be confused: *Cathayeia* Ohwi (1931; extant *Flacourtiaceae*) and *Cathaya* Chun & Kuang (1962; fossil *Pinaceae*) (Taxon 36: 429. 1987); *Cristella* Pat. (1887; *Fungi*) and *Christella* H. Lév. (1915; *Pteridophyta*) (Taxon 35: 551. 1986); *Coluria* R. Br. (1823; *Rosaceae*) and *Colura* (Dumort.) Dumort. (1835; *Hepaticae*) (Taxon 42: 433. 1993); *Acanthococcus* Hook. f. & Harv. (1845; *Rhodophyta*) and *Acanthococos* Barb. Rodr. (1900; *Palmae*) (Taxon 42: 433. 1993); *Rauia* Nees & Mart. (1823; *Rutaceae*) and *Rauhia* Traub (1957; *Amaryllidaceae*) (Taxon 42: 433. 1993).

53.6. When two or more homonyms have equal priority, the first of them that is adopted in an effectively published text (Art. 29-31) by an author who simultaneously rejects the other(s) is treated as having priority. Likewise, if an author in an effectively published text substitutes other names for all but one of these homonyms, the homonym for the taxon that is not renamed is treated as having priority.

Ex. 19. Linnaeus simultaneously published "10." *Mimosa cinerea* (Sp. Pl.: 517. 1753) and "25." *M. cinerea* (Sp. Pl.: 520. 1753). In 1759, he renamed species 10 *M. cineraria* L. and retained the name *M. cinerea* for species 25, so that the latter is treated as having priority over its homonym.

Ex. 20. Rouy & Foucaud (Fl. France 2: 30. 1895) published the name *Erysimum hieraciifolium* var. *longisiliquum,* with two different types, for two different taxa under different subspecies. Only one of these names can be maintained.

Note 3. A homonym renamed or rejected under Art. 53.6 remains legitimate and takes precedence over a later synonym of the same rank, should a transfer to another genus or species be effected.

Ex. 21. Mimosa cineraria L. (1759), based on *M. cinerea* L. (Sp. Pl.: 517 [non 520]. 1753; see Art. 53 Ex. 19), was transferred to *Prosopis* by Druce (1914) as *P. cineraria* (L.) Druce. However, the correct name in *Prosopis* would have been a combination based on *M. cinerea* had not that name been successfully proposed for rejection.

Article 54

54.1. Consideration of homonymy does not extend to the names of taxa not treated as plants, except as stated below:

(a) Later homonyms of the names of taxa once treated as plants are illegitimate, even though the taxa have been reassigned to a different group of organisms to which this *Code* does not apply.

(b) A name originally published for a taxon other than a plant, even if validly published under Art. 32-45 of this *Code,* is illegitimate if it becomes a homonym of a plant name when the taxon to which it applies is first treated as a plant (see also Art. 45.4).

Note 1. The *International code of nomenclature of bacteria* provides that a bacterial name is illegitimate if it is a later homonym of a name of a taxon of bacteria, fungi, algae, protozoa, or viruses.

Recommendation 54A

54A.1 Authors naming new taxa under this *Code* should, as far as is practicable, avoid using such names as already exist for zoological and bacteriological taxa.

Article 55

55.1. A name of a species or subdivision of a genus may be legitimate even if its epithet was originally placed under an illegitimate generic name (see also Art. 22.5).

Ex. 1. Agathophyllum Juss. (1789) is an illegitimate name, being a superfluous substitute for *Ravensara* Sonn. (1782). Nevertheless the name *A. neesianum* Blume (1851) is legitimate. Because Meisner (1864) cited *A. neesianum* as a synonym of his new *Mespilodaphne mauritiana* but did not adopt the epithet *neesiana, M. mauritiana* Meisn. is a superfluous name and hence illegitimate.

55.2. An infraspecific name may be legitimate even if its final epithet was originally placed under an illegitimate specific name (see also Art. 27.2).

Ex. 2. Agropyron japonicum var. *hackelianum* Honda (in Bot. Mag. (Tokyo) 41: 385. 1927) is legitimate, even though it was published under the illegitimate *A. japonicum* Honda (1927), a later homonym of *A. japonicum* (Miq.) P. Candargy (1901) (see also Art. 27 Ex. 1).

55.3. The names of species and of subdivisions of genera assigned to genera the names of which are conserved or sanctioned later homonyms, and which had earlier been assigned to the genera under the rejected homonyms, are legitimate under the conserved or sanctioned names without change of authorship or date if there is no other obstacle under the rules.

Ex. 3. Alpinia languas J. F. Gmel. (1791) and *Alpinia galanga* (L.) Willd. (1797) are to be accepted although *Alpinia* L. (1753), to which they were assigned by their authors, is rejected and the genus in which they are now placed is named *Alpinia* Roxb. (1810), nom. cons.

Article 56

56.1. Any name that would cause a disadvantageous nomenclatural change (Art. 14.1) may be proposed for rejection. A name thus rejected, or its basionym if it has one, is placed on a list of nomina utique rejicienda (App. V). Along with the listed names, all combinations based on them are similarly rejected, and none is to be used (see Rec. 50E.2).

56.2. The list of rejected names will remain permanently open for additions and changes. Any proposal for rejection of a name must be accompanied by a detailed statement of the cases both for and against its rejection, including considerations of typification. Such proposals must be submitted to the General Committee (see Div. III), which will refer them for examination to the committees for the various taxonomic groups (see also Art. 14.14 and Rec. 14A).

Article 57

57.1. A name that has been widely and persistently used for a taxon or taxa not including its type is not to be used in a sense that conflicts with current usage unless and until a proposal to deal with it under Art. 14.1 or 56.1 has been submitted and rejected.

Ex. 1. The name *Strophostyles helvola* (L.) Elliott was widely and persistently used from the mid-19th century for a taxon that Verdcourt (in Taxon 46: 357-359. 1997) reported did not include its type, which he found to be referable to *Macroptilium lathyroides* (L.) Urb., based on *Phaseolus lathyroides* L. (1763) and over which *P. helvolus,* the basionym of *S. helvola,* has priority. Verdcourt did not transfer the epithet *helvolus* to *Macroptilium* which would have conflicted with current usage, but proposed *P. helvolus* for conservation with a conserved type that he believed referred to the species to which the name *S. helvola* had been applied; the proposal was accepted. When Delgado-Salinas & Lavin (in Taxon 53: 839-841. 2004) later discovered that this first-conserved type applies to another species, *Strophostyles umbellata* (Muhl. ex Willd.) Britton, they also preserved current usage and proposed a new conserved type.

Article 58

58.1. The epithet in an illegitimate name if available may be used in a different combination, at the same or a different rank, if no other epithet is available from a name that has priority at that rank. The resulting name is then treated as new, either as a nomen novum with the same type as the illegitimate name (see also Art. 7.5 and Art. 33 Note 2) or as the name of a new taxon with a different type. Its priority does not date back to the publication of the illegitimate name.

Ex. 1. The name *Talinum polyandrum* Hook. (1855) is illegitimate, being a later homonym of *T. polyandrum* Ruiz & Pav. (1798). When Bentham, in 1863, transferred *T. polyandrum* Hook. to *Calandrinia,* he called it *C. polyandra.* This name has priority from 1863, and is cited as *C. polyandra* Benth., not *C. polyandra* "(Hook.) Benth."

Ex. 2. Hibiscus ricinifolius E. Mey. ex Harv. (1860) is illegitimate because *H. ricinoides* Garcke (1849) was cited in synonymy. When the epithet *ricinifolius* was combined at varietal rank under *H. vitifolius* by Hochreutiner (in Annuaire Conserv. Jard. Bot. Genève 4: 170. 1900) his name was legitimate and is treated as a nomen novum, typified by the type of *H. ricinoides,* that is to be cited as *H. vitifolius* var. *ricinifolius* Hochr., not "(E. Mey. ex Harv.) Hochr."

Ex. 3. When publishing *Collema tremelloides* var. *cyanescens,* Acharius (Syn. Meth. Lich.: 326. 1814) cited in synonymy *C. tremelloides* var. *caesium* Ach. (Lichenogr. Universalis: 656. 1810), a legitimate name at the same rank, thus rendering his new name superfluous and illegitimate. However, the epithet *cyanescens* was available for use in *Collema* at the rank of species, and the name *C. cyanescens* Rabenh. (1845), based on the same type, is legitimate. The correct author citation for *Leptogium cyanescens,* published by Körber (1855) by reference to *C. cyanescens* "Schaer.", is therefore (Rabenh.) Körb., not (Ach.) Körb. or (Schaer.) Körb. Körber ascribed the epithet *cyanescens* to Schaerer because this author was the first to use the epithet at specific rank in the name *Parmelia cyanescens* Schaerer (1842), which is however illegitimate being a later homonym of *P. cyanescens* (Pers.) Ach. (1803).

Note 1. In the case of re-use at the same rank of epithets of illegitimate superfluous names, the type of the name causing the original superfluity must be explicitly excluded.

Ex. 4. Menispermum villosum Lam. (1797) is an illegitimate superfluous name because *M. hirsutum* L. (1753) was cited in synonymy. The name *Cocculus villosus* DC. (1817), based on *M. villosum,* is also illegitimate since the type of *M. hirsutum* was not excluded and there was no obstacle to the use of the epithet *hirsutus* in *Cocculus.*

Ex. 5. Cenomyce ecmocyna Ach. (1810) is an illegitimate superfluous name for *Lichen gracilis* L. (1753), as is *Scyphophora ecmocyna* Gray (1821), based on *C. ecmocyna,* since the type of *L. gracilis* was not excluded and there was no obstacle to the use of the epithet *gracilis* in *Scyphophora.* However, when proposing the combination *Cladonia ecmocyna,* Leighton (1866) explicitly excluded that type and thereby published a new, legitimate name, *Cladonia ecmocyna* Leight.

Ex. 6. Diospyros discolor Willd. (1806) was illegitimate when published, because *Cavanillea philippensis* Desr. (1792) was cited as a synonym. *Embryopteris discolor,* based on *D. discolor* Willd., was published in 1837 by G. Don (Gen. Syst. 4: 41.), who clearly excluded *C. philippensis.* The name would, therefore, have been attributable to "G. Don" with priority from 1837. However, *D. discolor* is now a conserved name and no longer illegitimate, hence this provision no longer applies and the correct author citation is *Embryopteris discolor* (Willd.) G. Don, with priority from 1806.

CHAPTER VI. NAMES OF FUNGI WITH A PLEOMORPHIC LIFE CYCLE

Article 59

59.1. In non lichen-forming ascomycetous and basidiomycetous fungi (including *Ustilaginales*) with mitotic asexual morphs (anamorphs) as well as a meiotic sexual morph (teleomorph), the correct name covering the holomorph (i.e. the species in all its morphs) is the earliest legitimate name typified, or epitypified under Art. 59.7, by an element representing the teleomorph, i.e. the morph characterized by the production of asci/ascospores, basidia/basidiospores, teliospores, or other basidium-bearing organs.

Ex. 1. The name *Crocicreomyces guttiferae* Bat. & Peres (1964) was published for a lichen-forming fungus producing only an asexual morph. When it was recognized that *C. guttiferae* is conspecific with *Byssoloma aeruginescens* Vĕzda (1974), based on an ascospore-producing type, and that *Crocicreomyces* Bat. & Peres (1964) is synonymous with *Byssoloma* Trevis. (1853), Batista & Peres's epithet was correctly recombined as *B. guttiferae* (Bat. & Peres) Lücking & Sérus. (1998). As Art. 59 does not apply to lichen-forming fungi, no separate generic or specific names are available for use for the asexual morph.

59.2. For a binary name to qualify as a name of a holomorph, not only must its type specimen, or its epitype specimen under Art. 59.7, be teleomorphic, but also the protologue must include a description or diagnosis of this morph (or be so phrased that the possibility of reference to the teleomorph cannot be excluded) (see also Art. 59.7).

59.3. If these requirements are not fulfilled, the name is that of a form-taxon and is applicable only to the anamorph represented by its type, as described or referred to in the protologue. The accepted taxonomic disposition of the type of the name determines the application of the name, no matter whether the genus to which a subordinate taxon is assigned by the author(s) is holomorphic or anamorphic.

Ex. 2. The name *Ravenelia cubensis* Arthur & J. R. Johnst. (1918), based on a specimen bearing only uredinia (an anamorph), is a validly published and legitimate name of an anamorph, in spite of the attribution of the species to a holomorphic genus. It is legitimately combined with a generic name typified by an anamorph as *Uredo cubensis* (Arthur & J. R.

Johnst.) Cummins (1956). *Ravenelia cubensis* is not available for use inclusive of the teleomorph.

59.4. Irrespective of priority, names with a teleomorphic type or epitype (Art. 59.7) take precedence over names with only an anamorphic type when the types are judged to belong to the same holomorphic taxon. Priority of competing teleomorphic typified or epitypified names follows Principle III except that teleomorphic typified names published before 1 January 2007 take precedence over anamorphic typified names subsequently epitypified after 1 January 2007 by teleomorphs.

59.5. The provisions of this article shall not be construed as preventing the publication and use of binary names for form-taxa when it is thought necessary or desirable to refer to anamorphs alone.

Ex. 3. Because the teleomorph of *Gibberella stilboides* W. L. Gordon & C. Booth (1971) is only known from strains of the anamorph *Fusarium stilboides* Wollenw. (1924) mating in culture, and has not been found in nature, it may be thought desirable to use the name of the anamorph for the pathogen of *Coffea*.

Ex. 4. Cummins (1971), in *The rust fungi of cereals, grasses and bamboos,* found it to be neither necessary nor desirable to introduce new names of anamorphs under *Aecidium* Pers. : Pers. and *Uredo* Pers. : Pers., for the aecial and uredinial stages of species of *Puccinia* Pers. : Pers. of which the telial stage (teleomorph) was known.

Note 1. In the absence of existing legitimate names, specific or infraspecific names for anamorphs may be proposed at the time of publication of the name for the holomorphic fungus or later. The final epithets may, if desired, be identical, as long as they are not in homonymous combinations.

Ex. 5. The name *Penicillium brefeldianum* B. O. Dodge (1933), based on teleomorphic and anamorphic material, is a validly published and legitimate name of a holomorph, in spite of the attribution of the species to a generic name typified by an anamorph. It is legitimately combined with a holomorphic generic name as *Eupenicillium brefeldianum* (B. O. Dodge) Stolk & D. B. Scott (1967). *Penicillium brefeldianum* is not available for use in a restricted sense for the anamorph alone.

59.6. As long as there is direct and unambiguous evidence for the deliberate introduction of a new morph judged by the author(s) to be correlated with the morph typifying a purported basionym, and this evidence is strengthened by fulfilment of all requirements in Art. 32-45 for valid publication of a name of a new taxon, any indication such as "comb. nov." or "nom. nov." is regarded as a formal error, and the name introduced is treated as that of a new taxon, and attributed solely to the author(s) thereof. When only the requirements for valid publication of a new combination (Art. 33 and 34) have been fulfilled, the name is accepted as such and based, in accordance with Art. 7.4, on the type of the declared or implicit basionym.

Ex. 6. Mycosphaerella aleuritidis was published as "(Miyake) Ou comb. nov., syn. *Cerco-spora aleuritidis* Miyake" but with a Latin diagnosis of the teleomorph. The indication "comb. nov." is taken as a formal error, and *M. aleuritidis* S. H. Ou (1940) is accepted as a validly published new specific name for the holomorph, typified by the teleomorphic material described by Ou.

Ex. 7. Corticium microsclerotium was originally published as "(Matz) Weber, comb. nov., syn. *Rhizoctonia microsclerotia* Matz" with a description, only in English, of the teleomorph. Because of Art. 36, this may not be considered as the valid publication of the name of a new species, and so *C. microsclerotium* (Matz) G. F. Weber (1939) must be considered a validly published and legitimate new combination based on the specimen of the anamorph that typifies its basionym. *Corticium microsclerotium* G. F. Weber (1951), published with a Latin description and a teleomorphic type, is an illegitimate later homonym.

Ex. 8. Hypomyces chrysospermus Tul. (1860), presented as the name of a holomorph without the indication "comb. nov." but with explicit reference to *Mucor chrysospermus* (Bull.) Bull. and *Sepedonium chrysospermum* (Bull.) Fr., which are names of its anamorph, is not to be considered as a new combination but as the name of a newly described species, with a teleomorphic type.

59.7. Where a teleomorph has been discovered for a fungus previously known only as an anamorph and for which there is no existing legitimate name for the holomorph, an epitype exhibiting the teleomorph stage may be designated for the hitherto anamorphic name even when there is no hint of the teleomorph in the protologue of that name.

Recommendation 59A

59A.1. When a new morph of a fungus is described, it should be published either as a new taxon (e.g. gen. nov., sp. nov., var. nov.) the name of which has a teleomorphic type, or as a new anamorph (anam. nov.) the name of which has an anamorphic type.

59A.2. When in naming a new morph of a fungus the epithet of the name of a different, earlier described morph of the same fungus is used, the new name should be designated as the name of a new taxon or anamorph, as the case may be, but not as a new combination based on the earlier name.

59A.3. Authors should avoid the publication and use of binary names for anamorphs when the teleomorphic connection is firmly established and there is no practical need for separate names (as e.g. in rust fungi and members of the *Trichocomaceae*).

CHAPTER VII. ORTHOGRAPHY AND GENDER OF NAMES

SECTION 1. ORTHOGRAPHY

Article 60

60.1. The original spelling of a name or epithet is to be retained, except for the correction of typographical or orthographical errors and the standardizations imposed by Art. 60.5 (*u/v* or *i/j* used interchangeably), 60.6 (diacritical signs and ligatures), 60.8 (compounding forms), 60.9 (hyphens), 60.10 (apostrophes), 60.11 (terminations; see also Art. 32.7), and 60.12 (fungal epithets).

Ex. 1. Retention of original spelling: The generic names *Mesembryanthemum* L. (1753) and *Amaranthus* L. (1753) were deliberately so spelled by Linnaeus and the spelling is not to be altered to *"Mesembrianthemum"* and *"Amarantus"*, respectively, although these latter forms are philologically preferable (see Bull. Misc. Inform. Kew 1928: 113, 287. 1928). – *Phoradendron* Nutt. (1848) is not to be altered to *"Phoradendrum"*. – *Triaspis mozambica* A. Juss. (1843) is not to be altered to *"T. mossambica"*, as in Engler (Pflanzenw. Ost-Afrikas C: 232. 1895). – *Alyxia ceylanica* Wight (1848) is not to be altered to *"A. zeylanica"*, as in Trimen (Handb. Fl. Ceylon 3: 127. 1895). – *Fagus sylvatica* L. (1753) is not to be altered to *"F. silvatica"*. The classical spelling *silvatica* is recommended for adoption in the case of a new name (Rec. 60E), but the mediaeval spelling *sylvatica* is not an orthographical error. – *Scirpus cespitosus* L. (1753) is not to be altered to *"S. caespitosus"*.

*Ex. 2. Typographical errors: *Globba "brachycarpa"* Baker (1890) and *Hetaeria "alba"* Ridl. (1896) are typographical errors for *Globba trachycarpa* Baker and *Hetaeria alta* Ridl., respectively (see J. Bot. 59: 349. 1921).

Ex. 3. *"Torilis" taihasenzanensis* Masam. (in J. Soc. Trop. Agric. 6: 570. 1934) was a typographical error for *Trollius taihasenzanensis*, as noted on the errata slip inserted between pages 4 and 5 of the same volume.

Ex. 4. The misspelled *Indigofera "longipednnculata"* Y. Y. Fang & C. Z. Zheng (1983) is presumably a typographical error and is to be corrected to *I. longipedunculata*.

*Ex. 5. Orthographical error: *Gluta "benghas"* L. (1771), being an orthographical error for *G. renghas,* is cited as *G. renghas* L. (see Engler in Candolle & Candolle, Monogr. Phan. 4: 225. 1883); the vernacular name used as a specific epithet by Linnaeus is "renghas", not "benghas".

Note 1. Art. 14.11 provides for the conservation of an altered spelling of a name of a family, genus, or species.

Ex. 6. *Bougainvillea* (see App. III, *Spermatophyta, Dicotyledones*).

60.2. The words "original spelling" mean the spelling employed when the name was validly published. They do not refer to the use of an initial capital or lower-case letter, this being a matter of typography (see Art. 20.1 and 21.2, Rec. 60F).

60.3. The liberty of correcting a name is to be used with reserve, especially if the change affects the first syllable and, above all, the first letter of the name.

*Ex. 7. The spelling of the generic name *Lespedeza* Michx. (1803) is not to be altered, although it commemorates Vicente Manuel de Céspedes (see Rhodora 36: 130-132, 390-392. 1934). – *Cereus jamacaru* DC. (1828) may not be altered to *C.* *"mandacaru"*, even if *jamacaru* is believed to be a corruption of the vernacular name "mandacaru".

60.4. The letters *w* and *y,* foreign to classical Latin, and *k,* rare in that language, are permissible in Latin plant names. Other letters and ligatures foreign to classical Latin that may appear in Latin plant names, such as the German *ß* (double *s*), are to be transcribed.

60.5. When a name has been published in a work where the letters *u, v* or *i, j* are used interchangeably or in any other way incompatible with modern practices (e.g. one letter of a pair not being used in capitals, or not at all), those letters are to be transcribed in conformity with modern botanical usage.

Ex. 8. *Uffenbachia* Fabr. (1763), not *"Vffenbachia"; Taraxacum* Zinn (1757), not *"Taraxacvm"; Curculigo* Gaertn. (1788), not *"Cvrcvligo".*

Ex. 9. *"Geastrvm hygrometricvm"* and *"Vredo pvstvlata"* of Persoon (1801) are written, respectively, *Geastrum hygrometricum* Pers. and *Uredo pustulata* Pers.

Ex. 10. *Brachypodium "iaponicum"* of Miquel (1866) is written *Brachypodium japonicum* Miq.

60.6. Diacritical signs are not used in Latin plant names. In names (either new or old) drawn from words in which such signs appear, the signs are to be suppressed with the necessary transcription of the letters so modified; for example *ä, ö, ü* become, respectively, *ae, oe, ue; é, è, ê* become *e,* or sometimes *ae; ñ* becomes *n; ø* becomes *oe; å* becomes *ao.* The diaeresis, indicating that a vowel is to be pronounced separately from the preceding vowel (as in *Cephaëlis, Isoëtes*), is permissible; the ligatures *-æ-* and *-œ-,* indicating that the letters are pronounced together, are to be replaced by the separate letters *-ae-* and *-oe-.*

106

60.7. When changes in spelling by authors who adopt personal, geographic, or vernacular names in nomenclature are intentional latinizations, they are to be preserved, except when they concern *(a)* only the termination of epithets, to which Art. 60.11 applies, or *(b)* changes to personal names involving *(1)* omission of a final vowel or final consonant or *(2)* conversion of a final vowel to a different vowel, for which the final letter of the name is to be restored.

Ex. 11. *Clutia* L. (1753), *Gleditsia* L. (1753), and *Valantia* L. (1753), commemorating Cluyt, Gleditsch, and Vaillant, respectively, are not to be altered to*"Cluytia"*, *"Gleditschia"*, and*"Vaillantia"*; Linnaeus latinized the names of these botanists deliberately as Clutius, Gleditsius, and Valantius.

Ex. 12. *Abies alcoquiana* Veitch ex Lindl. (1861), commemorating "Rutherford Alcock Esq.", implies an intentional latinization of that name to Alcoquius. In transferring the epithet to *Picea,* Carrière (1867) deliberately changed the spelling to *"alcockiana"*. The resulting combination is nevertheless correctly cited as *P. alcoquiana* (Veitch ex Lindl.) Carrière (see Art. 61.4).

Ex. 13. *Abutilon glaziovii* K. Schum. (1891), *Desmodium bigelovii* A. Gray (1843), and *Rhododendron bureavii* Franch. (1887), commemorating A. F. M. Glaziou, J. Bigelow, and L. E. Bureau, respectively, are not to be changed to *A. "glazioui"*, *D. "bigelowii"*, or *R. "bureaui"*. In these three cases, the implicit latinizations Glaziovius, Bigelovius, and Bureavius result from conversion of a final consonant and do not affect merely the termination of the names.

Ex. 14. *Arnica chamissonis* Less. (1831) and *Tragus berteronianus* Schult. (1824), commemorating L. K. A. von Chamisso and C. L. G. Bertero, are not to be changed to *A. "chamissoi"* or *T. "berteroanus"*. The derivation of these epithets from the third declension genitive, a practice not now recommended in most cases (see Rec. 60C.2), involves the addition of letters to the personal name and does not affect merely the termination.

Ex. 15. *Acacia "brandegeana", Blandfordia "backhousii", Cephalotaxus "fortuni", Chenopodium "loureirei", Convolvulus "loureiri", Glochidion "melvilliorum", Hypericum "buckleii", Solanum "rantonnei",* and *Zygophyllum "billardierii"* were published to commemorate T. S. Brandegee, J. Backhouse, R. Fortune, J. de Loureiro, R. Melville and E. F. Melville, S. B. Buckley, V. Rantonnet, and J. J. H. de Labillardière (de la Billardière). The implicit latinizations are Brandegeus, Backhousius, Fortunus, Loureireus or Loureirus, Melvillius, Buckleius, Rantonneus, and Billardierius, but these are not acceptable under Art. 60.7. The names are correctly cited as *A. brandegeeana* I. M. Johnst. (1925), *B. backhousei* Gunn & Lindl. (1845), *Cephalotaxus fortunei* Hook. (1850), *Chenopodium loureiroi* Steud. (1840), *Convolvulus loureiroi* G. Don (1836), *G. melvilleorum* Airy Shaw (1971), *H. buckleyi* M. A. Curtis (1843), *S. rantonnetii* Carrière (1859), and *Z. billardierei* DC. (1824).

Note 2. The provisions of Art. 60.7, 60.11, and Rec. 60C deal with the latinization of names through their modification. This latinization is different from translation of names (e.g. Tabernaemontanus from Bergzabern) and from the use of an adjective indirectly derived from a personal name, which are thus not subject to modification under Art. 60.7 or 60.11.

Ex. 16. In *Wollemia nobilis* W. G. Jones & al. (1995), the use of the adjective *nobilis* is indirectly derived from the name of the discoverer David Noble. *Cladonia abbatiana* S. Steenroose (1991) honours the French lichenologist H. des Abbayes. In both cases the adjective is indirectly derived from a personal name. Since no typographical or orthographical error is present, the original spelling of those names may not be altered.

60.8. The use of a compounding form contrary to Rec. 60G in an adjectival epithet is treated as an error to be corrected.

Ex. 17. Candolle's *Pereskia "opuntiaeflora"* is to be cited as *P. opuntiiflora* DC. (1828), and *Myrosma "cannaefolia"* of the younger Linnaeus, as *M. cannifolia* L. f. (1782).

Ex. 18. *Cacalia "napaeafolia"* and *Senecio "napaeafolius"* are to be cited as *Cacalia napaeifolia* DC. (1838) and *Senecio napaeifolius* (DC.) Sch. Bip. (1845), respectively; the specific epithet refers to the resemblance of the leaves to those of the genus *Napaea* L. (not *"Napea"*), and the substitute (connecting) vowel *-i* should have been used instead of the genitive singular inflection *-ae*.

Ex. 19. However, in *Andromeda polifolia* L. (1753), the epithet is a pre-Linnaean plant name (*"Polifolia"* of Buxbaum) used in apposition and not an adjective; it is not to be altered to *"poliifolia"* (*Polium*-leaved).

60.9. The use of a hyphen in a compound epithet is treated as an error to be corrected by deletion of the hyphen, unless the epithet is formed of words that usually stand independently or the letters before and after the hyphen are the same, when a hyphen is permitted (see Art. 23.1 and 23.3).

Ex. 20. Hyphen to be omitted: *Acer pseudoplatanus* L. (1753), not *A. "pseudo-platanus";* *Eugenia costaricensis* O. Berg (1856), not *E. "costa-ricensis";* *Ficus neoëbudarum* Summerh. (1932), not *F. "neo-ebudarum";* *Lycoperdon atropurpureum* Vittad. (1842), not *L. "atro-purpureum";* *Croton ciliatoglandulifer* Ortega (1797), not *C. "ciliato-glandulifer";* *Scirpus* sect. *Pseudoëriophorum* Jurtzev (in Byull. Moskovsk. Obshch. Isp. Prir., Otd. Biol. 70(1): 132. 1965), not *S.* sect. *"Pseudo-eriophorum".*

Ex. 21. Hyphen to be maintained: *Aster novae-angliae* L. (1753), *Coix lacryma-jobi* L. (1753), *Arctostaphylos uva-ursi* (L.) Spreng. (1825), *Veronica anagallis-aquatica* L. (1753; Art. 23.3), *Athyrium austro-occidentale* Ching (1986).

Note 3. Art. 60.9 refers only to epithets (in combinations), not to names of genera or taxa in higher ranks; a generic name published with a hyphen can be changed only by conservation (Art. 14.11).

Ex. 22. *Pseudo-salvinia* Piton (1940) may not be changed to *"Pseudosalvinia";* whereas by conservation *"Pseudo-elephantopus"* was changed to *Pseudelephantopus* Rohr (1792).

60.10. The use of an apostrophe in an epithet is treated as an error to be corrected by deletion of the apostrophe.

Ex. 23. *Lycium "o'donellii", Cymbidium "i'ansoni",* and *Solanum tuberosum* var. *"muru'kewillu"* are to be corrected to *L. odonellii* F. A. Barkley (1953), *C. iansonii* Rolfe (1900), and *S. tuberosum* var. *murukewillu* Ochoa (in Phytologia 65: 112. 1988), respectively.

60.11. The use of a termination (for example *-i, -ii, -ae, -iae, -anus,* or *-ianus*) contrary to Rec. 60C.1 is treated as an error to be corrected (see also Art. 32.7). However, terminations of epithets formed in accordance with Rec. 60C.2 are not to be corrected.

Ex. 24. Rhododendron "potanini" Batalin (1892) must be corrected to *R. potaninii* since it commemorates G. N. Potanin, to whose name Rec. 60C.1 applies. However, *Phoenix theophrasti* Greuter (1967) must not be changed to *P. "theophrastii"* since it commemorates Theophrastus, to whose name Rec. 60C.2 applies.

Ex. 25. Rosa "pissarti" (Carrière in Rev. Hort. 1880: 314. 1880) is a typographical error for *R. "pissardi"* (see Rev. Hort. 1881: 190. 1881), which in its turn is treated as an error for *R. pissardii* Carrière (see Rec. 60C.1(b)).

Ex. 26. However, *Uladendron codesuri* Marc.-Berti (1971) is not to be changed to *U. "codesurii"* (as by Brenan in Index Kew., Suppl. 16. 1981), since the epithet does not commemorate a person but derives from an acronym (CODESUR, Comisión para el Desarrollo del Sur de Venezuela).

Ex. 27. Nigella degenii subsp. *barbro* Strid and *N. degenii* subsp. *jenny* Strid (in Opera Bot. 28: 58, 60. 1970) commemorate the wife and daughter of the author. These spellings are not to be changed since the personal names were not given Latin terminations to form the subspecific epithets.

Ex. 28. Asparagus tamaboki Yatabe (1893) and *Agropyron kamoji* Ohwi (1942) bear the Japanese vernacular names "tamaboki" and (in part) "kamojigusa" as their epithets and are therefore not correctable to *A. "tamabokii"* and *A. "kamojii"*.

Note 4. If the gender and/or number of a substantival epithet derived from a personal name is inappropriate for the sex and/or number of the person(s) whom the name commemorates, the termination is to be corrected in conformity with Rec. 60C.1.

Ex. 29. Rosa ×"toddii" was named by Wolley-Dod (in J. Bot. 69, Suppl.: 106. 1931) for "Miss E. S. Todd"; the name is to be corrected to *R. ×toddiae* Wolley-Dod.

Ex. 30. Astragalus "matthewsii", published by Podlech & Kirchhoff (in Mitt. Bot. Staatssamml. München 11: 432. 1974) to commemorate Victoria A. Matthews, is to be corrected to *A. matthewsiae* Podlech & Kirchhoff; it is not therefore a later homonym of *A. matthewsii* S. Watson (1883) (see Agerer-Kirchhoff & Podlech in Mitt. Bot. Staatssamml. München 12: 375. 1976).

Ex. 31. Codium "geppii" (Schmidt in Biblioth. Bot. 91: 50. 1923), which commemorates "A. & E. S. Gepp", is to be corrected to *C. geppiorum* O. C. Schmidt.

60.12. Epithets of fungus names derived from the generic name of an associated organism are to be spelled in accordance with the accepted spelling of that organism's name; other spellings are regarded as orthographical variants to be corrected (see Art. 61).

Ex. 32. Phyllachora "anonicola" (Chardón in Mycologia 32: 190. 1940) is to be altered to *P. annonicola* Chardón, since the spelling *Annona* is now accepted in preference to *"Anona".* –

Meliola "albizziae" (Hansford & Deighton in Mycol. Pap. 23: 26. 1948) is to be altered to *M. albiziae* Hansf. & Deighton, since the spelling *Albizia* is now accepted in preference to *"Albizzia"*.

Recommendation 60A

60A.1. When a new name or its epithet is to be derived from Greek, the transliteration to Latin should conform to classical usage.

60A.2. The Greek spiritus asper (rough breathing) should be transcribed in Latin as the letter *h*.

Ex. 1. Hyacinthus from ὑάκινθος.

Recommendation 60B

60B.1. When a new generic name, or epithet of a subdivision of a genus, is taken from the name of a person, it should be formed as follows:

(a) When the name of the person ends with a vowel, the letter *-a* is added (thus *Ottoa* after Otto; *Sloanea* after Sloane), except when the name ends with *-a,* when *-ea* is added (e.g. *Collaea* after Colla), or with *-ea* (as *Correa*), when no letter is added.

(b) When the name of the person ends with a consonant, the letters *-ia* are added, but when the name ends with *-er,* either of the terminations *-ia* and *-a* is appropriate (e.g. *Sesleria* after Sesler and *Kernera* after Kerner).

(c) In latinized personal names ending with *-us* this termination is dropped (e.g. *Dillenia* after Dillenius) before applying the procedure described under (a) and (b).

Note 1. The syllables not modified by these endings are unaffected unless they contain letters foreign to Latin plant names or diacritical signs (see Art. 60.6).

Note 2. More than one generic name, or epithet of a subdivision of a genus, may be based on the same personal name, e.g. by adding a prefix or suffix to that personal name or by using an anagram or abbreviation of it.

Ex. 1. Durvillaea Bory (1826) and *Urvillea* Kunth (1821); *Lapeirousia* Pourr. (1788) and *Peyrousea* DC. (1838); *Engleria* O. Hoffm. (1888), *Englerastrum* Briq. (1894), and *Englerella* Pierre (1891); *Bouchea* Cham. (1832) and *Ubochea* Baill. (1891); *Gerardia* L. (1753) and *Graderia* Benth. (1846); *Martia* Spreng. (1818) and *Martiusia* Schult. & Schult. f. (1822).

Recommendation 60C

60C.1. When personal names are given Latin terminations in order to form specific and infraspecific epithets formation of those epithets is as follows (but see Rec. 60C.2):

(a) If the personal name ends with a vowel or *-er,* substantival epithets are formed by adding the genitive inflection appropriate to the sex and number of the person(s) honoured (e.g. *scopoli-i* for Scopoli (m), *fedtschenko-i* for Fedtschenko (m), *fedtschenko-ae* for Fedtschenko (f), *glaziou-i* for Glaziou (m), *lace-ae* for Lace (f), *gray-i* for Gray (m), *hooker-orum* for the Hookers (m)), except when the name ends with *-a,* in which case adding *-e* (singular) or *-rum* (plural) is appropriate (e.g. *triana-e* for Triana (m), *pojarkova-e* for Pojarkova (f), *orlovskaja-e* for Orlovskaja (f)).

(b) If the personal name ends with a consonant (except *-er*), substantival epithets are formed by adding *-i-* (stem augmentation) plus the genitive inflection appropriate to the sex and number of the person(s) honoured (e.g. *lecard-ii* for Lecard (m), *wilson-iae* for Wilson (f), *verlot-iorum* for the Verlot brothers, *braun-iarum* for the Braun sisters, *mason-iorum* for Mason, father and daughter).

(c) If the personal name ends with a vowel, adjectival epithets are formed by adding *-an-* plus the nominative singular inflection appropriate to the gender of the generic name (e.g. *Cyperus heyne-anus* for Heyne, *Vanda lindley-ana* for Lindley, *Aspidium bertero-anum* for Bertero), except when the personal name ends with *-a* in which case *-n-* plus the appropriate inflection is added (e.g. *balansa-nus* (m), *balansa-na* (f), and *balansa-num* (n) for Balansa).

(d) If the personal name ends with a consonant, adjectival epithets are formed by adding *-i-* (stem augmentation) plus *-an-* (stem of adjectival suffix) plus the nominative singular inflection appropriate to the gender of the generic name (e.g. *Rosa webb-iana* for Webb, *Desmodium griffith-ianum* for Griffith, *Verbena hassler-iana* for Hassler).

Note 1. The hyphens in the above examples are used only to set off the total appropriate termination.

60C.2. Personal names already in Greek or Latin, or possessing a well-established latinized form, should be given their appropriate Latin genitive to form new substantival epithets (e.g. *alexandri* from Alexander or Alexandre, *augusti* from Augustus or August or Auguste, *martini* from Martinus or Martin, *linnaei* from Linnaeus, *martii* from Martius, *wislizeni* from Wislizenus, *edithae* from Editha or Edith, *elisabethae* from Elisabetha or Elisabeth, *murielae* from Muriela or Muriel, *conceptionis* from Conceptio or Concepción, *beatricis* from Beatrix or Béatrice, *hectoris* from Hector; but not *"cami"* from Edmond Camus or Aimée Camus). Treating modern family names, i.e. ones that do not have a well-established latinized form, as if they were in third declension should be avoided (e.g. *munronis* from Munro, *richardsonis* from Richardson).

60C.3. New epithets based on personal names that have a well-established latinized form should maintain the traditional use of that latinized form.

Ex. 1. In addition to the epithets in Rec. 60C.2, the following epithets commemorate personal names already in Latin or possessing a well-established latinized form: *(a)* second declension: *afzelii* based on Afzelius; *allemanii* based on Allemanius (Freire Allemão); *bauhini* based on Bauhinus (Bauhin); *clusii* based on Clusius; *rumphii* based on Rumphius (Rumpf); *solandri* based on Solandrus (Solander); *(b)* third declension: *bellonis* based on Bello; *brunonis* based on Bruno (Robert Brown); *chamissonis* based on Chamisso; *(c)* adjectives (see Art. 23.5): *afzelianus, clusianus, linnaeanus, martianus, rumphianus* and *brunonianus, chamissonianus.*

60C.4. In forming new epithets based on personal names the customary spelling of the personal name should not be modified unless it contains letters foreign to Latin plant names or diacritical signs (see Art. 60.4 and 60.6).

60C.5. In forming new epithets based on personal names prefixes and particles should be treated as follows:

(a) The Scottish patronymic prefix "Mac", "Mc", or "M'", meaning "son of", should be spelled "mac" and united with the rest of the name (e.g. *macfadyenii* after Macfadyen, *macgillivrayi* after MacGillivray, *macnabii* after McNab, *mackenii* after M'Ken).

(b) The Irish patronymic prefix "O" should be united with the rest of the name or omitted (e.g. *obrienii, brienianus* after O'Brien, *okellyi* after O'Kelly).

(c) A prefix consisting of an article (e.g. le, la, l', les, el, il, lo), or containing an article (e.g. du, de la, des, del, della), should be united to the name (e.g. *leclercii* after Le Clerc, *dubuyssonii* after DuBuysson, *lafarinae* after La Farina, *logatoi* after Lo Gato).

(d) A prefix to a family name indicating ennoblement or canonization should be omitted (e.g. *candollei* after de Candolle, *jussieui* after de Jussieu, *hilairei* after Saint-Hilaire, *remyi* after St Rémy); in geographical epithets, however, "St" is rendered as *sanctus* (m) or *sancta* (f) (e.g. *sancti-johannis,* of St John, *sanctae-helenae,* of St Helena).

(e) A German or Dutch prefix should be omitted (e.g. *iheringii* after von Ihering, *martii* after von Martius, *steenisii* after van Steenis, *strassenii* after zu Strassen, *vechtii* after van der Vecht), but when it is normally treated as part of the family name it should be included in the epithet (e.g. *vonhausenii* after Vonhausen, *vanderhoekii* after Vanderhoek, *vanbruntiae* after Van Brunt).

Recommendation 60D

60D.1. An epithet derived from a geographical name is preferably an adjective and usually takes the termination *-ensis, -(a)nus, -inus,* or *-icus.*

Ex. 1. Rubus quebecensis L. H. Bailey (from Quebec), *Ostrya virginiana* (Mill.) K. Koch (from Virginia), *Eryngium amorginum* Rech. f. (from Amorgos), *Fraxinus pennsylvanica* Marshall (from Pennsylvania).

Recommendation 60E

60E.1. The epithet in a new name should be written in conformity with the customary spelling of the word or words from which it is derived and in accordance with the accepted usage of Latin and latinization (see also Art. 23.5).

Ex. 1. *sinensis* (not *chinensis*).

Recommendation 60F

60F.1. All specific and infraspecific epithets should be written with an initial lower-case letter.

Recommendation 60G

60G.1. A compound name or an epithet which combines elements derived from two or more Greek or Latin words should be formed, as far as practicable, in accordance with classical usage. This may be stated as follows (see also Note 1):

(a) In a regular compound, a noun or adjective in non-final position appears as a compounding form generally obtained by

(1) removing the case ending of the genitive singular (Latin *-ae, -i, -us, -is;* transliterated Greek *-ou, -os, -es, -as, -ous* and its equivalent *-eos*) and

(2) before a consonant, adding a connecting vowel (*-i-* for Latin elements, *-o-* for Greek elements).

(3) Exceptions are common, and one should review earlier usages of a particular compounding form.

Ex. 1. The following are examples of the formation of a compound epithet derived from a generic name and another Greek or Latin word. The epithet meaning "having leaves like those of *Myrica*" is *myricifolia* (*Myric-*, connecting vowel *-i-* and ending *-folia*). The epithets *aquilegifolia* and *aquilegiaefolia* derived from the name *Aquilegia* must be changed to *aquilegiifolia* (*Aquilegi-*, connecting vowel *-i-* and ending *-folia*).

(b) In a pseudocompound, a noun or adjective in a non-final position appears as a word with a case ending, not as a modified stem. Examples are: *nidus-avis* (nest of bird), *Myos-otis* (ear of mouse), *albo-marginatus* (margined with white), etc. In epithets where tingeing is expressed, the modifying initial colour often is in the ablative because the preposition *e, ex,* is implicit, e.g. *atropurpureus* (blackish purple) from "ex atro purpureus" (purple tinged with black). Others have been deliberately introduced to reveal etymological differences when different word elements have the same compounding forms, such as *tubi-* from tube *(tubus, tubi)* or from trumpet *(tuba, tubae)* where *tubaeflorus* can only mean trumpet-flowered; also *carici-* is the compounding form from both papaya *(carica, caricae)* and sedge *(carex, caricis)* where *caricaefolius* can only mean papaya-leaved. The latter use of the genitive singular of the first declension for pseudocompounding is treated as an error to be corrected unless it makes an etymological distinction (see Art. 60.8).

Note 1. In forming some other apparently irregular compounds, classical usage is commonly followed.

Ex. 2. The compounding forms *hydro-* and *hydr- (Hydro-phyllum)* stem from water *(hydor, hydatos); calli- (Calli-stemon)* derive from the adjective beautiful *(kalos);* and *meli- (Meli-osma, Meli-lotus)* stem from honey *(mel, melitos).*

Note 2. The hyphens in the above examples are given solely for explanatory reasons. For the use of hyphens in generic names and in epithets see Art. 20.3, 23.1, and 60.9.

Recommendation 60H

60H.1. The etymology of new names or of epithets in new names should be given, especially when their meaning is not obvious.

Article 61

61.1. Only one orthographical variant of any one name is treated as validly published: the form that appears in the original publication, except as provided in Art. 60 (typographical or orthographical errors and standardizations), Art. 14.11 (conserved spellings), and Art. 32.7 (improper Latin terminations).

61.2. For the purpose of this *Code,* orthographical variants are the various spelling, compounding, and inflectional forms of a name or its final epithet (including typographical errors), only one nomenclatural type being involved.

61.3. If orthographical variants of a name appear in the original publication, the one that conforms to the rules and best suits the recommendations of Art. 60 is to be retained; otherwise the first author who, in an effectively published text (Art. 29-31), explicitly adopts one of the variants and rejects the other(s) must be followed.

61.4. The orthographical variants of a name are to be corrected to the validly published form of that name. Whenever such a variant appears in print, it is to be treated as if it were printed in its corrected form.

Note 1. In full citations it is desirable that the original form of a corrected orthographical variant of a name be added (Rec. 50F).

61.5. Confusingly similar names based on the same type are treated as orthographical variants. (For confusingly similar names based on different types, see Art. 53.3-5.)

Ex. 1. *"Geaster"* (Fries, 1829) and *Geastrum* Pers. (1794) : Pers. (1801) are similar names with the same type (see Taxon 33: 498. 1984); they are treated as orthographical variants despite the fact that they are derived from two different nouns, *aster (asteris)* and *astrum (astri).*

SECTION 2. GENDER

Article 62

62.1. A generic name retains the gender assigned by botanical tradition, irrespective of classical usage or the author's original usage. A generic name without a botanical tradition retains the gender assigned by its author (but see Art. 62.4).

Note 1. Botanical tradition usually maintains the classical gender of a Greek or Latin word, when this was well established.

*Ex. 1. In accordance with botanical tradition, *Adonis* L., *Atriplex* L., *Diospyros* L., *Hemerocallis* L., *Orchis* L., *Stachys* L., and *Strychnos* L. must be treated as feminine while *Lotus* L. and *Melilotus* Mill. must be treated as masculine. *Eucalyptus* L'Hér., which lacks a botanical tradition, retains the feminine gender assigned by its author. Although their ending suggests masculine gender, *Cedrus* Trew and *Fagus* L., like most other classical tree names, were traditionally treated as feminine and thus retain that gender; similarly, *Rhamnus* L. is feminine, despite the fact that Linnaeus assigned it masculine gender. *Phyteuma* L. (n), *Sicyos* L. (m), and *Erigeron* L. (m) are other names for which botanical tradition has re-established the classical gender despite another choice by Linnaeus.

62.2. Compound generic names take the gender of the last word in the nominative case in the compound. If the termination is altered, however, the gender is altered accordingly.

Ex. 2. Irrespective of the fact that *Parasitaxus* de Laub. (1972) was treated as masculine when published, its gender is feminine: it is a compound of which the last part coincides with the generic name *Taxus* L., which is feminine by botanical tradition (Art. 62.1).

Ex. 3. Compound generic names in which the termination of the last word is altered: *Stenocarpus* R. Br., *Dipterocarpus* C. F. Gaertn., and all other compounds ending in the Greek masculine *-carpos* (or *-carpus*), e.g. *Hymenocarpos* Savi, are masculine; those in *-carpa* or *-carpaea,* however, are feminine, e.g. *Callicarpa* L. and *Polycarpaea* Lam.; and those in *-carpon, -carpum,* or *-carpium* are neuter, e.g. *Polycarpon* L., *Ormocarpum* P. Beauv., and *Pisocarpium* Link.

(a) Compounds ending in *-botrys, -codon, -myces, -odon, -panax, -pogon, -stemon,* and other masculine words, are masculine.

Ex. 4. Irrespective of the fact that the generic names *Andropogon* L. and *Oplopanax* (Torr. & A. Gray) Miq. were originally treated as neuter by their authors, they are masculine.

(b) Compounds ending in *-achne, -chlamys, -daphne, -glochin, -mecon, -osma* (the modern transcription of the feminine Greek word οσμή, *osmē*), and other feminine words, are feminine. An exception is made in the case of names ending in *-gaster,* which strictly speaking ought to be feminine, but which are treated as masculine in accordance with botanical tradition.

Ex. 5. Irrespective of the fact that *Tetraglochin* Poepp., *Triglochin* L., *Dendromecon* Benth., and *Hesperomecon* Greene were originally treated as neuter, they are feminine.

(c) Compounds ending in *-ceras, -dendron, -nema, -stigma, -stoma,* and other neuter words, are neuter. An exception is made for names ending in *-anthos* (or *-anthus*), *-chilos (-chilus* or *-cheilos),* and *-phykos (-phycos* or *-phycus),* which ought to be neuter, since that is the gender of the Greek words ἄνθος, *anthos,* χεῖλος, *cheilos,* and φῦκος, *phykos,* but are treated as masculine in accordance with botanical tradition.

Ex. 6. Irrespective of the fact that *Aceras* R. Br. and *Xanthoceras* Bunge were treated as feminine when first published, they are neuter.

62.3. Arbitrarily formed generic names or vernacular names or adjectives used as generic names, of which the gender is not apparent, take the gender assigned to them by their authors. If the original author failed to indicate the gender, the next subsequent author may choose a gender, and that choice, if effectively published (Art. 29-31), is to be accepted.

Ex. 7. Taonabo Aubl. (1775) is feminine because Aublet's two species were *T. dentata* and *T. punctata.*

Ex. 8. Agati Adans. (1763) was published without indication of gender; feminine gender was assigned to it by Desvaux (in J. Bot. Agric. 1: 120. 1813), who was the first subsequent author to adopt the name in an effectively published text, and his choice is to be accepted.

Ex. 9. The original gender of *Manihot* Mill. (1754), as apparent from some of the species polynomials, was feminine, and *Manihot* is therefore to be treated as feminine.

62.4. Generic names ending in *-anthes, -oides* or *-odes* are treated as feminine and those ending in *-ites* as masculine, irrespective of the gender assigned to them by the original author.

Recommendation 62A

62A.1. When a genus is divided into two or more genera, the gender of the new generic name or names should be that of the generic name that is retained.

Ex. 1. When *Boletus* L. : Fr. is divided, the gender of the new generic names should be masculine: *Xerocomus* Quél. (1887), *Boletellus* Murrill (1909), etc.

DIVISION III. PROVISIONS FOR THE GOVERNANCE OF THE CODE

Div.III.1. The *Code* may be modified only by action of a plenary session of an International Botanical Congress on a resolution moved by the Nomenclature Section of that Congress[1].

Div.III.2. Permanent Nomenclature Committees are established under the auspices of the International Association for Plant Taxonomy. Members of these Committees are elected by an International Botanical Congress. The Committees have power to co-opt and to establish subcommittees; such officers as may be desired are elected.

(1) General Committee, composed of the secretaries of the other Committees, the rapporteur-général, the president and the secretary of the International Association for Plant Taxonomy, and at least 5 members to be appointed by the Nomenclature Section. The rapporteur-général is charged with the presentation of nomenclature proposals to the International Botanical Congress.

(2) Committee for Vascular Plants.

(3) Committee for Bryophyta.

(4) Committee for Fungi.

(5) Committee for Algae.

(6) Committee for Fossil Plants.

(7) Editorial Committee, charged with the preparation and publication of the *Code* in conformity with the decisions adopted by the International Botanical Congress. Chairman: the rapporteur-général of the previous Congress, who is charged with the general duties in connection with the editing of the *Code*.

---

[1] In the event that there should not be another International Botanical Congress, authority for the *International code of botanical nomenclature* shall be transferred to the International Union of Biological Sciences or to an organization at that time corresponding to it. The General Committee is empowered to define the machinery to achieve this.

Div.III.3. The Bureau of Nomenclature of the International Botanical Congress. Its officers are: *(1)* the president of the Nomenclature Section, elected by the organizing committee of the International Botanical Congress in question; *(2)* the recorder, appointed by the same organizing committee; *(3)* the rapporteur-général, elected by the previous Congress; *(4)* the vice-rapporteur, elected by the organizing committee on the proposal of the rapporteur-général.

Div.III.4. The voting on nomenclature proposals is of two kinds: *(a)* a preliminary guiding mail vote and *(b)* a final and binding vote at the Nomenclature Section of the International Botanical Congress.

Qualifications for voting:

(a) Preliminary mail vote:

> *(1)* The members of the International Association for Plant Taxonomy.
>
> *(2)* The authors of proposals.
>
> *(3)* The members of the Permanent Nomenclature Committees.

Note 1. No accumulation or transfer of personal votes is permissible.

(b) Final vote at the sessions of the Nomenclature Section:

> *(1)* All officially enrolled members of the Section. No accumulation or transfer of personal votes is permissible.
>
> *(2)* Official delegates or vice-delegates of the institutes appearing on a list drawn up by the Bureau of Nomenclature of the International Botanical Congress and submitted to the General Committee for final approval; such institutes are entitled to 1-7 votes, as specified on the list. No single institution, even in the wide sense of the term, is entitled to more than 7 votes. Transfer of institutional votes to specified vice-delegates is permissible, but no single person will be allowed more than 15 votes, personal vote included. Institutional votes may be deposited at the Bureau of Nomenclature to be counted in a specified way for specified proposals.[1]

---

[1] Prior to each International Botanical Congress any institution desiring to vote in the coming Nomenclature Section (and not listed as having been allocated a vote in the previous Nomenclature Section) should notify the Bureau of Nomenclature of the IBC of their wish to be allocated one or more votes and provide relevant information regarding the level of taxonomic activity in their institution.

APPENDIX I

NAMES OF HYBRIDS

Article H.1

H.1.1. Hybridity is indicated by the use of the multiplication sign × or by the addition of the prefix "notho-"[1] to the term denoting the rank of the taxon.

Article H.2

H.2.1. A hybrid between named taxa may be indicated by placing the multiplication sign between the names of the taxa; the whole expression is then called a hybrid formula.

Ex. 1. Agrostis L. × *Polypogon* Desf.; *Agrostis stolonifera* L. × *Polypogon monspeliensis* (L.) Desf.; *Salix aurita* L. × *S. caprea* L.; *Mentha aquatica* L. × *M. arvensis* L. × *M. spicata* L.; *Polypodium vulgare* subsp. *prionodes* (Asch.) Rothm. × subsp. *vulgare; Tilletia caries* (Bjerk.) Tul. × *T. foetida* (Wallr.) Liro.

Recommendation H.2A

H.2A.1. It is usually preferable to place the names or epithets in a formula in alphabetical order. The direction of a cross may be indicated by including the sexual symbols (♀: female; ♂: male) in the formula, or by placing the female parent first. If a non-alphabetical sequence is used, its basis should be clearly indicated.

Article H.3

H.3.1. Hybrids between representatives of two or more taxa may receive a name. For nomenclatural purposes, the hybrid nature of a taxon is indicated by placing the multiplication sign × before the name of an intergeneric hybrid or before the epithet in the name of an interspecific hybrid, or by prefixing the term "notho-" (optionally abbreviated "n-") to the term de-

---

[1] From the Greek νόθος, *nothos,* meaning hybrid.

noting the rank of the taxon (see Art. 3.2 and 4.4). All such taxa are designated nothotaxa.

Ex. 1. (The putative or known parentage is found in Art. H.2 Ex. 1.) ×*Agropogon* P. Fourn. (1934); ×*Agropogon littoralis* (Sm.) C. E. Hubb. (1946); *Salix* ×*capreola* Andersson (1867); *Mentha* ×*smithiana* R. A. Graham (1949); *Polypodium vulgare* nothosubsp. *mantoniae* (Rothm.) Schidlay (in Futák, Fl. Slov. 2: 225. 1966).

H.3.2. A nothotaxon cannot be designated unless at least one parental taxon is known or can be postulated.

H.3.3. For purposes of homonymy and synonymy the multiplication sign and the prefix "notho-" are disregarded.

Ex. 2. ×*Hordelymus* Bachteev & Darevsk. (1950) (= *Elymus* L. × *Hordeum* L.) is a later homonym of *Hordelymus* (Jess.) Harz (1885).

Note 1. Taxa which are believed to be of hybrid origin need not be designated as nothotaxa.

Ex. 3. The true-breeding tetraploid raised from the artificial cross *Digitalis grandiflora* L. × *D. purpurea* L. may, if desired, be referred to as *D. mertonensis* B. H. Buxton & C. D. Darl. (1931); *Triticum aestivum* L. (1753) is treated as a species although it is not found in nature and its genome has been shown to be composed of those of *T. dicoccoides* (Körn.) Körn., *T. speltoides* (Tausch) Gren. ex K. Richt., and *T. tauschii* (Coss.) Schmalh.; the taxon known as *Phlox divaricata* subsp. *laphamii* (A. W. Wood) Wherry (in Morris Arbor. Monogr. 3: 41. 1955) was believed by Levin (in Evolution 21: 92-108. 1967) to be a stabilized product of hybridization between *P. divaricata* L. subsp. *divaricata* and *P. pilosa* subsp. *ozarkana* Wherry; *Rosa canina* L. (1753), a polyploid believed to be of ancient hybrid origin, is treated as a species.

Recommendation H.3A

H.3A.1. The multiplication sign ×, indicating the hybrid nature of a taxon, should be placed so as to express that it belongs with the name or epithet but is not actually part of it. The exact amount of space, if any, between the multiplication sign and the initial letter of the name or epithet should depend on what best serves readability.

Note 1. The multiplication sign × in a hybrid formula is always placed between, and separate from, the names of the parents.

H.3A.2. If the multiplication sign is not available it should be approximated by a lower case letter "x" (not italicized).

Article H.4

H.4.1. When all the parent taxa can be postulated or are known, a nothotaxon is circumscribed so as to include all individuals (as far as they can be recognized) derived from the crossing of representatives of the stated parent taxa (i.e. not only the F_1 but subsequent filial generations and also back-crosses and combinations of these). There can thus be only one correct name corresponding to a particular hybrid formula; this is the earliest legitimate name (see Art. 6.3) in the appropriate rank (Art. H.5), and other names to which the same hybrid formula applies are synonyms of it.

Ex. 1. The names *Oenothera* ×*wienii* Renner ex Rostański (1977) and *O.* ×*drawertii* Renner ex Rostański (1966) are both considered to apply to the hybrid *O. biennis* L. × *O. villosa* Thunb.; the types of the two nothospecific names are known to differ by a whole gene complex; nevertheless, the later name is treated as a synonym of the earlier.

Note 1. Variation within nothospecies and nothotaxa of lower rank may be treated according to Art. H.12 or, if appropriate, according to the *International code of nomenclature for cultivated plants*.

Article H.5

H.5.1. The appropriate rank of a nothotaxon is that of the postulated or known parent taxa.

H.5.2. If the postulated or known parent taxa are of unequal rank the appropriate rank of the nothotaxon is the lowest of these ranks.

Note 1. When a taxon is designated by a name in a rank inappropriate to its hybrid formula, the name is incorrect in relation to that hybrid formula but may nevertheless be correct, or may become correct later (see also Art. 52 Note 3).

Ex. 1. The combination *Elymus* ×*laxus* (Fr.) Melderis & D. C. McClint. (1983), based on *Triticum laxum* Fr. (1842), was published for hybrids with the formula *E. farctus* subsp. *boreoatlanticus* (Simonet & Guin.) Melderis × *E. repens* (L.) Gould, so that the combination is in a rank inappropriate to the hybrid formula. It is, however, the correct name applicable to all hybrids between *E. farctus* (Viv.) Melderis and *E. repens*.

Ex. 2. Radcliffe-Smith incorrectly published the nothospecific name *Euphorbia* ×*cornubiensis* Radcl.-Sm. (1985) for *E. amygdaloides* L. × *E. characias* subsp. *wulfenii* (W. D. J. Koch) Radcl.-Sm., although the correct designation for hybrids between *E. amygdaloides* and *E. characias* L. is *E.* ×*martini* Rouy (1900); later, he remedied his mistake by publishing the combination *E.* ×*martini* nothosubsp. *cornubiensis* (Radcl.-Sm.) Radcl.-Sm. (in Taxon 35: 349. 1986). However, the name *E.* ×*cornubiensis* is potentially correct for hybrids with the formula *E. amygdaloides* × *E. wulfenii* W. D. J. Koch.

Recommendation H.5A

H.5A.1. When publishing a name of a new nothotaxon at the rank of species or below, authors should provide any available information on the taxonomic identity, at lower ranks, of the known or postulated parent plants of the type of the name.

Article H.6

H.6.1. A nothogeneric name (i.e. the name at generic rank for a hybrid between representatives of two or more genera) is a condensed formula or is equivalent to a condensed formula (but see Art. 11.9).

H.6.2. The nothogeneric name of a bigeneric hybrid is a condensed formula in which the names adopted for the parental genera are combined into a single word, using the first part or the whole of one, the last part or the whole of the other (but not the whole of both) and, optionally, a connecting vowel.

Ex. 1. ×*Agropogon* P. Fourn. (1934) (= *Agrostis* L. × *Polypogon* Desf.); ×*Gymnanacamptis* Asch. & Graebn. (1907) (= *Anacamptis* Rich. × *Gymnadenia* R. Br.); ×*Cupressocyparis* Dallim. (1938) (= *Chamaecyparis* Spach × *Cupressus* L.); ×*Seleniphyllum* G. D. Rowley (1962) (= *Epiphyllum* Haw. × *Selenicereus* (A. Berger) Britton & Rose).

Ex. 2. ×*Amarcrinum* Coutts (1925) is correct for *Amaryllis* L. × *Crinum* L., not "×*Crindonna*". The latter formula was proposed by Ragionieri (1921) for the same nothogenus, but was formed from the generic name adopted for one parent *(Crinum)* and a synonym *(Belladonna* Sweet) of the generic name adopted for the other *(Amaryllis)*. Being contrary to Art. H.6, it is not validly published under Art. 32.1(c).

Ex. 3. The name ×*Leucadenia* Schltr. (1919) is correct for *Leucorchis* E. Mey. × *Gymnadenia* R. Br., but if the generic name *Pseudorchis* Ség. is adopted instead of *Leucorchis*, ×*Pseudadenia* P. F. Hunt (1971) is correct.

Ex. 4. Boivin (1967) published ×*Maltea* for what he considered to be the intergeneric hybrid *Phippsia* (Trin.) R. Br. × *Puccinellia* Parl. As this is not a condensed formula, the name cannot be used for that intergeneric hybrid, for which the correct name is ×*Pucciphippsia* Tzvelev (1971). Boivin did, however, provide a Latin description and designate a type; consequently, *Maltea* B. Boivin is a validly published generic name and is correct if its type is treated as belonging to a separate genus, not to a nothogenus.

H.6.3. The nothogeneric name of an intergeneric hybrid derived from four or more genera is formed from the name of a person to which is added the termination -*ara;* no such name may exceed eight syllables. Such a name is regarded as a condensed formula.

Ex. 5. ×*Beallara* Moir (1970) (= *Brassia* R. Br. × *Cochlioda* Lindl. × *Miltonia* Lindl. × *Odontoglossum* Kunth).

H.6.4. The nothogeneric name of a trigeneric hybrid is either *(a)* a condensed formula in which the three names adopted for the parental genera are combined into a single word not exceeding eight syllables, using the whole or first part of one, followed by the whole or any part of another, followed by the whole or last part of the third (but not the whole of all three) and, optionally, one or two connecting vowels, or *(b)* a name formed like that of a nothogenus derived from four or more genera, i.e. from a personal name to which is added the termination -*ara.*

Ex. 6. ×*Sophrolaeliocattleya* Hurst (1898) (= *Cattleya* Lindl. × *Laelia* Lindl. × *Sophronitis* Lindl.); ×*Vascostylis* Takakura (1964) (= *Ascocentrum* Schltr. ex J. J. Sm. × *Rhynchostylis* Blume × *Vanda* W. Jones ex R. Br.); ×*Rodrettiopsis* Moir (1976) (= *Comparettia* Poepp. & Endl. × *Ionopsis* Kunth × *Rodriguezia* Ruiz & Pav.); ×*Devereuxara* Kirsch (1970) (= *Ascocentrum* Schltr. ex J. J. Sm. × *Phalaenopsis* Blume × *Vanda* W. Jones ex R. Br.).

Recommendation H.6A

H.6A.1. When a nothogeneric name is formed from the name of a person by adding the termination *-ara,* that person should preferably be a collector, grower, or student of the group.

Article H.7

H.7.1. The name of a nothotaxon which is a hybrid between subdivisions of a genus is a combination of an epithet, which is a condensed formula formed in the same way as a nothogeneric name (Art. H.6.2), with the name of the genus.

Ex. 1. *Ptilostemon* nothosect. *Platon* Greuter (in Boissiera 22: 159. 1973), comprising hybrids between *P.* sect. *Platyrhaphium* Greuter and *P.* sect. *Ptilostemon; P.* nothosect. *Plinia* Greuter (in Boissiera 22: 158. 1973), comprising hybrids between *P.* sect. *Platyrhaphium* and *P.* sect. *Cassinia* Greuter.

Article H.8

H.8.1. When the name or the epithet in the name of a nothotaxon is a condensed formula (Art. H.6 and H.7), the parental names used in its formation must be those which are correct for the particular circumscription, position, and rank accepted for the parental taxa.

Ex. 1. If the genus *Triticum* L. is interpreted on taxonomic grounds as including *Triticum* (s. str.) and *Agropyron* Gaertn., and the genus *Hordeum* L. as including *Hordeum* (s. str.) and *Elymus* L., then hybrids between *Agropyron* and *Elymus* as well as between *Triticum* (s. str.) and *Hordeum* (s. str.) are placed in the same nothogenus, ×*Tritordeum* Asch. & Graebn. (1902). If, however, *Agropyron* is separated generically from *Triticum,* hybrids between *Agropyron* and *Hordeum* (s. str. or s. lat.) are placed in the nothogenus ×*Agrohordeum* A. Camus (1927). Similarly, if *Elymus* is separated generically from *Hordeum,* hybrids between *Elymus* and *Triticum* (s. str. or s. lat.) are placed in the nothogenus ×*Elymotriticum* P. Fourn. (1935). If both *Agropyron* and *Elymus* are given generic rank, hybrids between them are placed in the nothogenus ×*Agroelymus* A. Camus (1927); ×*Tritordeum* is then restricted to hybrids between *Hordeum* (s. str.) and *Triticum* (s. str.), and hybrids between *Elymus* and *Hordeum* are placed in ×*Elyhordeum* Mansf. ex Tsitsin & Petrova (1955), a substitute name for ×*Hordelymus* Bachteev & Darevsk. (1950) non *Hordelymus* (Jess.)Harz (1885).

H.8.2. Names ending in *-ara* for nothogenera, which are equivalent to condensed formulae (Art. H.6.3-4), are applicable only to plants which are accepted taxonomically as derived from the parents named.

Ex. 2. If *Euanthe* Schltr. is recognized as a distinct genus, hybrids simultaneously involving its only species, *E. sanderiana* (Rchb.) Schltr., and the three genera *Arachnis* Blume, *Renanthera* Lour., and *Vanda* W. Jones ex R. Br. must be placed in ×*Cogniauxara* Garay & H. R. Sweet (1966); if, on the other hand, *E. sanderiana* is included in *Vanda,* the same hybrids are placed in ×*Holttumara* Holttum (1958) *(Arachnis × Renanthera × Vanda).*

Article H.9

H.9.1. In order to be validly published, the name of a nothogenus or of a nothotaxon with the rank of subdivision of a genus (Art. H.6 and H.7) must be effectively published (see Art. 29-31) with a statement of the names of the parent genera or subdivisions of genera, but no description or diagnosis is necessary, whether in Latin or in any other language.

Ex. 1. Validly published names: ×*Philageria* Mast. (1872), published with a statement of parentage, *Lapageria* Ruiz & Pav. × *Philesia* Comm. ex Juss.; *Eryngium* nothosect. *Alpestria* Burdet & Miège, pro sect. (in Candollea 23: 116. 1968), published with a statement of its parentage, *E.* sect. *Alpina* H. Wolff × *E.* sect. *Campestria* H. Wolff; ×*Agrohordeum* A. Camus (1927) (= *Agropyron* Gaertn. × *Hordeum* L.), of which ×*Hordeopyron* Simonet (1935, *"Hordeopyrum"*) is a later synonym.

Note 1. Since the names of nothogenera and nothotaxa with the rank of a subdivision of a genus are condensed formulae or treated as such, they do not have types.

Ex. 2. The name ×*Ericalluna* Krüssm. (1960) was published for plants which were thought to be the product of the cross *Calluna vulgaris* (L.) Hull × *Erica cinerea* L. If it is considered that these are not hybrids, but are variants of *E. cinerea,* the name ×*Ericalluna* Krüssm. remains available for use if and when known or postulated plants of *Calluna* Salisb. × *Erica* L. should appear.

Ex. 3. ×*Arabidobrassica* Gleba & Fr. Hoffm. (in Naturwissenschaften 66: 548. 1979), a nothogeneric name which was validly published with a statement of parentage for the result of somatic hybridization by protoplast fusion of *Arabidopsis thaliana* (L.) Heynh. with *Brassica campestris* L., is also available for intergeneric hybrids resulting from normal crosses between *Arabidopsis* Heynh. and *Brassica* L., should any be produced.

Note 2. However, names published merely in anticipation of the existence of a hybrid are not validly published under Art. 34.1(b).

Article H.10

H.10.1. Names of nothotaxa at the rank of species or below must conform with the provisions *(a)* in the body of the *Code* applicable to the same ranks (see Art. 40.1) and *(b)* in Art. H.3. Infringements of Art. H.3.1 are treated as errors to be corrected (see also Art. 11.9).

Ex. 1. The nothospecies name *Melampsora* ×*columbiana* G. Newc. (in Mycol. Res. 104: 271. 2000) was validly published, with a Latin description and designation of a holotype, for the hybrid between *M. medusae* Thüm. and *M. occidentalis* H. S. Jacks.

H.10.2. Taxa previously published as species or infraspecific taxa which are later considered to be nothotaxa may be indicated as such, without change of rank, in conformity with Art. 3 and 4 and by the application of Art. 50 (which also operates in the reverse direction).

H.10.3. The following are considered to be formulae and not true epithets: designations consisting of the epithets of the names of the parents combined in unaltered form by a hyphen, or with only the termination of one epithet changed, or consisting of the specific epithet of the name of one parent combined with the generic name of the other (with or without change of termination).

Ex. 2. The designation *Potentilla "atrosanguinea-pedata"* published by Maund (in Bot. Gard. 5: No. 385, t. 97. 1833) is considered to be a formula meaning *P. atrosanguinea* Lodd. ex D. Don × *P. pedata* Nestl.

Ex. 3. *Verbascum "nigro-lychnitis"* (Schiede, Pl. Hybr.: 40. 1825) is considered to be a formula, *V. lychnitis* L. × *V. nigrum* L.; the correct binary name for this hybrid is *V. ×schiedeanum* W. D. J. Koch (1844).

Ex. 4. The following names include true epithets (but see Rec. H.10A): *Acaena ×anserovina* Orchard (1969) (from *A. anserinifolia* (J. R. Forst. & G. Forst.) J. Armstr. and *A. ovina* A. Cunn.); *Micromeria ×benthamineolens* Svent. (1969) (from *M. benthamii* Webb & Berthel. and *M. pineolens* Svent.).

Note 1. Since the name of a nothotaxon at the rank of species or below has a type, statements of parentage play a secondary part in determining the application of the name.

Ex. 5. *Quercus ×deamii* Trel. (in Mem. Natl. Acad. Sci. 20: 14. 1924) when described was considered as the cross *Q. alba* L. × *Q. muehlenbergii* Engelm. However, progeny grown from acorns from the tree from which the type originated led Bartlett to conclude that the parents were in fact *Q. macrocarpa* Michx. and *Q. muehlenbergii.* If this conclusion is accepted, the name *Q. ×deamii* applies to *Q. macrocarpa* × *Q. muehlenbergii,* and not to *Q. alba* × *Q. muehlenbergii.*

Recommendation H.10A

H.10A.1. In forming epithets for names of nothotaxa at the rank of species and below, authors should avoid combining parts of the epithets of the names of the parents.

Recommendation H.10B

H.10B.1. When contemplating the publication of new names for hybrids between named infraspecific taxa, authors should carefully consider whether they are really needed, bearing in mind that formulae, though more cumbersome, are more informative.

Article H.11

H.11.1. The name of a nothospecies of which the postulated or known parent species belong to different genera is a combination of a nothospecific epithet with a nothogeneric name.

Ex. 1. ×*Heucherella tiarelloides* (Lemoine & É. Lemoine) H. R. Wehrh. is considered to have originated from the cross between a garden hybrid of *Heuchera* L. and *Tiarella cordifolia* L. (see Stearn in Bot. Mag. 165: ad t. 31. 1948). Its original name, *Heuchera* ×*tiarelloides* Lemoine & É. Lemoine (1912), is therefore incorrect.

Ex. 2. When *Orchis fuchsii* Druce was renamed *Dactylorhiza fuchsii* (Druce) Soó the name for its hybrid with *Coeloglossum viride* (L.) Hartm., ×*Orchicoeloglossum mixtum* Asch. & Graebn. (1907), became the basis of the necessary new combination ×*Dactyloglossum mixtum* (Asch. & Graebn.) Rauschert (1969).

H.11.2. The final epithet in the name of an infraspecific nothotaxon of which the postulated or known parental taxa are assigned to different species, may be placed subordinate to the name of a nothospecies (but see Rec. H.10B).

Ex. 3. *Mentha* ×*piperita* L. nothosubsp. *piperita* (= *M. aquatica* L. × *M. spicata* L. subsp. *spicata*); *Mentha* ×*piperita* nothosubsp. *pyramidalis* (Ten.) Harley (in Kew Bull. 37: 604. 1983) (= *M. aquatica* L. × *M. spicata* subsp. *tomentosa* (Briq.) Harley).

Article H.12

H.12.1. Subordinate taxa within nothospecies may be recognized without an obligation to specify parent taxa at the subordinate rank. In this case non-hybrid infraspecific categories of the appropriate rank are used.

Ex. 1. *Mentha* ×*piperita* f. *hirsuta* Sole; *Populus* ×*canadensis* var. *serotina* (R. Hartig) Rehder and *P.* ×*canadensis* var. *marilandica* (Poir.) Rehder (see also Art. H.4 Note 1).

Note 1. As there is no statement of parentage at the rank concerned there is no control of circumscription at this rank by parentage (compare Art. H.4).

Note 2. It is not feasible to treat subdivisions of nothospecies by the methods of both Art. H.10 and H.12.1 at the same rank.

H.12.2. Names published at the rank of nothomorph[1] are treated as having been published as names of varieties (see Art. 50).

---

[1] Pre-Sydney editions of the *Code* permitted only one rank under provisions equivalent to Art. H.12. That rank was equivalent to variety and the category was termed "nothomorph".

APPENDIX IIA

NOMINA FAMILIARUM ALGARUM, FUNGORUM, PTERIDO-PHYTORUM, ET FOSSILIUM CONSERVANDA ET REJICIENDA

In the following lists the nomina conservanda, arranged in alphabetical sequence within the major groups, have been inserted in the left column, in **bold-face italics**. Synonyms and earlier homonyms (nomina rejicienda) are listed in the right column.

(H) homonym (Art. 14.10; see also Art. 53), only the earliest being listed.

(=) taxonomic synonym (i.e. heterotypic synonym, based on a type different from that of the conserved name), to be rejected only in favour of the conserved name (Art. 14.6 and 14.7).

One name listed as conserved has no corresponding nomen rejiciendum because its conservation is no longer necessary (Art. 52.3; see Art. 14.13).

A. ALGAE

Acrochaetiaceae Melch. in Melchior & Werdermann (eds.), Engler's Syllabus, ed. 12, 1: 130. Mai-Dec 1954 [*Rhodoph.*].
Typus: *Acrochaetium* Nägeli

(=) *Rhodochortaceae* Nasr in Bull. Fac. Sci. Egypt. Univ. 26: 92. 1947.
Typus: *Rhodochorton* Nägeli

Bangiaceae Reinbold in Schriften Naturwiss. Vereins Schleswig-Holstein 9: 113. 1891 [*Rhodoph.*].
Typus: *Bangia* Lyngb.

(=) *Porphyraceae* Kütz., Phycol. General.: 382. 14-16 Sep 1843.
Typus: *Porphyra* C. Agardh (nom. cons.).

Chromulinaceae Engl. in Engler & Prantl, Nat. Pflanzenfam. 1(2): 570. Dec 1897 [*Chrysoph.*].
Typus: *Chromulina* Cienk.

(=) *Chrysomonadaceae* F. Stein, Organism. Infusionsthiere 3(1): x, 152. 1878.
Typus: *Chrysomonas* F. Stein

Cladophoraceae Wille in Warming, Haandb. Syst. Bot., ed. 2: 30. 1884 [*Chloroph.*].
Typus: *Cladophora* Kütz. (nom. cons.).

(=) *Pithophoraceae* Wittr. in Nova Acta Regiae Soc. Sci. Upsal., ser. 3, vol. extraord. (19): 47. 1877.
Typus: *Pithophora* Wittr.

(=) *Confervaceae* Dumort., Comment. Bot.: 71, 96. Nov (sero) - Dec (prim.) 1822.
Typus: *Conferva* L.

Euglenaceae Dujard., Hist. Nat. Zoophyt.: 347. 1841 [*Euglenoph.*].
Typus: *Euglena* Ehrenb.

(=) *Astasiaceae* Ehrenb., Symb. Phys., Zool.: [32]. 1831.
Typus: *Astasia* Ehrenb. (nom. rej.).

Eupodiscaceae Ralfs in Pritchard, Hist. Infus., ed. 4: 758, 842. 1861 [*Bacillarioph.*].
Typus: *Eupodiscus* Bailey (nom. cons.).

(H) *Eupodiscaceae* Kütz., Sp. Alg.: 134. 23-24 Jul 1849.
Typus: *Eupodiscus* Ehrenb. (nom. rej.).

Isochrysidaceae Bourrelly in Rev. Algol., Mém. Hors-Sér. 1: 227, 228. Oct-Dec 1957 [*Prymnesioph./Haptoph.*].
Typus: *Isochrysis* Parke

(=) *Ruttneraceae* Geitler in Int. Rev. Gesamten Hydrobiol. Hydrogr. 43: 108. 1943.
Typus: *Ruttnera* Geitler

Lomentariaceae Willkomm, Anleit. Stud. Bot. 2: 147. Feb 1854 [*Rhodoph.*].
Typus: *Lomentaria* Lyngb.

(=) *Chondrosiphonaceae* Kütz., Phycol. General.: 438. 14-16 Sep 1843.
Typus: *Chondrosiphon* Kütz.

Nemastomataceae F. Schmitz in Engler, Syllabus: 22. Apr 1892 [*Rhodoph.*].
Typus: *Nemastoma* J. Agardh (nom. cons.).

(=) *Gymnophlaeaceae* Kütz., Phycol. General.: 389, 390. 14-16 Sep 1843.
Typus: *Gymnophlaea* Kütz.

Ochromonadaceae Lemmerm. in Forschungsber. Biol. Stat. Plön 7: 105. 1899 [*Chrysoph.*].
Typus: *Ochromonas* Vysotskij

(=) *Dendromonadaceae* F. Stein, Organism. Infusionsthiere 3(1): x, 153. 1878.
Typus: *Dendromonas* F. Stein

(=) *Spumellaceae* Kent, Man. Infus.: 231. 1880.
Typus: *Spumella* Cienk.

Oscillatoriaceae Engl., Syllabus, ed. 2: 6. Mai 1898 [*Cyanoph.*].
Typus: *Oscillatoria* Vaucher ex Gomont

(=) *Lyngbyaceae* Hansg. in Nuova Notarisia 3: 1. 5 Jan 1892 (trib. *Lyngbyeae* Gomont in Ann. Sci. Nat., Bot., ser. 7, 15: 290. 1 Jan 1892).
Typus: *Lyngbya* C. Agardh ex Gomont

Plocamiaceae Kütz., Phycol. General.: 442, 449. 14-16 Sep 1843 [*Rhodoph.*].
Typus: *Plocamium* J. V. Lamour. (nom. cons.).

(=) *Thamnophoraceae* Decne. in Ann. Sci. Nat., Bot., ser. 2, 17: 359, 364. Jun 1842.
Typus: *Thamnophora* C. Agardh

Polyidaceae Kylin, Gatt. Rhodophyc.: 142, 166. 1956 [*Rhodoph.*].
Typus: *Polyides* C. Agardh

(=) *Spongiocarpaceae* Grev., Alg. Brit.: 68. Mar 1830.
Typus: *Spongiocarpus* Grev.

Protoperidiniaceae Balech in Publ. Espec. Inst. Esp. Oceanogr. 1: 77. 1988 (post 14 Nov) [*Dinoph.*].
Typus: *Protoperidinium* Bergh

(=) *Kolkwitziellaceae* Er. Lindem. in Engler & Prantl, Nat. Pflanzenfam., ed. 2, 2: 34, 71. 1928.
Typus: *Kolkwitziella* Er. Lindem.

(=) *Congruentidiaceae* J. Schiller in Rabenh. Krypt.-Fl., ed. 2, 10(3, 2): 320. Apr 1935.
Typus: *Congruentidium* T. H. Abé

(=) *Diplopsalidaceae* Matsuoka in Rev. Palaeobot. Palynol. 56: 98. Aug 1988.
Typus: *Diplopsalis* Bergh

Retortamonadaceae Wenrich in Trans. Amer. Microscop. Soc. 51: 233. Oct 1932 [*Trichomonadoph.*].
Typus: *Retortamonas* Grassi

(=) *Embadomonadaceae* A. G. Alexeev in Compt.-Rend. Séances Mém. Soc. Biol. 80: 358. 31 Mar 1917.
Typus: *Embadomonas* D. L. Mackinnon

(=) *Chilomastigaceae* Wenyon, Protozool.: 268, 287, 620. 1926.
Typus: *Chilomastix* A. G. Alexeev

Rhodomelaceae Horaninow, Char. Ess. Fam.: 238. 17 Jun 1847 [*Rhodoph.*].
Typus: *Rhodomela* C. Agardh (nom. cons.).

(=) *Rytiphlaeaceae* Decne. in Arch. Mus. Hist. Nat. 2: 161, 171. 1841.
Typus: *Rytiphlaea* C. Agardh

(=) *Heterocladiaceae* Decne. in Ann. Sci. Nat., Bot., ser. 2, 17: 359, 364. Jun 1842.
Typus: *Heterocladia* Decne.

(=) *Polyphacaceae* Decne. in Ann. Sci. Nat., Bot., ser. 2, 17: 359, 363. Jun 1842.
Typus: *Polyphacus* C. Agardh

(=) *Polysiphoniaceae* Kütz., Phycol. General.: 413, 416. 14-16 Sep 1843.
Typus: *Polysiphonia* Grev. (nom. cons.).

(=) *Amansiaceae* Kütz., Phycol. General.: 442, 447. 14-16 Sep 1843.
Typus: *Amansia* J. V. Lamour.

Siphonocladaceae Borzì, Stud. Algol. 1: 25. Jan-Jun 1883 [*Chloroph.*].
Typus: *Siphonocladus* F. Schmitz

Stigonemataceae Borzì in Nuova Notarisia 3: 43. 5 Apr 1892 (subtrib. *Stigonematinae* Bornet & Flahault in Ann. Sci. Nat., Bot., ser. 7, 5: 53. 1 Jan 1886) [*Cyanoph.*].
Typus: *Stigonema* C. Agardh ex Bornet & Flahault

(=) *Sirosiphonaceae* A. B. Frank in Leunis, Syn. Pflanzenk., ed. 3, 3: 215, 217. Aug 1886.
Typus: *Sirosiphon* Kütz. ex A. B. Frank

Tetrasporaceae Wittr. in Bih. Kongl. Svenska Vetensk.-Akad. Handl. 1(1): 28. 1872 [*Chloroph.*].
Typus: *Tetraspora* Link ex Desv.

(=) *Palmellaceae* Decne. in Ann. Sci. Nat., Bot., ser. 2, 17: 327, 333. Jun 1842.
Typus: *Palmella* Lyngb.

Thalassiosiraceae M. Lebour, Plankt. Diatoms N. Seas: 56. Mar 1930 [*Bacillarioph.*].
Typus: *Thalassiosira* Cleve

(=) *Lauderiaceae* Lemmerm. in Abh. Naturwiss. Vereine Bremen 16: 379. Sep 1899 (subtrib. *Lauderiinae* F. Schütt in Engler & Prantl, Nat. Pflanzenfam. 1(1b): 56, 82. Dec 1896).
Typus: *Lauderia* Cleve

(=) *Planktoniellaceae* Lemmerm. in Abh. Naturwiss. Vereine Bremen 16: 378. Sep 1899 (subtrib. *Planktoniellinae* F. Schütt in Engler & Prantl, Nat. Pflanzenfam. 1(1b): 55, 71. Dec 1896).
Typus: *Planktoniella* F. Schütt

Trentepohliaceae Hansg., Prodr. Algenfl. Böhmen 1: 37, 85. Jan-Apr 1886 [*Chloroph.*].
Typus: *Trentepohlia* Mart.

(=) *Byssaceae* Adans., Fam. Pl. 2: 1. Jul-Aug 1763.
Typus: *Byssus* L.

Vacuolariaceae Luther in Bih. Kongl. Svenska Vetensk. Akad. Handl. 24 (sect. 3, 13): 19. 1899 [*Raphidoph.*].
Typus: *Vacuolaria* Cienk.

(=) *Coelomonadaceae* Buetschli in Bronn, Kl. Ordn. Thier-Reichs 1: 819. 1884.
Typus: *Coelomonas* F. Stein

B. FUNGI

Corticiaceae Herter in Warnstorf & al., Krypt.-Fl. Brandenburg 6: 70. 30 Jan 1910.
Typus: *Corticium* Pers.

(=) *Cyphellaceae* Burnett, Outl. Bot.: 233. 1835.
Typus: *Cyphella* Fr. : Fr.

(=) *Peniophoraceae* Lotsy, Vortr. Bot. Stammesgesch. 1: 687, 689. Jan-Mar 1907.
Typus: *Peniophora* Cooke

(=) *Vuilleminiaceae* Maire ex Lotsy, Vortr. Bot. Stammesgesch. 1: 678. Jan-Mar 1907.
Typus: *Vuilleminia* Maire

Cortinariaceae R. Heim ex Pouzar in Česká Mykol. 37: 174. 28 Jul 1983.
Typus: *Cortinarius* (Pers.) Gray

(=) *Crepidotaceae* Singer in Lilloa 22: 584. 1951 (trib. *Crepidoteae* S. Imai in J. Fac. Agric. Hokkaido Univ. 43: 238. Aug 1938).
Typus: *Crepidotus* (Fr.) Staude

(=) *Galeropsidaceae* Singer in Bol. Soc. Argent. Bot. 10: 61. 15 Aug 1962.
Typus: *Galeropsis* Velen.

(=) *Thaxterogasteraceae* Singer in Bol. Soc. Argent. Bot. 10: 63. 1962.
Typus: *Thaxterogaster* Singer

(=) *Hebelomataceae* Locq., Fl. Mycol. 3: 146. 1977.
Typus: *Hebeloma* (Fr.) P. Kumm.

(=) *Inocybaceae* Jülich in Biblioth. Mycol. 85: 374. 1 Feb 1982.
Typus: *Inocybe* (Fr.) Fr.

(=) *Verrucosporaceae* Jülich in Biblioth. Mycol. 85: 393. 1 Feb 1982.
Typus: *Verrucospora* E. Horak

Gnomoniaceae G. Winter in Rabenh. Krypt.-Fl., ed. 2, 1(2): 334. Jul 1885.
Typus: *Gnomonia* Ces. & De Not.

(=) *Obryzaceae* Körber, Syst. Lich. Germ.: 427. Nov-Dec 1855.
Typus: *Obryzum* Wallr.

Helotiaceae Rehm in Rabenh. Krypt.-Fl., ed. 2, 1(3): 647. Jul 1892.
Typus: *Helotium* Pers., non Tode : Fr. [= *Cudoniella* Sacc.].

(=) *Bulgariaceae* Fr., Summa Veg. Scand.: 345, 357. 1849.
Typus: *Bulgaria* Fr. : Fr.

(=) *Cenangiaceae* Bonord., Handb. Mykol.: 26. 1851.
Typus: *Cenangium* Fr. : Fr.

(=) *Heterosphaeriaceae* Rehm in Rabenh. Krypt.-Fl., ed. 2, 1(3): 191, 198. Aug 1888.
Typus: *Heterosphaeria* Grev.

(=) *Cordieritaceae* Sacc., Syll. Fung. 8: 810. 20 Dec 1889.
Typus: *Cordierites* Mont.

Mycosphaerellaceae Lindau in Engler & Prantl, Nat. Pflanzenfam. 1(1): 421. Jun 1897.
Typus: *Mycosphaerella* Johanson

(=) *Ascosporaceae* Bonord., Handb. Mykol.: 62. 1851 (sero).
Typus: *Ascospora* Fr.

Physciaceae Zahlbr. A. Schneider, Text-book lichenol.: 151. 1897.
Typus: *Physcia* (Schreb.) Michx.

(=) *Caliciaceae* Chevall., Fl. Gén. Env. Paris I: 385. 5 Aug 1826.
Typus: *Calicium* Pers.

(=) *Pyxinaceae* J. W. Griff. & Henfr., Microgr. Dict., ed. 2: 422. Dec 1859.
Typus: *Pyxine* Fr.

Placynthiaceae E. Dahl in Meddel. Grønland 150(2): 49. 4 Feb 1950.
Typus: *Placynthium* (Ach.) Gray

(=) *Lecotheciaceae* Körb., Syst. Lich. Germ.: 397. Nov-Dec 1855.
Typus: *Lecothecium* Trevis.

(=) *Racoblennaceae* A. Massal. ex Stizenb. in Ber. Thätigk. St. Gallischen Naturwiss. Ges. 1861-1862: 142. 1862.
Typus: *Racoblenna* A. Massal.

Rhytismataceae Chevall., Fl. Gén. Env. Paris 1: 439. 5 Aug 1826.
Typus: *Rhytisma* Fr. : Fr.

(=) *Xylomataceae* Fr., Scleromyceti Sveciae 2: p. post titulum. 1820.
Typus: *Xyloma* Pers. : Fr.

Taphrinaceae Rostr. in Warming, Haandb. Syst. Bot., ed. 3: 141. 1891.
Typus: *Taphrina* Fr. : Fr.

(=) *Exoascaceae* G. Winter in Rabenh. Krypt.-Fl. 1(2): 3. Mar 1884.
Typus: *Exoascus* Fuckel

Trapeliaceae M. Choisy ex Hertel in Vortr. Gesamtgeb. Bot., ser. 2, 4: 181. 1970.
Typus: *Trapelia* M. Choisy

(=) *Saccomorphaceae* Elenkin in Ber. Biol. Süsswasser-Stat. Kaiserl. Naturf.-Ges. St. Petersburg 3: 193. 1912.
Typus: *Saccomorpha* Elenkin

Tricholomataceae R. Heim ex Pouzar in Česká Mykol. 37: 175. 28 Jul 1983.
Typus: *Tricholoma* (Fr.) Staude nom. cons.).

(=) *Hygrophoraceae* Lotsy, Vortr. Bot. Stammesgesch. 1: 706. Jan-Mar 1907.
Typus: *Hygrophorus* Fr.

(=) *Mycenaceae* Overeem, Icon. Fung. Malay. 14-15: 4. 1926.
Typus: *Mycena* (Pers.) Roussel

(=) *Hydnangiaceae* Gäum. & C. W. Dodge, Compar. Morph. Fungi: 485. 1928.
Typus: *Hydnangium* Wallr.

(=) *Physalacriaceae* Corner in Beih. Nova Hedwigia 33: 10. 1970.
Typus: *Physalacria* Peck

(=) *Amparoinaceae* Singer in Rev. Mycol. 40: 58. 1976.
Typus: *Amparoina* Singer

(=) *Dermolomataceae* Bon in Doc. Mycol. 9(35): 43. 1979.
Typus: *Dermoloma* (J. E. Lange) Herink

(=) *Macrocystidiaceae* Kühner in Bull. Mens. Soc. Linn. Lyon, Soc. Bot. 48: 172. Mar 1979.
Typus: *Macrocystidia* Joss.

(=) *Rhodotaceae* Kühner in Bull. Mens. Soc. Linn. Lyon, Soc. Bot. 49: 235. Apr 1980
Typus: *Rhodotus* Maire

(=) *Pleurotaceae* Kühner in Bull. Mens. Soc. Linn. Lyon, Soc. Bot. 49: 184. Mar 1980.
Typus: *Pleurotus* (Fr. : Fr.) P. Kumm. (nom. cons.).

(=) *Marasmiaceae* Kühner in Bull. Mens. Soc. Linn. Lyon, Soc. Bot. 49: 76. Feb 1980.
Typus: *Marasmius* Fr. (nom. cons.).

(=) *Hygrophoropsidaceae* Kühner in Bull. Mens. Soc. Linn. Lyon, Soc. Bot. 49: 414. Sep 1980.
Typus: *Hygrophoropsis* (J. Schröt.) Maire ex Martin-Sans

(=) *Biannulariaceae* Jülich in Biblioth. Mycol. 85: 356. 1 Feb 1982.
Typus: *Biannularia* Beck

(=) *Cyphellopsidaceae* Jülich in Biblioth. Mycol. 85: 362. 1 Feb 1982.
Typus: *Cyphellopsis* Donk

(=) *Fayodiaceae* Jülich in Biblioth. Mycol. 85: 367. 1 Feb 1982.
Typus: *Fayodia* Kühner

(=) *Laccariaceae* Jülich in Biblioth. Mycol. 85: 374. 1 Feb 1982.
Typus: *Laccaria* Berk. & Broome

(=) *Lentinaceae* Jülich in Biblioth. Mycol. 85: 376. 1 Feb 1982.
Typus: *Lentinus* Fr. : Fr.

(=) *Leucopaxillaceae* Jülich in Biblioth. Mycol. 85: 376. 1 Feb 1982.
Typus: *Leucopaxillus* Boursier
(=) *Lyophyllaceae* Jülich in Biblioth. Mycol. 85: 378. 1 Feb 1982.
Typus: *Lyophyllum* P. Karst.
(=) *Nyctalidaceae* Jülich in Biblioth. Mycol. 85: 381. 1 Feb 1982.
Typus: *Nyctalis* Fr.
(=) *Panellaceae* Jülich in Biblioth. Mycol. 85: 382. 1 Feb 1982.
Typus: *Panellus* P. Karst.
(=) *Resupinataceae* Jülich in Biblioth. Mycol. 85: 388. 1 Feb 1982.
Typus: *Resupinatus* (Nees) Gray
(=) *Squamanitaceae* Jülich in Biblioth. Mycol. 85: 390. 1 Feb 1982.
Typus: *Squamanita* Imbach
(=) *Termitomycetaceae* Jülich in Biblioth. Mycol. 85: 391. 1 Feb 1982.
Typus: *Termitomyces* R. Heim
(=) *Xerulaceae* Jülich in Biblioth. Mycol. 85: 394. 1 Feb 1982.
Typus: *Xerula* Maire

D. PTERIDOPHYTA

Adiantaceae Newman, Hist. Brit. Ferns: 5. 1-5 Feb 1840.
Typus: *Adiantum* L.

(=) *Parkeriaceae* Hook., Exot. Fl. 2: ad t. 147. Mar 1825.
Typus: *Parkeria* Hook.

Dicksoniaceae M. R. Schomb., Reis. Br.-Guiana 2: 1047. 7-10 Mar 1849 (*Dicksonieae* C. Presl [Tent. Pterid.] in Abh. Königl. Böhm. Ges. Wiss., ser. 4, 5: 133. 1836 [seors. impr.] ante 2 Dec]).
Typus: *Dicksonia* L'Hér.

(=) *Thyrsopteridaceae* C. Presl, Gefäß-bündel Farrn: 22, 38. 1847.
Typus: *Thyrsopteris* Kunze

Dryopteridaceae Herter in Revista Sudamer. Bot. 9: 15. Jun 1949.
Typus: *Dryopteris* Adans. (nom. cons.).

(=) *Peranemataceae* Ching in Sunyatsenia 5: 246. 30 Oct 1940 (nom. cons.).

Peranemataceae Ching in Sunyat-senia 5: 246. 30 Oct 1940 (trib. *Per-anemateae* C. Presl [Tent. Pterid.] in Abh. Königl. Böhm. Ges. Wiss., ser. 4, 5: 64. 1836 ([seors. impr.] ante 2 Dec)).
Typus: *Peranema* D. Don

(H)   *Peranemataceae* Buetschli in Bronn, Kl. Ordn. Thier-Reichs 1: 824. 1884, nom. illeg. [*Algae, Euglenoph.*].
Typus: *Peranema* Dujard., non D. Don

F. FOSSIL PLANTS (EXCL. DIATOMS)

Rhaetogonyaulacaceae G. Norris in Neues Jahrb. Geol. Paläontol., Abh. 155: 4. 12 May 1978 [*Dinoph.*].
Typus: *Rhaetogonyaulax* Sarjeant

(=)   *Shublikodiniaceae* V. D. Wiggins in Micropaleontology 19: 2. 4 Jun 1973.
Typus: *Shublikodinium* V. D. Wiggins

APPENDIX IIB

NOMINA FAMILIARUM BRYOPHYTORUM
ET SPERMATOPHYTORUM CONSERVANDA

The names of families printed in ***bold-face italics,*** arranged in alphabetical sequence within the major groups, are to be retained in all cases, with priority over unlisted synonyms (Art. 14.5) and homonyms (Art. 14.10).

When two listed names compete, the earlier must be retained unless the contrary is indicated or one of the competing names is listed in Art. 18.5. For any family including the type of an alternative family name, one or the other of these alternative names is to be used.

By a decision of the Vienna Congress (2005), the starting-point date (Art. 13.1) for suprageneric names in *Sphagnaceae, Hepaticae* and *Spermatophyta* is Jussieu's *Genera plantarum* (4 Aug 1789). For details of the changes arising from this and other decisions taken in Vienna see the Preface to this *Code* (p. xiii).

C. BRYOPHYTA

Bryoxiphiaceae Besch. in J. Bot. (Morot) 6: 183. 16 Mai 1892 [*Musci*].
Typus: *Bryoxiphium* Mitt. (nom. cons.).

Ditrichaceae Limpr. in Rabenh. Krypt.-Fl., ed. 2, 4(1): 482. Oct 1887 [*Musci*].
Typus: *Ditrichum* Hampe (nom. cons.).

Entodontaceae Kindb., Gen. Eur. N.-Amer. Bryin.: 7. 1897 [*Musci*].
Typus: *Entodon* Müll. Hal.

Ephemeraceae J. W. Griff. & Henfr., Microgr. Dict.: 235. 1 Apr 1855 [*Musci*].
Typus: *Ephemerum* Hampe (nom. cons.).

Eustichiaceae Broth. in Engler & Prantl, Nat. Pflanzenfam., ed. 2, 10: 420. Mai-Jun 1924 [*Musci*].
Typus: *Eustichia* Brid.

Fossombroniaceae Hazsl., Magyar Birodalom Moh-Flórája: 20, 36. 30 Mai - 9 Jun 1885 [*Hepat.*].
Typus: *Fossombronia* Raddi

Lejeuneaceae Rostovzev, Morfol. Sist. Pechen. Mkhov: 94. 4-11 Nov 1913 [*Hepat.*].
Typus: *Lejeunea* Lib. (nom. cons.).

Porellaceae Cavers in New Phytol. 9: 292. 1910 [*Hepat.*].
Typus: *Porella* L.

Pottiaceae Hampe in Bot. Zeitung (Berlin) 11: 329. 6 Mai 1853 (subtrib. *Pottiinae* Müll. Hal, Syn. Musc. Frond. 1: 546. Mar 1849) [*Musci*].
Typus: *Pottia* Ehrh. ex Fürnr. (nom. cons.).

Sematophyllaceae Broth. in Engler & Prantl, Nat. Pflanzenfam. 1(3): 706. 7 Mar 1905 [*Musci*].
Typus: *Sematophyllum* Mitt.

E. SPERMATOPHYTA

Abietaceae Gray, Nat. Arr. Brit. Pl. 2: 222, 223. 10 Jan 1822.
Typus: *Abies* Mill.

Note: If this family is united with *Pinaceae*, the name *Abietaceae* is rejected in favour of *Pinaceae*.

Acanthaceae Juss., Gen. Pl.: 102. 4 Aug 1789.
Typus: *Acanthus* L.

Aceraceae Juss., Gen. Pl.: 250. 4 Aug 1789.
Typus: *Acer* L.

Achariaceae Harms in Engler & Prantl, Nat. Pflanzenfam., Nachtr. 1: 256. 4 Oct 1897.
Typus: *Acharia* Thunb.

Achatocarpaceae Heimerl in Engler & Prantl, Nat. Pflanzenfam., ed. 2, 16c: 174. 28 Apr 1934.
Typus: *Achatocarpus* Triana

Actinidiaceae Gilg & Werderm. in Engler & Prantl, Nat. Pflanzenfam., ed. 2, 21: 36. 30 Jul 1925.
Typus: *Actinidia* Lindl.

Note: If this family is united with *Saurauiaceae*, the name *Actinidiaceae* must be used.

Adoxaceae E. Mey., Preuss. Pfl.-Gatt.: 198. 1-7 Sep 1839.
Typus: *Adoxa* L.

Aextoxicaceae Engl. & Gilg in Engler, Syllabus, ed. 8: 250. Jan-Feb 1920.
Typus: *Aextoxicon* Ruiz & Pav.

Agavaceae Dumort., Anal. Fam. Pl.: 57, 58. 1829.
Typus: *Agave* L.

Aizoaceae Martinov, Tekhno-Bot. Slovar: 15. 3 Aug 1820.
Typus: *Aizoon* L.

Akaniaceae Stapf in Bull. Misc. Inform. Kew 1912: 380. 13 Dec 1912.
Typus: *Akania* Hook. f.

Alangiaceae DC., Prodr. 3: 203. Mar (med.) 1828.
Typus: *Alangium* Lam. (nom. cons.).

Alismataceae Vent., Tabl. Règne Vég. 2: 157. 5 Mai 1799.
Typus: *Alisma* L.

Alliaceae Borkh., Bot. Wörterb. 1: 15. 1797.
Typus: *Allium* L.

Alsinaceae Bartl. in Bartling & Wendland, Beitr. Bot. 2: 159. Dec 1825.
Typus: *Alsine* L. [= *Stellaria* L.].

Alstroemeriaceae Dumort., Anal. Fam. Pl.: 57, 58. 1829.
Typus: *Alstroemeria* L.

Altingiaceae Lindl., Veg. Kingd.: 253. 14-28 Mar 1846.
Typus: *Altingia* Noronha

Amaranthaceae Juss., Gen. Pl.: 87. 4 Aug 1789.
Typus: *Amaranthus* L.

Amaryllidaceae J. St.-Hil., Expos. Fam. Nat. 1: 134. Feb-Apr 1805.
Typus: *Amaryllis* L. (nom. cons.).

Amborellaceae Pichon in Bull. Mus. Natl. Hist. Nat., sér. 2, 20: 384. 25 Oct 1948.
Typus: *Amborella* Baill.

Ambrosiaceae Bercht. & J. Presl, Přir. Rostlin: 254. Jan-Apr 1820.
Typus: *Ambrosia* L.

Amygdalaceae Marquis, Esq. Règne Vég.: 49. 15-22 Jul 1820.
Typus: *Amygdalus* L.

Anacardiaceae R. Br. in Tuckey, Narr. Exped. Zaire: 431. 5 Mar 1818.
Typus: *Anacardium* L.

Ancistrocladaceae Planch. ex Walp. in Ann. Bot. Syst. 2: 175. 15-16 Dec 1851.
Typus: *Ancistrocladus* Wall. (nom. cons.).

Annonaceae Juss., Gen. Pl.: 283. 4 Aug 1789.
Typus: *Annona* L.

Apiaceae Lindl., Intr. Nat. Syst. Bot., ed. 2: 21. 13 Jun 1836; nom. alt.: *Umbelliferae.*
Typus: *Apium* L.

Apocynaceae Juss., Gen. Pl.: 143. 4 Aug 1789.
Typus: *Apocynum* L.

Aponogetonaceae Planch. in Bot. Mag.: ad t. 4894. 1 Jan 1856.
Typus: *Aponogeton* L. f. (nom. cons.).

Apostasiaceae Lindl., Nix. Pl.: 22. 17 Sep 1833.
Typus: *Apostasia* Blume

Aquifoliaceae Bercht. & J. Presl, Přir. Rostlin 2(109*): [438], 440. 1825.
Typus: *Aquifolium* Mill., nom. illeg. (*Ilex* L.).

Araceae Juss., Gen. Pl.: 23. 4 Aug 1789.
Typus: *Arum* L.

Araliaceae Juss., Gen. Pl.: 217. 4 Aug 1789.
Typus: *Aralia* L.

Araucariaceae Henkel & W. Hochst., Syn. Nadelhölz.: xvii, 1. 17-18 Jan 1865.
Typus: *Araucaria* Juss.

Arecaceae Bercht. & J. Presl, Přir. Rostlin: 266. Jan-Apr 1820; nom. alt.: *Palmae.*
Typus: *Areca* L.

Aristolochiaceae Juss., Gen. Pl.: 72. 4 Aug 1789.
Typus: *Aristolochia* L.

Asclepiadaceae Borkh., Bot. Wörterb. 1: 31. 1797.
Typus: *Asclepias* L.

Asparagaceae Juss., Gen. Pl.: 40. 4 Aug 1789.
Typus: *Asparagus* L.

Asteraceae Bercht. & J. Presl, Přir. Rostlin: 254. Jan-Apr 1820; nom. alt.: *Compositae.*
Typus: *Aster* L.

Asteranthaceae R. Knuth in Engler, Pflanzenr. 105: 1. 22 Aug 1939.
Typus: *Asteranthos* Desf.

Austrobaileyaceae Croizat in Cact. Succ. J. (Los Angeles) 15: 64. Mai 1943 (subfam. *Austrobaileyoideae* Croizat in J. Arnold Arbor. 21: 404. 24 Jul 1940).
Typus: *Austrobaileya* C. T. White

Avicenniaceae Miq. in Lehmann, Pl. Preiss. 1: 353. 14-16 Aug 1845 ([un-ranked] *Avicennieae* Endl., Ench. Bot.: 314. 15-21 Aug 1841).
Typus: *Avicennia* L.

Balanitaceae M. Roem., Fam. Nat. Syn. Monogr. 1: 26. 14 Sep - 15 Oct 1846.
Typus: *Balanites* Delile (nom. cons.).

Balanopaceae Benth. & Hook. f., Gen. Pl. 3: v, 341. 7 Feb 1880.
Typus: *Balanops* Baill.

Balanophoraceae Rich. in Mém. Mus. Hist. Nat. 8: 429. Nov 1822.
Typus: *Balanophora* J. R. Forst. & G. Forst.

Balsaminaceae A. Rich. in Bory, Dict. Class. Hist. Nat. 2: 173. 31 Dec 1822.
Typus: *Balsamina* Mill. [= *Impatiens* L.].

Barbeyaceae Rendle in Oliver, Fl. Trop. Afr. 6(2): 14. Mar 1916.
Typus: *Barbeya* Schweinf.

Barringtoniaceae DC. ex F. Rudolphi, Syst. Orb. Veg.: 56. 5-12 Jul 1830.
Typus: *Barringtonia* J. R. Forst. & G. Forst. (nom. cons.).

Basellaceae Raf., Fl. Tellur. 3: 44. Nov-Dec 1837.
Typus: *Basella* L.

Bataceae Mart. ex Perleb, Clav. Class.: 17. 18-24 Mar 1838.
Typus: *Batis* P. Browne

Begoniaceae C. Agardh, Aphor. Bot.: 200. 13 Jun 1824.
Typus: *Begonia* L.

Berberidaceae Juss., Gen. Pl.: 286. 4 Aug 1789.
Typus: *Berberis* L.

Betulaceae Gray, Nat. Arr. Brit. Pl. 2: 222, 243. 10 Jan 1822.
Typus: *Betula* L.
Note: If this family is united with *Corylaceae*, the name *Betulaceae* must be used.

Bignoniaceae Juss., Gen. Pl.: 137. 4 Aug 1789.
Typus: *Bignonia* L. (nom. cons.).

Bixaceae Kunth, Malvac., Büttner., Tiliac.: 17. 20 Apr 1822.
Typus: *Bixa* L.

Bombacaceae Kunth, Malvac., Büttner., Tiliac.: 5. 20 Apr 1822.
Typus: *Bombax* L. (nom. cons.).

Boraginaceae Juss., Gen. Pl.: 128. 4 Aug 1789.
Typus: *Borago* L.

Brassicaceae Burnett, Outl. Bot.: 854, 1093, 1123. Feb 1835; nom. alt.: *Cruciferae*.
Typus: *Brassica* L.

Bretschneideraceae Engl. & Gilg in Engler, Syllabus, ed. 9 & 10: 218. 6 Nov 1924.
Typus: *Bretschneidera* Hemsl.

Bromeliaceae Juss., Gen. Pl.: 49. 4 Aug 1789.
Typus: *Bromelia* L.

Brunelliaceae Engl. in Engler & Prantl, Nat. Pflanzenfam., Nachtr. 1: 182. 2 Aug 1897.
Typus: *Brunellia* Ruiz & Pav.

Bruniaceae R. Br. ex DC., Prodr. 2: 43. Nov (med.) 1825.
Typus: *Brunia* Lam. (nom. cons.).

Brunoniaceae Dumort., Anal. Fam. Pl.: 19, 21. 1829.
Typus: *Brunonia* Sm. ex R. Br.

Buddlejaceae K. Wilh., Samenpflanzen: 90. 12 Nov 1910.
Typus: *Buddleja* L.

Burmanniaceae Blume, Enum. Pl. Javae: 27. Oct-Dec 1827.
Typus: *Burmannia* L.

Burseraceae Kunth in Ann. Sci. Nat. (Paris) 2: 346. Jun 1824.
Typus: *Bursera* Jacq. ex L. (nom. cons.).

Butomaceae Mirb., Hist. Nat. Pl. 8: 194. 1804.
Typus: *Butomus* L.

Buxaceae Dumort., Comment. Bot.: 54. Nov (sero) - Dec (prim.) 1822.
Typus: *Buxus* L.

Byblidaceae Domin in Acta Bot. Bohem. 1: 3. 1922 (subfam. *Byblidoideae* Engl. & Gilg in Engler, Syllabus, ed. 7: 329. Oct 1912 - Mar 1913).
Typus: *Byblis* Salisb.

Byttneriaceae R. Br. in Flinders, Voy. Terra Austr. 2: 540. 19 Jul 1814.
Typus: *Byttneria* Loefl. (nom. cons.).

Cabombaceae Rich. ex A. Rich. in Bory, Dict. Class. Hist. Nat. 2: 608. 31 Dec 1822.
Typus: *Cabomba* Aubl.

Cactaceae Juss., Gen. Pl.: 310. 4 Aug 1789.
Typus: *Cactus* L. (nom. rej.) (*Mammillaria* Haw., nom. cons.).

Caesalpiniaceae R. Br. in Flinders, Voy. Terra Austr. 2: 551. 19 Jul 1814.
Typus: *Caesalpinia* L.

Callitrichaceae Link, Enum. Hort. Berol. Alt. 1: 7. 16 Mar - 30 Jun 1821.
Typus: *Callitriche* L.

Calycanthaceae Lindl. in Bot. Reg.: ad t. 404. 1 Oct 1819.
Typus: *Calycanthus* L. (nom. cons.).

Calyceraceae R. Br. ex Rich. in Mém. Mus. Hist. Nat. 6: 74. Nov 1820.
Typus: *Calycera* Cav. (nom. cons.).

Campanulaceae Juss., Gen. Pl.: 163. 4 Aug 1789.
Typus: *Campanula* L.

Canellaceae Mart., Nov. Gen. Sp. Pl. 3: 168, 170. Sep 1832.
Typus: *Canella* P. Browne (nom. cons.).

Cannabaceae Martinov, Tekhno-Bot. Slovar: 99. 3 Aug 1820.
Typus: *Cannabis* L.

Cannaceae Juss., Gen. Pl.: 62. 4 Aug 1789.
Typus: *Canna* L.

Capparaceae Juss., Gen. Pl.: 242. 4 Aug 1789.
Typus: *Capparis* L.

Caprifoliaceae Juss., Gen. Pl.: 210. 4 Aug 1789.
Typus: *Caprifolium* Mill. (*Lonicera* L.).

Cardiopteridaceae Blume, Rumphia 3: 205. Jun 1847.
Typus: *Cardiopteris* Wall. ex Royle

Caricaceae Dumort., Anal. Fam. Pl.: 37, 42. 1829.
Typus: *Carica* L.

Cartonemataceae Pichon in Notul. Syst. (Paris) 12: 219. 25 Mai 1946.
Typus: *Cartonema* R. Br.

Caryocaraceae Voigt, Hort. Suburb. Calcutt.: 88. Aug-Dec 1845.
Typus: *Caryocar* L.

Caryophyllaceae Juss., Gen. Pl.: 299. 4 Aug 1789.
Typus: *Caryophyllus* Mill., non L. (*Dianthus* L.).

Cassythaceae Bartl. ex Lindl., Nix. Pl.: 15. 17 Sep 1833.
Typus: *Cassytha* L.

Casuarinaceae R. Br. in Flinders, Voy. Terra Austr. 2: 571. 19 Jul 1814.
Typus: *Casuarina* L.

Celastraceae R. Br. in Flinders, Voy. Terra Austr. 2: 554. 19 Jul 1814.
Typus: *Celastrus* L.

Note: If this family is united with *Hippocrateaceae,* the name *Celastraceae* must be used.

Centrolepidaceae Endl., Gen. Pl.: 119. Dec 1836.
Typus: *Centrolepis* Labill.

Cephalotaceae Dumort., Anal. Fam. Pl.: 59, 61. 1829.
Typus: *Cephalotus* Labill. (nom. cons.).

Cephalotaxaceae Neger, Nadelhölzer: 23, 30. 28 Oct 1907.
Typus: *Cephalotaxus* Siebold & Zucc. ex Endl.

Ceratophyllaceae Gray, Nat. Arr. Brit. Pl. 2: 395, 554. 10 Jan 1822.
Typus: *Ceratophyllum* L.

Cercidiphyllaceae Engl., Syllabus, ed. 5: 126. 20-22 Jul 1907.
Typus: *Cercidiphyllum* Siebold & Zucc.

Chenopodiaceae Vent., Tabl. Règne Vég. 2: 253. 5 Mai 1799.
Typus: *Chenopodium* L.

Chloranthaceae R. Br. ex Sims in Bot. Mag.: ad t. 2190. 1 Nov 1820.
Typus: *Chloranthus* Sw.

Chrysobalanaceae R. Br. in Tuckey, Narr. Exped. Zaire: 433. 5 Mar 1818.
Typus: *Chrysobalanus* L.

Cichoriaceae Juss., Gen. Pl.: 168. 4 Aug 1789.
Typus: *Cichorium* L.

Circaeasteraceae Hutch., Fam. Fl. Pl. 1: 98. 15 Jan 1926.
Typus: *Circaeaster* Maxim.

Cistaceae Juss., Gen. Pl.: 294. 4 Aug 1789.
Typus: *Cistus* L.

Clethraceae Klotzsch in Linnaea 24: 12. Mai 1851.
Typus: *Clethra* L.

Clusiaceae Lindl., Intr. Nat. Syst. Bot., ed. 2: 74. 13 Jun 1836; nom. alt.: *Guttiferae*.
Typus: *Clusia* L.

Cneoraceae Vest, Anleit. Stud. Bot.: 267. 1818.
Typus: *Cneorum* L.

Cochlospermaceae Planch. in London J. Bot. 6: 305. Jun-Jul 1847.
Typus: *Cochlospermum* Kunth (nom. cons.).

Colchicaceae DC., Essai Propr. Méd. Pl.: 56. Jul 1804.
Typus: *Colchicum* L.

Columelliaceae D. Don in Edinburgh New Philos. J. 6: 46, 49. Oct-Dec 1828.
Typus: *Columellia* Ruiz & Pav. (nom. cons.).

Combretaceae R. Br., Prodr.: 351. 27 Mar 1810.
Typus: *Combretum* Loefl. (nom. cons.).

Commelinaceae Mirb., Hist. Nat. Pl. 8: 177. 1804.
Typus: *Commelina* L.

Compositae Giseke, Prael. Ord. Nat. Pl.: 538. Apr 1792; nom. alt.: *Asteraceae*.
Typus: *Aster* L.

Connaraceae R. Br. in Tuckey, Narr. Exped. Zaire: 431. 5 Mar 1818.
Typus: *Connarus* L.

Convolvulaceae Juss., Gen. Pl.: 132. 4 Aug 1789.
Typus: *Convolvulus* L.

Cordiaceae R. Br. ex Dumort., Anal. Fam. Pl.: 20, 25. 1829.
Typus: *Cordia* L.

Coriariaceae DC., Prodr. 1: 739. Jan (med.) 1824.
Typus: *Coriaria* L.

Cornaceae Bercht. & J. Presl, Přir. Rostlin 2(23*): [91], 92. 1825.
Typus: *Cornus* L.

Corsiaceae Becc. in Malesia 1: 238. Sep 1878.
Typus: *Corsia* Becc.

Corylaceae Mirb., Elém. Physiol. Vég. Bot. 2: 906. 24-30 Jun 1815.
Typus: *Corylus* L.

Note: If this family is united with *Betulaceae,* the name *Corylaceae* is rejected in favour of *Betulaceae.*

Corynocarpaceae Engl. in Engler & Prantl, Nat. Pflanzenfam., Nachtr. 1: 215. 4 Oct 1897.
Typus: *Corynocarpus* J. R. Forst. & G. Forst.

Crassulaceae J. St.-Hil., Expos. Fam. Nat. 2: 123. Feb-Apr 1805.
Typus: *Crassula* L.

Crossosomataceae Engl. in Engler & Prantl, Nat. Pflanzenfam., Nachtr. 1: 185. 2 Aug 1897.
Typus: *Crossosoma* Nutt.

Cruciferae Juss., Gen. Pl.: 237. 4 Aug 1789; nom. alt.: *Brassicaceae.*
Typus: *Brassica* L.

Crypteroniaceae A. DC., Prodr. 16(2): 677. Jul (med.) 1868.
Typus: *Crypteronia* Blume

Cucurbitaceae Juss., Gen. Pl.: 393. 4 Aug 1789.
Typus: *Cucurbita* L.

Cunoniaceae R. Br. in Flinders, Voy. Terra Austr. 2: 548. 19 Jul 1814.
Typus: *Cunonia* L. (nom. cons.).

Cupressaceae Gray, Nat. Arr. Brit. Pl. 2: 222, 225. 10 Jan 1822.
Typus: *Cupressus* L.

Cuscutaceae Dumort., Anal. Fam. Pl.: 20, 25. 1829.
Typus: *Cuscuta* L.

Cyanastraceae Engl. in Bot. Jahrb. Syst. 28: 357. 22 Mai 1900.
Typus: *Cyanastrum* Oliv.

Cycadaceae Pers., Syn. Pl. 2: 630. Sep 1807.
Typus: *Cycas* L.

Cyclanthaceae Poit. ex A. Rich. in Bory, Dict. Class. Hist. Nat. 5: 222. 15 Mai 1824.
Typus: *Cyclanthus* Poit. ex A. Rich.

Cymodoceaceae Vines, Stud. Text-Book Bot.: 553. Mar 1895.
Typus: *Cymodocea* K. D. Koenig (nom. cons.).

Cynomoriaceae Endl. ex Lindl., Nix. Pl.: 23. 17 Sep 1833.
Typus: *Cynomorium* L.

Cyperaceae Juss., Gen. Pl.: 26. 4 Aug 1789.
Typus: *Cyperus* L.

Cyrillaceae Lindl., Veg. Kingd.: 445. 14-28 Mar 1846.
Typus: *Cyrilla* Garden ex L.

Daphniphyllaceae Müll. Arg. in Candolle, Prodr. 16(1): 1. Nov (med.) 1869.
Typus: *Daphniphyllum* Blume

Datiscaceae Dumort., Anal. Fam. Pl.: 13, 14. 1829.
Typus: *Datisca* L.

Degeneriaceae I. W. Bailey & A. C. Sm. in J. Arnold Arbor. 23: 357. 15 Jul 1942.
Typus: *Degeneria* I. W. Bailey & A. C. Sm.

Desfontainiaceae Endl. in Pfeiffer, Nomencl. Bot. 1: 1037. 21 Feb 1873 ([un-
ranked] *Desfontainieae* Endl., Ench.
Bot.: 336. 15-21 Aug 1841).
Typus: *Desfontainia* Ruiz & Pav.

Dialypetalanthaceae Rizzini & Occhi-
oni in Lilloa 17: 253. 30 Dec 1948.
Typus: *Dialypetalanthus* Kuhlm.

Diapensiaceae Lindl., Intr. Nat. Syst.
Bot., ed. 2: 233. 13 Jun 1836.
Typus: *Diapensia* L.

Dichapetalaceae Baill. in Martius, Fl.
Bras. 12(1): 365. 1 Apr 1886.
Typus: *Dichapetalum* Thouars

Dichondraceae Dumort., Anal. Fam.
Pl.: 20, 24. 1829.
Typus: *Dichondra* J. R. Forst. & G.
Forst.

Diclidantheraceae J. Agardh, Theoria
Syst. Pl.: 195. Apr-Sep 1858.
Typus: *Diclidanthera* Mart.

Didiereaceae Radlk. in Engler &
Prantl, Nat. Pflanzenfam. 3(5): 462.
ante Mai 1896.
Typus: *Didierea* Baill.

Dilleniaceae Salisb., Parad. Lond.: ad
t. 73. 1 Jun 1807.
Typus: *Dillenia* L.

Dioncophyllaceae Airy Shaw in Kew
Bull. [6] 1951: 333. 26 Jan 1952.
Typus: *Dioncophyllum* Baill.

Dioscoreaceae R. Br., Prodr.: 294. 27
Mar 1810.
Typus: *Dioscorea* L.

Dipentodontaceae Merr. in Brittonia
4: 69, 73. 16 Dec 1941.
Typus: *Dipentodon* Dunn

Dipsacaceae Juss., Gen. Pl.: 194. 4 Aug
1789.
Typus: *Dipsacus* L.

Dipterocarpaceae Blume, Bijdr. Fl.
Ned. Ind.: 222. 20 Sep - 7 Dec 1825.
Typus: *Dipterocarpus* C. F. Gaertn.

Dodonaeaceae Kunth ex Small, Fl.
S.E. U.S.: 724, 737. 22 Jul 1903.
Typus: *Dodonaea* Mill.

Donatiaceae B. Chandler in Notes Roy.
Bot. Gard. Edinburgh 5: 44. Nov 1911.
Typus: *Donatia* J. R. Forst. & G.
Forst. (nom. cons.).

Dracaenaceae Salisb., Gen. Pl.: 73.
Apr-Mai 1866.
Typus: *Dracaena* Vand. ex L.

Droseraceae Salisb., Parad. Lond.: ad
t. 95. 1 Feb 1808.
Typus: *Drosera* L.

Dysphaniaceae Pax in Bot. Jahrb. Syst.
61: 230. 15 Jun 1927.
Typus: *Dysphania* R. Br.

Ebenaceae Gürke in Engler & Prantl,
Nat. Pflanzenfam. 4(1): 153. Dec 1891.
Typus: *Ebenus* Kuntze, non L. (*Maba*
J. R. Forst. & G. Forst.).

Ehretiaceae Mart., Nov. Gen. Sp. Pl.
2: 136, 138. Jan-Jun 1827.
Typus: *Ehretia* P. Browne

Elaeagnaceae Juss., Gen. Pl.: 74. 4
Aug 1789.
Typus: *Elaeagnus* L.

Elaeocarpaceae Juss. in Candolle, Es-
sai Propr. Méd. Pl., ed. 2: 87. 4-11
Mai 1816.
Typus: *Elaeocarpus* L.

Elatinaceae Dumort., Anal. Fam. Pl.:
44, 49. 1829.
Typus: *Elatine* L.

Empetraceae Hook. & Lindl. in Hooker, Fl. Scot. [2]: 297. 10 Mai 1821.
Typus: *Empetrum* L.

Epacridaceae R. Br., Prodr.: 535. 27 Mar 1810.
Typus: *Epacris* Cav. (nom. cons.).

Ephedraceae Dumort., Anal. Fam. Pl.: 11, 12. 1829.
Typus: *Ephedra* L.

Ericaceae Juss., Gen. Pl.: 159. 4 Aug 1789.
Typus: *Erica* L.

Eriocaulaceae Martinov, Tekhno-Bot. Slovar: 237. 3 Aug 1820.
Typus: *Eriocaulon* L.

Erythropalaceae Planch. ex Miq., Fl. Ned. Ind. 1(1): 704. 10 Jul 1856.
Typus: *Erythropalum* Blume

Erythroxylaceae Kunth in Humboldt & al., Nov. Gen. Sp. 5, ed. 4°: 175; ed. f°: 135. 25 Feb 1822.
Typus: *Erythroxylum* P. Browne

Escalloniaceae R. Br. ex Dumort., Anal. Fam. Pl.: 35, 37. 1829.
Typus: *Escallonia* Mutis ex L. f.

Eucommiaceae Engl., Syllabus, ed. 6: 145. Jun-Dec 1909.
Typus: *Eucommia* Oliv.

Eucryphiaceae Gay in Bot. Zeitung (Berlin) 6: 130. 18 Feb 1848.
Typus: *Eucryphia* Cav.

Euphorbiaceae Juss., Gen. Pl.: 384. 4 Aug 1789.
Typus: *Euphorbia* L.

Eupomatiaceae Orb., Dict. Univ. Hist. Nat. 5: 511. 29 Mar 1845.
Typus: *Eupomatia* R. Br.

Eupteleaceae K. Wilh., Samenpflanzen: 17. 12 Nov 1910.
Typus: *Euptelea* Siebold & Zucc.

Fabaceae Lindl., Intr. Nat. Syst. Bot., ed. 2: 148. 13 Jun 1836; nom. alt.: *Leguminosae, Papilionaceae.*
Typus: *Faba* Mill. [= *Vicia* L.].

Fagaceae Dumort., Anal. Fam. Pl.: 11, 12. 1829.
Typus: *Fagus* L.

Flacourtiaceae Rich. ex DC., Prodr. 1: 255. Jan (med.) 1824.
Typus: *Flacourtia* Comm. ex L'Hér.

Note: If this family is united with *Samydaceae,* the name *Flacourtiaceae* must be used.

Flagellariaceae Dumort., Anal. Fam. Pl.: 59, 60. 1829.
Typus: *Flagellaria* L.

Fouquieriaceae DC., Prodr. 3: 349. Mar (med.) 1828.
Typus: *Fouquieria* Kunth

Francoaceae A. Juss. in Ann. Sci. Nat. (Paris) 25: 9. Jan 1832.
Typus: *Francoa* Cav.

Frankeniaceae Desv. in Gérardin & Desvaux, Dict. Rais. Bot.: 188. 12-19 Apr 1817.
Typus: *Frankenia* L.

Fumariaceae Marquis, Esq. Règne Vég.: 50. 15-22 Jul 1820.
Typus: *Fumaria* L.

Garryaceae Lindl. in Edwards's Bot. Reg.: ad t. 1686. 1 Jul 1834.
Typus: *Garrya* Douglas ex Lindl.

Geissolomataceae A. DC., Prodr. 14: 491. Oct (med.) 1856.
Typus: *Geissoloma* Lindl. ex Kunth

Gentianaceae Juss., Gen. Pl.: 141. 4 Aug 1789.
Typus: *Gentiana* L.

Geosiridaceae Jonker in Receuil Trav. Bot. Néerl. 36 [Meded. Bot. Mus. Herb. Rijks Univ. Utrecht 60]: 477. [seors. impr.] 18 Mai 1939.
Typus: *Geosiris* Baill.

Geraniaceae Juss., Gen. Pl.: 268. 4 Aug 1789.
Typus: *Geranium* L.

Gesneriaceae Rich. & Juss. in Candolle, Essai Propr. Méd. Pl., ed. 2: 192. 4-11 Mai 1816.
Typus: *Gesneria* L.

Ginkgoaceae Engl. in Engler & Prantl, Nat. Pflanzenfam., Nachtr. 1: 19. 16 Jul 1897.
Typus: *Ginkgo* L.

Globulariaceae DC. in Lamarck & Candolle, Fl. Franç., ed. 3, 3: 427. 17 Sep 1805.
Typus: *Globularia* L.

Gnetaceae Blume, Nov. Pl. Expos.: 23. Aug-Dec 1833.
Typus: *Gnetum* L.

Gomortegaceae Reiche in Ber. Deutsch. Bot. Ges. 14: 232. 19 Aug 1896.
Typus: *Gomortega* Ruiz & Pav.

Gonystylaceae Tiegh. in Just's Bot. Jahresber. 21(2): 389, 390. Jul-Dec 1896.
Typus: *Gonystylus* Teijsm. & Binn.

Goodeniaceae R. Br., Prodr.: 573. 27 Mar 1810.
Typus: *Goodenia* Sm.

Gramineae Juss., Gen. Pl.: 28. 4 Aug 1789; nom. alt.: *Poaceae.*
Typus: *Poa* L.

Greyiaceae Hutch., Fam. Fl. Pl. 1: 202. 15 Jan 1926.
Typus: *Greyia* Hook. & Harv.

Grossulariaceae DC. in Lamarck & Candolle, Fl. Franç., ed. 3, 4(2): 405. 17 Sep 1805.
Typus: *Grossularia* Mill.

Grubbiaceae Endl. ex Meisn., Pl. Vasc. Gen. 1: 323; 2: 239. 18-24 Jul 1841.
Typus: *Grubbia* P. J. Bergius

Gunneraceae Meisn., Pl. Vasc. Gen. 1: 345, 346; 2: 257. 13-15 Feb 1842.
Typus: *Gunnera* L.

Guttiferae Juss., Gen. Pl.: 255. 4 Aug 1789; nom. alt.: *Clusiaceae.*
Typus: *Clusia* L.

Gyrostemonaceae A. Juss. in Orbigny, Dict. Univ. Hist. Nat. 6: 450. 1845 ([unranked] *Gyrostemoneae* Endl., Ench. Bot.: 509. 15-21 Aug 1841).
Typus: *Gyrostemon* Desf.

Haemodoraceae R. Br., Prodr.: 299. 27 Mar 1810.
Typus: *Haemodorum* Sm.

Haloragaceae R. Br. in Flinders, Voy. Terra Austr. 2: 549. 19 Jul 1814.
Typus: *Haloragis* J. R. Forst. & G. Forst.

Hamamelidaceae R. Br. in Abel, Narr. Journey China: 374. 15 Aug 1818.
Typus: *Hamamelis* L.

Heliotropiaceae Schrad. in Commentat. Soc. Regiae Sci. Gott. Recent. 4: 192. Dec 1819.
Typus: *Heliotropium* L.

Hernandiaceae Blume, Bijdr. Fl. Ned. Ind.: 550. 24 Jan 1826.
Typus: *Hernandia* L.

Heteropyxidaceae Engl. & Gilg in Engler, Syllabus, ed. 8: 281. Jan-Feb 1920.
Typus: *Heteropyxis* Harv. (nom. cons.).

Himantandraceae Diels in Bot. Jahrb. Syst. 55: 126. 27 Nov 1917.
Typus: *Himantandra* F. Muell. ex Diels [= *Galbulimima* F. M. Bailey].

Hippocastanaceae A. Rich., Bot. Méd.: 680. 14-21 Jun 1823.
Typus: *Hippocastanum* Mill. (*Aesculus* L.).

Hippocrateaceae Juss. in Ann. Mus. Natl. Hist. Nat. 18: 486. Jul-Aug 1811.
Typus: *Hippocratea* L.
Note: If this family is united with *Celastraceae,* the name *Hippocrateaceae* is rejected in favour of *Celastraceae*.

Hippuridaceae Vest, Anleit. Stud. Bot.: 265, 278. 1818.
Typus: *Hippuris* L.

Hoplestigmataceae Gilg in Engler, Syllabus, ed. 9 & 10: 322. 6 Nov 1924.
Typus: *Hoplestigma* Pierre

Humbertiaceae Pichon in Notul. Syst. (Paris) 13: 23. 15 Dec 1947.
Typus: *Humbertia* Lam.

Humiriaceae A. Juss. in Saint-Hilaire, Fl. Bras. Merid. 2: 87. 10 Oct 1829.
Typus: *Humiria* Aubl. (nom. cons.).

Hydnoraceae C. Agardh, Aphor. Bot.: 88. 21 Dec 1821.
Typus: *Hydnora* Thunb.

Hydrangeaceae Dumort., Anal. Fam. Pl.: 36, 38. 1829.
Typus: *Hydrangea* L.

Hydrocharitaceae Juss., Gen. Pl.: 67. 4 Aug 1789.
Typus: *Hydrocharis* L.

Hydrocotylaceae Bercht. & J. Presl, Přir. Rostlin: 258. Jan-Apr 1820.
Typus: *Hydrocotyle* L.

Hydrophyllaceae R. Br. in Bot. Reg.: ad t. 242. 1 Dec 1817.
Typus: *Hydrophyllum* L.

Hydrostachyaceae Engl. in Bot. Jahrb. Syst. 20: 136. 16 Nov 1894 (trib. *Hydrostachyeae* Tul. in Ann. Sci. Nat., Bot., sér. 3, 11: 91. Feb 1849).
Typus: *Hydrostachys* Thouars

Hypericaceae Juss., Gen. Pl.: 254. 4 Aug 1789.
Typus: *Hypericum* L.

Hypoxidaceae R. Br. in Flinders, Voy. Terra Austr. 2: 576. 19 Jul 1814.
Typus: *Hypoxis* L.

Icacinaceae Miers in Ann. Mag. Nat. Hist., ser. 2, 8: 174. Sep 1851 (trib. *Icacineae* Benth. in Ann. Mag. Nat. Hist. 7: 215. Mai 1841).
Typus: *Icacina* A. Juss.

Illecebraceae R. Br., Prodr.: 413. 27 Mar 1810.
Typus: *Illecebrum* L.

Illiciaceae A. C. Sm. in Sargentia 7: 8. 28 Nov 1947 (trib. *Illicieae* DC., Prodr. 1: 77. Jan (med.) 1824).
Typus: *Illicium* L.

Iridaceae Juss., Gen. Pl.: 57. 4 Aug 1789.
Typus: *Iris* L.

Irvingiaceae Exell & Mendonça, Consp. Fl. Angol. 1: 279, 395. 20 Aug 1951 (subfam. *Irvingioideae* Engl. in Engler & Drude, Veg. Erde 9(3,1): 765. Feb-Sep 1915).
Typus: *Irvingia* Hook. f.

Iteaceae J. Agardh, Theoria Syst. Pl.: 151. Apr-Sep 1858.
Typus: *Itea* L.

Ixonanthaceae Planch. ex Miq., Fl. Ned. Ind. 1(2): viii, 494. 23 Dec 1858 - 6 Oct 1859.
Typus: *Ixonanthes* Jack

Juglandaceae DC. ex Perleb, Vers. Arzneikr. Pfl.: 143. Mai 1818.
Typus: *Juglans* L.

Julianiaceae Hemsl. in J. Bot. 44: 379. Oct 1906.
Typus: *Juliania* Schltdl., non La Llave & Lex. (*Amphipterygium* Standl.).

Juncaceae Juss., Gen. Pl.: 43. 4 Aug 1789.
Typus: *Juncus* L.

Juncaginaceae Rich., Démonstr. Bot.: ix. Mai 1808.
Typus: *Juncago* Ség. (*Triglochin* L.).
Note: If this family is united with *Potamogetonaceae,* the name *Juncaginaceae* is rejected in favour of *Potamogetonaceae.*

Koeberliniaceae Engl. in Engler & Prantl, Nat. Pflanzenfam. 3(6): 319. 14 Mai 1895.
Typus: *Koeberlinia* Zucc.

Krameriaceae Dumort., Anal. Fam. Pl.: 20, 23. 1829.
Typus: *Krameria* L.

Labiatae Juss., Gen. Pl.: 110. 4 Aug 1789; nom. alt.: *Lamiaceae.*
Typus: *Lamium* L.

Lacistemataceae Mart., Nov. Gen. Sp. Pl. 1: 154, 158. Jan-Mar 1826.
Typus: *Lacistema* Sw.

Lactoridaceae Engl. in Engler & Prantl, Nat. Pflanzenfam. 3(2): 19. 27 Jan 1888.
Typus: *Lactoris* Phil.

Lamiaceae Martinov, Tekhno-Bot. Slovar: 355. 3 Aug 1820; nom. alt.: *Labiatae.*
Typus: *Lamium* L.

Lardizabalaceae R. Br. in Trans. Linn. Soc. London 13: 212. 23 Mai - 21 Jun 1821.
Typus: *Lardizabala* Ruiz & Pav.

Lauraceae Juss., Gen. Pl.: 80. 4 Aug 1789.
Typus: *Laurus* L.

Lecythidaceae A. Rich. in Bory, Dict. Class. Hist. Nat. 9: 259. 25 Feb 1825.
Typus: *Lecythis* Loefl.

Leeaceae Dumort., Anal. Fam. Pl.: 21, 27. 1829.
Typus: *Leea* D. Royen ex L. (nom. cons.).

Leguminosae Juss., Gen. Pl.: 345. 4 Aug 1789; nom. alt.: *Fabaceae.*
Typus: *Faba* Mill. [= *Vicia* L.].

Leitneriaceae Benth. & Hook. f., Gen. Pl. 3: vi, 396. 7 Feb 1880.
Typus: *Leitneria* Chapm.

Lemnaceae Martinov, Tekhno-Bot. Slovar: 362. 3 Aug 1820.
Typus: *Lemna* L.

Lennoaceae Solms in Abh. Naturf. Ges. Halle 11: 174. 7 Jan 1870.
Typus: *Lennoa* La Llave & Lex.

Lentibulariaceae Rich. in Poiteau & Turpin, Fl. Paris. 1, ed. 4°: 26; ed. f°: 23. 1808.
Typus: *Lentibularia* Ség. (*Utricularia* L.).

Lepidobotryaceae J. Léonard in Bull. Jard. Bot. Etat 20: 38. Jun 1950.
Typus: *Lepidobotrys* Engl.

Lilaeaceae Dumort., Anal. Fam. Pl.: 62, 65. 1829.
Typus: *Lilaea* Bonpl.

Liliaceae Juss., Gen. Pl.: 48. 4 Aug 1789.
Typus: *Lilium* L.

Limnanthaceae R. Br. in London Edinburgh Philos. Mag. & J. Sci. 3: 70, 71. Jul 1833.
Typus: *Limnanthes* R. Br. (nom. cons.).

Limoniaceae Ser., Fl. Pharm.: 456. 1851.
Typus: *Limonium* Mill. (nom. cons.).

Linaceae DC. ex Perleb, Vers. Arzneikr. Pfl.: 107. Mai 1818.
Typus: *Linum* L.

Lissocarpaceae Gilg in Engler, Syllabus, ed. 9 & 10: 324. 6 Nov 1924.
Typus: *Lissocarpa* Benth.

Loasaceae Juss. in Ann. Mus. Natl. Hist. Nat. 5: 21. 1804.
Typus: *Loasa* Adans.

Lobeliaceae Juss. in Bonpland, Descr. Pl. Malmaison: [19]. 17-24 Sep 1813.
Typus: *Lobelia* L.

Loganiaceae R. Br. ex Mart., Nov. Gen. Sp. Pl. 2: 133. Jan-Jun 1827.
Typus: *Logania* R. Br. (nom. cons.).

Loranthaceae Juss. in Ann. Mus. Natl. Hist. Nat. 12: 292. 1808.
Typus: *Loranthus* Jacq. (nom. cons.).

Lowiaceae Ridl., Fl. Malay Penins. 4: 291. 1 Dec 1924.
Typus: *Lowia* Scort. [= *Orchidantha* N. E. Br.].

Lythraceae J. St.-Hil., Expos. Fam. Nat. 2: 175. Feb-Apr 1805.
Typus: *Lythrum* L.

Magnoliaceae Juss., Gen. Pl.: 280. 4 Aug 1789.
Typus: *Magnolia* L.

Malaceae Small, Fl. S.E. U.S.: 495, 529. 22 Jul 1903.
Typus: *Malus* Mill.

Malesherbiaceae D. Don in Edinburgh New Philos. J. 2: 321. Jan-Mar 1827.
Typus: *Malesherbia* Ruiz & Pav.

Malpighiaceae Juss., Gen. Pl.: 252. 4 Aug 1789.
Typus: *Malpighia* L.

Malvaceae Juss., Gen. Pl.: 271. 4 Aug 1789.
Typus: *Malva* L.

Marantaceae R. Br. in Flinders, Voy. Terra Austr. 2: 575. 19 Jul 1814.
Typus: *Maranta* L.

Marcgraviaceae Bercht. & J. Presl, Přir. Rostlin: 218. Jan-Apr 1820.
Typus: *Marcgravia* L.

Martyniaceae Horan., Char. Ess. Fam.: 130. 30 Jun 1847.
Typus: *Martynia* L.

Mayacaceae Kunth in Abh. Königl. Akad. Wiss. Berlin 1840: 93. 1842.
Typus: *Mayaca* Aubl.

Medusagynaceae Engl. & Gilg in Engler, Syllabus, ed. 9 & 10: 280. 6 Nov 1924.
Typus: *Medusagyne* Baker

Medusandraceae Brenan in Kew Bull. [7] 1952: 228. 25 Jul 1952.
Typus: *Medusandra* Brenan

Melanthiaceae Batsch ex Borkh., Bot. Wörterb. 2: 8. 1797.
Typus: *Melanthium* L.

Melastomataceae Juss., Gen. Pl.: 328. 4 Aug 1789.
Typus: *Melastoma* L.

Meliaceae Juss., Gen. Pl.: 263. 4 Aug 1789.
Typus: *Melia* L.

Melianthaceae Horan., Prim. Lin. Syst. Nat.: 100. 2 Nov 1834.
Typus: *Melianthus* L.

Menispermaceae Juss., Gen. Pl.: 284. 4 Aug 1789.
Typus: *Menispermum* L.

Menyanthaceae Dumort., Anal. Fam. Pl.: 20, 25. 1829.
Typus: *Menyanthes* L.

Mesembryanthemaceae Philib., Intr. Bot. 3: 268. 28 Jan 1800.
Typus: *Mesembryanthemum* L. (nom. cons.).

Mimosaceae R. Br. in Flinders, Voy. Terra Austr. 2: 551. 19 Jul 1814.
Typus: *Mimosa* L.

Misodendraceae J. Agardh, Theoria Syst. Pl.: 236. Apr-Sep 1858.
Typus: *Misodendrum* Banks ex DC.

Mitrastemonaceae Makino in Bot. Mag. (Tokyo) 25: 252. 20 Dec 1911.
Typus: *Mitrastemon* Makino

Molluginaceae Bartl. in Bartling & Wendland, Beitr. Bot. 2: 158. Dec 1825.
Typus: *Mollugo* L.

Monimiaceae Juss. in Ann. Mus. Natl. Hist. Nat. 14: 133. 1809.
Typus: *Monimia* Thouars

Monotropaceae Nutt., Gen. N. Amer. Pl. 1: 272. 14 Jul 1818.
Typus: *Monotropa* L.
Note: If this family is united with *Pyrolaceae,* the name *Monotropaceae* is rejected in favour of *Pyrolaceae.*

Montiniaceae Nakai, Chosakuronbun Mokuroku [Ord. Fam. Trib. Nov.]: 243. 20 Jul 1943.
Typus: *Montinia* Thunb.

Moraceae Gaudich. in Trinius, Gen. Pl.: 13. 1835.
Typus: *Morus* L.

Moringaceae Martinov, Tekhno-Bot. Slovar: 404. 3 Aug 1820.
Typus: *Moringa* Adans.

Musaceae Juss., Gen. Pl.: 61. 4 Aug 1789.
Typus: *Musa* L.

Myoporaceae R. Br., Prodr.: 514. 27 Mar 1810.
Typus: *Myoporum* G. Forst.

Myricaceae Rich. ex Kunth in Humboldt & al., Nov. Gen. Sp. 2, ed. 4°: 16. 28 Apr 1817.
Typus: *Myrica* L.

Myristicaceae R. Br., Prodr.: 399. 27 Mar 1810.
Typus: *Myristica* Gronov. (nom. cons.).

Myrothamnaceae Nied. in Engler & Prantl, Nat. Pflanzenfam. 3(2a): 103. 9 Mar 1891.
Typus: *Myrothamnus* Welw.

Myrsinaceae R. Br., Prodr.: 532. 27 Mar 1810.
Typus: *Myrsine* L.

Myrtaceae Juss., Gen. Pl.: 322. 4 Aug 1789.
Typus: *Myrtus* L.

Najadaceae Juss., Gen. Pl.: 18. 4 Aug 1789.
Typus: *Najas* L.

Nelumbonaceae A. Rich. in Bory, Dict. Class. Hist. Nat. 11: 492. 10 Feb 1827.
Typus: *Nelumbo* Adans.

Nepenthaceae Dumort., Anal. Fam. Pl.: 14, 16. 1829.
Typus: *Nepenthes* L.

Neuradaceae Kostel., Allg. Med.-Pharm. Fl.: 1476. Jan-Oct 1835.
Typus: *Neurada* L.

Nolanaceae Bercht. & J. Presl, Přir. Rostlin: 244. Jan-Apr 1820.
Typus: *Nolana* L. ex L. f.

Nyctaginaceae Juss., Gen. Pl.: 90. 4 Aug 1789.
Typus: *Nyctago* Juss., nom. illeg. (*Mirabilis* L.).

Nymphaeaceae Salisb. in Ann. Bot. (König & Sims) 2: 70. Jun 1805.
Typus: *Nymphaea* L. (nom. cons.).

Nyssaceae Juss. ex Dumort., Anal. Fam. Pl.: 13. 1829.
Typus: *Nyssa* L.

Ochnaceae DC. in Nouv. Bull. Sci. Soc. Philom. Paris 2: 208. Jan 1811.
Typus: *Ochna* L.

Octoknemaceae Tiegh. in Solereder, Syst. Anat. Dicot. Ergänz.: 83, 84. 9 Mai 1908.
Typus: *Octoknema* Pierre

Olacaceae R. Br. in Tuckey, Narr. Exped. Zaire: 452. 5 Mar 1818.
Typus: *Olax* L.

Oleaceae Hoffmanns. & Link, Fl. Portug. 1: 62. 1 Sep 1809.
Typus: *Olea* L.

Oliniaceae Harv. & Sond., Fl. Cap. 2: ix. 15-31 Oct 1862.
Typus: *Olinia* Thunb. (nom. cons.).

Onagraceae Juss., Gen. Pl.: 317. 4 Aug 1789.
Typus: *Onagra* Mill. (*Oenothera* L.).

Opiliaceae Valeton, Crit. Overz. Olacin.: 136. 7 Jul 1886.
Typus: *Opilia* Roxb.

Orchidaceae Juss., Gen. Pl.: 64. 4 Aug 1789.
Typus: *Orchis* L.

Orobanchaceae Vent., Tabl. Règne Vég. 2: 292. 5 Mai 1799.
Typus: *Orobanche* L.

Oxalidaceae R. Br. in Tuckey, Narr. Exped. Zaire: 433. 5 Mar 1818.
Typus: *Oxalis* L.

Paeoniaceae Raf., Anal. Nat.: 176. Apr-Jul 1815.
Typus: *Paeonia* L.

Palmae Juss., Gen. Pl.: 37. 4 Aug 1789; nom. alt.: *Arecaceae.*
Typus: *Areca* L.

Pandaceae Engl. & Gilg in Engler, Syllabus, ed. 7: 223. Oct 1912 - Mar 1913.
Typus: *Panda* Pierre

Pandanaceae R. Br., Prodr.: 340. 27 Mar 1810.
Typus: *Pandanus* Parkinson

Papaveraceae Juss., Gen. Pl.: 235. 4 Aug 1789.
Typus: *Papaver* L.

Papilionaceae Giseke, Prael. Ord. Nat. Pl.: 415. Apr 1792; nom. alt.: *Fabaceae.*
Typus: *Faba* Mill. [= *Vicia* L.].

Parnassiaceae Martinov, Tekhno-Bot. Slovar: 456. 3 Aug 1820.
Typus: *Parnassia* L.

Passifloraceae Juss. ex Roussel, Fl. Calvados, ed. 2: 334. 30 Mai 1806.
Typus: *Passiflora* L.

Pedaliaceae R. Br., Prodr.: 519. 27 Mar 1810.
Typus: *Pedalium* D. Royen ex L.

Penaeaceae Sweet ex Guill. in Bory, Dict. Class. Hist. Nat. 13: 171. 1 Mar 1828.
Typus: *Penaea* L.

Pentaphragmataceae J. Agardh, Theoria Syst. Pl.: 95. Apr-Sep 1858.
Typus: *Pentaphragma* Wall. ex G. Don

Pentaphylacaceae Engl. in Engler & Prantl, Nat. Pflanzenfam., Nachtr. 1: 214. 4 Oct 1897.
Typus: *Pentaphylax* Gardner & Champ.

Penthoraceae Rydb. ex Britton, Man. Fl. N. States: 475. 16 Oct 1901.
Typus: *Penthorum* L.

Peridiscaceae Kuhlm. in Arq. Serv. Florest. 3: 4. 1950.
Typus: *Peridiscus* Benth.

Periplocaceae Schltr. in Schumann & Lauterbach, Nachtr. Fl. Schutzgeb. Südsee: 351. Nov 1905.
Typus: *Periploca* L.

Petermanniaceae Hutch., Fam. Fl. Pl. 2: 113. 20 Aug 1934.
Typus: *Petermannia* F. Muell. (nom. cons.).

Petrosaviaceae Hutch., Fam. Fl. Pl. 2: 36. 20 Aug 1934.
Typus: *Petrosavia* Becc.

Philesiaceae Dumort., Anal. Fam. Pl.: 53, 54, 97. 1829.
Typus: *Philesia* Comm. ex Juss.

Philydraceae Link, Enum. Hort. Berol. Alt. 1: 5. 16 Mar - 30 Jun 1821.
Typus: *Philydrum* Banks ex Gaertn.

Phrymaceae Schauer in Candolle, Prodr. 11: 520. 25 Nov 1847.
Typus: *Phryma* L.

Phytolaccaceae R. Br. in Tuckey, Narr. Exped. Zaire: 454. 5 Mar 1818.
Typus: *Phytolacca* L.

Picrodendraceae Small in J. New York Bot. Gard. 18: 184. Aug 1917.
Typus: Picrodendron Griseb. (nom. cons.).

Pinaceae Spreng. ex F. Rudolphi, Syst. Orb. Veg.: 35. 5-12 Jul 1830.
Typus: *Pinus* L.
Note: If this family is united with *Abietaceae,* the name *Pinaceae* must be used.

Piperaceae Giseke, Prael. Ord. Nat. Pl.: 123. Apr 1792.
Typus: *Piper* L.

Pittosporaceae R. Br. in Flinders, Voy. Terra Austr. 2: 542. 19 Jul 1814.
Typus: *Pittosporum* Banks ex Sol. (nom. cons.).

Plantaginaceae Juss., Gen. Pl.: 89. 4 Aug 1789.
Typus: *Plantago* L.

Platanaceae T. Lestib., Botanogr. Elém.: 526. 12-19 Jun 1826.
Typus: *Platanus* L.

Plumbaginaceae Juss., Gen. Pl.: 92. 4 Aug 1789.
Typus: *Plumbago* L.

Poaceae Barnhart in Bull. Torrey Bot. Club 22: 7. 15 Jan 1895 (trib. *Poeae* R. Br. in Flinders, Voy. Terra Austr. 2: 583. 19 Jul 1814); nom. alt.: *Gramineae.*
Typus: *Poa* L.

Podocarpaceae Endl., Syn. Conif.: 203. Mai-Jun 1847.
Typus: *Podocarpus* L'Hér. ex Pers. (nom. cons.).

Podophyllaceae DC., Syst. Nat. 1: 126. 1-15 Nov 1817.
Typus: *Podophyllum* L.

Podostemaceae Rich. ex Kunth in Humboldt & al., Nov. Gen. Sp. 1, ed. 4°: 246. 4-11 Mai 1816.
Typus: *Podostemum* Michx.

Polemoniaceae Juss., Gen. Pl.: 136. 4 Aug 1789.
Typus: *Polemonium* L.

Polygalaceae Hoffmanns. & Link, Fl. Portug. 1: 62. 1 Sep 1809.
Typus: *Polygala* L.

Polygonaceae Juss., Gen. Pl.: 82. 4 Aug 1789.
Typus: *Polygonum* L. (nom. cons.).

Pontederiaceae Kunth in Humboldt & al., Nov. Gen. Sp. 1, ed. 4°: 265. 4-11 Mai 1816.
Typus: *Pontederia* L.

Portulacaceae Juss., Gen. Pl.: 312. 4 Aug 1789.
Typus: *Portulaca* L.

Posidoniaceae Vines, Stud. Text-Book Bot.: 553. Mar 1895.
Typus: *Posidonia* K. D. Koenig (nom. cons.).

Potamogetonaceae Bercht. & J. Presl, Přir. Rostlin 1(7*): [1], 3. 1823.
Typus: *Potamogeton* L.
Note: If this family is united with *Juncaginaceae,* the name *Potamogetonaceae* must be united.

Primulaceae Batsch ex Borkh., Bot. Wörterb. 2: 240. 1797.
Typus: *Primula* L.

Proteaceae Juss., Gen. Pl.: 78. 4 Aug 1789.
Typus: *Protea* L. (nom. cons.).

Pterostemonaceae Small in N. Amer. Fl. 22(1-6): 2. 22 Mai 1905.
Typus: *Pterostemon* Schauer

Punicaceae Bercht. & J. Presl, Přir. Rostlin 2(94*): [378]. 1825.
Typus: *Punica* L.

Pyrolaceae Lindl., Syn. Brit. Fl.: 175. 16 Mar 1829.
Typus: *Pyrola* L.
Note: If this family is united with *Monotropaceae,* the name *Pyrolaceae* must be used.

Quiinaceae Choisy ex Engl. in Martius, Fl. Bras. 12(1): 475-476. 1 Apr 1888.
Typus: *Quiina* Aubl.

Rafflesiaceae Dumort., Anal. Fam. Pl.: 13, 14. 1829.
Typus: *Rafflesia* R. Br.

Ranunculaceae Juss., Gen. Pl.: 231. 4 Aug 1789.
Typus: *Ranunculus* L.

Rapateaceae Dumort., Anal. Fam. Pl.: 60, 62. 1829.
Typus: *Rapatea* Aubl.

Resedaceae Martinov, Tekhno-Bot. Slovar: 541. 3 Aug 1820.
Typus: *Reseda* L.

Restionaceae R. Br., Prodr.: 243. 27 Mar 1810.
Typus: *Restio* Rottb. (nom. cons.).

Rhamnaceae Juss., Gen. Pl.: 376. 4 Aug 1789.
Typus: *Rhamnus* L.

Rhizophoraceae Pers., Syn. Pl. 2: 2. Nov 1806.
Typus: *Rhizophora* L.

Rhoipteleaceae Hand.-Mazz. in Repert. Spec. Nov. Regni Veg. 30: 75. 15 Feb 1932.
Typus: *Rhoiptelea* Diels & Hand.-Mazz.

Roridulaceae Martinov, Tekhno-Bot. Slovar: 549. 3 Aug 1820.
Typus: *Roridula* Burm. f. ex L.

Rosaceae Juss., Gen. Pl.: 334. 4 Aug 1789.
Typus: *Rosa* L.

Rubiaceae Juss., Gen. Pl.: 196. 4 Aug 1789.
Typus: *Rubia* L.

Ruppiaceae Horan., Prim. Lin. Syst. Nat.: 46. 2 Nov 1834.
Typus: *Ruppia* L.

Ruscaceae M. Roem., Handb. Allg. Bot. 3: 446. 1840.
Typus: *Ruscus* L.

Rutaceae Juss., Gen. Pl.: 296. 4 Aug 1789.
Typus: *Ruta* L.

Sabiaceae Blume, Mus. Bot. 1: 368. 1851.
Typus: *Sabia* Colebr.

Salicaceae Mirb., Elém. Physiol. Vég. Bot. 2: 905. 24-30 Jun 1815.
Typus: *Salix* L.

Salvadoraceae Lindl., Intr. Nat. Syst. Bot., ed. 2: 269. 13 Jun 1836.
Typus: *Salvadora* L.

Samydaceae Vent. in Mém. Cl. Sci. Math. Inst. Natl. France 1807(2): 149. 1808.
Typus: *Samyda* Jacq. (nom. cons.).
Note: If this family is united with *Flacourtiaceae,* the name *Samydaceae* is rejected in favour of *Flacourtiaceae.*

Santalaceae R. Br., Prodr.: 350. 27 Mar 1810.
Typus: *Santalum* L.

Sapindaceae Juss., Gen. Pl.: 246. 4 Aug 1789.
Typus: *Sapindus* L. (nom. cons.).

Sapotaceae Juss., Gen. Pl.: 151. 4 Aug 1789.
Typus: *Sapota* Mill. (*Achras* L., nom. rej.) [= *Manilkara* Adans.(nom. cons.)].

Sarcolaenaceae Caruel in Atti Reale Accad. Lincei, Mem. Cl. Sci. Fis., ser. 3, 10: 226, 248. 1881.
Typus: *Sarcolaena* Thouars

Sarcospermataceae H. J. Lam in Bull. Jard. Bot. Buitenzorg, sér. 3, 7: 248. Feb 1925.
Typus: *Sarcosperma* Hook. f.

Sargentodoxaceae Stapf ex Hutch., Fam. Fl. Pl. 1: 100. 15 Jan 1926.
Typus: *Sargentodoxa* Rehder & E. H. Wilson

Sarraceniaceae Dumort., Anal. Fam. Pl.: 53. 1829.
Typus: *Sarracenia* L.

Saurauiaceae Griseb., Grundr. Syst. Bot.: 98. 1-2 Jun 1854.
Typus: *Saurauia* Willd. (nom. cons.).
Note: If this family is united with *Actinidiaceae,* the name *Saurauiaceae* is rejected in favour of *Actinidiaceae.*

Saururaceae Rich. ex T. Lestib., Botanogr. Elém.: 453. 12-19 Jun 1826.
Typus: *Saururus* L.

Saxifragaceae Juss., Gen. Pl.: 308. 4 Aug 1789.
Typus: *Saxifraga* L.

Scheuchzeriaceae F. Rudolphi, Syst. Orb. Veg.: 28. 5-12 Jul 1830.
Typus: *Scheuchzeria* L.

Schisandraceae Blume, Fl. Javae 32-33: 3. 25 Jun 1830.
Typus: *Schisandra* Michx. (nom. cons.).

Scrophulariaceae Juss., Gen. Pl.: 117. 4 Aug 1789.
Typus: *Scrophularia* L.

Scyphostegiaceae Hutch., Fam. Fl. Pl. 1: 229. 15 Jan 1926.
Typus: *Scyphostegia* Stapf

Scytopetalaceae Engl. in Engler & Prantl, Nat. Pflanzenfam., Nachtr. 1: 242. 4 Oct 1897.
Typus: *Scytopetalum* Pierre ex Engl.

Selaginaceae Choisy [Mém. Sélag.: 19] in Mém. Soc. Phys. Genève 2: 89. 1823.
Typus: *Selago* L.

Simaroubaceae DC. in Nouv. Bull. Sci. Soc. Philom. Paris 2: 209. Jan 1811.
Typus: *Simarouba* Aubl. (nom. cons.).

Siphonodontaceae Gagnep. & Tardieu in Notul. Syst. (Paris) 14: 102. 25 Jul 1951 (subfam. *Siphonodontoideae* Croizat in Lilloa 13: 41. 29 Dec 1947).
Typus: *Siphonodon* Griff.

Smilacaceae Vent., Tabl. Règne Vég. 2: 146. 5 Mai 1799.
Typus: *Smilax* L.

Solanaceae Juss., Gen. Pl.: 124. 4 Aug 1789.
Typus: *Solanum* L.

Sonneratiaceae Engl. in Engler & Prantl, Nat. Pflanzenfam., Nachtr. 1: 261. 4 Oct 1897.
Typus: *Sonneratia* L. f. (nom. cons.).

Sparganiaceae Hanin, Cours Bot.: 400. 16-23 Apr 1811.
Typus: *Sparganium* L.

Sphenocleaceae T. Baskerv., Affin. Pl.: 110. 1839 (subfam. *Sphenocleoideae* Lindl., Intr. Nat. Syst. Bot., ed. 2: 238. 13 Jun 1836).
Typus: *Sphenoclea* Gaertn. (nom. cons.).

Stachyuraceae J. Agardh, Theoria Syst. Pl.: 152. Apr-Sep 1858.
Typus: *Stachyurus* Siebold & Zucc.

Stackhousiaceae R. Br. in Flinders, Voy. Terra Austr. 2: 555. 19 Jul 1814.
Typus: *Stackhousia* Sm.

Staphyleaceae Martinov, Tekhno-Bot. Slovar: 598. 3 Aug 1820.
Typus: *Staphylea* L.

Stemonaceae Caruel in Nuovo Giorn. Bot. Ital. 10: 94. Apr 1878.
Typus: *Stemona* Lour.

Stenomeridaceae J. Agardh, Theoria Syst. Pl.: 66. Apr-Sep 1858.
Typus: *Stenomeris* Planch.

Sterculiaceae Vent. in Salisbury, Parad. Lond.: ad t. 69. 1 Mai 1807.
Typus: *Sterculia* L.

Stilbaceae Kunth, Handb. Bot.: 393. 21-28 Sep 1831.
Typus: *Stilbe* P. J. Bergius

Strasburgeriaceae Tiegh. in Solereder, Syst. Anat. Dicot. Ergänz.: 45. 9 Mai 1908.
Typus: *Strasburgeria* Baill.

Strelitziaceae Hutch., Fam. Fl. Pl. 2: 72. 20 Aug 1934.
Typus: *Strelitzia* Aiton

Stylidiaceae R. Br., Prodr.: 565. 27 Mar 1810.
Typus: *Stylidium* Sw. ex Willd. (nom. cons.).

Styracaceae DC. & Spreng. [transl. Jameson], Elem. Philos. Pl.: 140. Jul 1821.
Typus: *Styrax* L.

Note: If this family is united with *Symplocaceae,* the name *Styracaceae* must be used.

Surianaceae Arn. in Wight & Arnott, Prodr. Fl. Ind. Orient.: 360. 10 Oct 1834.
Typus: *Suriana* L.

Symplocaceae Desf. in Mém. Mus. Hist. Nat. 6: 9. 1820.
Typus: *Symplocos* Jacq.

Note: If this family is united with *Styracaceae,* the name *Symplocaceae* is rejected in favour of *Styracaceae.*

Taccaceae Dumort., Anal. Fam. Pl.: 57, 58. 1829.
Typus: *Tacca* J. R. Forst. & G. Forst. (nom. cons.).

Tamaricaceae Link, Enum. Hort. Berol. Alt. 1: 291. 16 Mar - 30 Jun 1821.
Typus: *Tamarix* L.

Taxaceae Gray, Nat. Arr. Brit. Pl. 2: 222, 226. 10 Jan 1822.
Typus: *Taxus* L.

Taxodiaceae Saporta in Ann. Sci. Nat., Bot., sér. 5, 4: 44. 1865.
Typus: *Taxodium* Rich.

Tecophilaeaceae Leyb. in Bonplandia 10: 370. 1862.
Typus: *Tecophilaea* Bertero ex Colla

Tetracentraceae A. C. Sm. in J. Arnold Arbor. 26: 135. 16 Apr 1945.
Typus: *Tetracentron* Oliv.

Tetragoniaceae Lindl., Intr. Nat. Syst. Bot., ed. 2: 209. 13 Jun 1836.
Typus: *Tetragonia* L.

Theaceae Mirb. in Bot. Reg.: ad t. 112. 1 Mai 1816.
Typus: *Thea* L. [= *Camellia* L.].

Theligonaceae Dumort., Anal. Fam. Pl.: 15, 17. 1829.
Typus: *Theligonum* L.

Theophrastaceae D. Don in Edwards's Bot. Reg.: ad t. 1764. 1 Jun 1835.
Typus: *Theophrasta* L.

Thismiaceae J. Agardh, Theoria Syst. Pl.: 99. Apr-Sep 1858.
Typus: *Thismia* Griff.

Thurniaceae Engl., Syllabus, ed. 5: 94. 20-22 Jul 1907.
Typus: *Thurnia* Hook. f.

Thymelaeaceae Juss., Gen. Pl.: 76. 4 Aug 1789.
Typus: *Thymelaea* Mill. (nom. cons.).

Tiliaceae Juss., Gen. Pl.: 289. 4 Aug 1789.
Typus: *Tilia* L.

Tovariaceae Pax in Engler & Prantl, Nat. Pflanzenfam. 3(2): 207. Mar 1891.
Typus: *Tovaria* Ruiz & Pav. (nom. cons.).

Trapaceae Dumort., Anal. Fam. Pl.: 36, 39. 1829.
Typus: *Trapa* L.

Tremandraceae R. Br. ex DC., Prodr. 1: 343. Jan (med.) 1824.
Typus: *Tremandra* R. Br. ex DC.

Trichopodaceae Hutch., Fam. Fl. Pl. 2: 143. 20 Aug 1934.
Typus: *Trichopus* Gaertn.

Tricyrtidaceae Takht., Divers. Classific. Fl. Pl.: 482. 24 Apr 1997.
Typus: *Tricyrtis* Wall.

Trigoniaceae A. Juss. in Orbigny, Dict. Univ. Hist. Nat. 12: 670. 7 Jul 1849.
Typus: *Trigonia* Aubl.

Trilliaceae Chevall., Fl. Gén. Env. Paris 2: 297. 1827.
Typus: *Trillium* L.

Trimeniaceae Gibbs, Fl. Arfak Mts.: 135. Jul 1917.
Typus: *Trimenia* Seem. (nom. cons.).

Triuridaceae Gardner in Trans. Linn. Soc. London 19: 160. 5 Aug 1843.
Typus: *Triuris* Miers

Trochodendraceae Eichler in Flora 48: 14. 18 Jan 1865.
Typus: *Trochodendron* Siebold & Zucc.

Tropaeolaceae Juss. ex DC., Prodr. 1: 683. Jan (med.) 1824.
Typus: *Tropaeolum* L.

Turneraceae Kunth ex DC., Prodr. 3: 345. Mar (med.) 1828.
Typus: *Turnera* L.

Typhaceae Juss., Gen. Pl.: 25. 4 Aug 1789.
Typus: *Typha* L.

Ulmaceae Mirb., Elém. Physiol. Vég. Bot. 2: 905. 24-30 Jun 1815.
Typus: *Ulmus* L.

Umbelliferae Juss., Gen. Pl.: 218. 4 Aug 1789; nom. alt.: *Apiaceae.*
Typus: *Apium* L.

Urticaceae Juss., Gen. Pl.: 400. 4 Aug 1789.
Typus: *Urtica* L.

Uvulariaceae A. Gray ex Kunth, Enum. Pl. 4: 199. 17-19 Jul 1843.
Typus: *Uvularia* L.

Vacciniaceae DC. ex Perleb, Vers. Arzneikr. Pfl.: 228. Mai 1818.
Typus: *Vaccinium* L.

Valerianaceae Batsch, Tab. Affin. Regni Veg.: 227. 2 Mai 1802.
Typus: *Valeriana* L.

Velloziaceae J. Agardh, Theoria Syst. Pl.: 9. Apr-Sep 1858.
Typus: *Vellozia* Vand.

Verbenaceae J. St.-Hil., Expos. Fam. Nat. 1: 245. Feb-Apr 1805.
Typus: *Verbena* L.

Violaceae Batsch, Tab. Affin. Regni Veg.: 57. 2 Mai 1802.
Typus: *Viola* L.

Vitaceae Juss., Gen. Pl.: 267. 4 Aug 1789.
Typus: *Vitis* L.

Vochysiaceae A. St.-Hil. in Mém. Mus. Hist. Nat. 6: 265. 1820.
Typus: *Vochysia* Aubl. (nom. cons.).

Walleriaceae H. Huber ex Takht. in Bot. Zhurn. (Moscow & Leningrad) 79(12): 65. 7-28 Feb 1995 (subfam. *Wallerioideae* R. Dahlgren in Monogr. Syst. Bot. Missouri Bot. Gard. 25: 75. Jun 1988).
Typus: *Walleria* J. Kirk.

Welwitschiaceae Caruel in Nuovo Giorn. Bot. Ital. 11: 17, 18. 2 Jan 1879.
Typus: *Welwitschia* Hook. f. (nom. cons.).

Winteraceae R. Br. ex Lindl., Intr. Nat. Syst. Bot.: 26. Sep 1830.
Typus: *Wintera* Murray, nom. illeg. (*Drimys* J. R. Forst. & G. Forst., nom. cons.).

Xanthorrhoeaceae Dumort., Anal. Fam. Pl.: 60, 62, 103. 1829.
Typus: *Xanthorrhoea* Sm.

Xyridaceae C. Agardh, Aphor. Bot.: 158. 23 Mai 1823.
Typus: *Xyris* L.

Zannichelliaceae Chevall., Fl. Gén. Env. Paris 2: 256. 1827.
Typus: *Zannichellia* L.

Zingiberaceae Martinov, Tekhno-Bot. Slovar: 682. 3 Aug 1820.
Typus: *Zingiber* Mill. (nom. cons.).

Zosteraceae Dumort., Anal. Fam. Pl.: 65, 66. 1829.
Typus: *Zostera* L.

Zygophyllaceae R. Br. in Flinders, Voy. Terra Austr. 2: 545. 19 Jul 1814.
Typus: *Zygophyllum* L.

APPENDIX III

NOMINA GENERICA CONSERVANDA ET REJICIENDA

In the following lists the nomina conservanda have been inserted in the left column, in **bold-face italics**. Synonyms and earlier homonyms (nomina rejicienda) are listed in the right column. Conserved names are listed alphabetically within the major groups:

A. *Algae*
1. *Bacillariophyceae* (incl. fossil diatoms)
2. *Bodonophyceae*
3. *Chlorophyceae*
4. *Chrysophyceae*
5. *Cyanophyceae*
6. *Dinophyceae*
7. *Euglenophyceae*
8. *Phaeophyceae*
9. *Rhodophyceae*
10. *Trichomonadophyceae*
11. *Xanthophyceae*

B. *Fungi*

C. *Bryophyta*
1. *Hepaticae*
2. *Musci*

D. *Pteridophyta*

E. *Spermatophyta*
1. *Gymnospermae*
2. *Monocoyledones*
3. *Dicotyledones*

F. Fossil plants (excl. diatoms)

orth. cons.   orthographia conservanda, spelling to be conserved (Art. 14.11).

typ. cons.   typus conservandus, type to be conserved (Art. 14.9; see also Art. 14.3 and 10.4); as by Art. 14.8, listed types of conserved names may not be changed even if they are not explicitly designated as "typ. cons."

typ. des.   typi designatio, designation of type (Art. 10.5); used with names that became nomenclatural synonyms by type designation.

vide   see; usually followed by a reference to the author and place of publication of first type designation (Art. 10.5); also used (as "etiam vide", see also) for cross-reference to another relevant entry.

(H)   homonym (Art. 14.10; see also Art. 53), only the earliest being listed.

(≡)   nomenclatural synonym (i.e., homotypic synonym, based on the same nomenclatural type as the conserved name; Art. 14.4), usually only the earliest legitimate one being listed (but more than one in some cases in which homotypy results from type designation).

             Nomenclatural synonyms of rejected names, when they exist, are cited instead of the type. Nomenclatural synonyms that are part of a type entry are placed in parentheses (round brackets).

(=)   taxonomic synonym (i.e., heterotypic synonym, based on a type different from that of the conserved name), to be rejected only in favour of the conserved name (Art. 14.6 and 14.7).

Some type citations are followed by an indication of heterotypic synonymy (the supposedly correct name and its basionym, if any), reflecting current taxonomic opinion and in no way binding for nomenclatural purposes.

Some names listed as conserved have no corresponding nomina rejicienda because they were conserved solely to maintain a particular type, because evidence after their conservation may have indicated that conservation was unnecessary (see Art. 14.13), or because they were conserved to eliminate doubt about their legitimacy.

A. ALGAE

A1. BACILLARIOPHYCEAE (INCL. FOSSIL DIATOMS)

Acanthoceras Honigm. in Arch. Hydrobiol. Planktonk. 5: 76. 16 Oct 1909.
Typus: *A. magdeburgense* Honigm.

(H) *Acanthoceras* Kütz. in Linnaea 15: 731. Feb-Mar 1842 [*Rhodoph.*].
Typus: *A. shuttleworthianum* Kütz.

Actinella F. W. Lewis in Proc. Acad. Nat. Sci. Philadelphia 1863: 343. 1864.
Typus: *A. punctata* F. W. Lewis [Foss.].

(H) *Actinella* Pers., Syn. Pl. 2: 469. Sep 1807 [*Dicot.: Comp.*].
≡ *Actinea* Juss. 1803.

Arachnoidiscus H. Deane pat. ex Shadbolt in Trans. Roy. Microscop. Soc. London 3: 49. 1852.
Typus: *A. japonicus* Shadbolt ex A. Pritch. (Hist. Infus., ed. 3: 319. 1852).

(H) *Arachnodiscus* Bailey ex Ehrenb. in Ber. Bekanntm. Verh. Königl. Preuss. Akad. Wiss. Berlin 1849: 63. 1849 [*Bacillarioph.*].
≡ *Hemiptychus* Ehrenb. 1848 (nom. rej. sub *Arachnoidiscus*).

(=) *Hemiptychus* Ehrenb. in Ber. Bekanntm. Verh. Königl. Preuss. Akad. Wiss. Berlin 1848: 7. 1848.
Typus: *H. ornatus* Ehrenb.

Aulacodiscus Ehrenb. in Ber. Bekanntm. Verh. Königl. Preuss. Akad. Wiss. Berlin 1844: 73. 1844.
Typus: *A. crux* Ehrenb. [Foss.].

(=) *Tripodiscus* Ehrenb., Lebende Thierart. Kreidebild.: 50. 1840.
Typus: *T. germanicus* Ehrenb. (*T. argus* Ehrenb., nom. alt.).

(=) *Pentapodiscus* Ehrenb. in Ber. Bekanntm. Verh. Königl. Preuss. Akad. Wiss. Berlin 1843: 165. 1843.
Typus: *P. germanicus* Ehrenb.

(=) *Tetrapodiscus* Ehrenb. in Ber. Bekanntm. Verh. Königl. Preuss. Akad. Wiss. Berlin 1843: 165. 1843.
Typus: *T. germanicus* Ehrenb.

Auricula Castrac. in Atti Accad. Pontif. Sci. Nuovi Lincei 26: 407. 1873.
Typus: *A. amphitritis* Castrac.

(H) *Auricula* Hill, Brit. Herb.: 98. 31 Mar 1756 [*Dicot.: Primul.*].
Typus: non designatus.

Brebissonia Grunow in Verh. K.K. Zool.-Bot. Ges. Wien 10: 512. 1860.
Typus: *B. boeckii* (Ehrenb.) O'Meara (in Proc. Roy. Irish Acad., ser. 2, 2(Sci.): 338. Oct 1875) (*Cocconema boeckii* Ehrenb.).

(H) *Brebissonia* Spach, Hist. Nat. Vég. 4: 401. 11 Apr 1835 [*Dicot.: Onagr.*].
Typus (vide Spach in Ann. Sci. Nat., Bot., ser. 2, 4: 175. 1835): *B. microphylla* (Kunth) Spach (*Fuchsia microphylla* Kunth).

Cerataulina H. Perag. ex F. Schütt in Engler & Prantl, Nat. Pflanzenfam. 1(1b): 95. Dec (sero) 1896.
Typus: *C. bergonii* (H. Perag.) F. Schütt (*Cerataulus bergonii* H. Perag.) [= *C. pelagica* (Cleve) Hendey (*Zygoceras pelagicum* Cleve)].

(=) *Syringidium* Ehrenb. in Ber. Bekanntm. Verh. Königl. Preuss. Akad. Wiss. Berlin 1845: 357. 1845.
Typus: *S. bicorne* Ehrenb.

Coscinodiscus Ehrenb. in Abh. Königl. Akad. Wiss. Berlin, Phys. Abh., 1838: 128. 1839.
Typus: *C. argus* Ehrenb. (typ. cons.) [Foss.].

Cyclotella (Kütz.) Bréb., Consid. Diatom.: 19. 1838, (*Frustulia* subg. *Cyclotella* Kütz. in Linnaea 8: 535. 1834).
Typus: *C. tecta* Håk. & R. Ross. (in Taxon 33: 529. 20 Aug 1984) (typ. cons.) [= *C. distinguenda* Hust.].

Cymatopleura W. Sm. in Ann. Mag. Nat. Hist., ser. 2, 7: 12. Jan 1851.
Typus: *C. solea* (Bréb.) W. Sm. (*Cymbella solea* Bréb.) [= *C. librile* (Ehrenb.) Pant. (*Navicula librile* Ehrenb.)].

(=) *Sphinctocystis* Hassall, Hist. Brit. Freshwater Alg. 1: 436. Jul-Dec 1845.
Typus: *S. librile* (Ehrenb.) Hassall (*Navicula librile* Ehrenb.).

Cymbella C. Agardh, Consp. Diatom.: 1. 4 Dec 1830.
Typus: *C. cymbiformis* C. Agardh (typ. cons.).

Diatoma Bory, Dict. Class. Hist. Nat. 5: 461. 15 Mai 1824.
Typus: *D. vulgaris* Bory (typ. cons.).

(H) *Diatoma* Lour., Fl. Cochinch.: 290, 295. Sep 1790 [*Dicot.: Rhizophor.*].
Typus: *D. brachiata* Lour.

Diatomella Grev. in Ann. Mag. Nat. Hist., ser. 2, 15: 259. Apr 1855.
Typus: *D. balfouriana* Grev.

(=) *Disiphonia* Ehrenb., Mikrogeologie: 260. 1854.
Typus: *D. australis* Ehrenb. [Foss.].

Didymosphenia Mart. Schmidt in Schmidt, Atlas Diatom.-Kunde: t. 214, f. 1-12. Mar 1899.
Typus: *D. geminata* (Lyngb.) Mart. (etiam vide *Gomphonema* [*Bacillarioph.*]).

(≡) *Dendrella* Bory, Dict. Class. Hist. Nat. 5: 393. 15 Mai 1824 (typ. des. in Regnum Veg. 3: 70. 1952).

(=) *Diomphala* Ehrenb. in Ber. Bekanntm. Verh. Königl. Preuss. Akad. Wiss. Berlin 1842: 336. 1843.
Typus: *D. clava-herculis* Ehrenb. [Foss.].

Eupodiscus Bailey in Smithsonian Contr. Knowl. 2(8): 39. 1851.
Typus: *E. radiatus* Bailey (typ. cons.).

(H) *Eupodiscus* Ehrenb. in Ber. Bekanntm. Verh. Königl. Preuss. Akad. Wiss. Berlin 1844: 73. 1844 [*Bacillarioph.*].
≡ *Tripodiscus* Ehrenb. 1840 (nom. rej. sub *Aulacodiscus*).

Fragilariopsis Hust. in Schmidt's Atlas Diatom.-Kunde, expl. t. 299, f. 9-14. Aug 1913.
Typus: *F. antarctica* Hust. *(Fragilaria antarctica* Castrac., non A. F. Schwarz).

(=) *Pseudo-eunotia* Grunow in Van Heurck, Syn. Diatom. Belgique, expl. t. 35, f. 22. Mai-Jun 1881.
Typus: *P. doliolus* (G. C. Wall.) Grunow *(Synedra doliolus* G. C. Wall.).

Frustulia Rabenh., Süssw.-Diatom.: 50. Mar-Mai 1853.
Typus: *F. saxonica* Rabenh. (typ. cons.).

(H) *Frustulia* C. Agardh, Syst. Alg.: xiii, 1. Mai-Sep 1824 [*Bacillarioph.*].
Typus (vide Round & al., Diatoms: 690. 1990): *F. obtusa* (Lyngb.) C. Agardh *(Echinella obtusa* Lyngb.).

Gomphonema Ehrenb. in Abh. Königl. Akad. Wiss. Berlin, Phys. Kl., 1831: 87. 1832.
Typus: *G. acuminatum* Ehrenb. (typ. cons.).

(H) *Gomphonema* C. Agardh, Syst. Alg.: xvi, 11. Mai-Sep 1824 [*Bacillarioph.*].
≡ *Didymosphenia* Mart. Schmidt 1899 (nom. cons.).

Gyrosigma Hassall, Hist. Brit. Freshwater Alg. 1: 435. Jul-Dec 1845.
Typus: *G. hippocampus* Hassall, nom. illeg. (*Navicula hippocampus* Ehrenb., nom. illeg., *Frustulia attenuata* Kütz., *Gyrosigma attenuatum* (Kütz.) Rabenh.).

(=) *Scalptrum* Corda in Alman. Carlsbad 5: 193. 1835.
Typus: *S. striatum* Corda

Hantzschia Grunow in Monthly Microscop. J. 18: 174. 1 Oct 1877. Typus: *H. amphioxys* (Ehrenb.) Grunow (*Eunotia amphioxys* Ehrenb.).

(H) *Hantzschia* Auersw. in Hedwigia 2: 60. 1862 [*Fungi*]. Typus: *H. phycomyces* Auersw.

Hemiaulus Heib., Krit. Overs. Danske Diatom.: 45. 4 Jun 1863. Typus: *H. proteus* Heib. [Foss.].

(H) *Hemiaulus* Ehrenb. in Ber. Bekanntm. Verh. Königl. Preuss. Akad. Wiss. Berlin 1844: 199. 1844 [*Bacillarioph.*]. Typus: *H. antarcticus* Ehrenb.

Licmophora C. Agardh in Flora 10: 628. 28 Oct 1827. Typus: *L. argentescens* C. Agardh

(=) *Styllaria* Drap. ex Bory, Dict. Class. Hist. Nat. 2: 129. 31 Dec 1822. Typus: *S. paradoxa* (Lyngb.) Bory (Hist. Nat. Zooph.: 709. 1827) (*Echinella paradoxa* Lyngb.).

(=) *Exilaria* Grev., Scott. Crypt. Fl.: ad t. 289. Apr 1827. Typus: *E. flabellata* Grev.

Melosira C. Agardh, Syst. Alg.: xiv, 8. Mai-Sep 1824 (*'Meloseira'*) (orth. cons.). Typus: *M. nummuloides* C. Agardh

(=) *Lysigonium* Link in Nees, Horae Phys. Berol.: 4. 1-8 Feb 1820. Typus (vide Regnum Veg. 3: 71. 1952): *Conferva moniliformis* O. F. Müll.

Nitzschia Hassall, Hist. Brit. Freshwater Alg. 1: 435. Jul-Dec 1845. Typus: *N. elongata* Hassall, nom. illeg. (*Bacillaria sigmoidea* Nitzsch, *N. sigmoidea* (Nitzsch) W. Sm.).

(≡) *Sigmatella* Kütz., Alg. Aq. Dulc. Germ.: No. 2. Jan-Feb 1833.

(=) *Homoeocladia* C. Agardh in Flora 10: 629. 28 Oct 1827. Typus: *H. martiana* C. Agardh

Pantocsekia Grunow ex Pant., Beitr. Foss. Bacill. Ung. 1: 47. 1886. Typus: *P. clivosa* Grunow ex Pant. [Foss.].

(H) *Pantocsekia* Griseb. ex Pant. in Oesterr. Bot. Z. 23: 267. Sep 1873 [*Dicot.: Convolvul.*]. Typus: *P. illyrica* Griseb. ex Pant.

Peronia Bréb. & Arn. ex Kitton in Quart. J. Microscop. Sci., ser. 2, 8: 16. 1868. Typus: *P. erinacea* Bréb. & Arn. ex Kitton, nom. illeg. (*Gomphonema fibula* Bréb. ex Kütz., *P. fibula* (Bréb. ex Kütz.) R. Ross).

(H) *Peronia* Redouté, Liliac.: ad t. 342. 15 Nov 1811 [*Monocot.: Marant.*]. Typus: *P. stricta* F. Delaroche

Pinnularia Ehrenb. in Ber. Bekanntm. Verh. Königl. Preuss. Akad. Wiss. Berlin 1843: 45. 1843.
Typus: *P. viridis* (Nitzsch) Ehrenb. (*Bacillaria viridis* Nitzsch) (typ. cons.).

(H) *Pinnularia* Lindl. & Hutton, Foss. Fl. Gr. Brit. 2: [81], t. 111. 1834.
Typus: *P. capillacea* Lindl. & Hutton [Foss.].

(=) *Stauroptera* Ehrenb. in Ber. Bekanntm. Verh. Königl. Preuss. Akad. Wiss. Berlin 1843: 45. 1843.
Typus: *S. semicruciata* Ehrenb. [Foss.].

Pleurosigma W. Sm. in Ann. Mag. Nat. Hist., ser. 2, 9: 2. Jan 1852.
Typus: *P. angulatum* (E. J. Quekett) W. Sm. (*Navicula angulata* E. J. Quekett) (typ. cons.: Northern Ireland, Belfast, Aug 1849, *Smith* (BM No. 23671)).

(=) *Scalptrum* Corda in Alman. Carlsbad 5: 193. 1835.
Typus: *S. striatum* Corda

(=) *Gyrosigma* Hassall, Hist. Brit. Freshwater Alg. 1: 435. Jul-Dec 1845 (nom. cons.).

(=) *Endosigma* Bréb. in Orbigny, Dict. Univ. Hist. Nat. 11: 418, 419. 1848.
Typus: non designatus.

Podocystis Bailey in Smithsonian Contr. Knowl. 7(3): 11. 1854.
Typus: *P. americana* Bailey

(H) *Podocystis* Fr., Summa Veg. Scand.: 512. 1849 [*Fungi*].
Typus (vide Laundon in Mycol. Pap. 99: 14. 1965): *P. capraearum* (DC.) Fr. (*Uredo capraearum* DC.).

(=) *Euphyllodium* Shadbolt in Trans. Roy. Microscop. Soc. London, ser. 2, 2: 14. 1854.
Typus: *E. spathulatum* Shadbolt

Rhabdonema Kütz., Kieselschal. Bacill.: 126. 7-9 Nov 1844.
Typus: *R. minutum* Kütz.

(=) *Tessella* Ehrenb., Zus. Erkenntn. Organis.: 23. 1836.
Typus: *T. catena* Ehrenb. [Foss.].

Rhizosolenia Brightw. in Quart. J. Microscop. Sci. 6: 94. Apr 1858.
Typus: *R. styliformis* Brightw.

(H) *Rhizosolenia* Ehrenb. in Abh. Königl. Akad. Wiss. Berlin, Phys. Abh., 1841: 402. 1843 [*Bacillarioph.*].
Typus: *R. americana* Ehrenb. [Foss.]

Rhopalodia O. Müll. in Bot. Jahrb. Syst. 22: 57. 19 Nov 1895.
Typus: *R. gibba* (Ehrenb.) O. Müll. (*Navicula gibba* Ehrenb.).

(=) *Pyxidicula* Ehrenb. in Abh. Königl. Akad. Wiss. Berlin, Phys. Abh., 1833: 295. 1834.
Typus: *P. operculata* (C. Agardh) Ehrenb. (*Frustulia operculata* C. Agardh).

Skeletonema Grev. in Trans. Roy. Microscop. Soc. London, ser. 2, 13: 43. Jul 1865.
Typus: *S. costatum* (Grev.) Cleve (*Melosira costata* Grev.) (typ. cons.).

Skeletonemopsis P. A. Sims in Diatom Res. 9: 408. 1995.
Typus: *S. barbadensis* (Grev.) P. A. Sims (*Skeletonema barbadense* Grev.) [Foss.].

Staurophora Mereschk. in Beih. Bot. Central. 15: 20. 1903
Typus: *S. amphioxys* (W. Greg.) D. G. Mann *(Stauroneis amphioxys* W. Greg.).

(H) *Staurophora* Willd. in Ges. Naturf. Freunde Berlin Mag. Neuesten Entdeck. Gesammten Naturk. 3: 101. Apr-Jun 1809 [*Hepat.*].
Typus: *S. pulchella* Willd., nom. illeg. (*Marchantia cruciata* L.).

Tetracyclus Ralfs in Ann. Mag. Nat. Hist. 12: 105. Aug 1843.
Typus: *T. lacustris* Ralfs

(=) *Biblarium* Ehrenb. in Ber. Bekanntm. Verh. Königl. Preuss. Akad. Wiss. Berlin 1843: 47. 1843.
Typus: *B. glans* (Ehrenb.) Ehrenb. (*Navicula glans* Ehrenb.) [Foss.].

Thalassiothrix Cleve & Grun. in Kongl. Svenska Vetenskapsakad. Handl., ser. 4, 17(2): 108. 1880.
Typus: *T. longissima* Cleve & Grun. (typ. cons.).

A2. BODONOPHYCEAE

Karotomorpha B. V. Travis in Trans. Amer. Microscop. Soc. 53: 277. Jul 1934.
Typus: *K. bufonis* (Dobell) B. V. Travis (*Monocercomonas bufonis* Dobell).

(≡) *Tetramastix* A. G. Alexeev in Compt.-Rend. Séances Mém. Soc. Biol. 79: 1076. 2 Dec 1916.

165

A3. CHLOROPHYCEAE

Acetabularia J. V. Lamour. in Nouv. Bull. Sci. Soc. Philom. Paris 3: 185. Dec 1812. Typus: *A. acetabulum* (L.) P. C. Silva (in Univ. Calif. Publ. Bot. 25: 255. 9 Jun 1952) (*Madrepora acetabulum* L.).

(≡) *Acetabulum* Boehm. in Ludwig, Def. Gen. Pl., ed. 3: 504. 1760.

Anadyomene J. V. Lamour. in Nouv. Bull. Sci. Soc. Philom. Paris 3: 187. Dec 1812 (*'Anadyomena'*) (orth. cons.). Typus: *A. flabellata* J. V. Lamour. [= *A. stellata* (Wulfen) C. Agardh (*Ulva stellata* Wulfen)].

Aphanochaete A. Braun, Betracht. Erschein. Verjüng. Natur: 196. 1850. Typus: *A. repens* A. Braun [= *A. confervicola* (Nägeli) Rabenh. (*Herposteiron confervicola* Nägeli)].

(=) *Herposteiron* Nägeli in Kützing, Sp. Alg.: 424. 23-24 Jul 1849. Typus: *H. confervicola* Nägeli

Bambusina Kütz., Sp. Alg.: 188. 23-24 Jul 1849. Typus: *B. brebissonii* Kütz., nom. illeg. (*Didymoprium borreri* Ralfs, *B. borreri* (Ralfs) Cleve).

(=) *Gymnozyga* Ehrenb. ex Kütz., Sp. Alg.: 188. 23-24 Jul 1849. Typus: *G. moniliformis* Ehrenb. ex Kütz.

Chaetomorpha Kütz., Phycol. Germ.: 203. 14-16 Aug 1845. Typus: *C. melagonium* (Weber & Mohr) Kütz. (*Conferva melagonium* Weber & Mohr).

(=) *Chloronitum* Gaillon in Cuvier, Dict. Sci. Nat. 53: 389. 1828. Typus (vide Silva in Univ. Calif. Publ. Bot. 25: 270. 1952): *C. aereum* (Dillwyn) Gaillon (*Conferva aerea* Dillwyn).

(=) *Spongopsis* Kütz., Phycol. General.: 261. 14-16 Sep 1843. Typus: *S. mediterranea* Kütz.

Chlamydomonas Ehrenb. in Abh. Königl. Akad. Wiss. Berlin, Phys. Abh., 1833: 288. 1834 *('Chlamidomonas')* (orth. cons.).
Typus: *C. pulvisculus* (O. F. Müll.) Ehrenb. (*Monas pulvisculus* O. F. Müll.).

(=) *Protococcus* C. Agardh, Syst. Alg.: xvii, 13. Mai-Sep 1824.
Typus (vide Drouet & Daily in Butler Univ. Bot. Stud. 12: 167. 1956): *P. nivalis* (F. A. Bauer) C. Agardh (*Uredo nivalis* F. A. Bauer).

(=) *Sphaerella* Sommerf. in Mag. Naturvidensk. 4: 252. 1824.
Typus (vide Hazen in Mem. Torrey Bot. Club 6: 238. 1899): *S. nivalis* (Bauer) Sommerf. (*Uredo nivalis* Bauer)

Chlorella Beij. in Bot. Zeitung (Berlin) 48: 758. 21 Nov 1890.
Typus: *C. vulgaris* Beij.

(=) *Zoochlorella* K. Brandt in Verh. Physiol. Ges. Berlin 1881-1882: 24. 1881; in Arch. Anat. Physiol., Physiol. Abt. 1881: 571. 2 Dec 1881.
Typus: *Z. conductrix* K. Brandt

Chlorococcum Menegh. in Mem. Reale Accad. Sci. Torino, ser. 2, 5: 24. 1842.
Typus: *C. infusionum* (Schrank) Menegh. (*Lepra infusionum* Schrank) (typ. cons.).

(H) *Chlorococcum* Fr., Syst. Orb. Veg.: 356. Dec 1825 [*Chloroph.*].
≡ *Protococcus* C. Agardh 1824 (nom. rej. sub *Chlamydomonas*).
≡ *Sphaerella* Sommerf. 1824 (nom. rej. sub *Chlamydomonas*).

Chloromonas Gobi in Bot. Zap. 15: 232, 255. 1899-1900.
Typus: *C. reticulata* (Gorozh.) Gobi (*Chlamydomonas reticulata* Gorozh.).

(H) *Chloromonas* Kent, Man. Infus.: 369, 401. 1881 [*Euglenoph.*].
≡ *Cryptoglena* Ehrenb. 1832.

(=) *Tetradonta* Korshikov in Russk. Arh. Protistol. 4: 183, 195. 1925.
Typus: *T. variabilis* Korshikov

(=) *Platychloris* Pascher, Süsswasserflora 4: 138, 331. Jan-Mar 1927.
Typus: *P. minima* Pascher (*Chlamydomonas minima* Pascher, non P. A. Dang.).

167

Cladophora Kütz., Phycol. General.: 262. 14-16 Sep 1843.
Typus: *C. oligoclona* (Kütz.) Kütz. (*Conferva oligoclona* Kütz.).

(=) *Conferva* L., Sp. Pl.: 1164. 1 Mai 1753.
Typus (vide Bonnem. in J. Phys. Chim. Hist. Nat. Arts 94: 198. 1822): *C. rupestris* L.

(=) *Annulina* Link in Nees, Horae Phys. Berol.: 4. 1-8 Feb 1820.
Typus (vide Silva in Univ. Calif. Publ. Bot. 25: 270. 1952): *A. glomerata* (L.) Nees (Horae Phys. Berol.: [index]. 1820) (*Conferva glomerata* L.).

Cladophoropsis Børgesen in Overs. Kongel. Danske Vidensk. Selsk. Forh. Medlemmers Arbeider 1905: 288. 10 Jun 1905.
Typus: *C. membranacea* (C. Agardh) Børgesen (*Conferva membranacea* C. Agardh).

(=) *Spongocladia* Aresch. in Öfvers. Förh. Kongl. Svenska Vetensk.-Akad. 10: 202. 1853.
Typus: *S. vaucheriiformis* Aresch.

Coleochaete Bréb. in Ann. Sci. Nat., Bot., ser. 3, 1: 29. Jan 1844.
Typus: *C. scutata* Bréb.

(=) *Phyllactidium* Kütz., Phycol. General.: 294. 14-16 Sep 1843.
Typus (vide Meneghini in Atti Riunione Sci. Ital. 6: 457. 1845): *P. pulchellum* Kütz.

Enteromorpha Link in Nees, Horae Phys. Berol.: 5. 1-8 Feb 1820.
Typus: *E. intestinalis* (L.) Nees (Horae Phys. Berol.: [index]. 1-8 Feb 1820.) (*Ulva intestinalis* L.).

(≡) *Splaknon* Adans., Fam. Pl. 2: 13, 607. Jul-Aug 1763 (typ. des.: Silva in Univ. Calif. Publ. Bot. 25: 294. 1952).

Gloeococcus A. Braun, Betracht. Erschein. Verjüng. Natur: 169. 1850.
Typus: *G. minor* A. Braun (typ. cons.).

(H) *Gloiococcus* Shuttlew. in Biblioth. Universelle Genève, ser. 2, 25: 405. Feb 1840 [*Algae*].
Typus: *G. grevillei* (C. Agardh) Shuttlew. (*Haematococcus grevillei* C. Agardh).

Gongrosira Kütz., Phycol. General.: 281. 14-16 Sep 1843.
Typus: *G. sclerococcus* Kütz., nom. illeg. (*Stereococcus viridis* Kütz., *G. viridis* (Kütz.) De Toni).

(≡) *Stereococcus* Kütz. in Linnaea 8: 379. 1833.

Haematococcus Flot. in Nov. Actorum Acad. Caes. Leop.-Carol. Nat. Cur. 20: 413. 1844.
Typus: *H. pluvialis* Flot. (typ. cons.) [= *H. lacustris* (Gir.-Chantr.) Rostaf. (*Volvox lacustris* Gir.-Chantr.)].

(H)   *Haematococcus* C. Agardh, Icon. Alg. Eur.: ad t. 22. 1830 [*Cyanoph.*].
Typus (vide Morren in Nouv. Mém. Acad. Roy. Sci. Bruxelles 14(7): 9. 1841): *H. sanguineus* (C. Agàrdh) C. Agardh (*Palmella sanguinea* C. Agardh).

(=)   *Disceraea* Morren & C. Morren in Nouv. Mém. Acad. Roy. Sci. Bruxelles 14(5): 37. 1841.
Typus: *D. purpurea* Morren & C. Morren

Halimeda J. V. Lamour. in Nouv. Bull. Sci. Soc. Philom. Paris 3: 186. Dec 1812 (*'Halimedea'*) (orth. cons.).
Typus: *H. tuna* (J. Ellis & Sol.) J. V. Lamour. (Hist. Polyp. Corall.: 309. 1816) (*Corallina tuna* J. Ellis & Sol.).

(≡)   *Sertularia* Boehm. in Ludwig, Def. Gen. Pl., ed. 3: 504. 1760 (typ. des.: Silva in Univ. Calif. Publ. Bot. 25: 294. 1952).

Hydrodictyon Roth, Bemerk. Crypt. Wassergew.: 48. Feb-Aug 1797.
Typus: *H. reticulatum* (L.) Bory (Dict. Class. Hist. Nat. 6: 506. 9 Oct 1824) (*Conferva reticulata* L.).

(≡)   *Reticula* Adans., Fam. Pl. 2: 3, 598. Jul-Aug 1763.

Microspora Thur. in Ann. Sci. Nat., Bot., ser. 3, 14: 221. Oct 1850.
Typus: *M. floccosa* (Vaucher) Thur. (*Prolifera floccosa* Vaucher).

(H)   *Microspora* Hassall in Ann. Mag. Nat. Hist. 11: 363. Mai 1843 [*Chloroph.*].
Typus: non designatus.

Mougeotia C. Agardh, Syst. Alg.: xxvi, 83. Mai-Sep 1824.
Typus: *M. genuflexa* (Roth) C. Agardh (*Conferva genuflexa* Roth).

(H)   *Mougeotia* Kunth in Humboldt & al., Nov. Gen. Sp. 5, ed. f°: 253; ed. 4°: 326. 1823 [*Dicot.: Stercul.*].
Typus: non designatus.

(≡)   *Serpentinaria* Gray, Nat. Arr. Brit. Pl. 1: 299. 1 Nov 1821 (typ. des.: Silva in Univ. Calif. Publ. Bot. 25: 252. 1952).

(=)   *Agardhia* Gray, Nat. Arr. Brit. Pl. 1: 279, 299. 1 Nov 1821.
Typus: *A. caerulescens* (Sm.) Gray (*Conferva caerulescens* Sm.).

Prasiola Menegh. in Nuovi Saggi Imp. Regia Accad. Sci. Padova 4: 360. 1838.
Typus: *P. crispa* (Lightf.) Kütz. (Phycol. General.: 295. 14-16 Sep 1843) (*Ulva crispa* Lightf.).

(=) *Humida* Gray, Nat. Arr. Brit. Pl. 1: 278, 281. 1 Nov 1821.
Typus (vide Drouet in Acad. Nat. Sci. Philadelphia Monogr. 15: 312. 1968): *H. muralis* (Dillwyn) Gray (*Conferva muralis* Dillwyn).

Schizogonium Kütz., Phycol. General.: 245. 14-16 Sep 1843.
Typus: *S. murale* (Dillwyn) Kütz. (*Conferva muralis* Dillwyn).

(≡) *Humida* Gray, Nat. Arr. Brit. Pl. 1: 278, 281. 1 Nov 1821.

Sirogonium Kütz., Phycol. General.: 278. 14-16 Sep 1843.
Typus: *S. sticticum* (Sm.) Kütz. (*Conferva stictica* Sm.).

(≡) *Choaspis* Gray, Nat. Arr. Brit. Pl. 1: 279, 299. 1 Nov 1821

Sphaerozosma Ralfs, Brit. Desmid: 65. 1 Jan 1848.
Typus: *S. vertebratum* Ralfs

(H) *Sphaerozosma* Corda, Icon. Fung. 5: 27. Jun 1842 [*Fungi*].
≡ *Sphaerosoma* Klotzsch 1839.

Spirogyra Link in Nees, Horae Phys. Berol.: 5. 1-8 Feb 1820.
Typus: *S. porticalis* (O. F. Müll.) Dumort. (Comment. Bot.: 99. Nov (sero) - Dec (prim.) 1822) (*Conferva porticalis* O. F. Müll.).

(=) *Conjugata* Vaucher, Hist. Conferv. Eau Douce: 3, 37. Mar 1803.
Typus (vide Bonnemaison in J. Phys. Chim. Hist. Nat. Arts 94: 195. 1822): *C. princeps* Vaucher

Stigeoclonium Kütz., Phycol. General.: 253. 14-16 Sep 1843 *('Stygeoclonium')* (orth. cons.).
Typus: *S. tenue* (C. Agardh) Kütz. (*Draparnaldia tenuis* C. Agardh).

(=) *Myxonema* Fr., Syst. Orb. Veg.: 343. Dec 1825.
Typus (vide Hazen in Mem. Torrey Bot. Club 11: 193. 1902): *M. lubricum* (Dillwyn) Fr. (Fl. Scan.: 329. 1835) (*Conferva lubrica* Dillwyn).

Struvea Sond. in Bot. Zeitung (Berlin) 3: 49. 24 Jan 1845.
Typus: *S. plumosa* Sond.

(H) *Struvea* Rchb., Deut. Bot. Herb.-Buch, Syn.: 222, 236. Jul 1841 [*Gymnosp.: Tax.*].
≡ *Torreya* Arn. 1838 (nom. cons.).

Trentepohlia Mart., Fl. Crypt. Erlang.: 351. Jun 1817.
Typus: *T. aurea* (L.) Mart. (*Byssus aurea* L.).

(H) *Trentepohlia* Roth, Catal. Bot. 2: 73. 1800 [*Dicot.: Cruc.*].
Typus: non designatus.

(=) *Byssus* L., Sp. Pl.: 1168. 1 Mai 1753.
Typus (vide Fries, Stirp. Agri Femsion.: 42. 1825): *B. jolithus* L.

Ulva L., Sp. Pl.: 1163. 1 Mai 1753.
Typus: *U. lactuca* L. (typ. cons.).

Urospora Aresch. in Nova Acta Regiae Soc. Sci. Upsal., ser. 3, 6(2): 15. 1866.
Typus: *U. mirabilis* Aresch.

(=) *Hormiscia* Fr., Fl. Scan.: 326. 1835. Typus (vide Silva in Univ. Calif. Publ. Bot. 25: 270. 1952): *H. penicilliformis* (Roth) Fr. (*Conferva penicilliformis* Roth).

(=) *Codiolum* A. Braun, Alg. Unicell.: 19. Apr-Oct 1855.
Typus: *C. gregarium* A. Braun

Zygnema C. Agardh in Liljeblad, Utkast Sv. Fl., ed. 3: 492, 595. 1816.
Typus: *Z. cruciatum* (Vaucher) C. Agardh (*Conjugata cruciata* Vaucher).

(=) *Lucernaria* Roussel, Fl. Calvados, ed. 2: 20, 84. 1806.
Typus: *L. pellucida* Roussel

Zygogonium Kütz., Phycol. General.: 280. 14-16 Sep 1843.
Typus: *Z. ericetorum* (Roth) Kütz. (*Conferva ericetorum* Roth).

(≡) *Leda* Bory, Dict. Class. Hist. Nat. 1: 595. 27 Mai 1822 (typ. des.: Silva in Univ. Calif. Publ. Bot. 25: 253. 1952).

A4. CHRYSOPHYCEAE

Anthophysa Bory, Dict. Class. Hist. Nat. 1: 427, 597. 27 Mai 1822 (*'Anthophysis'*) (orth. cons.).
Typus: *A. muelleri* Bory, nom. illeg. (*Volvox vegetans* O. F. Müll., *A. vegetans* (O. F. Müll.) F. Stein).

Hydrurus C. Agardh, Syst. Alg.: xviii, 24. Mai-Sep 1824.
Typus: *H. vaucheri* C. Agardh, nom. illeg. (*Conferva foetida* Vill., *H. foetidus* (Vill.) Trevis.).

(≡) *Carrodorus* Gray, Nat. Arr. Brit. Pl. 1: 318, 350. 1 Nov 1821.

(=) *Cluzella* Bory, Dict. Class. Hist. Nat. 3: 14. 6 Sep 1823.
Typus: *C. myosurus* (Ducluz.) Bory (Dict. Class. Hist. Nat. 4: 234. 27 Dec 1823) (*Batrachospermum myosurus* Ducluz.).

A5. CYANOPHYCEAE

Anabaena Bory ex Bornet & Flahault in Ann. Sci. Nat., Bot., ser. 7, 7: 180, 224. 1 Jan 1886 .
Typus: *A. oscillarioides* Bory ex Bornet & Flahault

(H) *Anabaena* A. Juss., Euphorb. Gen.: 46. 21 Feb 1824 [*Dicot.: Euphorb.*].
Typus: *A. tamnoides* A. Juss.

Aphanothece Nägeli in Neue Denkschr. Allg. Schweiz. Ges. Gesammten Naturwiss. 10(7): 59. 1849.
Typus: *A. microscopica* Nägeli

(=) *Coccochloris* Spreng., Mant. Prim. Fl. Hal.: 14. 4 Jul 1807.
Typus: *C. stagnina* Spreng.

Gloeocapsa Kütz., Phycol. General.: 173. 14-16 Sep 1843.
Typus: *G. atrata* Kütz., nom. illeg. (*Microcystis atra* Kütz.).

(=) *Bichatia* Turpin in Mém. Mus. Hist. Nat. 16: 163. 1828.
Typus: *B. vesiculinosa* Turpin

Homoeothrix (Thur. ex Bornet & Flahault) Kirchn. in Engler & Prantl, Nat. Pflanzenfam. 1(1a): 85, 87. Aug 1898 (*Calothrix* sect. *Homoeothrix* Thur. ex Bornet & Flahault in Ann. Sci. Nat., Bot., ser. 7, 3: 345, 347. 1 Jan 1886).
Typus: *Calothrix juliana* Bornet & Flahault (*H. juliana* (Bornet & Flahault) Kirchn.).

(=) *Amphithrix* Bornet & Flahault in Ann. Sci. Nat., Bot., ser. 7, 3: 340, 343. 1 Jan 1886.
Typus (vide Geitler in Engler & Prantl, Nat. Pflanzenfam., ed. 2, 1b: 175. 1942): *A. janthina* Bornet & Flahault
(=) *Tapinothrix* Sauv. in Bull. Soc. Bot. France 39: cxxiii. 1892.
Typus: *T. bornetii* Sauv.

Leptolyngbya Anagn. & Komárek in Arch. Hydrobiol., Suppl. 80: 390. Mar 1988.
Typus: *L. boryana* (Gomont) Anagn. & Komárek (*Plectonema boryanum* Gomont) (typ. cons.).

(=) *Spirocoleus* (Möbius ex Kirchn.) Crow in Trans. Amer. Microscop. Soc. 46: 147. Apr 1927 (*Lyngbya* sect. *Spirocoleus* Möbius ex Kirchn. in Engler & Prantl, Nat. Pflanzenfam. 1(1a): 67. Aug 1898).
Typus: *S. lagerheimii* (Gomont) Möbius ex Crow (*Lyngbya lagerheimii* Gomont).

Lyngbya C. Agardh ex Gomont in Ann. Sci. Nat., Bot., ser. 7, 16: 95, 118. 1 Jan 1892.
Typus: *L. confervoides* C. Agardh ex Gomont

(H) *Lyngbyea* Sommerf., Suppl. Fl. Lapp.: 189. 1826 [*Bacillarioph.*].
Typus: non designatus.

Microchaete Thur. ex Bornet & Flahault in Ann. Sci. Nat., Bot., ser. 7, 5: 82, 83. 1 Jan 1886.
Typus: *M. grisea* Thur. ex Bornet & Flahault

(H) *Microchaete* Benth., Pl. Hartw.: 209. Nov 1845 [*Dicot.: Comp.*].
Typus (vide Pfeiffer, Nomencl. Bot. 2: 304. 1874): *M. pulchella* (Kunth) Benth. (*Cacalia pulchella* Kunth).

Microcystis Lemmerm., Krypt.-Fl. Brandenburg 3: 45, 72. 4 Mar 1907.
Typus: *M. aeruginosa* (Kütz.) Lemmerm. (*Micraloa aeruginosa* Kütz.) (typ. cons.).

(H) *Microcystis* Kütz. in Linnaea 8: 372. 1833 [*Euglenoph.*].
Typus (vide Drouet & Daily in Butler Univ. Bot. Stud. 12: 152. 1956): *M. noltei* (C. Agardh) Kütz. (*Haematococcus noltei* C. Agardh).

Nodularia Mert. ex Bornet & Flahault in Ann. Sci. Nat., Bot., ser. 7, 7: 180, 243. 1 Jan 1886.
Typus: *N. spumigena* Mert. ex Bornet & Flahault

(H) *Nodularia* Link ex Lyngb., Tent. Hydrophytol. Dan.: xxx, 99. Apr-Aug 1819 [*Rhodoph.*].
≡ *Lemanea* Bory 1808 (nom. cons.).

Rivularia C. Agardh ex Bornet & Flahault in Ann. Sci. Nat., Bot., ser. 7, 3: 341; 4: 345. 1 Jan 1886.
Typus: *R. atra* Roth ex Bornet & Flahault

(H) *Rivularia* Roth, Catal. Bot. 1: 212. Jan-Feb 1797 [*Chloroph.*]
Typus (vide Hazen in Mem. Torrey Bot. Club 11: 210. 1902): *R. cornudamae* Roth

Trichodesmium Ehrenb. ex Gomont in Ann. Sci. Nat., Bot., ser. 7, 16: 96, 193. 1 Jan 1892.
Typus: *T. erythraeum* Ehrenb. ex Gomont

(H) *Trichodesmium* Chevall., Fl. Gén. Env. Paris 1: 382. 5 Aug 1826 [*Fungi*].
≡ *Graphiola* Poit. 1824.

A6. DINOPHYCEAE

Abedinium Loebl. & A. R. Loebl. in Stud. Trop. Oceanogr. 3: 1, 14. Jun 1966.
Typus: *A. dasypus* (Cachon & Cachon-Enj.) Loebl. & A. R. Loebl. (*Leptophyllus dasypus* Cachon & Cachon-Enj.).

(≡) *Leptophyllus* Cachon & Cachon-Enj. in Bull. Inst. Océanogr. 62 (1292): 7. Feb 1964.

Amphilothus Kof. ex Poche in Arch. Protistenk. 30: 264. 12 Sep 1913.
Typus: *A. elegans* (F. Schütt) Er. Lindem. (in Engler & Prantl, Nat. Pflanzenfam., ed. 2, 2: 69. 1928) (*Amphitholus elegans* F. Schütt).

(≡) *Amphitholus* F. Schütt in Ergebn. Plankt.-Exped. Humboldt-Stiftung IV.M.a.A: 34. Jan 1895.

Dinamoebidium Pascher in Arch. Protistenk. 37: 31. 30 Aug 1916.
Typus: *D. varians* (Pascher) Pascher (*Dinamoeba varians* Pascher).

(≡) *Dinamoeba* Pascher in Arch. Protistenk. 36: 118. 8 Jan 1916.

Dogelodinium Loebl. & A. R. Loebl. in Stud. Trop. Oceanogr. 3: 1, 27. Jun 1966.
Typus: *D. ovoides* (Cachon) Loebl. & A. R. Loebl. (*Collinella ovoides* Cachon).

(≡) *Collinella* Cachon in Ann. Sci. Nat., Zool., ser. 12, 6: 49. Apr-Jun 1964.

Gyrodinium Kof. & Swezy in Mem. Univ. Calif. 5: 273. 28 Jun 1921.
Typus: *G. spirale* (Bergh) Kof. & Swezy (*Gymnodinium spirale* Bergh).

(≡) *Spirodinium* F. Schütt in Engler & Prantl, Nat. Pflanzenfam. 1(1b): 3, 5. 1896.

Keppenodinium Loebl. & A. R. Loebl. in Stud. Trop. Oceanogr. 3: 1, 38. Jun 1966.
Typus: *K. mycetoides* (Cachon) Loebl. & A. R. Loebl. (*Hollandella mycetoides* Cachon).

(≡) *Hollandella* Cachon in Ann. Sci. Nat., Zool., ser. 12, 6: 53. Apr-Jun 1964.

Latifascia Loebl. & A. R. Loebl. in Stud. Trop. Oceanogr. 3: 1, 38. Jun 1966.
Typus: *L. inaequalis* (Kof. & Skogsb.) Loebl. & A. R. Loebl. (*Heteroschisma inaequale* Kof. & Skogsb.).

(≡) *Heteroschisma* Kof. & Skogsb. in Mem. Mus. Comp. Zool. Harvard Coll. 51: 36. Dec 1928.

Sphaeripara Poche in Arch. Naturgesch. 77, Suppl. 1: 80. Sep 1911.
Typus: *S. catenata* (Neresh.) Loebl. & A. R. Loebl. (in Stud. Trop. Oceanogr. 3: 56. Jun 1966) (*Lohmannia catenata* Nehresh.).

(≡) *Lohmannia* Neresh. in Biol. Zentralbl. 23: 757. 1 Nov 1903.

A7. EUGLENOPHYCEAE

Anisonema Dujard., Hist. Nat. Zoo-phyt.: 327, 344. 1841.
Typus: *A. acinus* Dujard.

(H)   *Anisonema* A. Juss., Euphorb. Gen.: 19. 21 Feb 1824 [*Dicot.: Euphorb.*].
Typus: *A. reticulatum* (Poir.) A. Juss. (*Phyllanthus reticulatus* Poir.).

Astasia Dujard., Hist. Nat. Zoophyt.: 353, 356. 1841.
Typus: *A. limpida* Dujard. (typ. cons.).

(H)   *Astasia* Ehrenb. in Ann. Phys. Chem. 94: 508. 1830 [*Euglenoph.*].
Typus (vide Silva in Taxon 9: 20. 1960): *A. haematodes* Ehrenb.

Lepocinclis Perty in Mitth. Naturf. Ges. Bern 1849: 28. 15 Feb 1849.
Typus: *L. globulus* Perty

(=)   *Crumenula* Dujard. in Ann. Sci. Nat., Zool., ser. 2, 5: 204, 205. Apr 1836.
Typus: *C. texta* Dujard.

Phacus Dujard., Hist. Nat. Zoo-phyt.: 327, 334. 1841.
Typus: *P. longicauda* (Ehrenb.) Dujard. (*Euglena longicauda* Ehrenb.) (typ. cons.).

(H)   *Phacus* Nitzsch in Ersch & Gruber, Allg. Encycl. Wiss. Künste, Sect. 1, 16: 69. 1827 [*Euglenoph.*].
≡ *Virgulina* Bory 1823.

A8. PHAEOPHYCEAE

Agarum Dumort., Comment. Bot.: 102. Nov (sero) - Dec (prim.) 1822.
Typus: *A. clathratum* Dumort. (*Fucus agarum* S. G. Gmel.).

(H)   *Agarum* Link in Neues J. Bot. 3(1,2): 7. Apr 1809 [*Rhodoph.*].
Typus: *A. rubens* (L.) Link (*Fucus rubens* L.).

Alaria Grev., Alg. Brit.: xxxix, 25. Mar 1830.
Typus: *A. esculenta* (L.) Grev. (*Fucus esculentus* L.).

(≡)   *Musaefolia* Stackh. in Mém. Soc. Imp. Naturalistes Moscou 2: 53, 66. 1809.

Ascophyllum Stackh. in Mém. Soc. Imp. Naturalistes Moscou 2: 54, 66. 1809 (*'Ascophylla'*) (orth. cons.).
Typus: *A. laevigatum* Stackh., nom. illeg. (*Fucus nodosus* L., *A. nodosum* (L.) Le Jol.).

(≡)   *Nodularius* Roussel, Fl. Calvados, ed. 2, 93. 1806 (typ. des.: Silva in Univ. Calif. Publ. Bot. 25: 299. 1952).

175

Asteronema Delépine & Asensi in Bull. Soc. Bot. France 122: 296. 18 Nov 1975.
Typus: *A. australe* Delépine & Asensi

(H)   *Asteronema* Trevis., Nomencl. Alg.: 46. post 1 Mar 1845 [*Fungi*].
Typus: *A. microscopicum* (Kütz.) Trevis. (*Asterothrix microscopica* Kütz.).

Carpomitra Kütz., Phycol. General.: 343. 14-16 Sep 1843.
Typus: *C. cabrerae* (Clemente) Kütz. (*Fucus cabrerae* Clemente).

(≡)   *Dichotomocladia* Trevis. in Atti Riunione Sci. Ital. 4: 333. 15 Aug 1843.

(=)   *Chytraphora* Suhr in Flora 17: 721. 14 Dec 1834.
Typus: *C. filiformis* Suhr

Chordaria C. Agardh, Syn. Alg. Scand.: xii. Mai-Dec 1817.
Typus: *C. flagelliformis* (O. F. Müll.) C. Agardh (*Fucus flagelliformis* O. F. Müll.).

(H)   *Chordaria* Link in Neues J. Bot. 3(1,2): 8. Apr 1809 [*Phaeoph.*].
≡ *Chorda* Stackh. 1797.

Cystophora J. Agardh in Linnaea 15: 3. Apr 1841.
Typus: *C. retroflexa* (Labill.) J. Agardh (*Fucus retroflexus* Labill.).

(=)   *Blossevillea* Decne. in Bull. Acad. Roy. Sci. Bruxelles 7(1): 410. 1840 (*'Blosvillea'*).
Typus (vide Silva in Univ. Calif. Publ. Bot. 25: 279. 1952): *B. torulosa* (R. Br. ex Turner) Decne. (in Ann. Sci. Nat., Bot., ser. 2, 17: 331. Jun 1842) (*Fucus torulosus* R. Br. ex Turner).

Cystoseira C. Agardh, Spec. Alg. 1: 50. Jan-Apr 1820.
Typus: *C. concatenata* (L.) C. Agardh (*Fucus concatenatus* L.) [= *C. foeniculacea* (L.) Grev. (*Fucus foeniculaceus* L.)].

(=)   *Gongolaria* Boehm. in Ludwig, Def. Gen. Pl., ed. 3: 503. 1760.
Typus: *Fucus abies-marina* S. G. Gmel.

(=)   *Baccifer* Roussel, Fl. Calvados, ed. 2: 94. 1806.
Typus: *Fucus baccatus* S. G. Gmel.

(=)   *Abrotanifolia* Stackh. in Mém. Soc. Imp. Naturalistes Moscou 2: 56, 81. 1809.
Typus (vide Papenfuss in Hydrobiologia 2: 184. 1950): *A. loeflingii* Stackh. (*Fucus abrotanifolius* L.).

(=)   *Ericaria* Stackh. in Mém. Soc. Imp. Naturalistes Moscou 2: 56, 80. 1809.
Typus (vide Papenfuss in Hydrobiologia 2: 185. 1950): *Fucus ericoides* L.

Desmarestia J. V. Lamour. in Ann. Mus. Natl. Hist. Nat. 20: 43. 1813.
Typus: *D. aculeata* (L.) J. V. Lamour. (*Fucus aculeatus* L.).

(≡) *Hippurina* Stackh. in Mém. Soc. Imp. Naturalistes Moscou 2: 59, 89. 1809 (typ. des.: Silva in Univ. Calif. Publ. Bot. 25: 257. 1952).

(=) *Herbacea* Stackh. in Mém. Soc. Imp. Naturalistes Moscou 2: 58, 89. 1809. Typus: *H. ligulata* Stackh. (*Fucus ligulatus* Lightf., non S. G. Gmel.).

(=) *Hyalina* Stackh. in Mém. Soc. Imp. Naturalistes Moscou 2: 58, 88. 1809. Typus: *H. mutabilis* Stackh., nom. illeg. (*Fucus viridis* O. F. Müll.).

Desmotrichum Kütz., Phycol. Germ.: 244. 14-16 Aug 1845.
Typus: *D. balticum* Kütz.

(H) *Desmotrichum* Blume, Bijdr.: 329. 20 Sep-7 Dec 1825 [*Monocot.: Orchid.*]. ≡ *Flickingeria* A. D. Hawkes 1961.

(=) *Diplostromium* Kütz., Phycol. General.: 298. 14-16 Sep 1843.
Typus (vide Silva in Univ. Calif. Publ. Bot. 25: 257. 1952): *D. tenuissimum* (C. Agardh) Kütz. (*Zonaria tenuissima* C. Agardh).

Dictyopteris J. V. Lamour. in Nouv. Bull. Sci. Soc. Philom. Paris 1: 332. Mai 1809.
Typus: *D. polypodioides* (DC.) J. V. Lamour. (*Fucus polypodioides* Desf., non S. G. Gmel., *Ulva polypodioides* DC.).

(≡) *Granularius* Roussel, Fl. Calvados, ed. 2: 90. 1806 (typ. des.: Silva in Regnum Veg. 101: 745. 1979).

(=) *Neurocarpus* F. Weber & D. Mohr in Beitr. Naturk. 1: 300. 15 Nov 1805-1806.
Typus: *N. membranaceus* (Stackh.) F. Weber & D. Mohr (*Polypodoidea membranacea* Stackh.; *Fucus membranaceus* Stackh., non Burm. f.).

Dictyosiphon Grev., Alg. Brit.: xliii, 55. Mar 1830.
Typus: *D. foeniculaceus* (Huds.) Grev. (*Conferva foeniculacea* Huds.) (etiam vide *Scytosiphon* [*Phaeoph.*]).

Dictyota J. V. Lamour. in J. Bot. (Desvaux) 2: 38. 1-8 Apr 1809.
Typus: *D. dichotoma* (Huds.) J. V. Lamour. (*Ulva dichotoma* Huds.) (typ. cons.).

Ectocarpus Lyngb., Tent. Hydrophytol. Dan.: xxxi, 130. Apr-Aug 1819.
Typus: *E. siliculosus* (Dillwyn) Lyngb. (*Conferva siliculosa* Dillwyn).

(=) *Colophermum* Raf., Précis Découv. Somiol.: 49. Jun-Dec 1814.
Typus: *C. floccosum* Raf.

Elachista Duby, Bot. Gall.: 972. Mai 1830 *('Elachistea')* (orth. cons.).
Typus: *E. scutellata* Duby, nom. illeg. (*Conferva scutulata* Sm., *E. scutulata* (Sm.) Aresch.).

(=) *Opospermum* Raf., Précis Découv. Somiol.: 48. Jun-Dec 1814.
Typus: *O. nigrum* Raf.

Halidrys Lyngb., Tent. Hydrophytol. Dan.: xxix, 37. Apr-Aug 1819.
Typus: *H. siliquosa* (L.) Lyngb. (*Fucus siliquosus* L.) (typ. cons.).

(H) *Halidrys* Stackh. in Mém. Soc. Imp. Naturalistes Moscou 2: 53, 62. 1809 (typ. des.: Papenfuss in Hydrobiologia 2: 186. 1950) [*Phaeoph.*].
≡ *Fucus* L. 1753 (typ. des.: De Toni in Flora 74: 173. 1891).

(≡) *Siliquarius* Roussel, Fl. Calvados, ed. 2: 94. 1806.

Hesperophycus Setch. & N. L. Gardner in Univ. Calif. Publ. Bot. 4: 127. 26 Aug 1910.
Typus: *H. californicus* P. C. Silva (in Taxon 39: 5. 22 Feb 1990) (typ. cons.).

Himanthalia Lyngb., Tent. Hydrophytol. Dan.: xxix, 36. Apr-Aug 1819.
Typus: *H. lorea* (L.) Lyngb. (*Fucus loreus* L.) [= *H. elongata* (L.) Gray (*Fucus elongatus* L.)].

(≡) *Funicularius* Roussel, Fl. Calvados, ed. 2: 91. 1806.

(=) *Lorea* Stackh. in Mém. Soc. Imp. Naturalistes Moscou 2: 60, 94. 1809.
Typus: *L. elongata* (L.) Stackh. (*Fucus elongatus* L.).

Hormosira (Endl.) Menegh. in Nuovi Saggi Imp. Regia Accad. Sci. Padova 4: 368. 1838 (*Cystoseira* sect. *Hormosira* Endl., Gen. Pl.: 10. Aug 1836).
Typus: *Fucus moniliformis* Labill., non Esper [= *H. banksii* (Turner) Decne. (*Fucus banksii* Turner)].

(≡) *Moniliformia* J. V. Lamour. in Bory, Dict. Class. Hist. Nat. 7: 71. 5 Mar 1825.

Laminaria J. V. Lamour. in Ann. Mus. Natl. Hist. Nat. 20: 40. 1813.
Typus: *L. digitata* (Huds.) J. V. Lamour. (*Fucus digitatus* Huds.).

(=) *Saccharina* Stackh. in Mém. Soc. Imp. Naturalistes Moscou 2: 53, 65. 1809.
Typus (vide Silva in Univ. Calif. Publ. Bot. 25: 259. 1952): *S. plana* Stackh. (*Fucus saccharinus* L.).

Leptonematella P. C. Silva in Taxon 8: 63. 12 Mar 1959.
Typus: *L. fasciculata* (Reinke) P. C. Silva (*Leptonema fasciculatum* Reinke).

Padina Adans., Fam. Pl. 2: 13, 586. Jul-Aug 1763.
Typus: *P. pavonica* (L.) J. V. Lamour. (Hist. Polyp. Corall.: 304. 1816) (*Fucus pavonicus* L.).

Petalonia Derbès & Solier in Ann. Sci. Nat., Bot., ser. 3, 14: 265. Nov 1850.
Typus: *P. debilis* (C. Agardh) Derbès & Solier (*Laminaria debilis* C. Agardh).

(=) *Fasciata* Gray, Nat. Arr. Brit. Pl. 1: 383. 1 Nov 1821.
Typus (vide Silva in Univ. Calif. Publ. Bot. 25: 299. 1952): *F. attenuata* Gray, nom. illeg. (*Fucus fascia* O. F. Müll.).

Pylaiella Bory, Dict. Class. Hist. Nat. 4: 393. 27 Dec 1823 *('Pilayella')* (orth. cons.).
Typus: *P. littoralis* (L.) Kjellm. *(Conferva littoralis* L.).

Saccorhiza Bach. Pyl., Fl. Terre-Neuve: 23. 22 Jan 1830.
Typus: *Laminaria bulbosa* (Huds.) J. Agardh (Spec. Alg. 1: 138. Apr-Sep 1848) (*Fucus bulbosus* Huds.) [= *S. polyschides* (Lightf.) Batters (*Fucus polyschides* Lightf.)].

(=) *Polyschidea* Stackh. in Mém. Soc. Imp. Naturalistes Moscou 2: 53, 65. 1809.
Typus (vide Papenfuss in Hydrobiologia 2: 189. 1950): *Fucus polyschides* Lightf.

Sargassum C. Agardh, Spec. Alg. 1: 1. Jan-Apr 1820.
Typus: *S. bacciferum* (Turner) C. Agardh (*Fucus bacciferus* Turner).

(=) *Acinaria* Donati, Essai Hist. Nat. Mer Adriat.: 26, 33. Jan-Mar 1758.
Typus: *Sargassum donatii* (Zanardini) Kütz. (*S. vulgare* var. *donatii* Zanardini).

Scytosiphon C. Agardh, Spec. Alg. 1: 160. Jan-Apr 1820.
Typus: *S. lomentaria* (Lyngb.) Link (Handbuch 3: 232. 1833) (*Chorda lomentaria* Lyngb.) (typ. cons.).

(H) *Scytosiphon* C. Agardh, Disp. Alg. Suec.: 24. 11 Dec 1811 (typ. des.: Silva in Regnum Veg. 8: 205. 1956) [*Phaeoph.*].
≡ *Dictyosiphon* Grev. 1830 (nom. cons.).

Spermatochnus Kütz., Phycol. Ge-neral.: 334. 14-16 Sep 1843.
Typus: *S. paradoxus* (Roth) Kütz. (*Conferva paradoxa* Roth) (typ. cons.).

Stilophora J. Agardh in Linnaea 15: 6. Apr 1841.
Typus: *S. rhizodes* (Turner) J. Ag-ardh (*Fucus rhizodes* Turner) (typ. cons.).

(H) *Stilophora* C. Agardh in Flora 10: 642. 7 Nov 1827 [*Phaeoph.*].
≡ *Hydroclathrus* Bory 1825.

Zonaria C. Agardh, Syn. Alg. Scand.: xx. Mai-Dec 1817.
Typus: *Z. flava* C. Agardh (typ. cons.) [= *Z. tournefortii* (Lamour.) Mont. (*Fucus tournefortii* Lamour.)].

(H) *Zonaria* Drap. ex F. Weber & D. Mohr in Beitr. Naturk. 1: 247-253. 15 Nov 1805-1806 [*Phaeoph.*].
≡ *Padina* Adans. 1763 (nom. cons.).

A9. RHODOPHYCEAE

Ahnfeltia Fr., Fl. Scan.: 309. 1836.
Typus: *A. plicata* (Huds.) Fr. (*Fucus plicatus* Huds.) (typ. cons.).

Areschougia Harv. in Trans. Roy. Irish Acad. 22 (Sci.): 554. 1855.
Typus: *A. laurencia* (Hook. f. & Harv.) Harv. (*Thamnocarpus laurencia* Hook. f. & Harv.).

(H) *Areschougia* Menegh. in Giorn. Bot. Ital. 1(1,1): 293. Mai-Jun 1844 [*Phaeoph.*].
Typus (vide Silva in Univ. Calif. Publ. Bot. 25: 283. 1952): *A. stellaris* (Aresch.) Menegh. (*Elachista stellaris* Aresch.).

Audouinella Bory, Dict. Class. Hist. Nat. 3: 340. 6 Sep 1823 (*'Auduinella'*) (orth. cons.).
Typus: *A. miniata* Bory [= *A. hermannii* (Roth) Duby (*Conferva hermannii* Roth)].

180

Bostrychia Mont. in Sagra, Hist. Phys. Cuba, Bot. Pl. Cell.: 39. 1842 (sero).
Typus: *B. scorpioides* (Huds.) Mont. (in Orbigny, Dict. Univ. Hist. Nat. 2: 661. 1842) (*Fucus scorpioides* Huds.).

(H) *Bostrychia* Fr., Syst. Mycol. 1: lii. 1 Jan 1821 [*Fungi*].
Typus: *B. chrysosperma* (Pers. : Fr.) Fr. (*Sphaeria chrysosperma* Pers. : Fr.).

(≡) *Amphibia* Stackh. in Mém. Soc. Imp. Naturalistes Moscou 2: 58, 89. 1809.

Botryocladia (J. Agardh) Kylin in Acta Univ. Lund., ser. 2, sect. 2, 27(11): 17. Mai-Oct 1931 (*Chrysymenia* sect. *Botryocladia* J. Agardh, Spec. Gen. Ord. Alg. 2: 214. Jan-Jun 1851).
Typus: *Chrysymenia uvaria* J. Agardh, nom. illeg. (*B. uvaria* Kylin, nom. illeg., *Fucus botryoides* Wulfen, *B. botryoides* (Wulfen) Feldmann).

(=) *Myriophylla* Holmes in Ann. Bot. (London) 8: 340. Sep 1894.
Typus: *M. beckeriana* Holmes

Calliblepharis Kütz., Phycol. General.: 403. 14-16 Sep 1843.
Typus: *C. ciliata* (Huds.) Kütz. (*Fucus ciliatus* Huds.).

(≡) *Ciliaria* Stackh. in Mém. Soc. Imp. Naturalistes Moscou 2: 54, 70. 1809 (typ. des.: Papenfuss in Hydrobiologia 2: 191. 1950).

Caloglossa (Harv.) G. Martens in Flora 52: 234, 237. 25 Mai 1869 (*Delesseria* subg. *Caloglossa* Harv. in Smithsonian Contr. Knowl. 5(5): 98. Mar. 1853).
Typus: *Delesseria leprieurii* Mont. (*C. leprieurii* (Mont.) G. Martens).

(=) *Apiarium* Durant, Alg. New York: 18. 1850.
Typus: *A. apicula* Durant.

Catenella Grev., Alg. Brit.: lxiii, 166. Mar 1830.
Typus: *C. opuntia* (Gooden. & Woodw.) Grev. (*Fucus opuntia* Gooden. & Woodw.) [= *C. caespitosa* (With.) L. M. Irvine (*Ulva caespitosa* With.)].

(=) *Clavatula* Stackh. in Mém. Soc. Imp. Naturalistes Moscou 2: 95, 97. 1809.
Typus: *C. caespitosa* Stackh. (*Fucus caespitosus* Stackh., non Forssk.).

Ceramium Roth, Catal. Bot. 1: 146. Jan-Feb 1797.
Typus: *C. virgatum* Roth (typ. cons.) [= *Ceramium rubrum* (Huds.) C. Agardh (*Conferva rubra* Huds.)].

(H) *Ceramion* Adans., Fam. Pl. 2: 13, 535. Jul-Aug 1763 [*Rhodoph.*].
≡ *Ceramianthemum* Donati ex Léman 1817 (nom. rej. sub *Gracilaria*).

Chondria C. Agardh, Syn. Alg. Scand.: xviii. Mai-Dec 1817.
Typus: *C. tenuissima* (With.) C. Agardh (*Fucus tenuissimus* With.).

(≡) *Dasyphylla* Stackh., Nereis Brit., ed. 2: ix, xi. 1816 (ante Aug).
Typus (vide Papenfuss in Hydrobiologia 2: 192. 1950): *D. woodwardii* Stackh. (*Fucus dasyphyllus* Woodw.).

Chylocladia Grev. in Hooker, Brit. Fl., ed. 4, 2(1): 256, 297. 1833.
Typus: *C. kaliformis* (With.) Grev. (*Fucus kaliformis* With.) [= *C. verticillata* (Lightf.) Bliding (*Fucus verticillatus* Lightf.)].

(≡) *Kaliformis* Stackh. in Mém. Soc. Imp. Naturalistes Moscou 2: 56, 78. 1809 (typ. des.: Papenfuss in Hydrobiologia 2: 198. 1950).

Corynomorpha J. Agardh in Acta Univ. Lund. 8 (sect. 3, 6): 3. 1872.
Typus: *C. prismatica* (J. Agardh) J. Agardh (*Dumontia prismatica* J. Agardh).

(≡) *Prismatoma* (J. Agardh) Harv., Index Gen. Alg.: 11. Jul-Aug 1860 (*Acrotilus* subg. *Prismatoma* J. Agardh, Spec. Gen. Ord. Alg. 2: 193. Jan-Jun 1851).

Cryptopleura Kütz., Phycol. General.: 444. 14-16 Sep 1843.
Typus: *C. lacerata* (S. G. Gmel.) Kütz. (*Fucus laceratus* S. G. Gmel.) [= *C. ramosa* (Huds.) Kylin ex Newton (*Ulva ramosa* Huds.)].

(H) *Cryptopleura* Nutt. in Trans. Amer. Philos. Soc., ser. 2, 7: 431. 2 Apr 1841 [*Dicot.: Comp.*].
Typus: *C. californica* Nutt.

(≡) *Papyracea* Stackh. in Mém. Soc. Imp. Naturalistes Moscou 2: 56, 76. 1809 (typ. des.: Papenfuss in Index Nom. Gen.: No. 00816. 1955).

Dasya C. Agardh, Syst. Alg.: xxxiv, 211. Mai-Sep 1824 (*'Dasia'*) (orth. cons.).
Typus: *D. pedicellata* (C. Agardh) C. Agardh (*Sphaerococcus pedicellatus* C. Agardh) [= *D. baillouviana* (S. G. Gmel.) Mont. (*Fucus baillouviana* S. G. Gmel.)].

(≡) *Baillouviana* Adans., Fam. Pl. 2: 13, 523. Jul-Aug 1763.
Typus: *Fucus baillouviana* S. G. Gmel.

Delesseria J. V. Lamour. in Ann. Mus. Natl. Hist. Nat. 20: 122. 1813.
Typus: *D. sanguinea* (Huds.) J. V. Lamour. (*Fucus sanguineus* Huds.).

(≡) *Hydrolapatha* Stackh. in Mém. Soc. Imp. Naturalistes Moscou 2: 54, 67. 1809 (typ. des.: Papenfuss in Hydrobiologia 2: 196. 1950).

Dudresnaya P. Crouan & H. Crouan in Ann. Sci. Nat., Bot., ser. 2, 3: 98. Feb 1835.
Typus: *D. coccinea* (C. Agardh) P. Crouan & H. Crouan (*Mesogloia coccinea* C. Agardh) (typ. cons.) [= *D. verticillata* (With.) Le Jol. (*Ulva verticillata* With.)].

(H) *Dudresnaya* Bonnem. in J. Phys. Chim. Hist. Nat. Arts 94: 180. Apr 1822 [*Phaeoph.*].
Typus: *Alcyonidium vermiculatum* (Sm.) J. V. Lamour. (*Rivularia vermiculata* Sm.).

(=) *Borrichius* Gray, Nat. Arr. Brit. Pl. 1: 317, 330. 1 Nov 1821.
Typus: *B. gelatinosus* Gray, nom. illeg. (*Ulva verticillata* With.).

Erythrotrichia Aresch. in Nova Acta Regiae Soc. Sci. Upsal., ser. 2, 14: 435. 1850.
Typus: *E. ceramicola* (Lyngb.) Aresch. (*Conferva ceramicola* Lyngb.) [= *E. carnea* (Dillwyn) J. Agardh (*Conferva carnea* Dillwyn)].

(≡) *Goniotrichum* Kütz., Phycol. General.: 244. 14-16 Sep 1843.

(=) *Porphyrostromium* Trevis., Sagg. Algh. Coccot.: 100. 1848.
Typus: *P. boryi* Trevis., nom. illeg. (*Porphyra boryana* Mont.).

Falklandiella Kylin, Gatt. Rhodoph.: 391. 1956.
Typus: *F. harveyi* (Hook. f.) Kylin (*Ptilota harveyi* Hook. f.).

Furcellaria J. V. Lamour. in Ann. Mus. Natl. Hist. Nat. 20: 45. 1813.
Typus: *F. lumbricalis* (Huds.) J. V. Lamour. (*Fucus lumbricalis* Huds.).

(=) *Fastigiaria* Stackh. in Mém. Soc. Imp. Naturalistes Moscou 2: 59, 90. 1809.
Typus (vide Papenfuss in Hydrobiologia 2: 194. 1950): *F. linnaei* Stackh., nom. illeg. (*Fucus fastigiatus* L.).

Gastroclonium Kütz., Phycol. General.: 441. 14-16 Sep 1843.
Typus: *G. ovale* Kütz., nom. illeg. (*Fucus ovalis* Huds., nom. illeg., *Fucus ovatus* Huds., *G. ovatum* (Huds.) Papenf.).

(=) *Sedoidea* Stackh. in Mém. Soc. Imp. Naturalistes Moscou 2: 57, 83. 1809.
Typus (vide Papenfuss in Hydrobiologia 2: 202. 1950): *Fucus sedoides* Gooden. & Woodw., nom. illeg. (*Fucus vermicularis* S. G. Gmel.).

Gelidium J. V. Lamour. in Ann. Mus. Natl. Hist. Nat. 20: 128. 1813.
Typus: *G. corneum* (Huds.) J. V. Lamour. (*Fucus corneus* Huds.).

(≡) *Cornea* Stackh. in Mém. Soc. Imp. Naturalistes Moscou 2: 57, 83. 1809 (typ. des.: Papenfuss in Hydrobiologia 2: 191-192. 1950).

Gracilaria Grev., Alg. Brit.: liv, 121. Mar 1830.
Typus: *G. compressa* (C. Agardh) Grev. (*Sphaerococcus compressus* C. Agardh) (typ. cons.).

(=) *Ceramianthemum* Donati ex Léman in Cuvier, Dict. Sci. Nat., 7: 421. 1817.
Typus (vide Ardissone in Mem. Soc. Crittog. Ital. 1: 240-241. 1883): *Fucus bursa-pastoris* S. G. Gmel.

(=) *Plocaria* Nees, Horae Phys. Berol.: 42. 1-8 Feb 1820.
Typus: *P. candida* Nees

Grateloupia C. Agardh, Spec. Alg. 1: 221. Oct 1822.
Typus: *G. filicina* (J. V. Lamour.) C. Agardh (*Delesseria filicina* J. V. Lamour.).

(H) *Grateloupia* Bonnem. in J. Phys. Chim. Hist. Nat. Arts 94: 189. Apr 1822 [*Rhodoph.*].
Typus: *Conferva arbuscula* Dillwyn

Griffithsia C. Agardh, Syn. Alg. Scand.: xxviii. Mai-Dec 1817 (*'Griffitsia'*) (orth. cons.).
Typus: *G. corallina* C. Agardh, nom. illeg. (*Conferva corallina* Murray, nom. illeg., *Conferva corallinoides* L., *G. corallinoides* (L.) Trevis.).

Halymenia C. Agardh, Syn. Alg. Scand.: xix, 35. Mai-Dec 1817.
Typus: *H. floresia* (Clemente) C. Agardh (*Fucus floresius* Clemente) (typ. cons.).

Helminthocladia J. Agardh, Spec. Gen. Ord. Alg. 2: 412. Jan-Jun 1852.
Typus: *H. purpurea* (Harv.) J. Agardh (*Mesogloia purpurea* Harv.) [= *H. calvadosii* (J. V. Lamour. ex Turpin) Setchell (*Dumontia calvadosii* J. V. Lamour. ex Turpin)].

(H) *Helminthocladia* Harv., Gen. S. Afr. Pl.: 396. Aug-Dec 1838 [*Phaeoph.*].
≡ *Mesogloia* C. Agardh 1817.

Helminthora J. Agardh, Spec. Gen. Ord. Alg. 2: 415. Jan-Jun 1852.
Typus: *H. divaricata* (C. Agardh) J. Agardh (*Mesogloia divaricata* C. Agardh).

(H) *Helminthora* Fr., Syst. Orb. Veg.: 341. Dec 1825 [*Rhodoph.*].
Typus: *H. multifida* (F. Weber & D. Mohr) Fr. (Fl. Scan.: 311. 1835) (*Rivularia multifida* F. Weber & D. Mohr).

Heterosiphonia Mont., Prodr. Gen. Phyc.: 4. 1 Aug - 10 Sep 1842.
Typus: *H. berkeleyi* Mont.

(=) *Ellisius* Gray, Nat. Arr. Brit. Pl. 1: 317, 333. 1 Nov 1821.
Typus (vide Silva in Univ. Calif. Publ. Bot. 25: 290. 1952): *E. coccineus* (Huds.) Gray (*Conferva coccinea* Huds.).

Hildenbrandia Nardo in Isis (Oken) 1834: 676. 1834 *('Hildbrandtia')* (orth. cons.).
Typus: *H. prototypus* Nardo

Husseya J. Agardh, Spec. Gen. Ord. Alg. 3(4): 123. 2 Mar 1901.
Typus: *H. australis* J. Agardh [= *H. rubra* (Harv.) P. C. Silva (*Chondria rubra* Harv.)].

(H) *Husseia* Berk. in London J. Bot. 6: 508. 1847 [*Fungi*].
Typus (vide Clements & Shear, Gen. Fung., ed. 2: 353. 1931): *H. insignis* Berk.

(=) *Rhododactylis* J. Agardh, Spec. Gen. Ord. Alg. 3(1): 566. Aug 1876.
Typus (vide Kylin in Acta Univ. Lund., ser. 2, sect. 2, 28(8): 48. 1932): *R. rubra* (Harv.) J. Agardh (*Chondria rubra* Harv.).

Iridaea Bory, Dict. Class. Hist. Nat. 9: 15. 25 Feb 1826 (*Iridaea, 'Iridea'*) (orth. cons.).
Typus: *I. cordata* (Turner) Bory (*Fucus cordatus* Turner) (typ. cons.).

(H) *Iridea* Stackh., Nereis Brit., ed. 2: ix, xii. 1816 (ante Aug) [*Phaeoph.*].
≡ *Hyalina* Stackh. 1809 (nom. rej. sub *Desmarestia*).

Laurencia J. V. Lamour. in Ann. Mus. Natl. Hist. Nat. 20: 130. 1813.
Typus: *L. obtusa* (Huds.) J. V. Lamour. (*Fucus obtusus* Huds.).

(=) *Osmundea* Stackh. in Mém. Soc. Imp. Naturalistes Moscou 2: 56, 79. 1809.
Typus (vide Silva in Univ. Calif. Publ. Bot. 25: 292. 1952): *O. expansa* Stackh., nom. illeg. (*Fucus osmunda* S. G. Gmel.).

Lemanea Bory in Ann. Mus. Natl. Hist. Nat. 12: 178. 1808.
Typus: *L. corallina* Bory, nom. illeg. (*Conferva fluviatilis* L., *L. fluviatilis* (L.) C. Agardh).

(≡) *Apona* Adans., Fam. Pl. 2: 2, 519. Jul-Aug 1763.

Lenormandia Sond. in Bot. Zeitung (Berlin) 3: 54. 24 Jan 1845.
Typus: *L. spectabilis* Sond.

(H) *Lenormandia* Delise in Desmazières, Pl. Crypt. N. France: No. 1144. 1841 [*Fungi*].
Typus: *L. jungermanniae* Delise

Lithothamnion Heydr. in Ber. Deutsch. Bot. Ges. 15: 412. 7 Sep 1897.
Typus: *L. muelleri* Lenorm. ex Rozanov (in Mém. Soc. Sci. Nat. Cherbourg 12: 101. 1866) (typ. cons.).

(H) *Lithothamniun* Phil. in Arch. Naturgesch. 3(1): 387. 1837 [*Rhodoph.*].
Typus (vide Lemoine in Ann. Inst. Océanogr. 2(2): 66. 1911): *L. byssoides* (Lam.) Phil. (*Nullipora byssoides* Lam.).

Martensia Hering in Ann. Mag. Nat. Hist. 8: 92. Oct 1841.
Typus: *M. elegans* Hering

(H) *Martensia* Giseke, Prael. Ord. Nat. Pl.: 202, 207, 249. Apr 1792 [*Monocot.: Zingiber.*].
Typus: *M. aquatica* (Retz.) Giseke (*Heritiera aquatica* Retz.).

Nemastoma J. Agardh, Alg. Mar. Medit.: 89. 9 Apr 1842 (*'Nemostoma'*) (orth. cons.).
Typus: *N. dichotomum* J. Agardh

Neurocaulon Zanardini ex Kütz., Sp. Alg.: 744. 23-24 Jul 1849.
Typus: *N. foliosum* (Menegh.) Zanardini ex Kütz. (*Iridaea foliosa* Menegh.) (typ. cons.).

Nitophyllum Grev., Alg. Brit.: xlvii, 77. Mar 1830.
Typus: *N. punctatum* (Stackh.) Grev. (*Ulva punctata* Stackh.).

(=) *Scutarius* Roussel, Fl. Calvados, ed. 2: 91. 1806.
Typus (vide Silva in Univ. Calif. Publ. Bot. 25: 268. 1952): *Fucus ocellatus* J. V. Lamour.

Odonthalia Lyngb., Tent. Hydrophytol. Dan.: xxix, 9. Apr-Aug 1819.
Typus: *O. dentata* (L.) Lyngb. (*Fucus dentatus* L.).

(≡) *Fimbriaria* Stackh. in Mém. Soc. Imp. Naturalistes Moscou 2: 95, 96. 1809 (typ. des.: Silva, Univ. Calif. Publ. Bot. 25: 269. 1952).

Phacelocarpus Endl. & Diesing in Bot. Zeitung (Berlin) 3: 289. 25 Apr 1845.
Typus: *P. tortuosus* Endl. & Diesing

(=) *Ctenodus* Kütz., Phycol. General.: 407. 14-16 Sep 1843.
Typus: *C. labillardierei* (Mert. ex Turner) Kütz. (*Fucus labillardierei* Mert. ex Turner).

Phyllophora Grev., Alg. Brit.: lvi, 135. Mar 1830.
Typus: *P. crispa* (Huds.) P. S. Dixon (in Bot. Not. 117: 63. 31 Mar 1964) (*Fucus crispus* Huds.) (typ. cons.).

(≡) *Epiphylla* Stackh., Nereis Brit., ed. 2: x, xii. 1816 (ante Aug).
(=) *Membranifolia* Stackh. in Mém. Soc. Imp. Naturalistes Moscou 2: 55, 75. 1809.
Typus (vide Papenfuss in Hydrobiologia 2: 198. 1950): *M. lobata* Stackh., nom. illeg. (*Fucus membranifolius* Gooden. & Woodw., nom. illeg., *Fucus pseudoceranoides* S. G. Gmel.).

Phymatolithon Foslie in Kongel. Norske Vidensk. Selsk. Skr. (Trondheim) 1898(2): 4. 14 Oct 1898.
Typus: *P. polymorphum* Foslie, nom. illeg. (*Millepora polymorpha* L., nom. illeg., *Millepora calcarea* Pall., *P. calcareum* (Pall.) W. H. Adey & D. L. McKibbin).

(≡) *Apora* Gunnerus in Kongel. Norske Vidensk. Selsk. Skr. (Copenhagen) 4: 72. 1768.

Platoma Schousb. ex F. Schmitz in Nuova Notarisia 5: 627. Apr 1894 (gend. neut. cons.).
Typus: *P. cyclocolpum* (Mont.) F. Schmitz (*Halymenia cyclocolpa* Mont.).

Pleonosporium Nägeli in Sitzungsber. Königl. Bayer. Akad. Wiss. München 1861(2): 326, 339. 1862.
Typus: *P. borreri* (Sm.) Nägeli (*Conferva borreri* Sm.).

Plocamium J. V. Lamour. in Ann. Mus. Natl. Hist. Nat. 20: 137. 1813.
Typus: *P. vulgare* J. V. Lamour., nom. illeg. (*Fucus cartilagineus* L., *P. cartilagineum* (L.) P. S. Dixon).

(≡) *Nereidea* Stackh. in Mém. Soc. Imp. Naturalistes Moscou 2: 58, 86. 1809 (typ. des.: Silva in Univ. Calif. Publ. Bot. 25: 264. 1952).

Plumaria F. Schmitz in Nuova Notarisia 7: 5. Jan 1896.
Typus: *P. elegans* (Bonnem.) F. Schmitz (*Ptilota elegans* Bonnem.).

(H) *Plumaria* Heist. ex Fabr., Enum.: 207. 1759 [*Monocot.: Cyper.*].
≡ *Eriophorum* L. 1753.

Polyneura (J. Agardh) Kylin in Acta Univ. Lund., ser. 2, sect. 2, 20(6): 33. 1924 (*Nitophyllum* subg. *Polyneura* J. Agardh, Spec. Gen. Ord. Alg. 3(3): 51. 1898).
Typus: *Nitophyllum hilliae* (Grev.) Grev. (*Delesseria hilliae* Grev., *P. hilliae* (Grev.) Kylin).

(H) *Polyneura* J. Agardh in Acta Univ. Lund. 35 (sect. 2, 4): 60. 1899 [*Rhodoph.*].
Typus: *P. californica* J. Agardh

Polysiphonia Grev., Scott. Crypt. Fl.: ad t. 90. 1 Dec 1823.
Typus: *P. urceolata* (Dillwyn) Grev. (*Conferva urceolata* Dillwyn) (typ. cons.).

(=) *Grammita* Bonnem. in J. Phys. Chim. Hist. Nat. Arts 94: 186. Apr 1822.
Typus: *Conferva fucoides* Huds.

(=) *Vertebrata* Gray, Nat. Arr. Brit. Pl. 1: 317, 338. 1 Nov 1821.
Typus: *V. fastigiata* Gray, nom. illeg. (*Conferva polymorpha* L.).

(=) *Gratelupella* Bory, Dict. Class. Hist. Nat. 3: 340. 6 Sep 1823.
Typus (vide Bory, Dict. Class. Hist. Nat. 7: 481. 1825): *Ceramium brachygonium* Lyngb.

Porphyra C. Agardh, Syst. Alg.: xxxii, 190. Mai-Sep 1824.
Typus: *P. purpurea* (Roth) C. Agardh (*Ulva purpurea* Roth, nom. cons.).

(H) *Porphyra* Lour., Fl. Cochinch.: 63, 69. Sep 1790 [*Dicot.: Verben.*].
Typus: *P. dichotoma* Lour.

(=) *Phyllona* Hill, Hist. Pl., ed. 2: 79. 1773.
Typus: non designatus.

Porphyridium Nägeli in Neue Denkschr. Allg. Schweiz. Ges. Gesammten Naturwiss. 10(7): 71, 138. 1849.
Typus: *P. cruentum* (Gray) Nägeli (*Olivia cruenta* Gray).

(=) *Chaos* Bory ex Desm., Cat. Pl. Omises Botanogr. Belgique: 1. Mar 1823.
Typus: *C. sanguinarius* Bory ex Desm., nom. illeg. (*Phytoconis purpurea* Bory).

(=) *Sarcoderma* Ehrenb. in Ann. Phys. Chem. 94: 504. 1830.
Typus: *S. sanguineum* Ehrenb.

Prionitis J. Agardh, Spec. Gen. Ord. Alg. 2: 185. Jan-Jun 1851.
Typus: *P. ligulata* J. Agardh [= *P. lanceolata* (Harv.) Harv. (*Gelidium lanceolatum* Harv.)].

(H) *Prionitis* Adans., Fam. Pl. 2: 499, 594. Jul-Aug 1763 [*Dicot.: Umbell.*].
≡ *Falcaria* Fabr. 1759 (nom. cons.).

Ptilota C. Agardh, Syn. Alg. Scand.: xix, 39. Mai-Dec 1817.
Typus: *P. gunneri* P. C. Silva, C. A. Maggs, & M. H. Hommersand *(Fucus ptilotus* Gunnerus) (typ. cons.).

Rhodochorton Nägeli in Sitzungsber. Bayer. Akad. Wiss. München 1861(2): 326, 355. 1862.
Typus: *R. purpureum* (Lightf.) Rosenv. (*Byssus purpurea* Lightf.) (typ. cons.).

Rhodomela C. Agardh, Spec. Alg. 1: 368. Oct 1822.
Typus: *R. subfusca* (Woodw.) C. Agardh (*Fucus subfuscus* Woodw.).

(=)　*Fuscaria* Stackh. in Mém. Soc. Imp. Naturalistes Moscou 2: 59, 93. 1809.
Typus: *F. variabilis* Stackh., nom. illeg. (*Fucus variabilis* With., nom. illeg., *Fucus confervoides* Huds.).

Rhodophyllis Kütz. in Bot. Zeitung (Berlin) 5: 23. 8 Jan 1847.
Typus: *R. bifida* Kütz., nom. illeg. (*Fucus bifudus* Turner, non S. G. Gmel., *Bifida divaricata* Stackh., *R. divaricata* (Stackh.) Papenf.).

(≡)　*Bifida* Stackh. in Mém. Soc. Imp. Naturalistes Moscou 2: 95, 97. 1809 (typ. des.: Silva in Univ. Calif. Publ. Bot. 25: 264. 1952).

(=)　*Inochorion* Kütz., Phycol. General.: 443. 14-16 Sep 1843.
Typus: *I. dichotomum* Kütz.

Rhodymenia Grev., Alg. Brit.: xlviii, 84. Mar 1830 *('Rhodomenia')* (orth. cons.).
Typus: *R. palmetta* (J. V. Lamour.) Grev. (*Delesseria palmetta* J. V. Lamour.) [= *R. pseudopalmata* (J. V. Lamour.) P. C. Silva (*Fucus pseudopalmatus* J. V. Lamour.)].

Schizymenia J. Agardh, Spec. Gen. Ord. Alg. 2: 158, 169. Jan-Jun 1851.
Typus: *S. dubyi* (Chauv.) J. Agardh (*Halymenia dubyi* Chauv.) (typ. cons.).

Suhria J. Agardh ex Endl., Gen. Pl., Suppl. 3: 41. Oct 1843.
Typus: *S. vittata* (L.) J. Agardh ex Endl. (*Fucus vittatus* L.).

(=)　*Chaetangium* Kütz., Phycol. General.: 392. 14-16 Sep 1843.
Typus: *C. ornatum* (L.) Kütz. (*Fucus ornatus* L.).

Vidalia J. Agardh, Spec. Gen. Ord. Alg. 2: 1117. Jan-Aug 1863. Typus: *V. spiralis* (J. V. Lamour.) J. Agardh (*Delesseria spiralis* J. V. Lamour.).

(=) *Volubilaria* J. V. Lamour. ex Bory, Dict. Class. Hist. Nat. 16: 630. 30 Oct 1830. Typus: *V. mediterranea* J. V. Lamour. ex Bory, nom. illeg. (*Fucus volubilis* L.).

(=) *Spirhymenia* Decne. in Arch. Mus. Hist. Nat. (Paris) 2: 177. 1841. Typus: *S. serrata* (Suhr) Decne. (in Ann. Sci. Nat., Bot., ser. 2, 17: 358. Jun 1842) (*Carpophyllum serratum* Suhr, '*denticulatum*' lapsu).

(=) *Epineuron* Harv. in London J. Bot. 4: 532. 1845. Typus (vide Silva in Univ. Calif. Publ. Bot. 25: 293. 1952): *E. colensoi* Hook. f. & Harv. (in London J. Bot. 4: 532. 1845).

A10. TRICHOMONADOPHYCEAE

Chilomastix A. G. Alexeev in Arch. Zool. Exp. Gén. 46: xi. 26 Dec 1910. Typus: *C. caulleryi* (A. G. Alexeev) A. G. Alexeev (*Macrostoma caulleryi* A. G. Alexeev).

(≡) *Macrostoma* A. G. Alexeev in Compt.-Rend. Séances Mém. Soc. Biol. 67: 200. 17 Jul 1909.

A11. XANTHOPHYCEAE

Botrydiopsis Borzì in Boll. Soc. Ital. Microscop. 1: 69. 1889. Typus: *B. arhiza* Borzì

(H) *Botrydiopsis* Trevis., Nomencl. Alg.: 70. 1845 [*Plantae*]. Typus: *B. vulgaris* (Bréb.) Trevis. (*Botrydina vulgaris* Bréb.).

Centritractus Lemmerm. in Ber. Deutsch. Bot. Ges. 18: 274. 24 Jul 1900 ('*Centratractus*') (orth. cons.). Typus: *C. belonophorus* (Schmidle) Lemmerm. (*Schroederia belonophora* Schmidle).

Monodus Chodat in Beitr. Krypto-
gamenfl. Schweiz 4(2): 185. 1913.
Typus: *M. acuminata* (Gerneck) Chodat (*Chlorella acuminata* Gerneck)
(typ. cons.).

Ophiocytium Nägeli in Neue Denkschr. Allg. Schweiz. Ges. Gesammten Naturwiss. 10(7): 87. 1849.
Typus: *O. apiculatum* Nägeli [= *O. cochleare* (Eichw.) A. Braun (*Spiro-gyra cochlearis* Eichw.)].

(=) *Spirodiscus* Ehrenb. in Abh. Königl.
Akad. Wiss. Berlin, Phys. Kl., 1831:
68. 1832.
Typus: *S. fulvus* Ehrenb.

Tetraedriella Pascher in Arch. Protistenk. 69: 423. 5 Feb 1930.
Typus: *T. acuta* Pascher

(=) *Polyedrium* Nägeli in Neue Denkschr. Allg. Schweiz. Ges. Gesammten Naturwiss. 10(7): 83. 1849.
Typus: *P. tetraëdricum* Nägeli

B. FUNGI

Agaricus L., Sp. Pl.: 1171. 1 Mai
1753.
Typus: *A. campestris* L. : Fr. (typ.
cons.).

Aleurodiscus Rabenh. ex J. Schröt.
in Cohn, Krypt.-Fl. Schlesien 3(1):
429. 2 Jun 1888.
Typus: *A. amorphus* (Pers. : Fr.) J.
Schröt. (*Peziza amorpha* Pers. : Fr.).

(=) *Cyphella* Fr., Syst. Mycol. 2: 201.
1822 : Fr., ibid.
Typus: *C. digitalis* (Alb. & Schwein.)
Fr.

Alternaria Nees, Syst. Pilze: 72.
1816-1817.
Typus: *A. tenuis* Nees (*Torula alter-nata* Fr. : Fr., *A. alternata* (Fr. : Fr.)
Keissl.).

Amanita Pers., Tent. Disp. Meth.
Fung.: 65. 14 Oct - 31 Dec 1797.
Typus: *A. muscaria* (L. : Fr.) Pers.
(*Agaricus muscarius* L. : Fr.).

(H) *Amanita* Boehm., Defin. Gen. Pl.:
490. 1760 (typ. des.: Earle in Bull.
New York Bot. Gard. 5: 382. 1909;
Donk in Beih. Nova Hedwigia 5: 20.
1962) [*Fungi*].
≡ *Agaricus* L. 1753 (nom. cons.).

191

Amaurochaete Rostaf., Vers. Syst. Mycetozoen: 8. 1873.
Typus: *A. atra* (Alb. & Schwein.) Rostaf. (*Lycogala atrum* Alb. & Schwein.).

(=) *Lachnobolus* Fr., Syst. Orb. Veg.: 148. 1825.
Typus: *L. cribrosus* Fr.

Amanitopsis Roze in Bull. Soc. Bot. France 23: 50, 51. 1876 (post 11 Feb).
Typus: *A. vaginata* (Bull. : Fr.) Roze (*Agaricus vaginatus* Bull. : Fr.).

(≡) *Vaginarius* Roussel, Fl. Calvados, ed. 2: 59. 1806 (typ. des.: Donk in Beih. Nova Hedwigia 5: 292. 1962).
(=) *Vaginata* Gray, Nat. Arr. Brit. Pl. 1: 601. 1 Nov 1821.
Typus: *V. livida* (Pers.) Gray (*Amanita livida* Pers.).

Amphisphaeria Ces. & De Not. in Comment. Soc. Crittog. Ital. 1: 223. Jan 1863.
Typus: *A. umbrina* (Fr.) De Not. (*Sphaeria umbrina* Fr.) (typ. cons.).

Anema Nyl. ex Forssell, Beitr. Gloeolich.: 40, 91. Mai-Jun 1885.
Typus: *A. decipiens* (A. Massal.) Forssell (*Omphalaria decipiens* A. Massal.).

(≡) *Omphalaria* A. Massal., Framm. Lichenogr.: 13. 1855.
Typus: *O. decipiens* A. Massal.

Anisomeridium (Müll. Arg.) M. Choisy, Icon. Lich. Univ.: 3. Jan 1928 (*Arthopyrenia* sect. *Anisomeridium* Müll. Arg. in Flora 66: 290. 21 Jun 1883).
Typus: *Arthopyrenia xylogena* Müll. Arg. [nomen sub *Anisomeridium* deest].

(=) *Microthelia* Körb., Syst. Lich. Germ.: 372. Jan-Mar 1855.
Typus (vide Fries, Gen. Heterolich. Eur.: 111. 1861): *M. micula* Körb., nom. illeg. (*Verrucaria biformis* Borrer, *Anisomeridium biforme* (Borrer) R. C. Harris).
(=) *Ditremis* Clem., Gen. Fungi: 41, 173. Jun-Dec 1909.
Typus: *D. inspersa* (Müll. Arg.) Clem. (*Pleurotrema inspersum* Müll. Arg.).

Anzia Stizenb. in Flora 44: 393. 7 Jul 1861.
Typus: *A. colpodes* (Ach.) Stizenb. (*Lichen colpodes* Ach.).

(=) *Chondrospora* A. Massal. in Atti Reale Ist. Veneto Sci. Lett. Arti, ser. 3, 5: 248. 1860.
Typus: *C. semiteres* (Mont. & Bosch) A. Massal. (*Parmelia semiteres* Mont. & Bosch).

Aposphaeria Sacc. in Michelia 2: 4. 25 Apr 1880.
Typus: *A. pulviscula* (Sacc.) Sacc. (*Phoma pulviscula* Sacc.).

(H) *Aposphaeria* Berk., Outl. Brit. Fungol.: 315. Aug-Dec 1860 [*Fungi*].
Typus: *A. complanata* (Tode : Fr.) Berk. (*Sphaeria complanata* Tode : Fr.).

Arthonia Ach. in Neues J. Bot. 1(3): 3. Jan 1806.
Typus: *A. radiata* (Pers.) Ach. (*Opegrapha radiata* Pers.).

(=) *Coniocarpon* DC. in Lamarck & Candolle, Fl. Franç., ed. 3, 2: 323. 17 Sep 1805.
Typus: *C. cinnabarinum* DC.

Arthopyrenia A. Massal., Ric. Auton. Lich. Crost.: 165. Jun-Dec 1852.
Typus: *A. cerasi* (Schrad.) A. Massal. (*Verrucaria cerasi* Schrad.) (typ. cons.).

Arthrorhaphis Th. Fr., Lich. Arct.: 203. Mai-Dec 1860.
Typus: *A. flavovirescens* (A. Massal.) Th. Fr. (*Raphiospora flavovirescens* A. Massal.) [= *A. citrinella* (Ach.) Poelt (*Lichen citrinellus* Ach.)].

(≡) *Raphiospora* A. Massal., Alc. Gen. Lich.: 11. Mai-Aug 1853 (typ. des.: Jørgensen & Santesson in Taxon 42: 881. 1993).

Aschersonia Mont. in Ann. Sci. Nat., Bot., ser. 3, 10: 121. Aug 1848.
Typus: *A. taitensis* Mont.

(H) *Aschersonia* Endl., Gen. Pl., Suppl. 2: 103. Mar-Jun 1842 [*Fungi*].
Typus: *A. crustacea* (Jungh.) Endl. (*Laschia crustacea* Jungh.).

Aspicilia A. Massal., Ric. Auton. Lich. Crost.: 36. Jun-Dec 1852.
Typus: [specimen] *"Urceolaria cinerea* β *alba"*, Schaerer, Lich. Helv. Exsicc., ed. 2, 6: No. 127 (VER) (typ. cons.) [= *A. cinerea* (L.) Körb. (*Lichen cinereus* L.)].

(=) *Circinaria* Link in Neues J. Bot. 3(1-2): 5. Apr 1809.
Typus: *Urceolaria hoffmannii* (Ach.) Ach., nom. illeg. (*Verrucaria contorta* Hoffm.).

(=) *Sagedia* Ach. in Kongl. Vetensk. Acad. Nya Handl. 30: 164. Jul-Sep 1809.
Typus (vide Laundon & Hawksworth in Taxon 37: 478. 1988): *S. zonata* Ach.

(=) *Sphaerothallia* T. Nees in Nova Acta Phys.-Med. Acad. Caes. Leop.-Carol. Nat. Cur. 15(2): 360. 1831.
Typus (vide Choisy in Bull. Soc. Bot. France 76: 525. 1929): *Lecanora esculenta* (Pall.) Eversm. (*Lichen esculentus* Pall.).

(=) *Chlorangium* Link in Bot. Zeitung (Berlin) 7: 731. 12 Oct 1849.
Typus: *C. jussufii* (Link) Link (*Placodium jussufii* Link).

Bacidina Vězda in Folia Geobot. Phytotax. 25: 431. 4 Jan 1991.
Typus: *B. phacodes* (Körb.) Vězda (*Bacidia phacodes* Körb.

(=) *Lichingoldia* D. Hawksw. & Poelt in Pl. Syst. Evol. 154: 203. 30 Jul 1986.
Typus: *L. gyalectiformis* D. Hawksw. & Poelt

(=) *Woessia* D. Hawksw. & Poelt in Pl. Syst. Evol. 154: 207. 30 Jul 1986.
Typus: *W. fusarioides* D. Hawksw. & al.

Badimia Vězda in Folia Geobot. Phytotax. 21: 206. 2 Jun 1986.
Typus: *B. dimidiata* (Bab. ex Leight.) Vězda (*Lecanora dimidiata* Bab. ex Leight.).

(=) *Pseudogyalecta* Vězda in Folia Geobot. Phytotax. 10: 408. 27 Dec 1975.
Typus: *P. verrucosa* Vězda

Boletus L., Sp. Pl.: 1176. 1 Mai 1753.
Typus: *B. edulis* Bull. : Fr. (typ. cons.).

Buellia De Not. in Giorn. Bot. Ital. 2(1,1): 195. 1846.
Typus: *B. disciformis* (Fr.) Mudd (Man. Brit. Lich.: 216. 1861) (*Lecidia parasema* var. *disciformis* Fr.).

(=) *Gassicurtia* Fée, Essai Crypt. Ecorc.: xlvi. 4 Dec 1824.
Typus: *G. coccinea* Fée (Essai Crypt. Ecorc.: 100. Mai-Oct 1825).

Caloplaca Th. Fr., Lich. Arct.: 218. Mai-Dec 1860.
Typus: *C. cerina* (Ehrh. ex Hedw.) Th. Fr. (*Lichen cerinus* Ehrh. ex Hedw.).

(=) *Gasparrinia* Tornab., Lichenogr. Sicul.: 27. 1849.
Typus: *G. murorum* (Hoffm.) Tornab. (*Lichen murorum* Hoffm.).

(=) *Pyrenodesmia* A. Massal. in Atti Reale Ist. Veneto Sci. Lett. Arti, ser. 2, 3 (Appunt. 4): 119. 1853.
Typus: *P. chalybaea* (Fr.) A. Massal. (*Parmelia chalybaea* Fr.).

(=) *Xanthocarpia* A. Massal. & De Not. in Massalongo, Alc. Gen. Lich.: 11. Mai-Aug 1853.
Typus: *X. ochracea* (Schaer.) A. Massal. & De Not. (*Lecidea ochracea* Schaer.)

Calvatia Fr., Summa Veg. Scand.: 442. 1849.
Typus: *C. craniiformis* (Schwein.) Fr. ex De Toni (*Bovista craniiformis* Schwein.).

(=) *Omalycus* Raf., Précis Découv. Somiol. Jun-Dec 1814.
Typus: *O. violacinus* Raf.

(=) *Langermannia* Rostk. in Sturm, Deutschl. Fl., sect. 3, 5: 23. 3-9 Nov 1839.
Typus: *L. gigantea* (Batsch : Pers.) Rostk. (*Lycoperdon giganteum* Batsch : Pers.).

(=) *Hippoperdon* Mont. in Ann. Sci. Nat., Bot., ser. 2, 17: 121. Feb 1842.
Typus: *H. crucibulum* Mont.

Candida Berkhout, Schimmelgesl. Monilia: 41. 1923.
Typus: *C. vulgaris* Berkhout

(=) *Syringospora* Quinq. in Arch. Physiol. Norm. Pathol. 1: 293. 1868.
Typus: *S. robinii* Quinq., nom. illeg. (*Oidium albicans* C. P. Robin).

(=) *Parendomyces* Queyrat & Laroche in Bull. & Mém. Soc. Méd. Hôp. Paris, ser. 3, 28: 136. 1909.
Typus: *P. albus* Queyrat & Laroche

(=) *Parasaccharomyces* Beurm. & Gougerot in Tribune Méd. (Paris) 42: 502. 7 Aug 1909.
Typus: non designatus.

(=) *Pseudomonilia* A. Geiger in Centralbl. Bakteriol., 2. Abth. 27: 134. 1 Jun 1910.
Typus: *P. albomarginata* A. Geiger

Catinaria Vain. in Acta Soc. Fauna Fl. Fenn. 53(1): 143. 4 Nov - 2 Dec 1922.
Typus: *C. atropurpurea* (Schaer.) Vězda & Poelt (*Lecidea sphaeroides* var. *atropurpurea* Schaer.) (typ. cons.).

(≡) *Biatorina* A. Massal., Ric. Auton. Lich. Crost.: 134. Jun-Dec 1852 (typ. des.: Santesson in Symb. Bot. Upsal. 12: 428. 1952).

Ceratiomyxa J. Schröt. in Engl. & Prantl, Nat. Pflanzenfam. I(I): 16. Sep 1889.
Typus: *C. mucida* (Pers.) J. Schröt. (*Isaria mucida* Pers.).

(=) *Famintzinia* Hazsl. in Oesterr. Bot. Z. 27: 85. 1877.
Typus: *F. porioides* (Alb. & Schwein.) Hazsl. (*Ceratium porioides* Alb. & Schwein.).

Cetraria Ach., Methodus: 292. Jan-Apr 1803.
Typus: *C. islandica* (L.) Ach. (*Lichen islandicus* L.).

(≡) *Platyphyllum* Vent., Tabl. Règne Vég. 2: 34. 5 Mai 1799.

Ceuthospora Grev., Scott. Crypt. Fl.: ad t. 253-254. Sep 1826.
Typus: *C. lauri* (Grev.) Grev. (*Cryptosphaeria lauri* Grev.).

(H) *Ceuthospora* Fr., Syst. Orb. Veg.: 119. Dec 1825 [*Fungi*].
Typus: *C. phaeocomes* (Rebent. : Fr.) Fr. (*Sphaeria phaeocomes* Rebent. : Fr.).

Chlorociboria Seaver ex C. S. Ramamurthi & al. in Mycologia 49: 857. 28 Mar 1958.
Typus: *C. aeruginosa* (Pers. : Fr.) Seaver ex C. S. Ramamurthi & al. (*Peziza aeruginosa* Pers. : Fr.) (typ. cons.).

Chlorophyllum Massee in Kew Bull. 1898: 136. 1898.
Typus: *C. esculentum* Massee [= *C. molybdites* (G. Mey. : Fr.) Massee (*Agaricus molybdites* G. Mey. : Fr.)].

(=) *Endoptychum* Czern. in Bull. Soc. Imp. Naturalistes Moscou 18(2, III): 146. 1845.
Typus: *E. agaricoides* Czern.

Chondropsis Nyl. ex Cromb. in J. Linn. Soc., Bot. 17: 397. 1879.
Typus: *C. semiviridis* (Nyl.) Cromb. (*Parmeliopsis semiviridis* Nyl.).

(H) *Chondropsis* Raf., Fl. Tellur. 3: 29, 97. Nov-Dec 1837 [*Dicot.: Gentian.*].
Typus: *C. trinervis* (L.) Raf. (*Chironia trinervis* L.).

Chrysothrix Mont. in Ann. Sci. Nat., Bot., ser. 3, 18: 312. Nov 1852.
Typus: *C. noli-tangere* Mont., nom. illeg. (*Peribotryon pavonii* Fr. : Fr., *C. pavonii* (Fr. : Fr.) J. R. Laundon).

(≡) *Peribotryon* Fr., Syst. Mycol. 3: 287. 1832.

(=) *Pulveraria* Ach., Methodus: 1. Jan-Apr 1803.
Typus (vide Laundon in Taxon 30: 663. 1981): *P. chlorina* (Ach.) Ach. (*Lichen chlorinus* Ach.).

Cistella Quél., Enchir. Fung.: 319. Jan-Jun 1886.
Typus: *C. dentata* (Pers. : Fr.) Quél. (*Peziza dentata* Pers. : Fr.).

(H) *Cistella* Blume, Bijdr.: 293. 20 Sep - 7 Dec 1825 [*Monocot.: Orchid.*].
Typus: *C. cernua* (Willd.) Blume (*Malaxis cernua* Willd.).

Cladonia P. Browne, Civ. Nat. Hist. Jamaica: 81. 10 Mar 1756.
Typus: *C. subulata* (L.) F. H. Wigg. (Prim. Fl. Holsat.: 90. 29 Mar 1780) (*Lichen subulatus* L.).

Clavaria L., Sp. Pl.: 1182. 1 Mai 1753.
Typus: *C. fragilis* Holmsk. (Beata Ruris 1: 7. 1790) : Fr. (Syst. Mycol. 1: 484. 1 Jan 1821).

Collema F. H. Wigg., Prim. Fl. Holsat.: 89. 29 Mar 1780.
Typus: *C. lactuca* (Weber) F. H. Wigg. (*Lichen lactuca* Weber).

(=) *Gabura* Adans., Fam. Pl. 2: 6, 560. Jul-Aug 1763.
Typus: *Lichen fascicularis* L

(=) *Kolman* Adans., Fam. Pl. 2: 7, 542. Jul-Aug 1763.
Typus: *Lichen nigrescens* Huds..

Collybia (Fr.) Staude, Schwämme Mitteldeutschl.: xxviii, 119. 1857 (*Agaricus* "trib." *Collybia* Fr., Syst. Mycol. 1: 9, 129. 1 Jan 1821).
Typus: *Agaricus tuberosus* Bull. : Fr. (*C. tuberosa* (Bull. : Fr.) P. Kumm.).

(=) *Gymnopus* Pers. ex Gray, Nat. Arr. Brit. Pl. 1: 604. 1 Nov 1821.
Typus: *G. fusipes* (Bull. : Fr.) Gray (*Agaricus fusipes* Bull. : Fr.).

Coniothyrium Corda, Icon. Fung. 4: 38. Sep 1840.
Typus: *C. palmarum* Corda

(=) *Clisosporium* Fr., Novit. Fl. Suec.: 80. 18 Dec 1819 : Fr., Syst. Mycol. 1: xlvii. 1 Jan 1821.
Typus: *C. lignorum* Fr. : Fr.

Conocybe Fayod in Ann. Sci. Nat., Bot., ser. 7, 9: 357. Jun 1889.
Typus: *C. tenera* (Schaeff. : Fr.) Fayod (*Agaricus tener* Schaeff. : Fr.).

(=) *Raddetes* P. Karst. in Hedwigia 26: 112. Mai-Jun 1887.
Typus: *R. turkestanicus* P. Karst.

(=) *Pholiotina* Fayod in Ann. Sci. Nat., Bot., ser. 7, 9: 359. Jun 1889.
Typus: *P. blattaria* (Fr. : Fr.) Fayod (*Agaricus blattarius* Fr. : Fr.).

(=) *Pholiotella* Speg. in Bol. Acad. Nac. Ci. 11: 412. 1889.
Typus: *P. blattariopsis* Speg.

Coprinopsis P. Karst. in Acta Soc. Fauna Fl. Fenn. 2(1): 27. Oct-Dec 1881.
Typus: *C. friesii* (Quél.) P. Karst. (*Coprinus friesii* Quél.).

(=) *Pselliophora* P. Karst. in Bidrag Kännedom Finlands Natur Folk 32: 528. Jul-Dec 1879.
Typus: *P. atramentaria* (Bull. : Fr.) P. Karst. (*Agaricus atramentarius* Bull. : Fr.).

Cordyceps Fr., Observ. Mycol. 2 (revis.): 316. 1824.
Typus: *C. militaris* (L. : Fr.) Fr. (*Clavaria militaris* L. : Fr.).

Cortinarius (Pers.) Gray, Nat. Arr. Brit. Pl. 1: 627. 1 Nov 1821 *('Cortinaria')* (orth. cons.) (*Agaricus* sect. *Cortinarius* Pers., Syn. Meth. Fung.: 276. 31 Dec 1801).
Typus: *Agaricus violaceus* L. : Fr. (*C. violaceus* (L. : Fr.) Gray).

Corynespora Güssow in Z. Pflanzenkrankh. 16: 10. 24 Jan 1906.
Typus: *C. mazei* Güssow

(=) *Coccosporium* Corda in Sturm, Deutschl. Fl., Abth. 3, 3: 49. 1831.
Typus: *C. maculiforme* Corda

Craterellus Pers., Mycol. Eur. 2: 4. Jan-Jul 1825 *('Cratarellus')* (orth. cons.).
Typus: *C. cornucopioides* (L. : Fr.) Pers. (*Peziza cornucopioides* L. : Fr.) (etiam vide *Pezicula*).

(H) *Craterella* Pers. in Neues Mag. Bot. 1: 112. Apr-Aug 1794 [*Fungi*].
Typus: *C. pallida* Pers.

(≡) *Trombetta* Adans., Fam. Pl. 2: 6, 613. Jul-Aug 1763 (typ. des.: Kuntze, Revis. Gen. Pl. 2: 873. 1891).

Cribraria Pers. in Neues Mag. Bot. 1: 91. 1794.
Typus: *C. rufescens* Pers.

(H) *Cribraria* Schrad. ex J. F. Gmel. Syst. Nat. 2: 1471. 1792.
Typus: *C. pallida* Schrad. ex J. F. Gmel.

Crocynia (Ach.) A. Massal. in Atti Reale Ist. Veneto Sci. Lett. Arti, ser. 3, 5: 251. 1860 (*Lecidea* sect. *Crocynia* Ach., Lichenogr. Universalis: 217. 1810).
Typus: *Lecidea gossypina* (Sw.) Ach. (*Lichen gossypinus* Sw., *C. gossypina* (Sw.) A. Massal.).

(≡) *Symplocia* A. Massal., Neagen. Lich.: 4. Nov-Dec 1854.

Cryphonectria (Sacc.) Sacc., Syll. Fung. 17: 783. 25 Mai 1905 (*Nectria* subg. *Cryphonectria* Sacc., Syll. Fung. 2: 507. 1883).
Typus: *Nectria parasitica* (Murrill) Sacc. (*Diaporthe parasitica* Murrill, *C. parasitica* (Murrill) M. E. Barr) (typ. cons.).

Cryptococcus Vuill. in Rev. Gen. Sci. Pures Appl. 12: 741. 1901.
Typus: *C. neoformans* (San Felice) Vuill. (*Saccharomyces neoformans* San Felice) (typ. cons.*:* No. 72042 (BPI) e cult. No. CBS 132).

(H) *Cryptococcus* Kütz. in Linnaea 8: 365. 1833 [*Fungi*].
Typus: *C. mollis* Kütz.

Cryptosphaeria Ces. & De Not. in Comment. Soc. Crittog. Ital. 4: 231. 1853.
Typus: *C. millepunctata* Grev. (Fl. Edin.: 360. 18 Mar 1824) (typ. cons.).

(H) *Cryptosphaeria* Grev., Scott. Crypt. Fl.: ad t. 13. Sep 1822 [*Fungi*].
Typus: *C. taxi* (Sowerby : Fr.) Grev. (*Sphaeria taxi* Sowerby : Fr.).

Cryptothecia Stirt. in Proc. Roy. Philos. Soc. Glasgow 10: 164. 1876.
Typus: *C. subnidulans* Stirt.

(=) *Myriostigma* Kremp., Lich. Foliicol.: 22. 1874.
Typus: *M. candidum* Kremp.

Cylindrocarpon Wollenw. in Phyto-pathology 3: 225. 2 Oct 1913.
Typus: *C. cylindroides* Wollenw.

(=) *Fusidium* Link in Ges. Naturf. Freun-de Berlin Mag. Neuesten Entdeck. Gesammten Naturk. 3: 8. Jan-Mar 1809 : Fr., Syst. Mycol. 1: xl. 1 Jan 1821.
Typus: *F. candidum* Link : Fr.

Daldinia Ces. & De Not. in Comment. Soc. Crittog. Ital. 1: 197. Jan 1863.
Typus: *D. concentrica* (Bolton : Fr.) Ces. & De Not. (*Sphaeria concentrica* Bolton : Fr.).

(≡) *Peripherostoma* Gray, Nat. Arr. Brit. Pl. 1: 513. 1 Nov 1821 (per typ. des.).
(≡) *Stromatosphaeria* Grev., Fl. Edin.: lxxiii, 355. 18 Mar 1824 (per typ. des.).

Debaryomyces Lodder & Kreger in Kreger, Yeasts, ed. 3: 130, 145. 1984.
Typus: *D. hansenii* (Zopf) Lodder & Kreger (*Saccharomyces hansenii* Zopf).

(H) *Debaryomyces* Klöcker in Compt.-Rend. Trav. Carlsberg Lab. 7: 273. 1909 [*Fungi*].
Typus: *D. globosus* Klöcker
(≡) *Debaryozyma* Van der Walt & Johannsen in Persoonia 10: 147. 28 Dec 1978.

Didymosphaeria Fuckel in Jahrb. Nassauischen Vereins Naturk. 23-24: 140. 19 Feb - 24 Nov 1870.
Typus: [specimen] *"Amphisphaeria epidermidis"*, Germany, "in sylva Hostrichiensi", *Fuckel* in Fungi Rhen. No. 1770 (S) (typ. cons.) [= *D. futilis* (Berk. & Broome) Rehm (*Sphaeria futilis* Berk. & Broome)].

Dothidea Fr. Observ. Mycol. 2: 347.
Apr-Mai 1818. : Fr. Syst. Mycol. 1:
LII. 1 Jan 1821.
Typus: *D. sambuci* (Pers. : Fr.) Fr. :
Fr. (typ. cons.).

Dothiora Fr., Summa Veg. Scand.:
418. 1849.
Typus: *D. pyrenophora* (Fr. : Fr.) Fr.
(*Dothidea pyrenophora* Fr. : Fr.).

(H) *Dothiora* Fr., Fl. Scan.: 347. 1837
[*Fungi*].
Typus: *Variolaria melogramma* Bull.
: Fr.

Drechslera S. Ito in Proc. Imp.
Acad. Japan 6: 355. 1930.
Typus: *D. tritici-vulgaris* (Y. Nisik.)
S. Ito ex S. Hughes (*Helmintho-
sporium tritici-vulgaris* Y. Nisik.).

(=) *Angiopoma* Lév. in Ann. Sci. Nat.,
Bot., ser. 2, 16: 235. Oct 1841.
Typus: *A. campanulatum* Lév.

Encoelia (Fr.) P. Karst. in Bidrag
Kännedom Finlands Natur Folk 19:
18, 217. 1871 (*Peziza* "trib." *En-
coelia* Fr., Syst. Mycol. 2: 74. 1822).
Typus: *Peziza furfuracea* Roth : Fr.
(*E. furfuracea* (Roth : Fr.) P. Karst.).

(≡) *Phibalis* Wallr., Fl. Crypt. Germ. 2:
445. Feb-Mar 1833.

Epidermophyton Sabour. in Arch.
Méd. Exp. Anat. Pathol. 19: 754.
Nov 1907.
Typus: *E. inguinale* Sabour.

(H) *Epidermidophyton* E. Lang in Vier-
teljahresschr. Dermatol. Syph. 11:
263. 1879 [*Fungi*].
Typus: non designatus.

Eutypella (Nitschke) Sacc. in Atti
Soc. Veneto-Trentino Sci. Nat. Pa-
dova 4: 80. Oct 1875 (*Valsa* subg.
Eutypella Nitschke, Pyrenomyc.
Germ.: 163. Jan 1870).
Typus: *Valsa sorbi* (Alb. & Schwein.
: Fr.) Fr. (*Sphaeria prunastri* var.
sorbi Alb. & Schwein. : Fr., *E. sorbi*
(Alb. & Schwein. : Fr.) Sacc.).

(=) *Scoptria* Nitschke, Pyrenomyc. Germ.:
83. Jan 1867.
Typus: *S. isariphora* Nitschke

Fusicladium Bonord., Handb. Mykol.: 80. 1851.
Typus: *F. virescens* Bonord.

(=) *Spilocaea* Fr., Novit. fl. svec. 5
(cont.): 79. 18 Dec 1819 : Fr. Syst
mycol. 1: XXXIX. 1 Jan 1821.
Typus: *S. pomi* Fr.
(=) *Cycloconium* Castagne, Cat. Pl. Marseille: 220. Oct 1845.
Typus: *C. olagineum* Castagne

Gautieria Vittad., Monogr. Tuber-
ac.: 25. 1831.
Typus: *G. morchelliformis* Vittad.

(H) *Gautiera* Raf., Med. Fl. 1: 202. 11
Jan 1828 [*Dicot.: Eric.*].
≡ *Gaultheria* L. 1753.

Gloeophyllum P. Karst. in Bidrag
Kännedom Finlands Natur Folk 37:
x, 79. 1882 *('Gleophyllum')* (orth.
cons.).
Typus: *G. sepiarium* (Wulfen : Fr.)
P. Karst. (*Agaricus sepiarius* Wulfen : Fr.).

(≡) *Serda* Adans., Fam. Pl. 2: 11, 604.
Jul-Aug 1763 (typ. des.: Donk in
Persoonia 1: 279. 1960).
(≡) *Sesia* Adans., Fam. Pl. 2: 10, 604.
Jul-Aug 1763 (typ. des.: Donk in
Persoonia 1: 280. 1960).
(=) *Ceratophora* Humb., Fl. Friberg.:
112. 1793.
Typus: *C. fribergensis* Humb.

Guignardia Viala & Ravaz in Bull.
Soc. Mycol. France 8: 63. 22 Mai
1892.
Typus: *G. bidwellii* (Ellis) Viala &
Ravaz (*Sphaeria bidwellii* Ellis) (typ.
cons.).

Gyalidea Lettau ex Vĕzda in Folia
Geobot. Phytotax. Bohemoslov. 1:
312. 1966.
Typus: *G. lecideopsis* (A. Massal.)
Lettau ex Vĕzda (*Gyalecta lecide-
opsis* A. Massal.).

(=) *Solorinella* Anzi, Cat. Lich. Sondr.:
37. Aug 1860.
Typus: *S. asteriscus* Anzi
(=) *Aglaothecium* Groenh. in Persoonia
3: 349. 20 Apr 1962.
Typus: *A. saxicola* Groenh.

Gyalideopsis Vĕzda in Folia Geobot. Phytotax. 7: 204. 7 Aug 1972.
Typus: *G. peruviana* Vĕzda

(=) *Diploschistella* Vain. in Ann. Univ.
Fenn. Åbo., A, 2(3): 26. 1926.
Typus: *D. urceolata* Vain.

Gymnoderma Nyl. in Flora 43: 546.
21 Sep 1860.
Typus: *G. coccocarpum* Nyl.

(H) *Gymnoderma* Humb., Fl. Friberg.:
109. 1793 [*Fungi*].
Typus: *G. sinuatum* Humb.

Gyrodon Opat. in Arch. Naturgesch. 2(1): 5. 1836.
Typus: *G. sistotremoides* Opat., nom. illeg. (*Boletus sistotremoides* Fr., non Alb. & Schwein., *Boletus sistotrema* Fr. : Fr., *G. sistotrema* (Fr. : Fr.) P. Karst. [= *G. lividus* (Bull. : Fr.) P. Karst. (*Boletus lividus* Bull. : Fr.)].

(=) *Anastomaria* Raf., Ann. Nat.: 16. Mar-Jul 1820.
Typus: *A. campanulata* Raf.

Gyromitra Fr., Summa Veg. Scand.: 346. 1849.
Typus: *G. esculenta* (Pers. : Fr.) Fr. (*Helvella esculenta* Pers. : Fr.).

(=) *Gyrocephalus* Pers. in Mém. Soc. Linn. Paris 3: 77. 1824.
Typus: *G. aginnensis* Pers., nom. Illeg. (*Helvella sinuosa* Brond.).

Haematomma A. Massal., Ric. Auton. Lich. Crost.: 32. Jun-Dec 1852.
Typus: *H. vulgare* A. Massal. (typ. cons.).

Helminthosporium Link in Ges. Naturf. Freunde Berlin Mag. Neuesten Entdeck. Gesammten Naturk. 3: 10. Jan-Mar 1809 *('Helmisporium')* (orth. cons.).
Typus: *H. velutinum* Link : Fr.

Hemitrichia Rostaf., Vers. Syst. Mycetozoen: 14. 1873.
Typus: *H. clavata* (Pers.)Rostaf. in Fuckel (*Trichia clavata* Pers.).

(=) *Hyporhamma* Corda, Icon. Fung. 6: 13. Oct 1854.
Typus: *H. reticulatum* (Pers.)Corda (*Trichia reticulata* Pers.).

Hexagonia Fr., Epicr. Syst. Mycol.: 496. 1838 *('Hexagona')* (orth. cons.).
Typus: *H. hirta* (P. Beauv. : Fr.) Fr. (*Favolus hirtus* P. Beauv. : Fr.) (typ. cons.).

(H) *Hexagonia* Pollini, Hort. Veron. Pl.: 35. 1816 [*Fungi*].
Typus: *H. mori* Pollini

Hirneola Fr. in Kongl. Vetensk. Acad. Handl. 1848: 144. 1848.
Typus: *H. nigra* Fr., nom. illeg. (*Peziza nigricans* Sw. : Fr., *H. nigricans* (Sw. : Fr.) P. W. Graff).

(H) *Hirneola* Fr., Syst. Orb. Veg.: 93. Dec 1825 [*Fungi*].
≡ *Mycobonia* Patouillard 1894 (nom. cons.).

(=) *Laschia* Fr. in Linnaea 5: 533. Oct 1830 : Fr., Syst. Mycol. 3 (index): 107. 1832.
Typus: *L. delicata* Fr. : Fr.

Hyaloscypha Boud. in Bull. Soc.
Mycol. France 1: 118. Mai 1885.
Typus: *H. vitreola* (P. Karst.) Boud.
(*Peziza vitreola* P. Karst.) (typ. cons.).

Hydnum L., Sp. Pl.: 1178. 1 Mai 1753.
Typus: *H. repandum* L. : Fr. (typ.
cons.).

Hymenochaete Lév. in Ann. Sci.
Nat., Bot., ser. 3, 5: 150. 1846.
Typus: *H. rubiginosa* (Dicks. : Fr.)
Lév. (*Helvella rubiginosa* Dicks. : Fr.).

(H) *Hymenochaeta* P. Beauv. ex T.
Lestib., Essai Cypér.: 43. 29 Mar
1819 [*Monocot.: Cyper.*].
Typus: non designatus.

(=) *Cyclomyces* Kunze ex Fr. in Linnaea
5: 512. 1830. (: Fr., Syst. Mycol. 3,
Index: 80. 1832)
Typus: *C. fuscus* Kunze ex Fr. : Fr.

Hyphodontia J. Erikss. in Symb.
Bot. Upsal. 16(1): 101. 1958.
Typus: *H. pallidula* (Bres.) J. Erikss.
(*Gonatobotrys pallidula* Bres.) (typ.
cons.).

(=) *Xylodon* (Pers.) Gray, Nat. Arr. Brit.
Pl. 1: 649. 1 Nov 1821 (*Sistotrema*
sect. *Xylodon* Pers., Syn. Meth. Fung.:
552. 31 Dec 1801).
Typus: *Sistotrema quercinum* Pers. :
Fr. (*X. quercinus* (Pers. : Fr.) Gray).

(=) *Grandinia* Fr., Epicr. Syst. Mycol.:
527. 1838.
Typus (vide Clements & Shear, Gen.
Fung., ed. 2: 346. 1931): *Thelephora*
granulosa Pers. : Fr.

(=) *Lyomices* P. Karst. in Rev. Mycol.
(Toulouse) 3(9): 23. 1881.
Typus: *L. serus* (Pers.) P. Karst.
(*Hydnum serum* Pers.).

(=) *Kneiffiella* P. Karst. in Bidrag Kän-
nedom Finlands Natur Folk 48: 371.
Jan-Sep 1889.
Typus: *K. barba-jovis* (Bull. : Fr.) P.
Karst. (*Hydnum barba-jovis* Bull.
: Fr.).

(=) *Schizopora* Velen., České Houby:
638. 1922.
Typus: *S. laciniata* Velen.

(=) *Chaetoporellus* Bondartsev & Singer in Mycologia 36: 67. 1 Feb 1944.
Typus: *C. latitans* (Bourdot & Galzin) Bondartsev & Singer (*Poria la-*
titans Bourdot & Galzin).

Hypholoma (Fr.) P. Kumm., Führ. Pilzk.: 21, 72. Jul-Aug 1871 (*Agaricus* "trib." *Hypholoma* Fr., Syst. Mycol. 1: 11, 287. 1 Jan 1821).
Typus: *Agaricus fascicularis* Fr. : Fr. (*H. fasciculare* (Fr. : Fr.) P. Kumm.) (typ. cons.).

Hypoderma De Not. in Giorn. Bot. Ital. 2(1,2): 13. 1847.
Typus: *H. rubi* (Pers. : Fr.) DC. ex Chevall. (*Hysterium rubi* Pers. : Fr.) (typ. cons.).

(H)   *Hypoderma* DC. in Lamarck & Candolle, Fl. Franç., ed. 3, 2: 304. 17 Sep 1805 (typ. des.: Cannon & Minter in Taxon 32: 580. 1983) [*Fungi*]. ≡ *Lophodermium* Chevall. 1826 (nom. cons.).

Hypoxylon Bull., Hist. Champ. France: 168. 1791.
Typus: *H. coccineum* Bull. [= *H. fragiforme* (Pers. : Fr.) Kickx (*Sphaeria fragiformis* Pers. : Fr.)].

(H)   *Hypoxylon* Adans., Fam. Pl. 2: 9, 616. Jul-Aug 1763 [*Fungi*].
Typus (vide Donk in Regnum Veg. 34: 16. 1964): *Xylaria polymorpha* (Pers. : Fr.) Grev. (*Sphaeria polymorpha* Pers. : Fr.).

(=)   *Sphaeria* Haller, Hist. Stirp. Helv. 3: 120. 25 Mar 1768 : Fr., Syst. Mycol. 1: lii. 1 Jan 1821.
Typus (vide Donk in Regnum Veg. 34: 16. 1964): *Sphaeria fragiformis* Pers. : Fr.

Icmadophila Trevis. in Rivista Period. Lav. Regia Accad. Sci. Padova 1: 267. 1853-1853.
Typus: *I. aeruginosa* (Scop.) Trevis. (*Lichen aeruginosus* Scop.).

(≡)   *Tupia* L. Marchand in Bijdr. Natuurk. Wetensch. 5: 191. 1830.

Isaria Pers. in Neues Mag. Bot. 1; 121. 1794 : Fr., Syst. Mycol. 1: xlviii. 1 Jan 1821.
Typus: *I. farinosa* (Holm) Fr. : Fr. (*Ramaria farinosa* Holm) (typ. cons.).

Karstenia Fr. in Acta Soc. Fauna Fl. Fenn. 2(6): 166. Jan-Jun 1885.
Typus: *K. sorbina* (P. Karst.) Fr. (*Propolis sorbina* P. Karst.).

(H)   *Karstenia* Göpp. in Nova Acta Phys.-Med. Acad. Caes. Leop.-Carol. Nat. Cur. 17, Suppl.: 451. 18 Nov 1836 [Foss.].
Typus: non designatus.

Kluyveromyces Van der Walt in Antonie van Leeuwenhoek Ned. Tijdschr. Hyg. 22: 271. 1956.
Typus: *K. marxianus* (E. C. Hansen) Van der Walt (*Saccharomyces marxianus* E. C. Hansen) (typ. cons.).

Lachnocladium Lév., Considér. Mycol.: 108. 1846.
Typus: *L. brasiliense* (Lév.) Pat. (*Eriocladus brasiliensis* Lév.).

(≡)   *Eriocladus* Lév. in Ann. Sci. Nat., Bot., ser. 3, 5: 158. 1846.

Lactarius Pers., Tent. Disp. Meth. Fung.: 63. 14 Oct - 31 Dec 1797 (*'Lactaria'*) (orth. cons.).
Typus: *L. piperatus* (L. : Fr.) Pers. (*Agaricus piperatus* L. : Fr.).

Laetinaevia Nannf. in Nova Acta Regiae Soc. Sci. Upsal., ser. 4, 8(2): 190. 1932 (post 5 Feb).
Typus: *L. lapponica* (Nannf.) Nannf. (*Naevia lapponica* Nannf.).

(=)   *Myridium* Clem., Gen. Fung.: 67. Jul-Oct 1909.
Typus: *M. myriosporum* (W. Phillips & Harkn.) Clem. (*Orbilia myriospora* W. Phillips & Harkn.).

Laetiporus Murrill in Bull. Torrey Bot. Club 31: 607. 26 Nov 1904.
Typus: *L. speciosus* Murrill, nom. illeg. (*Polyporus sulphureus* Bull. : Fr., *L. sulphureus* (Bull. : Fr.) Murrill).

(=)   *Cladoporus* (Pers.) Chevall., Fl. Gén. Env. Paris 1: 260. 656. 5 Aug 1826 (*Polyporus* [unranked] *Cladoporus* Pers., Mycol. Eur. 2: 122. Jan-Jul 1825).
Typus: *C. fulvus* Chevall., nom. illeg. (*Polyporus ramosus* Bull.).

Lecanactis Körb., Syst. Lich. Germ.: 275. Jan 1855.
Typus: *L. abietina* (Ach.) Körb. (*Lichen abietinus* Ach.) (typ. cons.).

(H)   *Lecanactis* Eschw., Syst. Lich.: 14, 25. 1824 [*Fungi*].
Typus: *L. lobata* Eschw.

(=)   *Pyrenotea* Fr., Syst. Mycol. 1: xxiii. 1 Jan 1821 : Fr., ibid.
Typus: *P. incrustans* (Ach.) Fr. (in Kongl. Vetensk. Acad. Handl. 1821: 332. 1821) (*Cyphelium incrustans* Ach.).

Lepiota (Pers.) Gray, Nat. Arr. Brit. Pl. 1: 601. 1 Nov 1821 (*Agaricus* sect. *Lepiota* Pers., Tent. Disp. Meth. Fung.: 68. 14 Oct - 31 Dec 1797).
Typus: *Agaricus colubrinus* Pers., non Bull. (*A. clypeolarius* Bull. : Fr., *L. clypeolaria* (Bull. : Fr.) P. Kumm.) (typ. cons.).

Lepraria Ach., Methodus: 3. Jan-Apr 1803.
Typus: *L. incana* (L.) Ach. (*Byssus incana* L.).

(≡)   *Pulina* Adans., Fam. Pl. 2: 3, 595. Jul-Aug 1763 (typ. des.: Laundon in Taxon 12: 37. 1963).

(≡)   *Conia* Vent., Tabl. Règne Vég. 2: 32. 5 Mai 1799 (typ. des.: Laundon in Taxon 12: 37. 1963).

Leptoglossum P. Karst. in Bidrag Kännedom Finlands Natur Folk 32: xvii, 242. Jul-Dec 1879.
Typus: *L. muscigenum* (Bull. : Fr.) P. Karst. (*Agaricus muscigenus* Bull. : Fr.).

(=)   *Boehmia* Raddi, Sp. Nov. Fung. Firenze: 15. 1806.
Typus: *B. muscoides* Raddi

Leptorhaphis Körb., Syst. Lich. Germ.: 371. Jan-Mar 1855.
Typus: *L. oxyspora* (Nyl.) Körb. (*Verrucaria oxyspora* Nyl.).

(=)   *Endophis* Norman, Conat. Praem. Gen. Lich.: 28. 1852.
Typus: non designatus.

Leptosphaeria Ces. & De Not. in Comment. Soc. Crittog. Ital. 1: 234. Jan 1863.
Typus: *L. doliolum* (Pers. : Fr.) Ces. & De Not. (*Sphaeria doliolum* Pers. : Fr.).

(≡)   *Bilimbiospora* Auersw. in Rabenhorst, Fungi Europaei, ed. 2: No. 261 [in sched. corr.]. 1861.

(=)   *Nodulosphaeria* Rabenh., Herb. Mycol., ed. 2: No. 725. 1858 (nom. cons.).

Letharia (Th. Fr.) Zahlbr. in Hedwigia 31: 36. Jan-Apr 1892 (*Evernia* [unranked] *Letharia* Th. Fr., Lichenogr. Scand. 1: 32. 1871).
Typus: *Evernia vulpina* (L.) Ach. (*Lichen vulpinus* L., *Letharia vulpina* (L.) Hue).

(≡) *Chlorea* Nyl. in Mém. Soc. Sci. Nat. Cherbourg 3: 170. Jun 1855.

Lichina C. Agardh, Syn. Alg. Scand.: xii, 9. Mai-Dec 1817.
Typus: *L. pygmaea* (Lightf.) C. Agardh (*Fucus pygmaeus* Lightf.).

(=) *Pygmaea* Stackh. in Mém. Soc. Imp. Naturalistes Moscou 2: 60, 95. 1809.
Typus: *Fucus lichenoides* J. F. Gmel.

Lopadium Körb., Syst. Lich. Germ.: 210. Jan 1855.
Typus: *L. pezizoideum* (Ach.) Körb. (*Lecidea pezizoidea* Ach.).

(=) *Brigantiaea* Trevis., Spighe e Paglie: 7. Jul 1853.
Typus: *B. tricolor* (Mont.) Trevis. (*Biatora tricolor* Mont.).

Lophiostoma Ces. & De Not. in Comment. Soc. Crittog. Ital. 1: 219. Jan 1863.
Typus: *L. macrostomum* (Tode : Fr.) Ces. & De Not. (*Sphaeria macrostoma* Tode : Fr.).

(≡) *Platysphaera* Dumort., Comment. Bot.: 87. Nov (sero) - Dec (prim.) 1822.

Lophodermium Chevall., Fl. Gén. Env. Paris 1: 435. 5 Aug 1826.
Typus: *L. arundinaceum* (Schrad. : Fr.) Chevall. (*Hysterium arundinaceum* Schrad. : Fr.) (etiam vide *Hypoderma* [*Fungi*]).

Luttrellia Shearer in Mycologia 70: 692. 27 Jun 1978.
Typus: *L. estuarina* Shearer

(H) *Lutrellia* Khokhr. & Gornostaj in Azbukina & al., Vodorosli Griby Mhi Dal'n. Vost.: 80. 18 Apr 1978 [*Fungi*].
≡ *Exserohilum* K. J. Leonard & Suggs 1974.

Marasmius Fr., Fl. Scan.: 339. 1836.
Typus: *M. rotula* (Scop. : Fr.) Fr. (*Agaricus rotula* Scop. : Fr.).

(=) *Micromphale* Gray, Nat. Arr. Brit. Pl. 1: 621. 1 Nov 1821.
Typus: *M. venosa* (Pers.) Gray (*Agaricus venosus* Pers.).

Melanogaster Corda in Sturm, Deutschl. Fl., sect. 3, 3: 1. 1831.
Typus: *M. tuberiformis* Corda

(=) *Bullardia* Jungh. in Linnaea 5: 408. Jul 1830.
Typus: *B. inquinans* Jungh.

Melanoleuca Pat., Cat. Pl. Cell. Tunisie: 22. Jan-Apr 1897.
Typus: *M. vulgaris* (Pat.) Pat. (*Melaleuca vulgaris* Pat., *Agaricus melaleucus* Pers. : Fr., *Melanoleuca melaleuca* (Pers. : Fr.) Murrill).

(≡) *Psammospora* Fayod in Ann. Reale Accad. Agric. Torino 35: 91. 1893.

Melanospora Corda, Icon. Fung. 1: 24. Aug 1837.
Typus: *M. zamiae* Corda

(=) *Ceratostoma* Fr., Observ. Mycol. 2: 337. Apr-Mai 1818.
Typus (vide Fries, Summa Veg. Scand.: 396. 1849): *C. chioneum* (Fr. : Fr.) Fr. (*Sphaeria chionea* Fr. : Fr.).

(=) *Megathecium* Link in Abh. Königl. Akad. Wiss. Berlin, Phys. Kl. 1824: 176. 1826 (typ. des.: Cannon & Hawksworth in Taxon 32: 476. 1983). ≡ *Ceratostoma* Fr. 1818.

Micarea Fr., Syst. Orb. Veg.: 256. Dec 1825.
Typus: *M. prasina* Fr. (typ. cons.).

(H) *Micarea* Fr., Sched. Crit. Lich. Suec. Exsicc.: No. 97. 1825 (ante 7 Mai) [*Fungi*].
Typus: *Biatora fuliginea* Fr., nom. illeg. (*Lecidea fuliginea* Ach., nom. illeg., *L. icmalea* Ach.).

Mollisia (Fr.) P. Karst. in Bidrag Kännedom Finlands Natur Folk 19: 15, 189. 1871 (*Peziza* "trib." *Mollisia* Fr., Syst. Mycol. 2: 137. 1822).
Typus: *Peziza cinerea* Batsch : Fr. (*M. cinerea* (Batsch : Fr.) P. Karst.).

(=) *Tapesia* (Pers. : Fr.) Fuckel in Jahrb. Nassauischen Vereins Naturk. 23-24: 300. 19 Feb - 24 Nov 1870 (*Peziza* [unranked] *Tapesia* Pers., Mycol. Eur. 1: 270. 1 Jan - 14 Apr 1822 : Fr., Syst. Mycol. 2: 105. 1822).
Typus (vide Clements & Shear, Gen. Fungi: 325. 1931): *Peziza fusca* Pers. : Fr. (*T. fusca* (Pers. : Fr.) Fuckel).

Monilia Bonord., Handb. Mykol.: 76. 1851 (sero).
Typus: *M. cinerea* Bonord.

(H) *Monilia* Link in Ges. Naturf. Freunde Berlin Mag. Neuesten Entdeck. Gesammten Naturk. 3: 16. Jan-Mar 1809 : Fr., Syst. Mycol. 1: xlvi. 1 Jan 1821 [*Fungi*].
Typus: *M. antennata* (Pers. : Fr.) Pers.

Mucor Fresen., Beitr. Mykol.: 7. Aug 1850.
Typus: [icon] *'Mucor mucedo'*, Fresenius, Beitr. Mykol.: t. 1, f. 1-12. Aug 1850 (typ. cons.) [= *M. murorum* Naumov].

(H) *Mucor* L., Sp. Pl.: 1185. 1 Mai 1753 : Fr., Syst. Mycol. 3: 317. 1832 [*Fungi*].
Typus (vide Sumstine in Mycologia 2: 127. 1910): *M. mucedo* L. : Fr.

(=) *Hydrophora* Tode, Fungi Mecklenb. Sel. 2: 5. 1791 : Fr., Syst. Mycol. 3: 313. 1832.
Typus (vide Sumstine in Mycologia 2: 132. 1910): *H. stercorea* Tode : Fr.

Mutinus Fr., Summa Veg. Scand.: 434. 1849.
Typus: *M. caninus* (Schaeff. : Pers.) Fr. (*Phallus caninus* Schaeff. : Pers.).

(≡) *Cynophallus* (Fr. : Fr.) Corda, Icon. Fung. 5: 29. Jun 1842 (*Phallus* "trib." *Cynophallus* Fr., Syst. Mycol. 2: 282, 284. 1823 : Fr., ibid.).

(=) *Aedycia* Raf. in Med. Repos., ser. 2, 5: 358. Feb - Apr 1808.
Typus: *A. rubra* Raf.

(=) *Ithyphallus* Gray, Nat. Arr. Brit. Pl. 1: 675. 1 Nov 1821.
Typus: *I. inodorus* (Sowerby) Gray (*Phallus inodorus* Sowerby).

Mycoblastus Norman in Nyt Mag. Naturvidensk. 7: 236. 1853 (*'Mykoblastus'*) (orth. cons.).
Typus: *M. sanguinarius* (L.) Norman (*Lichen sanguinarius* L.).

Mycobonia Pat. in Bull. Soc. Mycol. France 10: 76. 30 Jun 1894.
Typus: *M. flava* (Sw. : Fr.) Pat. (*Peziza flava* Sw. : Fr.) (etiam vide *Hirneola* [*Fungi*]).

Mycoporum Flot. ex Nyl. in Mém. Soc. Sci. Nat. Cherbourg 3: 186. Jun 1855.
Typus: *M. elabens* Flot. ex Nyl.

(H) *Mycoporum* G. Mey., Nebenst. Beschaeft. Pflanzenk.: 327. Sep-Dec 1825 [*Fungi*].
Typus: *M. melinostigma* G. Mey.

Nectria (Fr.) Fr., Summa Veg. Scand.: 387. 1849 (*Hypocrea* sect. *Nectria* Fr., Syst. Orb. Veg.: 105. Dec 1825).
Typus: *Sphaeria cinnabarina* Tode : Fr. (*N. cinnabarina* (Tode : Fr.) Fr.) (typ. cons.).

(=) *Ephedrosphaera* Dumort., Comment. Bot.: 90. Nov (sero) - Dec (prim.) 1822.
Typus (vide Cannon & Hawksworth in Taxon 32: 477. 1983): *E. decolorans* (Pers.) Dumort. (*Sphaeria decolorans* Pers.).

(=) *Hydropisphaera* Dumort., Comment. Bot.: 89. Nov (sero) - Dec (prim.) 1822.
Typus: *H. peziza* (Tode : Fr.) Dumort. (*Sphaeria peziza* Tode : Fr.).

Nectriopsis Maire in Ann. Mycol. 9: 323. 10 Aug 1911.
Typus: *N. violacea* (Fr. : Fr.) Maire (*Sphaeria violacea* Fr. : Fr.).

(=) *Chrysogluten* Briosi & Farneti in Atti Ist. Bot. Univ. Pavia, ser. 2, 8: 117. 1904.
Typus: *C. biasolettianum* Briosi & Farneti

(=) *Dasyphthora* Clem., Gen. Fung.: 45, 173. Jul-Oct 1909.
Typus: *D. lasioderma* (Ellis)Clem. (*Nectria lasioderma* Ellis).

Nidularia Fr. in Fries & Nordholm, Symb. Gasteromyc. 1: 2. 22 Mai 1817.
Typus: *N. radicata* Fr.

(H) *Nidularia* Bull., Hist. Champ. France: 163. 1791 [*Fungi*].
Typus: *N. vernicosa* Bull.

Nodulosphaeria Rabenh., Herb. Mycol., ed. 2: No. 725. 1858.
Typus: [specimen] Rabenhorst, Herb. Mycol., ed. 2: No. 725 (S) (typ. cons.) [= *N. derasa* (Berk. & Broome) L. Holm (*Sphaeria derasa* Berk. & Broome)].

Ocellularia G. Mey., Nebenst. Beschaeft. Pflanzenk. 1: 327. Sep-Dec 1825.
Typus: *O. obturata* (Ach.) Spreng. (Syst. Veg. 4(1): 242. 1-7 Jan 1827) (*Thelotrema obturatum* Ach.).

(=) *Ascidium* Fée, Essai Crypt. Ecorc.: xlii. 4 Dec 1824.
Typus: *A. cinchonarum* Fée

Oidium Link in Willdenow, Sp. Pl. 6(1): 121. 1824.
Typus: *O. monilioides* (Nees : Fr.) Link (*Acrosporium monilioides* Nees : Fr.).

(H) *Oidium* Link in Ges. Naturf. Freunde Berlin Mag. Neuesten Entdeck. Gesammten Naturk. 3: 18. Jan-Mar 1809 : Fr., Syst. Mycol. 1: xlv. 1 Jan 1821 [*Fungi*].
Typus: *O. aureum* (Pers. : Fr.) Link (*Trichoderma aureum* (Pers. : Fr.) Pers.).

(≡) *Acrosporium* Nees, Syst. Pilze: 53. 1816-1817 : Fr., Syst. Mycol. 1: xlv. 1 Jan 1821.

Omphalina Quél., Enchir. Fung.: 42. Jan-Jun 1886.
Typus: *O. pyxidata* (Bull. : Fr.) Quél. (*Agaricus pyxidatus* Bull. : Fr.) (typ. cons.).

Opegrapha Ach. in Kongl. Vetensk. Acad. Nya Handl. 30: 97. 1809.
Typus: *O. vulgata* (Ach.) Ach. (*Lichen vulgatus* Ach.).

(H) *Opegrapha* Humb., Fl. Friberg.: 57. 1793 [*Fungi*].
Typus: *O. vulgaris* Humb., nom. illeg. (*Lichen scriptus* L.).

Panaeolus (Fr.) Quél. in Mém. Soc. Emul. Montbéliard, ser. 2, 5: 151. Aug-Dec 1872 (*Agaricus* subg. *Panaeolus* Fr., Summa Veg. Scand.: 297. 1849).
Typus: *Agaricus papilionaceus* Bull. : Fr. (*P. papilionaceus* (Bull. : Fr.) Quél.).

(≡) *Coprinarius* (Fr.) P. Kumm., Führ. Pilzk.: 20, 68. Jul-Aug 1871.

Panus Fr., Epicr. Syst. Mycol.: 396. 1838.
Typus: *P. conchatus* (Bull. : Fr.) Fr. (*Agaricus conchatus* Bull. : Fr.) (typ. cons.).

Parmelia Ach., Methodus: xxxiii, 153. Jan-Apr 1803.
Typus: *P. saxatilis* (L.) Ach. (*Lichen saxatilis* L.).

(≡) *Lichen* L., Sp. Pl.: 1140. 1 Mai 1753.

Parmeliopsis (Nyl.) Nyl. in Not. Sällsk. Fauna Fl. Fenn. Förh. 8: 121. Jun 1866 (*Parmelia* subg. *Parmeliopsis* Nyl. in Not. Sällsk. Fauna Fl. Fenn. Förh. 5: 130. Jun-Jul 1861). Typus: *Parmelia ambigua* (Wulfen) Ach. (*Lichen ambiguu*s Wulfen, *Parmeliopsis ambigua* (Wulfen) Nyl.) (typ. cons.).

Peccania A. Massal. ex Arnold in Flora 41: 93. 14 Feb 1858. Typus: *P. coralloides* (A. Massal.) A. Massal. (in Atti Reale Ist. Veneto Sci. Lett. Arti, ser. 3, 5: 335. 1860) (*Corinophoros coralloides* A. Massal.).

(≡) *Corinophoros* A. Massal. in Flora 39: 212. 14 Apr 1856.

Peltigera Willd., Fl. Berol. Prodr.: 347. 1787. Typus: *P. canina* (L.) Willd. (*Lichen caninus* L.).

(=) *Placodion* P. Browne ex Adans., Fam. Pl. 2: 7, 592. Jul-Aug 1763. Typus: non designatus.

Peridermium (Link) J. C. Schmidt & Kunze, Deutschl. Schwämme 6: 4. 1817 (*Hypodermium* subg. *Peridermium* Link in Ges. Naturf. Freunde Berlin Mag. Neuesten Entdeck. Gesammten Naturk. 7: 29. 1816). Typus: *Aecidium elatinum* Alb. & Schwein. (*P. elatinum* (Alb. & Schwein.) J. C. Schmidt & Kunze).

Pertusaria DC. in Lamarck & Candolle, Fl. Franç., ed. 3, 2: 319. 17 Sep 1805. Typus: *P. communis* DC., nom. illeg. (*Lichen verrucosus* Huds.) [= *P. pertusa* (L.) Tuck. (*Lichen pertusu*s L.)].

(=) *Lepra* Scop., Intr. Hist. Nat.: 61. Jan-Apr 1777. Typus: non designatus.

(=) *Variolaria* Pers. in Ann. Bot. (Usteri) 7: 23. 1794. Typus: *V. discoidea* Pers.

(=) *Leproncus* Vent., Tabl. Règne Vég. 2: 32. 5 Mai 1799. Typus: non designatus.

(=) *Isidium* Ach., Methodus: xxxiii, 136. Jan-Apr 1803. Typus: *I. corallinum* (L.) Ach. (*Lichen corallinus* L.).

Pezicula Tul. & C. Tul., Select. Fung. Carpol. 3: 182. 1865.
Typus: *P. carpinea* (Pers.) Sacc. (*Peziza carpinea* Pers.).

(H) *Pezicula* Paulet, Tab. Pl. Fung.: 24. 1791 [*Fungi*].
≡ *Craterellus* Pers. 1825 (nom. cons.) (typ. des.: Cannon & Hawksworth in Taxon 32: 478. 1983).

Phacidium Fr., Observ. Mycol. 1: 167. 1815.
Typus: *P. lacerum* Fr. : Fr. (typ. cons.).

Phaeocollybia R. Heim, Inocybe: 70. Mai-Jun 1931.
Typus: *P. lugubris* (Fr. : Fr.) R. Heim (*Agaricus lugubris* Fr. : Fr.).

(=) *Quercella* Velen., České Houby: 495. 1921.
Typus: *Q. aurantiaca* Velen.

Phaeotrema Müll. Arg. in Mém. Soc. Phys. Genève 29(8): 10. 1887.
Typus: *P. subfarinosum* (Fée) Müll. Arg. (*Pyrenula subfarinosa* Fée).

(=) *Asteristion* Leight. in Trans. Linn. Soc. London 27: 163. 1870.
Typus: *A. erumpens* Leight.

Phellinus Quél., Enchir. Fung.: 172. Jan-Jun 1886.
Typus: *P. igniarius* (L. : Fr.) Quél. (*Boletus igniarius* L. : Fr.).

(=) *Mison* Adans., Fam. Pl. 2: 10, 578. Jul-Aug 1763 [*Fungi*].
Typus: non designatus.

Phillipsia Berk. in J. Linn. Soc., Bot. 18: 388. 29 Apr 1881.
Typus: *P. domingensis* (Berk.) Berk. (*Peziza domingensis* Berk.).

(H) *Phillipsia* C. Presl in Sternberg, Vers. Fl. Vorwelt 2: 206. 1 Sep - 7 Oct 1838 [*Foss.*].
Typus: *P. harcourtii* (Witham) C. Presl (*Lepidodendron harcourtii* Witham).

Phlyctis (Wallr.) Flot. in Bot. Zeitung (Berlin) 8: 571. 2 Aug 1850 (*Peltigera* sect. *Phlyctis* Wallr., Fl. Crypt. Germ. 3: 553. 1831).
Typus: *Peltigera agelaea* (Ach.) Wallr. (*Lichen agelaeus* Ach., *Phlyctis agelaea* (Ach.) Flot.).

(H) *Phlyctis* Raf., Caratt. Nuov. Gen.: 91. 1810 [*Algae*].
Typus: non designatus.

Pholiota (Fr.) P. Kumm., Führ. Pilzk.: 22, 83. Jul-Aug 1871 (*Agaricus* "trib." *Pholiota* Fr., Syst. Mycol. 1: 240. 1 Jan 1821).
Typus: *Agaricus squarrosus* Batsch : Fr. (*P. squarrosa* (Batsch : Fr.) P. Kumm.).

(≡)   *Derminus* (Fr.) Staude, Schwämme Mitteldeutschl.: xxvi, 86. 1857.

Phoma Sacc. in Michelia 2: 4. 25 Apr 1880.
Typus: *P. herbarum* Westend. (in Bull. Acad. Roy. Sci. Belgique 19: 118. 1852) (typ. cons.).

(H)   *Phoma* Fr., Novit. Fl. Suec.: 80. 18 Dec 1819 : Fr., Syst. Mycol. 1: lii. 1 Jan 1821 [*Fungi*].
Typus: *P. pustula* (Pers. : Fr.) Fr. (*Sphaeria pustula* Pers. : Fr.).

Phomopsis (Sacc.) Bubák in Oesterr. Bot. Z. 55: 78. Feb 1905 (*Phoma* subg. *Phomopsis* Sacc., Syll. Fung. 3: 66. 1884).
Typus: *Phoma lactucae* Sacc. (*Phomopsis lactucae* (Sacc.) Bubák).

(H)   *Phomopsis* Sacc. & Roum. in Rev. Mycol. (Toulouse) 6: 32. Jan 1884 [*Fungi*].
Typus: *P. brassicae* Sacc. & Roum.

(=)   *Myxolibertella* Höhn. in Ann. Mycol. 1: 526. 10 Dec 1903.
Typus (vide Clements & Shear, Gen. Fung.: 359. 1931): *M. aceris* Höhn.

Phyllachora Nitschke ex Fuckel in Jahrb. Nassauischen Vereins Naturk. 23-24: 216. 19 Feb - 24 Nov 1870.
Typus: P. graminis (Pers. : Fr.) Fuckel (*Sphaeria graminis* Pers. : Fr.).

(H)   *Phyllachora* Nitschke ex Fuckel, Fungi Rhenani: No. 2056. 1867 [*Fungi*].
Typus: *P. agrostis* Fuckel

Phyllosticta Pers., Traité Champ. Comest.: 55, 147. 1818.
Typus: *P. convallariae* Pers.

Physconia Poelt in Nova Hedwigia 9: 30. Mai 1965.
Typus: *P. pulverulacea* Moberg (in Mycotaxon 8: 310. 13 Jan 1979) (typ. cons.).

Plectosporium M. E. Palm & al. in Mycologia 87: 398. 1995.
Typus: *P. tabacinus* (F. H. Beyma) M. E. Palm & al. (*Cephalosporium tabacinum* F. H. Beyma).

(=)   *Spermosporina* U. Braun in Cryptog. Bot. 4: 11. 1993.
Typus: *S. alismatis* (Oudem.) U. Braun (*Septoria alismatis* Oudem.).

215

Pleospora Rabenh. ex Ces. & De Not. in Comment. Soc. Crittog. Ital. 1: 217. Jan 1863.
Typus: *P. herbarum* (Fr.) Rabenh. ex Ces. & De Not. (*Sphaeria herbarum* Fr.).

(H) *Pleiospora* Harv., Thes. Cap. 1: 51. 1860 [*Dicot.: Legum.*].
Typus: *P. cajanifolia* Harv.

(=) *Clathrospora* Rabenh. in Hedwigia 1: 116. 1857.
Typus: *C. elynae* Rabenh.

Pleurotus (Fr.) P. Kumm., Führ. Pilzk.: 24, 104. Jul-Aug 1871 (*Agaricus* "trib." *Pleurotus* Fr., Syst. Mycol. 1: 178. 1 Jan 1821).
Typus: *Agaricus ostreatus* Jacq. : Fr. (*P. ostreatus* (Jacq. : Fr.) P. Kumm.).

(≡) *Pleuropus* (Pers.) Roussel, Fl. Calvados, ed. 2: 67. 1806 (typ. des.: Donk in Beih. Nova Hedwigia 5: 235. 1962).

(≡) *Crepidopus* Nees ex Gray, Nat. Arr. Brit. Pl. 1: 616. 1 Nov 1821 (per typ. des.).

(=) *Gelona* Adans., Fam. Pl. 2: 11, 561. Jul-Aug 1763.
Typus: non designatus.

(=) *Resupinatus* Gray, Nat. Arr. Brit. Pl. 1: 617. 1 Nov 1821.
Typus: *R. applicatus* (Batsch : Fr.) Gray (*Agaricus applicatus* Batsch : Fr.).

(=) *Pterophyllus* Lév. in Ann. Sci. Nat., Bot., ser. 3, 2: 178. 1844.
Typus: *P. bovei* Lév.

(=) *Hohenbuehelia* Schulzer in Verh. K.K. Zool.-Bot. Ges. Wien 16 (Abh.): 45. Jan-Mai 1866.
Typus: *H. petaloides* (Bull. : Fr.) Schulzer (*Agaricus petaloides* Bull. : Fr.).

Podospora Ces. in Rabenhorst, Klotzschii Herb. Mycol., ed. 2: No. 259 (vel 258). Jan-Jun 1856.
Typus: *P. fimiseda* (Ces. & De Not.) Niessl (in Hedwigia 22: 156. Oct 1883) (*Sordaria fimiseda* Ces. & De Not.) (typ. cons.: [specimen] Rabenhorst, Klotzschii Herb. Mycol., ed. 2: No. 259 (S).).

(=) *Schizothecium* Corda, Icon. Fung. 2: 29. Jul 1838.
Typus: *S. fimicola* Corda

Polyblastia A. Massal., Ric. Auton. Lich. Crost.: 147. Jun-Dec 1852.
Typus: *P. cupularis* A. Massal.

(=) *Sporodictyon* A. Massal. in Flora 35: 326. 7 Jun 1852.
Typus: *S. schaererianum* A. Massal.

Polysporina Vězda in Folia Geobot.
Phytotax. 13: 399. 29 Dec 1978.
Typus: *P. simplex* (Davies) Vězda
(*Lichen simplex* Davies) (etiam vide
Sarcogyne [*Fungi*]).

Porina Ach. in Kongl. Vetensk. Acad.
Nya Handl. 30: 158. Jul-Sep 1809.
Typus: *P. nucula* Ach. (Syn. Meth.
Lich.: 112. 1814) (typ. cons.).

Psathyrella (Fr.) Quél., Mém. Soc.
Emul. Montbéliard, ser. 2, 5: 152.
1872 (*Agaricus* subg. *Psathyrella* Fr.,
Summa veg. Scand. 297, 1849).
Typus: *Agaricus gracilis* Fr. : Fr. (*P.
gracilis* (Fr. : Fr.) Quél.) (typ. cons.).

Pseudocyphellaria Vain. in Acta
Soc. Fauna Fl. Fenn. 7(1): 182. 1-22
Nov 1890.
Typus: *P. aurata* (Ach.) Vain. (*Sticta aurata* Ach.).

($\equiv$) *Crocodia* Link, Handbuch 3: 177.
1833.

(=) *Stictina* Nyl., Syn. Meth. Lich. 1:
333. Apr 1860.
Typus (vide Clements & Shear, Gen.
Fung.: 322. 1931): *S. crocata* (L.)
Nyl. (*Lichen crocatus* L.).

(=) *Phaeosticta* Trevis., Lichenoth. Ve-
neta: No. 75. Apr 1869.
Typus (vide Choisy in Bull. Mens.
Soc. Linn. Lyon, ser. 2, 29: 125.
1960): *P. physciospora* (Nyl.) Trevis. (*Sticta fossulata* subsp. *physciospora* Nyl.).

(=) *Saccardoa* Trevis., Lichenoth. Ve-
neta: No. 75. Apr 1869.
Typus (vide Choisy in Bull. Mens.
Soc. Linn. Lyon, ser. 2, 29: 123.
1960): *S. crocata* (L.) Trevis. (*Lichen crocatus* L.).

(=) *Parmostictina* Nyl. in Flora 58: 363.
11 Aug 1875.
Typus: *Sticta hirsuta* Mont.

Pseudographis Nyl. in Mém. Soc. Sci. Nat. Cherbourg 3: 190. Jun 1855. Typus: *P. elatina* (Ach. : Fr.) Nyl. (*Lichen elatinus* Ach. : Fr.).

(=)   *Krempelhuberia* A. Massal., Geneac. Lich.: 34. 1-22 Sep 1854. Typus: *K. cadubriae* A. Massal.

Psora Hoffm., Deutschl. Fl. 2: 161. Feb-Apr 1796. Typus: *P. decipiens* (Hedw.) Hoffm. (*Lecidea decipiens* Hedw.).

(H)   *Psora* Hill, Veg. Syst. 4: 30. 1762 [*Dicot.: Comp.*]. Typus: non designatus.

(H)   *Psora* Hoffm., Descr. Pl. Cl. Crypt. 1: 37. 1789 [*Fungi*]. Typus: *P. caesia* Hoffm.

Pulvinula Boud. in Bull. Soc. Mycol. France 1: 107. Mai 1885. Typus: *P. convexella* (P. Karst.) Boud. (*Peziza convexella* P. Karst.).

(=)   *Pulparia* P. Karst. in Not. Sällsk. Fauna Fl. Fenn. Förh. 8: 205. 1866. Typus: *P. arctica* P. Karst.

Pycnoporus P. Karst. in Rev. Mycol. (Toulouse) 3(9): 18. 1881. Typus: *P. cinnabarinus* (Jacq. : Fr.) P. Karst. (*Boletus cinnabarinus* Jacq. : Fr.).

(≡)   *Xylometron* Paulet, Prosp. Traité Champ.: 29. 1808.

Pyrenopsis (Nyl.) Nyl., Syn. Meth. Lich. 1(1): 97. 8-14 Aug 1858 (*Synalissa* sect. *Pyrenopsis* Nyl. in Mém. Soc. Sci. Nat. Cherbourg 3: 164. Jun 1855). Typus: *P. fuscatula* Nyl. (typ. cons.).

Pyrenula Ach., Syn. Meth. Lich.: 117. 1814. Typus: *P. nitida* (Weigel) Ach. (*Sphaeria nitida* Weigel) (typ. cons.).

(H)   *Pyrenula* Ach. in Kongl. Vetensk. Acad. Nya Handl. 30: 160. 1809 [*Fungi*]. Typus: *P. margacea* (Wahlenb.) Ach. (*Thelotrema margaceum* Wahlenb.).

Pythium Pringsh. in Jahrb. Wiss. Bot. 1: 304. 1858. Typus: *P. monospermum* Pringsh.

(H)   *Pythium* Nees in Nova Acta Phys.-Med. Acad. Caes. Leop.-Carol. Nat. Cur. 11: 515. 1823 [*Fungi*]. Typus: non designatus.

(=)   *Artotrogus* Mont. in Gard. Chron. 5: 640. 1845. Typus: *A. hydnosporus* Mont.

Racodium Fr., Syst. Mycol. 3: 229.
1829.
Typus: *R. rupestre* Pers. : Fr.

(H) *Racodium* Pers. in Neues Mag. Bot.
1: 123. Apr-Aug 1794 : Fr., Syst.
Mycol. 1: xlvi. 1 Jan 1821 [*Fungi*].
Typus: *R. cellare* Pers. : Fr.

Ramalina Ach., Lichenogr. Universalis: 122, 598. 1810.
Typus: *R. fraxinea* (L.) Ach. (*Lichen
fraxineus* L.) (typ. cons.).

Ramaria Fr. ex Bonord., Handb.
Mykol.: 166. 1851 (sero).
Typus: *R. botrytis* (Pers. : Fr.) Ricken
(Vadem. Pilzfr.: 253. Mai-Jun 1918)
(*Clavaria botrytis* Pers. : Fr.).

(H) *Ramaria* Holmsk. ex Gray, Nat. Arr.
Brit. Pl. 1: 655. 1 Nov 1821 [*Fungi*].
Typus: non designatus.
(≡) *Cladaria* Ritgen, Aufeinanderfolge
Org. Gest.: 54. 1828.

Ramularia Unger, Exanth. Pflanzen.: 169. 1833.
Typus: *R. pusilla* Unger

(H) *Ramularia* Roussel, Fl. Calvados,
ed. 2: 98. 1806 (typ. des.: Agardh,
Spec. Alg. 1: 402. 1823) [*Chloroph.*].
≡ *Ulva* L. 1753 (nom. cons.).

Reticularia Bull., Herb. France: t.
326. 1787-1788.
Typus: [icon] *R. lycoperdon* Bull.,
Herb France: t. 446, fig. IV. 1790
(typ. cons.).

Rhabdospora (Durieu & Mont.)
Sacc., Syll. Fung. 3: 578. 15 Dec
1884 (*Septoria* sect. *Rhabdospora*
Durieu & Mont. in Durieu, Expl.
Sci. Algérie 1: 592. 1849).
Typus: *Septoria oleandri* Durieu &
Mont. (*R. oleandri* (Durieu & Mont.)
Sacc.).

(=) *Filaspora* Preuss in Linnaea 26:
718. Sep 1855.
Typus: *F. peritheciiformis* Preuss

Rhipidium Cornu in Bull. Soc. Bot.
France 18: 58. 1871 (post 24 Mar).
Typus: *R. interruptum* Cornu

(H) *Rhipidium* Wallr., Fl. Crypt. Germ.
2: 742. Feb-Mar 1833 [*Fungi*].
Typus: *R. stipticum* (Bull. : Fr.) Wallr.
(*Agaricus stipticus* Bull. : Fr.).

Rhizoctonia DC. in Lamarck & Candolle, Fl. Franç., ed. 3, 5: 110. 8 Oct 1815.
Typus: *R. solani* J. G. Kühn (typ. cons.).

Rhizopus Ehrenb. in Nova Acta Phys.-Med. Acad. Caes. Leop.-Carol. Nat. Cur. 10: 198. 1821.
Typus: *R. nigricans* Ehrenb., nom. illeg. (*Mucor stolonifer* Ehrenb. : Fr., *R. stolonifer* (Ehrenb. : Fr.) Vuill.).

(=) *Ascophora* Tode, Fungi Mecklenb. Sel. 1: 13. 1790 : Fr., Syst. Mycol. 3: 309. 1832.
Typus (vide Kirk in Taxon 35: 374. 1986): *A. mucedo* Tode : Fr.

Robillarda Sacc. in Michelia 2: 8. 25 Apr 1880.
Typus: *R. sessilis* (Sacc.) Sacc. (*Pestalotia sessilis* Sacc.).

(H) *Robillarda* Castagne, Cat. Pl. Marseille: 205. 1845 [*Fungi*].
Typus: *R. glandicola* Castagne

Roccella DC. in Lamarck & Candolle, Fl. Franç., ed. 3, 2: 334. 17 Sep 1805.
Typus: *R. fuciformis* (L.) DC. (*Lichen fuciformis* L.).

(=) *Thamnium* Vent., Tabl. Règne Vég. 2: 35. 5 Mai 1799.
Typus (vide Ahti in Taxon 33: 330. 1984): *T. roccella* (L.) J. St.-Hil. (Expos. Fam. Nat. 1: 21. Feb-Apr 1805) (*Lichen roccella* L.).

Rutstroemia P. Karst. in Bidrag Kännedom Finlands Natur Folk 19: 12, 105. 1871.
Typus: *R. firma* (Pers. : Fr.) P. Karst. (*Peziza firma* Pers. : Fr.) (typ. cons.).

Sarcogyne Flot. in Bot. Zeitung (Berlin) 9: 753, 759. 24 Oct 1851.
Typus: *S. corrugata* Flot.

(H) *Sarcogyne* Flot. in Bot. Zeitung (Berlin) 8: 381-382. 10 Mai 1850 (typ. des.: Jørgensen & Santesson in Taxon 42: 885. 1993) [*Fungi*].
≡ *Polysporina* Vězda 1978 (nom. cons.).

Schaereria Körb., Syst. Lich. Germ.: 232. Jan 1855.
Typus: [specimen] *"Schaereria lugubris"*, Falkenstein, *Krempelhuber* (M) (typ. cons.) [= *S. cinereorufa* (Schaer.) Th. Fr. (*Lecidea cinereorufa* Schaer.)].

Sclerotinia Fuckel in Jahrb. Nassau-ischen Vereins Naturk. 23-24: 330. 19 Feb - 24 Nov 1870.
Typus: *S. libertiana* Fuckel, nom. il-leg. (*Peziza sclerotiorum* Lib., *S. sclerotiorum* (Lib.) de Bary) (typ. cons.).

Scutellinia (Cooke) Lambotte, Fl. Mycol. Belge, Suppl. 1: 299. 1887 (*Peziza* subg. *Scutellinia* Cooke, Mycographia: 259. Feb 1879).
Typus: *Peziza scutellata* L. : Fr. (*S. scutellata* (L. : Fr.) Lambotte).

(=) *Patella* F. H. Wigg., Prim. Fl. Holsat.: 106. 29 Mar 1780.
Typus (vide Korf & Schumacher in Taxon 35: 378. 1986): *P. ciliata* F. H. Wigg.

Scutula Tul. in Ann. Sci. Nat., Bot., ser. 3, 17: 118. 1852.
Typus: [icon] *"Scutula wallrothii"* in Ann. Sci. Nat., Bot., ser. 3, 17: t. 14, f. 14-24. 1852 (typ. cons.) [= *S. epiblastematica* (Wallr.) Rehm (*Peziza epiblastematica* Wallr.)].

(H) *Scutula* Lour., Fl. Cochinch. 223, 235. Sep 1790 [*Dicot.: Melastomat.*].
Typus: non designatus.

Septobasidium Pat. in J. Bot. (Morot) 6: 63. 16 Feb 1892.
Typus: *S. velutinum* Pat.

(=) *Gausapia* Fr., Syst. Orb. Veg.: 302. Dec 1825.
Typus: *Thelephora pedicellata* Schwein.
(=) *Glenospora* Berk. & Desm. in J. Hort. Soc. London 4: 255. 1849.
Typus: *G. curtisii* Berk. & Desm.
(=) *Campylobasidium* Lagerh. ex F. Ludw., Lehrb. Nied. Krypt.: 474. Jul 1892.
Typus: non designatus.

Septoria Sacc., Syll. Fung. 3: 474. 15 Dec 1884.
Typus: *S. cytisi* Desm. (in Ann. Sci. Nat., Bot., ser. 3, 8: 24. Jul 1847).

(H) *Septaria* Fr., Novit. Fl. Suec.: 78. 18 Dec 1819 : Fr., Syst. Mycol. 1: xl. 1 Jan 1821 [*Fungi*].
Typus: *S. ulmi* Fr. : Fr.

Simocybe P. Karst. in Bidrag Kännedom Finlands Natur Folk 32: xxii, 416. Jul-Dec 1879.
Typus: *S. centunculus* (Fr. : Fr.) P. Karst. (*Agaricus centunculus* Fr. : Fr.) (typ. cons.).

Siphula Fr., Lichenogr. Eur. Reform.: 7, 406. Jun-Jul 1831.
Typus: *S. ceratites* (Wahlenb.) Fr. (*Baeomyces ceratites* Wahlenb.).

(H) *Siphula* Fr., Sched. Crit. Lich. Suec. Exsicc. 1: 3. 1824 [*Fungi*].
≡ *Dufourea* Ach. ex Luyk. 1809 (nom. rej. sub *Xanthoria*).

Sordaria Ces. & De Not. in Comment. Soc. Crittog. Ital. 1: 225. Jan 1863.
Typus: *S. fimicola* (Roberge ex Desm.) Ces. & De Not. (*Sphaeria fimicola* Roberge ex Desm.) (typ. cons.).

Sphaerophorus Pers. in Ann. Bot. (Usteri) 7: 23. 1794.
Typus: *S. coralloides* Pers., nom. illeg. (*Lichen globiferus* L.) (typ. cons.) [= *S. globosus* (Huds.) Vain., *Lichen globosus* Huds.].

Sphaeropsis Sacc. in Michelia 2: 105. 25 Apr 1880.
Typus: *S. visci* (Alb. & Schwein. : Fr.) Sacc. (*Sphaeria atrovirens* var. *visci* Alb. & Schwein. : Fr.).

(H) *Sphaeropsis* Lév. in Demidov, Voy. Russie Mér. 2: 112. 1842 [*Fungi*].
Typus: *S. conica* Lév.
(=) *Macroplodia* Westend. in Bull. Acad. Roy. Sci. Belgique, ser. 2, 2: 562. 1857.
Typus: *M. aquifolia* Westend.

Sphaerotheca Lév. in Ann. Sci. Nat., Bot., ser. 3, 15: 138. Mar 1851.
Typus: *S. pannosa* (Wallr. : Fr.) Lév. (*Alphitomorpha pannosa* Wallr. : Fr.).

(H) *Sphaerotheca* Desv. in Mém. Soc. Imp. Naturalistes Moscou 5: 68. 1817 [*Fungi*].
Typus: *S. albescens* Desv., nom. illeg. (*Aecidium thesii* Desv.).

Spongipellis Pat., Hyménomyc. Eur.: 140. Jan-Mar 1887.
Typus: *S. spumeus* (Sowerby : Fr.) Pat. (Boletus spumeus Sowerby : Fr.).

(=) *Somion* Adans., Fam. Pl. 2: 5, 606. Jul-Aug 1763.
Typus (vide Donk in Verh. Kon. Ned. Akad. Wetensch., Afd. Natuurk., Tweede Sect. 62: 175. 1974): *Hydnum occarium* Batsch : Fr.

Stagonospora (Sacc.) Sacc., Syll. Fung. 3: 445. 15 Dec 1884 (*Hendersonia* subg. *Stagonospora* Sacc. in Michelia 2: 8. 25 Apr 1880).
Typus: *Hendersonia paludosa* Sacc. & Speg. (*S. paludosa* (Sacc. & Speg.) Sacc.).

(=)   *Hendersonia* Berk. in Ann. Mag. Nat. Hist. 6: 430. 1841.
Typus: *H. elegans* Berk.

Staurothele Norman, Conat. Praem. Gen. Lich.: 28. 1852.
Typus: *S. clopima* (Wahlenb.) Th. Fr. (Lich. Arct.: 263. Mai-Dec 1860) (*Verrucaria clopima* Wahlenb.).

(=)   *Paraphysorma* A. Massal., Ric. Auton. Lich. Crost.: 116. Jun-Dec 1852.
Typus: *P. protuberans* (Schaer.) A. Massal. (*Parmelia cervina* var. *protuberans* Schaer.).

Stereocaulon Hoffm., Deutschl. Fl. 2: 128. Feb-Apr 1796.
Typus: *S. paschale* (L.) Hoffm. (*Lichen paschalis* L.).

(H)   *Stereocaulon* (Schreb.) Schrad., Spic. Fl. Germ.: 113. 16 Mai - 5 Jun 1794 (*Lichen* sect. *Stereocaulon* Schreb., Gen. Pl.: 768. Mai 1791) [*Fungi*].
Typus: *Lichen corallinus* L. (*S. corallinum* (L.) Schrad.).

Stilbella Lindau in Engler & Prantl, Nat. Pflanzenfam. 1(1**): 489. Sep 1900.
Typus: *S. erythrocephala* (Ditmar : Fr.) Lindau (*Stilbum erythrocephalum* Ditmar : Fr.).

(=)   *Botryonipha* Preuss in Linnaea 25: 79. Jun 1852.
Typus: *B. alba* Preuss

Telamonia (Fr.) Wünsche, Pilze: 87, 122. 1877 (*Agaricus* "trib." *Telamonia* Fr., Syst. Mycol. 1: 10, 210. 1 Jan 1821).
Typus: *Agaricus torvus* Fr. : Fr. (*T. torva* (Fr. : Fr.) Wünsche).

(≡)   *Raphanozon* P. Kumm., Führ. Pilzk.: 22. Jul-Aug 1871.

Thamnolia Ach. ex Schaer., Enum. Crit. Lich. Eur.: 243. Aug-Sep 1850.
Typus: *T. vermicularis* (Sw.) Ach. ex Schaer. (*Lichen vermicularis* Sw.).

(≡)   *Cerania* Ach. ex Gray, Nat. Arr. Brit. Pl. 1: 413. 1 Nov 1821.

Thecaphora Fingerh. in Linnaea 10: 230. Feb-Mai 1836.
Typus: *T. hyalina* Fingerh.

(=)   *Sorosporium* F. Rudolphi in Linnaea 4: 116. Jan 1829.
Typus: *S. saponariae* F. Rudolphi

Thelopsis Nyl. in Mém. Soc. Sci. Nat. Cherbourg 3: 194. Jun 1855.
Typus: *T. rubella* Nyl.

(=) *Sychnogonia* Körb., Syst. Lich. Germ.: 332. Jan-Mar 1855.
Typus: *S. bayrhofferi* Zwackh ex Körb.

Tholurna Norman in Flora 44: 409. 14 Jul 1861.
Typus: *T. dissimilis* (Norman) Norman (*Podocratera dissimilis* Norman).

(≡) *Podocratera* Norman in Förh. Skand. Naturf. Möte 1860: 426. 6-12 Apr 1861.

Tomentella Pers. ex Pat., Hyménomyc. Eur.: 154. Jan-Mar 1887.
Typus: *T. ferruginea* (Pers. : Fr.) Pat. (*Thelephora ferruginea* Pers. : Fr.).

(=) *Caldesiella* Sacc., Fungi Ital.: t. 125. Mai 1877.
Typus: *C. italica* Sacc.

(=) *Odontia* Pers. in Neues Mag. Bot. 1: 110. Apr-Aug 1794.
Typus (vide Banker in Bull. Torrey Bot. Club 29: 448. 1902): *O. ferruginea* Pers.

Toninia A. Massal., Ric. Auton. Lich. Crost.: 107. Jun-Dec 1852.
Typus: *T. cinereovirens* (Schaerer) A. Massal. (*Lecidea cinerovirens* Schaerer).

(=) *Thalloidima* A. Massal., Ric. Auton. Lich. Crost.: 95. Jun-Dec 1852.
Typus (vide Clements & Shear, Gen. Fungi, ed. 2: 319. 1931): *T. candidum* (Weber) A. Massal. (*Lichen candidus* Weber).

(=) *Skolekites* Norman, Conat. Praem. Gen. Lich.: 23. 1852 (typ. des.: Hafellner in Nova Hedwigia, Beih. 79: 264. 1984).
≡ *Thalloidima* A. Massal. (nom. rej.)

Trapelia M. Choisy in Bull. Soc. Bot. France 76 523. 1929.
Typus: *T. coarctata* (Turner ex Sm.) M. Choisy (*Lichen coarctatus* Turner ex Sm.).

(=) *Discocera* A. L. Sm. & Ramsb. in Trans. Brit. Mycol. Soc. 6: 48. 17 Aug 1917.
Typus: *D. lichenicola* A. L. Sm. & Ramsb.

Tremella Pers. in Neues Mag. Bot. 1: 111. Apr-Aug 1794.
Typus: *T. mesenterica* Schaeff. : Fr. (Syst. Mycol. 2: 210. 1822) (typ. cons.).

Tricholoma (Fr.) Staude, Schwämme Mitteldeutschl.: xxviii, 125. 1857 (*Agaricus* "trib." *Tricholoma* Fr., Syst. Mycol. 1: 9, 36. 1 Jan 1821). Typus: *Agaricus flavovirens* Alb. & Schwein. : Fr. (*T. flavovirens* (Alb. & Schwein. : Fr.) S. Lundell).

(H) *Tricholoma* Benth. in Candolle, Prodr. 10: 426. 8 Apr 1846 [*Dicot.: Scrophular.*]. Typus: *T. elatinoides* Benth.

Trypethelium Spreng., Anleit. Kenntn. Gew. 3: 350. 28 Mar 1804. Typus: *T. eluteriae* Spreng.

(=) *Bathelium* Ach., Methodus: 111. Jan-Apr 1803. Typus: *B. mastoideum* Afzel. ex Ach.

Tubercularia Tode, Fungi Mecklenb. Sel. 1: 18. 1790. Typus: *T. vulgaris* Tode : Fr.

Umbilicaria Hoffm., Descr. Pl. Cl. Crypt. 1: 8. 1789. Typus: *U. hyperborea* (Ach.) Hoffm. (*Lichen hyperboreus* Ach.) (typ. cons.).

(H) *Umbilicaria* Heist. ex Fabr., Enum.: 42. 1759 [*Dicot.: Boragin.*]. ≡ *Omphalodes* Mill. 1754.

Urocystis Rabenh. ex Fuckel in Jahrb. Nassauischen Vereins Naturk. 23-24: 41. 19 Feb - 24 Nov 1870. Typus: *U. occulta* (Wallr.) Fuckel (*Erysibe occulta* Wallr.).

(=) *Polycystis* Lév. in Ann. Sci. Nat., Bot., ser. 3, 5: 269. 1846. Typus: *P. pompholygodes* (Schltdl.) Lév. (*Caeoma pompholygodes* Schltdl.).

(=) *Tuburcinia* Fr., Syst. Mycol. 3: 439. 1832 : Fr., ibid. Typus: *T. orobanches* (Mérat) Fr. (*Rhizoctonia orobanches* Mérat).

Uromyces (Link) Unger, Exanth. Pflanzen.: 277. 1832 (*Hypodermium* subg. *Uromyces* Link in Ges. Naturf. Freunde Berlin Mag. Neuesten Entdeck. Gesammten Naturk. 7: 28. 1816). Typus: *Uredo appendiculata* Pers. : Pers. (*Uromyces appendiculatus* (Pers. : Pers.) Unger).

(=) *Coeomurus* Link ex Gray, Nat. Arr. Brit. Pl. 1: 541. 1 Nov 1821. Typus: *C. phaseolorum* (R. Hedw. ex DC.) Gray (*Puccinia phaseolorum* R. Hedw. ex DC.).

(=) *Pucciniola* L. Marchand in Bijdr. Natuurk. Wetensch. 4: 47. 1829. Typus: *P. diadelphiae* L. Marchand

Valsa Fr., Summa Veg. Scand.: 410. 1849. Typus: *V. ambiens* (Pers. : Fr.) Fr. (*Sphaeria ambiens* Pers. : Fr.).

(H) *Valsa* Adans., Fam. Pl. 2: 9, 617. Jul-Aug 1763 [*Fungi*]. Typus (vide Cannon & Hawksworth in Taxon 32: 478. 1983): *Sphaeria disciformis* Hoffm.

Venturia Sacc., Syll. Fung. 1: 586. 13 Jun 1882.
Typus: *V. inaequalis* (Cooke) G. Winter (in Thümen, Mycoth. Univ.: No. 261. 1875) (*Sphaerella inaequalis* Cooke).

(H) *Venturia* De Not. in Giorn. Bot. Ital. 1(1,1): 332. Mai-Jun 1844 [*Fungi*].
Typus: *V. rosae* De Not.

Verrucaria Schrad., Spic. Fl. Germ.: 108. 16 Mai - 5 Jun 1794.
Typus: *V. rupestris* Schrad.

(H) *Verrucaria* Scop., Intr. Hist. Nat.: 61. Jan-Apr 1777 [*Fungi*].
Typus: *Baeomyces roseus* Pers.

Verticillium Nees, Syst. Pilze: 57. 1816 [-1817].
Typus: *V. dahliae* Kleb. in Mycol. Centralbl. 3: 66. 1913 (holotype specimen in HBG) (typ. cons.).

Volutella Fr., Syst. Mycol. 3: 458, 466. 1832.
Typus: *V. ciliata* (Alb. & Schwein. : Fr.) Fr. (*Tubercularia ciliata* Alb. & Schwein. : Fr.) (typ. cons.).

Volvariella Speg. in Anales Mus. Nac. Hist. Nat. Buenos Aires 6: 119. 4 Apr 1899.
Typus: *V. argentina* Speg.

(=) *Volvarius* Roussel, Fl. Calvados, ed. 2: 59. 1806.
Typus (vide Earle in Bull. New York Bot. Gard. 5: 395, 449. 1909): *Agaricus volvaceus* Bull. : Fr.

Xanthoparmelia (Vain.) Hale in Phytologia 28: 485. 1974 (*Parmelia* sect. *Xanthoparmelia* Vain. in Acta Soc. Fauna Fl. Fenn. 7(7): 60. 1890).
Typus: *X. conspersa* (Ach.) Hale (*Lichen consperus* Ach.).

(=) *Chondriopsis* Nyl. ex Cromb. in J. Linn. Soc. Bot. 17: 397. 1879.
Typus: *C. semiviridis* (Nyl.) Cromb. (*Parmeliopsis semiviridis* Nyl.).

Xanthoria (Fr.) Th. Fr., Lich. Arct.: 166. Mai-Dec 1860 (*Parmelia* [unranked] *Xanthoria* Fr., Syst. Orb. Veg.: 243. Dec 1825).
Typus: *Parmelia parietina* (L.) Ach. (*Lichen parietinus* L., *X. parietina* (L.) Th. Fr.).

(≡) *Blasteniospora* Trevis., Tornab. Blasteniosp.: 2. Feb 1853.

(=) *Dufourea* Ach. ex Luyk., Tent. Hist. Lich.: 93. 21 Dec 1809.
Typus (vide De Notaris in Giorn. Bot. Ital. 2(1,1): 224. 1846): *D. flammea* (L. f.) Ach. (Lichenogr. Universalis: 103, 524. 1810) (*Lichen flammeus* L. f.).

Xerocomus Quél. in Mougeot & Ferry, Fl. Vosges, Champ.: 477. 1887. Typus: *X. subtomentosus* (L. : Fr.) Quél. (*Boletus subtomentosus* L. : Fr.).

(≡)   *Versipellis* Quél., Enchir. Fung.: 157. Jan-Jun 1886.

Xeromphalina Kühner & Maire in Konrad & Maubl., Icon. select. fung. 6: 236. Mar 1934 (fasc. VIII) ('Xeromphalia') (orth. cons.). Typus: *X. campanella* (Batsch: Fr.) Kühner & Maire (*Agaricus campanella* Batsch: Fr.)

Xylaria Hill ex Schrank, Baier. Fl. 1: 200. 1789. Typus: *X. hypoxylon* (L. : Fr.) Grev. (Fl. Edin.: 355. 18 Mar 1824) (*Clavaria hypoxylon* L. : Fr.) (typ. cons.).

C. BRYOPHYTA

C1. HEPATICAE

Acrolejeunea (Spruce) Schiffn. in Engler & Prantl, Nat. Pflanzenfam. 1(3): 119, 128. Sep 1893 (*Lejeunea* subg. *Acrolejeunea* Spruce in Trans. & Proc. Bot. Soc. Edinburgh 15: 74, 115. Apr 1884). Typus: *Lejeunea torulosa* (Lehm. & Lindenb.) Spruce (*Jungermannia torulosa* Lehm. & Lindenb., *A. torulosa* (Lehm. & Lindenb.) Schiffn.).

(H)   *Acro-lejeunea* Steph. in Bot. Gaz. 15: 286. Nov 1890 [*Hepat.*]. Typus: *A. parviloba* Steph.

Adelanthus Mitt. in J. Proc. Linn. Soc., Bot. 7: 243. 5 Apr 1864. Typus: *A. falcatus* (Hook.) Mitt. (*Jungermannia falcata* Hook.).

(H)   *Adelanthus* Endl., Gen. Pl.: 1327. Oct 1840 [*Dicot.: Icacin.*]. Typus: *A. scandens* (Thunb.) Endl. ex Baill. (*Cavanilla scandens* Thunb.).

Asterella P. Beauv. in Cuvier, Dict. Sci. Nat. 3: 257. 30 Jan 1805. Typus: *A. tenella* (L.) P. Beauv. (*Marchantia tenella* L.) (typ. cons.).

Bazzania Gray, Nat. Arr. Brit. Pl. 1:
704, 775. 1 Nov 1821 *('Bazzanius')*
(orth. cons.).
Typus: *B. trilobata* (L.) Gray (*Jungermannia trilobata* L.).

Calypogeia Raddi, Jungermanniogr.
Etrusca: 31. 1818 *('Calypogeja')*
(orth. cons.).
Typus: *C. fissa* (L.) Raddi (*Mnium fissum* L., nom. cons.) (etiam vide *Mnium* [*Musci*]).

Cephaloziella (Spruce) Schiffn. in
Engler & Prantl, Nat. Pflanzenfam.
1(3): 98. Sep 1893 (*Cephalozia* subg.
Cephaloziella Spruce, Cephalozia:
23, 62. Oct-Dec 1882).
Typus: *Cephalozia divaricata* (Sm.)
Dumort. (*Jungermannia divaricata*
Sm., *Cephaloziella divaricata* (Sm.)
Schiffn.).

(=) *Dichiton* Mont., Syll. Gen. Sp.
Crypt. 52. Feb 1856.
Typus: *D. perpusillus* Mont., nom.
illeg. (*Jungermannia calyculata* Durieu & Mont., *D. calyculatus* (Durieu & Mont.) Trevis.).

Chiloscyphus Corda in Naturalientausch 12 [Opiz, Beitr. Naturgesch.]:
651. Sep 1829 *('Cheilocyphos')* (orth.
cons.).
Typus: *C. polyanthos* (L.) Corda (*Jungermannia polyanthos* L.).

Conocephalum Hill, Gener. Nat.
Hist., ed. 2, 2: 118. 1773 *('Conicephala')* (orth. cons.).
Typus: *C. conicum* (L.) Dumort.
(Comment. Bot.: 115. Nov (sero) -
Dec (prim.) 1822) *('Conocephalus conicus')* (*Marchantia conica* L.).

Diplophyllum (Dumort.) Dumort.,
Recueil Observ. Jungerm.: 15. 1835
(*Jungermannia* sect. *Diplophyllum*
Dumort., Syll. Jungerm. Europ.: 44.
1831).
Typus: *Jungermannia albicans* L.
(*D. albicans* (L.) Dumort.).

(H) *Diplophyllum* Lehm. in Ges. Naturf.
Freunde Berlin Mag. Neuesten Entdeck. Gesammten Naturk. 8: 310.
1818 [*Dicot.: Scrophular.*].
≡ *Oligospermum* D. Y. Hong 1984.

Gymnomitrion Corda in Natural-
ientausch 12 [Opiz, Beitr. Natur-
gesch.]: 651. Sep 1829.
Typus: *G. concinnatum* (Lightf.)
Corda (*Jungermannia concinnata*
Lightf.).

(≡)   *Cesius* Gray, Nat. Arr. Brit. Pl. 1:
705. 1 Nov 1821.

Haplomitrium Nees, Naturgesch.
Eur. Leberm. 1: 109. 15 Sep - 15
Dec 1833.
Typus: *H. hookeri* (Sm.) Nees (*Jun-
germannia hookeri* Sm.).

(≡)   *Scalius* Gray, Nat. Arr. Brit. Pl. 1:
704. 1 Nov 1821.

Heteroscyphus Schiffn. in Oesterr.
Bot. Z. 60: 171. Mai 1910.
Typus: *H. aselliformis* (Reinw. &
al.) Schiffn. (*Jungermannia aselliformis* Reinw. & al.).

(≡)   *Gamoscyphus* Trevis. in Mem. Reale
Ist. Lombardo Sci., Ser. 3, Cl. Sci.
Mat. 4: 422. 1877.

Jubula Dumort., Comment. Bot.:
112. Nov (sero) - Dec (prim.) 1822.
Typus: *J. hutchinsiae* (Hook.) Dumort. (*Jungermannia hutchinsiae*
Hook.) (typ. cons.).

Lejeunea Lib. in Ann. Gén. Sci.
Phys. 6: 372. 1820 (*'Lejeunia'*) (orth.
cons.).
Typus: *L. serpillifolia* Lib. (non *Jungermannia serpillifolia* Scop. 1772,
nec *Jungermannia serpyllifolia* Dicks.
1801) [= *L. cavifolia* (Ehrh.) Lindb.
(*Jungermannia cavifolia* Ehrh.)].

Lembidium Mitt. in Hooker, Handb.
N. Zeal. Fl.: 754. 1867.
Typus: *L. nutans* (Hook. f. & Tay-
lor) A. Evans (in Trans. Connecticut
Acad. Arts 8: 266. 1892) (*Junger-
mannia nutans* Hook. f. & Taylor).

(H)   *Lembidium* Körb., Syst. Lich. Germ.:
358. Jan-Mar 1855 [*Fungi*].
Typus: *L. polycarpum* Körb.

Lepidozia (Dumort.) Dumort., Recueil Observ. Jungerm.: 19. 1835 (*Pleuroschisma* sect. *Lepidozia* Dumort., Syll. Jungerm. Europ.: 69. 1831).
Typus: *Pleuroschisma reptans* (L.) (*Jungermannia reptans* L., *L. reptans* (L.) Dumort.) (etiam vide *Mastigophora* [*Hepat.*]).

Lethocolea Mitt. in Hooker, Handb. N. Zeal. Fl.: 751, 753. 1867.
Typus: *L. drummondii* Mitt., nom. illeg. (*Gymnanthe drummondii* Mitt., nom. illeg., *Podanthe squamata* Taylor, *L. squamata* (Taylor) E. A. Hodgs.).

(≡) *Podanthe* Taylor in London J. Bot. 5: 413. 1846.

Lopholejeunea (Spruce) Schiffn. in Engler & Prantl, Nat. Pflanzenfam. 1(3): 119, 129. Sep 1893 (*Lejeunea* subg. *Lopholejeunea* Spruce in Trans. & Proc. Bot. Soc. Edinburgh 15: 74, 119. Apr 1884).
Typus: *Lejeunea sagrana* (Mont.) Gottsche & al. (*Phragmicoma sagrana* Mont., *Lopholejeunea sagrana* (Mont.) Schiffn.).

(H) *Lopho-lejeunea* Steph. in Bot. Gaz. 15: 285. Nov 1890 [*Hepat.*].
Typus: *L. multilacera* Steph.

Mannia Opiz in Naturalientausch 12 [Opiz, Beitr. Naturgesch.]: 646. Sep 1829.
Typus: *M. michelii* Opiz, nom. illeg. (*Grimaldia dichotoma* Raddi, nom. illeg., *Marchantia androgyna* L., *Mannia androgyna* (L.) A. Evans).

(≡) *Cyathophora* Gray, Nat. Arr. Brit. Pl. 1: 678, 683. 1 Nov 1821.

Marchesinia Gray, Nat. Arr. Brit. Pl. 1: 679 (*'Marchesinius'*), 689, 817 (*'Marchesinus'*). 1 Nov 1821 (orth. cons.).
Typus: *M. mackaii* (Hook.) Gray (*Jungermannia mackaii* Hook.).

Mastigophora Nees, Naturgesch. Eur. Leberm. 3: 89. Apr 1838.
Typus: *M. woodsii* (Hook.) Nees (*Jungermannia woodsii* Hook.).

(H)   *Mastigophora* Nees, Naturgesch. Eur. Leberm. 1: 95, 101. 15 Sep - 15 Dec 1833 [*Hepat.*].
≡ *Lepidozia* (Dumort.) Dumort. 1835 (nom. cons.).

Mylia Gray, Nat. Arr. Brit. Pl. 1: 693. 1 Nov 1821 (*'Mylius'*) (orth. cons.).
Typus: *M. taylorii* (Hook.) Gray (*Jungermannia taylorii* Hook.).

Nardia Gray, Nat. Arr. Brit. Pl. 1: 694. 1 Nov 1821 (*'Nardius'*) (orth. cons.).
Typus: *N. compressa* (Hook.) Gray (*Jungermannia compressa* Hook.).

Pallavicinia Gray, Nat. Arr. Brit. Pl. 1: 775. 1 Nov 1821 (*'Pallavicinius'*) (orth. cons.).
Typus: *P. lyellii* (Hook.) Carruth. (in J. Bot. 3: 302. 1 Oct 1865) (*Jungermannia lyellii* Hook.).

Pellia Raddi, Jungermanniogr. Etrusca: 38. 1818.
Typus: *P. fabroniana* Raddi, nom. illeg. (*Jungermannia epiphylla* L., *P. epiphylla* (L.) Corda).

(≡)   *Merkia* Borkh., Tent. Disp. Pl. German.: 156. Apr 1792 (typ. des.: Grolle in Taxon 24: 693. 1975).

Plagiochasma Lehm. & Lindenb. in Lehmann, Nov. Stirp. Pug. 4: 13. Feb-Mar 1832.
Typus: *P. cordatum* Lehm. & Lindenb.

(=)   *Aytonia* J. R. Forst. & G. Forst., Char. Gen. Pl.: 74. 29 Nov 1775.
Typus: *A. rupestris* J. R. Forst. & G. Forst.

Plagiochila (Dumort.) Dumort., Recueil Observ. Jungerm.: 14. 1835 (*Radula* sect. *Plagiochila* Dumort., Syll. Jungerm. Europ.: 42. 1831).
Typus: *Radula asplenioides* (L.) Dumort. (*Jungermannia asplenioides* L., *P. asplenioides* (L.) Dumort.).

(=)   *Carpolepidum* P. Beauv., Fl. Oware 1: 21. Jun 1805.
Typus (vide Bonner, Index Hepat. 3: 526. 1963): *C. dichotomum* P. Beauv.

Radula Dumort., Comment. Bot.: 112. Nov (sero) - Dec (prim.) 1822. Typus: *R. complanata* (L.) Dumort. (*Jungermannia complanata* L.).

(≡) *Martinellius* Gray, Nat. Arr. Brit. Pl. 1: 690. 1 Nov 1821 (per typ. des.).

Reboulia Raddi in Opusc. Sci. 2: 357. Nov 1818-1819 (prim.) *('Rebouillia')* (orth. cons.). Typus: *R. hemisphaerica* (L.) Raddi (*Marchantia hemisphaerica* L.).

Riccardia Gray, Nat. Arr. Brit. Pl. 1: 679, 683. 1 Nov 1821 *('Riccardius')* (orth. cons.). Typus: *R. multifida* (L.) Gray (*Jungermannia multifida* L.).

Riccia L., Sp. Pl.: 1138. 1 Mai 1753. Typus: *R. glauca* L. (typ. cons.).

Saccogyna Dumort., Comment. Bot.: 113. Nov (sero) - Dec (prim.) 1822. Typus: *S. viticulosa* (L.) Dumort. (*Jungermannia viticulosa* L.).

(≡) *Lippius* Gray, Nat. Arr. Brit. Pl. 1: 679, 706. 1 Nov 1821.

Scapania (Dumort.) Dumort., Recueil Observ. Jungerm.: 14. 1835 (*Radula* sect. *Scapania* Dumort., Syll. Jungerm. Europ.: 38. 1831). Typus: *Radula undulata* (L.) Dumort. (*Jungermannia undulata* L., *S. undulata* (L.) Dumort.) (typ. cons.).

Solenostoma Mitt. in J. Linn. Soc., Bot. 8: 51. 30 Jun 1864. Typus: *S. tersum* (Nees) Mitt. (*Jungermannia tersa* Nees).

(=) *Gymnoscyphus* Corda in Sturm, Deutschl. Fl., sect. 2, Heft 26-27: 158. 1-7 Mar 1835. Typus: *G. repens* Corda

Taxilejeunea (Spruce) Schiffn. in Engler & Prantl, Nat. Pflanzenfam. 1(3): 118, 125. Sep 1893 (*Lejeunea* subg. *Taxilejeunea* Spruce in Trans. & Proc. Bot. Soc. Edinburgh 15: 77, 212. Apr 1884). Typus: *Lejeunea chimborazensis* Spruce (*T. chimborazensis* (Spruce) Steph.).

(H) *Taxilejeunea* Steph. in Hedwigia 28: 262. Jul-Aug 1889 [*Hepat.*]. Typus: *T. convexa* Steph.

Telaranea Spruce ex Schiffn. in Engler & Prantl, Nat. Pflanzenfam. 1(3): 103. Sep 1893   .
Typus: *T. chaetophylla* (Spruce) Schiffn. (*Lepidozia chaetophylla* Spruce).

(=) *Arachniopsis* Spruce, Cephalozia 84. Oct-Dec 1882.
Typus (vide R. M. Schust., Nova Hedwigia 10: 34. 1965): *A. coactilis* Spruce.

Trachylejeunea (Spruce) Schiffn. in Engler & Prantl, Nat. Pflanzenfam. 1(3): 119, 126. Sep 1893 (*Lejeunea* subg. *Trachylejeunea* Spruce in Trans. & Proc. Bot. Soc. Edinburgh 15: 76, 180. Apr 1884).
Typus: *Lejeunea acanthina* Spruce (*T. acanthina* (Spruce) Schiffn.).

(H) *Trachylejeunea* Steph. in Hedwigia 28: 262. Jul-Aug 1889 [*Hepat.*].
Typus: *T. elegantissima* Steph.

Treubia K. I. Goebel in Ann. Jard. Bot. Buitenzorg 9: 1. 1890 (ante 1 Oct).
Typus: *T. insignis* K. I. Goebel

Trichocolea Dumort., Comment. Bot.: 113. Nov (sero) - Dec (prim.) 1822 (*'Thricholea'*) (orth. cons.).
Typus: *T. tomentella* (Ehrh.) Dumort. (*Jungermannia tomentella* Ehrh.)

C2. MUSCI

Acidodontium Schwägr., Sp. Musc. Frond. Suppl. 2(2): 152. Mai 1827.
Typus: *A. kunthii* Schwägr., nom. illeg. (*Bryum megalocarpum* Hook., *A. megalocarpum* (Hook.) Renauld & Cardot).

Aloina Kindb. in Bih. Kongl. Svenska Vetensk.-Akad. Handl. 6(19): 22. Mar-Aug 1882.
Typus: *A. aloides* (W. D. J. Koch ex Schultz) Kindb. (in Bih. Kongl. Svenska Vetensk.-Akad. Handl. 7(9): 136. Jan-Jun 1883) (*Trichostomum aloides* W. D. J. Koch ex Schultz).

(=) *Aloidella* (De Not.) Venturi in Comment. Fauna Veneto Trentino 1(3): 124. 1 Jan 1868 (*Tortula* sect. *Aloidella* De Not., Musci Ital.: 3, 14. 1862).
Typus: non designatus.

Amblyodon P. Beauv. in Mag. Encycl. 5: 323. 21 Feb 1804 *('Amblyodum')* (orth. cons.).
Typus: *A. dealbatus* (Hedw.) P. Beauv. (in Cuvier, Dict. Sci. Nat. 2: 23. 12 Oct 1804) (*Meesia dealbata* Hedw.) (typ. cons.).

Amphidium Schimp., Coroll. Bryol. Eur.: 39. 1 Aug 1856.
Typus: *A. lapponicum* (Hedw.) Schimp. (*Anictangium lapponicum* Hedw.).

(H) *Amphidium* Nees in Sturm, Deutschl. Fl., Abt. 2, 5(17): 2. 25 Apr 1819 [*Musci*].
Typus: *A. pulvinatum* Nees

Anacolia Schimp., Syn. Musc. Eur., ed. 2: 513. 1876.
Typus: *A. webbii* (Mont.) Schimp. (*Glyphocarpa webbii* Mont.).

(=) *Glyphocarpa* R. Br. in Trans. Linn. Soc. London 12: 575. Aug 1819.
Typus: *G. capensis* R. Br.

Anoectangium Schwägr., Sp. Musc. Frond. Suppl. 1(1): 33. Jun-Jul 1811.
Typus: *A. compactum* Schwägr. (typ. cons.) [= *A. aestivum* (Hedw.) Mitt. (*Gymnostomum aestivum* Hedw.)].

(H) *Anictangium* Hedw., Sp. Musc. Frond.: 40. 1 Jan 1801 (typ. des.: Nees & Hornschuch in Nees & al., Bryol. Germ. 1: 90-91. 1823) [*Musci*].
≡ *Hedwigia* P. Beauv. 1804 (nom. cons.).

Aongstroemia Bruch & Schimp. in Bruch & al., Bryol. Europ. [1: 169]. Dec 1846 *('Angstroemia')* (orth. cons.).
Typus: *A. longipes* (Sommerf.) Bruch & Schimp. (*Weissia longipes* Sommerf.).

Atractylocarpus Mitt. in J. Linn. Soc., Bot. 12: 13. Jun 1869.
Typus: *A. alpinus* (Schimp. ex Milde) Lindb. (in Finland 1886(80): 2. 7 Apr 1886) (*Metzleria alpina* Schimp. ex Milde) (typ. cons.).

Atrichum P. Beauv. in Mag. Encycl. 5: 329. 21 Feb 1804.
Typus: *A. undulatum* (Hedw.) P. Beauv. (Prodr. Aethéogam.: 42. 10 Jan 1805) (*Polytrichum undulatum* Hedw.).

(≡) *Catharinea* Ehrh. ex D. Mohr, Observ. Bot.: 31. 1803 (post 19 Mar) (typ. des.: Pfeiffer, Nomencl. Bot. 1: 627. 1873).

Aulacomnium Schwägr., Sp. Musc. Frond. Suppl. 3(1): ad t. 215. Nov-Dec 1827 *('Aulacomnion')* (orth. cons.).
Typus: *A. androgynum* (Hedw.) Schwägr. (*Bryum androgynum* Hedw.).

(≡) *Gymnocephalus* Schwägr., Sp. Musc. Frond. Suppl. 1(2): 87. 1 Jan - 9 Mai 1816.
(=) *Arrhenopterum* Hedw., Sp. Musc. Frond.: 198. 1 Jan 1801.
Typus: *A. heterostichum* Hedw.

Barbula Hedw., Sp. Musc. Frond.: 115. 1 Jan 1801.
Typus: *B. unguiculata* Hedw.

(H) *Barbula* Lour., Fl. Cochinch.: 357, 366. Sep 1790 [*Dicot.: Verben.*].
Typus: *B. sinensis* Lour.

Bartramia Hedw., Sp. Musc. Frond.: 164. 1 Jan 1801.
Typus: *B. halleriana* Hedw. (typ. cons.).

(H) *Bartramia* L., Sp. Pl.: 389. 1 Mai 1753 [*Dicot.: Til.*].
Typus: *B. indica* L.

Bartramidula Bruch & Schimp. in Bruch & al., Bryol. Europ. [4: 55]. Apr 1846.
Typus: *B. wilsonii* Bruch & Schimp., nom. illeg. (*Glyphocarpa cernua* Wilson, *B. cernua* (Wilson) Lindb.).

(=) *Glyphocarpa* R. Br. in Trans. Linn. Soc. London 12: 575. Aug 1819.
Typus: *G. capensis* R. Br.

Braunia Bruch & Schimp. in Bruch & al., Bryol. Europ. [3: 159]. Apr. 1846.
Typus: *B. sciuroides* (Bals.-Criv. & De Not.) Bruch & Schimp. (*Anictangium sciuroides* Bals.-Criv. & De Not.)

(H) *Braunia* Hornsch. in Jahrb. Wiss. Krit. 1828(1): 467. Mar. 1828.
≡ *Neckera* Hedw. 1801 (nom. cons.).

Bryoxiphium Mitt. in J. Linn. Soc., Bot. 12: 24, 580. Jun 1869 *('Bryoziphium')* (orth. cons.).
Typus: *B. norvegicum* (Brid.) Mitt. (*Phyllogonium norvegicum* Brid.).

(≡) *Eustichium* Bruch & Schimp. in Bruch & al., Bryol. Europ. [2: 159]. Dec 1849.

Callicostella (Müll. Hal.) Mitt. in J. Proc. Linn. Soc., Suppl. Bot. 1: 66, 136. 21 Feb 1859 (*Hookeria* sect. *Callicostella* Müll. Hal., Syn. Musc. Frond. 2: 216. Jul 1851).
Typus: *Hookeria papillata* Mont. (*C. papillata* (Mont.) Mitt.).

(=) *Schizomitrium* Schimp. in Bruch & al., Bryol. Europ. [5: 59]. Jul 1851.
Typus (vide Crosby in Taxon 24: 355. 1975): *S. martianum* (Hornsch.) Crosby (*Hookeria martiana* Hornsch.).

Cinclidotus P. Beauv. in Mag. Encycl. 5: 319. 21 Feb 1804 (*'Ciccli-*
dotus') (orth. cons.).
Typus: *C. fontinaloides* (Hedw.) P.
Beauv. (Prodr. Aethéogam.: 52. 10
Jan 1805) (*Trichostomum fontinalo-*
ides Hedw.).

Crossidium Jur., Laubm.-Fl. Oes-   (=)   *Chloronotus* Venturi in Comment.
terr.-Ung.: 127. Apr-Jun 1882.   Fauna Veneto Trentino 1(3): 124. 1
Typus: *C. squamigerum* (Viv.) Jur.   Jan 1868.
(*Barbula squamigera* Viv.).   Typus: non designatus.

Cryptopodium Brid., Bryol. Univ. 1:   (≡)   *Cryptoseta* (Arn.) Kitt. in Mém. Soc.
xxxiv; 2: 30. 1827 (ante 21 Nov).   Linn. Paris 5: 283. Jul-Aug 1826
Typus: *C. bartramioides* (Hook.)   (*Bryum* [unranked] *Cryptoseta* Arn.,
Brid. (*Bryum bartramioides* Hook.).   Disp. Méth. Mousses: 43. 20 Dec
                                      1825 - 6 Feb 1826).

Cynodontium Bruch & Schimp. in   (H)   *Cynodontium* Brid., Muscol. Recent.
Schimper, Coroll. Bryol. Eur.: 12. 1   Suppl. 1: 155. 20 Apr 1806 [*Musci*].
Aug 1856.   ≡ *Cynontodium* Hedw. 1801 (nom.
Typus: *C. polycarpum* (Hedw.)   rej. sub *Distichium*).
Schimp. (*Fissidens polycarpus*
Hedw.) (typ. cons.).

Daltonia Hook. & Taylor, Muscol.
Brit.: 80. 1 Jan 1818.
Typus: *D. splachnoides* (Sm.) Hook.
& Taylor (*Neckera splachnoides* Sm.).

Desmotheca Lindb. in J. Linn. Soc.,   (≡)   *Cryptocarpon* Dozy & Molk., Musc.
Bot. 13: 184. 29 Mai 1872.   Frond. Archip. Ind.: 5. Aug-Sep 1844.
Typus: *D. apiculata* (Dozy & Molk.)
Lindb. ex Cardot (in Ann. Jard. Bot.
Buitenzorg, Suppl. 1: 11. 1897) (*Cryp-*
tocarpon apiculatum Dozy & Molk.).

Dicranella (Müll. Hal.) Schimp., Coroll. Bryol. Eur.: 13. 1 Aug 1856
(*Aongstroemia* sect. *Dicranella* Müll.
Hal., Syn. Musc. Frond. 1: 430. Oct
1848).
Typus: *Aongstroemia heteromalla*
(Hedw.) Müll. Hal. (*Dicranum hete-
romallum* Hedw., *Dicranella heteromalla* (Hedw.) Schimp.) (typ. cons.).

Dicranoloma (Renauld) Renauld in
Rev. Bryol. 28: 85. 1901 (*Leuco-
loma* subg. *Dicranoloma* Renauld,
Prodr. Fl. Bryol. Madagascar: 61.
1898).
Typus: *Leucoloma platyloma* (Besch.)
Renauld (*Dicranum platyloma* Besch.,
Dicranoloma platyloma (Besch.) Ren-
auld).

(=) *Megalostylium* Dozy & Molk., Musci Frond. Ined. Archip. Ind.: 145.
1848.
Typus: *M. brevisetum* Dozy & Molk.

Distichium Bruch & Schimp. in
Bruch & al., Bryol. Europ. [2: 153].
Apr 1846.
Typus: *D. capillaceum* (Hedw.) Bruch
& Schimp. (*Cynontodium capillac-
eum* Hedw.).

(=) *Cynontodium* Hedw., Sp. Musc.
Frond.: 57. 1 Jan 1801.
Typus: non designatus.

Ditrichum Timm ex Hampe in Flora
50: 181. 26 Apr 1867.
Typus: *D. homomallum* (Hedw.)
Hampe (*Didymodon homomallus*
Hedw.) (typ. cons.) [= *D. hetero-
mallum* (Hedw.) E. Britton (*Weissia
heteromalla* Hedw.)].

(H) *Ditrichum* Cass. in Bull. Sci. Soc.
Philom. Paris 1817: 33. Feb 1817
[*Dicot.: Comp.*].
Typus: *D. macrophyllum* Cass.

(≡) *Diaphanophyllum* Lindb. in Öfvers.
Förh. Kongl. Svenska Vetensk.-Ak-
ad. 19: 605. 1 Feb - 28 Mai 1863.

(=) *Aschistodon* Mont. in Ann. Sci. Nat.,
Bot., ser. 3, 4: 109. Aug 1845.
Typus: *A. conicus* Mont.

(=) *Lophiodon* Hook. f. & Wilson in London J. Bot. 3: 543. Sept-Oct 1844.
Typus: *L. strictus* Hook. f. & Wilson

(=) *Trichodon* Schimp., Coroll. Bryol.
Eur.: 36. 1 Aug 1856.
Typus: *T. cylindricus* (Hedw.) Schimp.
(*Trichostomum cylindricum* Hedw.).

Drepanocladus (Müll. Hal.) G. Roth in Hedwigia 38, Beibl.: (6). 28 Feb 1899 (*Hypnum* subsect. *Drepanocladus* Müll. Hal., Syn. Musc. Frond. 2: 321. Jul 1851).
Typus: *Hypnum aduncum* Hedw. (*D. aduncus* (Hedw.) Warnst.) (typ. cons.).

(H) *Drepanocladus* Müll. Hal. in Nuovo Giorn. Bot. Ital., ser. 2, 5: 203. 1898 [*Musci*].
Typus: *D. sinensi-uncinatus* Müll. Hal.

(=) *Drepano-hypnum* Hampe in Linnaea 37: 518. Oct 1872.
Typus: *D. fontinaloides* Hampe

Drummondia Hook. in Drummond, Musc. Amer.: No. 62. 1828.
Typus: *D. prorepens* (Hedw.) E. Britton (in Mem. Torrey Bot. Club 4: 180. 16 Apr 1894) (*Gymnostomum prorepens* Hedw.) (typ. cons.).

(≡) *Anodontium* Brid., Muscol. Recent. Suppl. 1: 41. 20 Apr 1806.

Ephemerella Müll. Hal., Syn. Musc. Frond. 1: 34. Feb 1848.
Typus: *E. pachycarpa* (Schwägr.) Müll. Hal. (*Phascum pachycarpum* Schwägr.).

(=) *Physedium* Brid., Bryol. Univ. 1: 51. Jan-Mar 1826.
Typus: *P. splachnoides* (Hornsch.) Brid. (*Phascum splachnoides* Hornsch.).

Ephemerum Hampe in Flora 20: 285. 14 Mai 1837.
Typus: *E. serratum* (Hedw.) Hampe (*Phascum serratum* Hedw.).

(H) *Ephemeron* Mill., Gard. Dict. Abr., ed. 4: [470]. 28 Jan 1754 [*Monocot.: Commelin.*].
≡ *Tradescantia* L. 1753.

Gollania Broth. in Engler & Prantl, Nat. Pflanzenfam. 1(3): 1044, 1054. 23 Jun 1908.
Typus: *G. neckerella* (Müll. Hal.) Broth. (*Hylocomium neckerella* Müll. Hal.) (typ. cons.).

Gymnostomum Nees & Hornsch. in Nees & al., Bryol. Germ. 1: 112, 153. 14 Feb - 15 Apr 1823.
Typus: *G. calcareum* Nees & Hornsch. (typ. cons.).

(H) *Gymnostomum* Hedw., Sp. Musc. Frond.: 30. 1 Jan 1801 [*Musci*].
Typus: non designatus.

Gyroweisia Schimp., Syn. Musc. Eur., ed. 2: 38. 1876.
Typus: *G. tenuis* (Hedw.) Schimp. (*Gymnostomum tenue* Hedw.) (typ. cons.).

(=) *Weisiodon* Schimp., Coroll. Bryol. Eur.: 9. 1 Aug 1856.
Typus: *W. reflexus* (Brid.) Schimp. (*Weissia reflexa* Brid.).

Haplohymenium Dozy & Molk., Musc. Frond. Ined. Archip. Ind.: 125. 1846.
Typus: *H. sieboldii* (Dozy & Molk.) Dozy & Molk. (*Leptohymenium sieboldii* Dozy & Molk.) (typ. cons.).

Hedwigia P. Beauv. in Mag. Encycl. 5: 304. 21 Feb 1804.
Typus: *H. ciliata* (Hedw.) P. Beauv. (*Anictangium ciliatum* Hedw.) (etiam vide *Anoectangium* [*Musci*]).

Helodium Warnst., Krypt.-Fl. Brandenburg 2: 675, 692. 9 Oct 1905.
Typus: *H. blandowii* (F. Weber & D. Mohr) Warnst. (*Hypnum blandowii* F. Weber & D. Mohr).

Holomitrium Brid., Bryol. Univ. 1: 226. Jan-Mar 1826 (*'Olomitrium'*) (orth. cons.).
Typus: *H. perichaetiale* (Hook.) Brid. (*Trichostomum perichaetiale* Hook.) (typ. cons.).

Homalia Brid., Bryol. Univ. 1: xlvi; 2: 807, 812. 1827 (ante 21 Nov) (*'Omalia'*) (orth. cons.).
Typus: *H. trichomanoides* (Hedw.) Brid. (*Leskea trichomanoides* Hedw.).

Hookeria Sm. in Smith & Sowerby, Engl. Bot.: ad t. 1902. 1 Jun 1808.
Typus: *H. lucens* (Hedw.) Sm. (*Hypnum lucens* Hedw.).

Hygroamblystegium Loeske, Moosfl. Harz.: 298. Jan-Mar 1903.
Typus: *H. irriguum* (Hook. & Wilson) Loeske (*Hypnum irriguum* Hook. & Wilson) [= *H. tenax* (Hedw.) Jenn. (*Hypnum tenax* Hedw.)].

(H) *Haplohymenium* Schwägr., Sp. Musc. Frond. Suppl. 3(2): ad t. 271. Jan 1829 [*Musci*].
Typus: *H. microphyllum* Schwägr. [= *Thuidium haplohymenium* (Harv.) A. Jaeger (*Hypnum haplohymenium* Harv.)].

(H) *Hedwigia* Sw., Prodr.: 4, 62. 20 Jun - 29 Jul 1788 [*Dicot.: Burser.*].
Typus: *H. balsamifera* Sw.

(H) *Helodium* Dumort., Fl. Belg.: 77. 1827 [*Dicot.: Umbell.*].
≡ *Helosciadium* W. D. J. Koch 1824.

(H) *Hookera* Salisb., Parad. Lond.: ad t. 98. 1 Mar 1808 [*Monocot.: Lil.*].
≡ *Brodiaea* Sm. 1810 (nom. cons.).

Hylocomium Schimp. in Bruch & al., Bryol. Europ. [5: 169]. 1852. Typus: *H. splendens* (Hedw.) Schimp. (*Hypnum splendens* Hedw.) (typ. cons.).

Hyophila Brid., Bryol. Univ. 1: 760. 1827 (ante 21 Nov). Typus: *H. javanica* (Nees & Blume) Brid. (*Gymnostomum javanicum* Nees & Blume) (typ. cons.).

Hypnum Hedw., Sp. Musc. Frond.: 236. 1 Jan 1801. Typus: *H. cupressiforme* Hedw. (typ. cons.).

Lepidopilum (Brid.) Brid., Bryol. Univ. 2: 267. 1827 (ante 21 Nov) (*Pilotrichum* [unranked] *Lepidopilum* Brid., Muscol. Recent. Suppl. 4: 141. 18 Dec 1818). Typus: *Pilotrichum scabrisetum* (Schwägr.) Brid. (*Neckera scabriseta* Schwägr., *L. scabrisetum* (Schwägr.) Steere).

(=) *Actinodontium* Schwägr., Sp. Musc. Frond. Suppl. 2(2): 75. Mai 1826. Typus: *A. ascendens* Schwägr.

Leptodon D. Mohr, Observ. Bot.: 27. 1803 (post 19 Mar). Typus: *L. smithii* (Dicks. ex Hedw.) F. Weber & D. Mohr (Index Mus. Pl. Crypt.: fol. 2, recto. Aug-Dec 1803) (*Hypnum smithii* Dicks. ex Hedw.).

Leptostomum R. Br. in Trans. Linn. Soc. London 10: 320. 7 Sep 1811. Typus: *L. inclinans* R. Br.

Leucoloma Brid., Bryol. Univ. 2: 218, 751. 1827 (ante 21 Nov). Typus: *L. bifidum* (Brid.) Brid. (*Hypnum bifidum* Brid.).

(≡) *Macrodon* Arn., Disp. Méth. Mousses: 51. 20 Dec 1825 - 6 Feb 1826.

(=) *Sclerodontium* Schwägr., Sp. Musc. Frond. Suppl. 2(1): 124. 1824. Typus: *S. pallidum* (Hook.) Schwägr. (*Leucodon pallidus* Hook.).

Meesia Hedw., Sp. Musc. Frond.: 173. 1 Jan 1801.
Typus: *M. longiseta* Hedw.

(H) *Meesia* Gaertn., Fruct. Sem. Pl. 1: 344. Dec 1788 [*Dicot.: Ochn.*].
Typus: *M. serrata* Gaertn.

Mittenothamnium Henn. in Hedwigia 41, Beibl.: 225. 15 Dec 1902.
Typus: *M. reptans* (Hedw.) Cardot (in Rev. Bryol. 40: 21. 1913) (*Hypnum reptans* Hedw.) (typ. cons.).

Mniobryum Limpr. in Rabenh. Krypt.-Fl., ed. 2, 4(2): 272. Jan 1892.
Typus: *M. carneum* Limpr., nom. illeg. (*Bryum delicatulum* Hedw., *M. delicatulum* (Hedw.) Dixon).

Mnium Hedw., Sp. Musc. Frond.: 188. 1 Jan 1801.
Typus: *M. hornum* Hedw. (typ. cons.).

(H) *Mnium* L., Sp. Pl.: 1109. 1 Mai 1753 (typ. des.: Proskauer in Taxon 12: 200. 1963) [*Hepat.*].
≡ *Calypogeia* Raddi 1818 (nom. cons.).

Muelleriella Dusén in Bot. Not. 1905: 304. 1905.
Typus: *M. crassifolia* (Hook. f. & Wilson) Dusén (*Orthotrichum crassifolium* Hook. f. & Wilson).

(H) *Muelleriella* Van Heurck, Treat. Diatom.: 435. Oct-Nov 1896 [*Bacillarioph.*].
Typus: *M. limbata* (Ehrenb.) Van Heurck (*Pyxidicula limbata* Ehrenb.).

Myrinia Schimp., Syn. Musc. Eur.: 482. Mar-Apr 1860.
Typus: *M. pulvinata* (Wahlenb.) Schimp. (*Leskea pulvinata* Wahlenb.).

(H) *Myrinia* Lilja, Fl. Sv. Odl. Vext., Suppl. 1: 25. 1840 [*Dicot.: Onagr.*].
Typus: *M. microphylla* Lilja

Neckera Hedw., Sp. Musc. Frond.: 200. 1 Jan 1801.
Typus: *N. pennata* Hedw.

(H) *Neckeria* Scop., Intr. Hist. Nat.: 313. Jan-Apr 1777 [*Dicot.: Papaver.*].
≡ *Capnoides* Mill. 1754 (nom. rej. sub *Corydalis*).

Oligotrichum DC. in Lamarck & Candolle, Fl. Franç., ed. 3, 2: 491. 17 Sep 1805.
Typus: *O. hercynicum* (Hedw.) DC. (*Polytrichum hercynicum* Hedw.) (typ. cons.).

Orthothecium Schimp. in Bruch & al., Bryol. Europ. [5: 105]. Jul 1851. Typus: *O. rufescens* (Brid.) Schimp. (*Hypnum rufescens* Brid.).

(H) *Orthothecium* Schott & Endl., Melet. Bot.: 31. 1832 [*Dicot.: Stercul.*]. Typus: *O. lhotskyanum* Schott & Endl.

Oxyrrhynchium (Schimp.) Warnst., Krypt.-Fl. Brandenburg 2: 764, 781. 9 Oct 1905 (*Eurhynchium* subg. *Oxyrrhynchium* Schimp. in Bruch & al., Bryol. Europ. [5: 224]. 1854). Typus: *Eurhynchium hians* (Hedw.) Sande Lac. (*Hypnum hians* Hedw., *O. hians* (Hedw.) Loeske) (typ. cons.).

Papillaria (Müll. Hal.) Lorentz, Moosstudien: 165. 1864 (*Neckera* subsect. *Papillaria* Müll. Hal., Syn. Musc. Frond. 2: 134. Sep 1850). Typus: *Neckera nigrescens* (Sw. ex Hedw.) Schwägr. (*Hypnum nigrescens* Sw. ex Hedw., *P. nigrescens* (Sw. ex Hedw.) A. Jaeger) (typ. cons.).

(H) *Papillaria* J. Kickx f., Fl. Crypt. Louvain: 73, 104. 20 Jun 1835 [*Fungi*].
≡ *Pycnothelia* Dufour 1821.

Pelekium Mitt. in J. Linn. Soc., Bot. 10: 176. 19 Mar 1868. Typus: *P. velatum* Mitt.

(=) *Lorentzia* Hampe in Flora 50: 75. 26 Feb 1867. Typus: *L. longirostris* Hampe (in Nuovo Giorn. Bot. Ital. 4: 288. 1872).

Platygyrium Schimp. in Bruch & al., Bryol. Europ. [5: 95]. Jul 1851. Typus: *P. repens* (Brid.) Schimp. (*Pterigynandrum repens* Brid.).

(=) *Pterigynandrum* Hedw., Sp. Musc. Frond.: 80. 1 Jan 1801. Typus (vide Schimper in Bruch & al., Bryol. Europ. [5: 121]. 1851): *P. filiforme* Hedw.

(=) *Leptohymenium* Schwägr., Sp. Musc. Frond. Suppl. 3(1): ad t. 246c. Apr-Dec 1828. Typus: *L. tenue* (Hook.) Schwägr. (*Neckera tenuis* Hook.).

Pleuridium Rabenh., Deutschl. Krypt.-Fl. 2(3): 79. Jul 1848. Typus: *P. subulatum* (Dicks. ex Hedw.) Rabenh. (*Phascum subulatum* Dicks. ex Hedw.) (typ. cons.).

(H) *Pleuridium* Brid., Muscol. Recent. Suppl. 4: 10. 18 Dec 1818 [*Musci*]. Typus (vide Pfeiffer, Nomencl. Bot. 2: 756. 1873): *P. alternifolium* (Dicks. ex Hedw.) Brid. (*Phascum alternifolium* Dicks. ex Hedw.).

Pleurozium Mitt. in J. Linn. Soc.,
Bot. 12: 22, 537. Jun 1869.
Typus: *P. schreberi* (Brid.) Mitt. (*Hyp-
num schreberi* Brid.) (typ. cons.).

Pottia Ehrh. ex Fürnr. in Flora 12(2),
Ergänzungsbl.: 10. Jul-Oct 1829.
Typus: *P. truncata* (Hedw.) Bruch
& Schimp. (in Bruch & al., Bryol.
Europ. [2: 37]. Aug 1843) (*Gymno-
stomum truncatum* Hedw.).

(=) *Anacalypta* Röhl. ex Leman in Cu-
vier, Dict. Sci. Nat. 2, Suppl.: 38. 12
Oct 1816.
Typus: *A. lanceolata* (Hedw.) Röhl.
ex Nees & Hornsch. (in Nees & al.,
Bryol. Germ. 2(2): 141. Jul-Oct 1831)
(*Encalypta lanceolata* Hedw.).

(=) *Physedium* Brid., Bryol. Univ. 1: 51.
Jan-Mar 1826.
Typus: *P. splachnoides* (Hornsch.)
Brid. (*Phascum splachnoides*
Hornsch.).

Pterygoneurum Jur., Laubm.-Fl. Oesterr.-Ung.: 95. Apr-Jun 1882 (*'Pteri-
goneurum'*) (orth. cons.).
Typus: *P. cavifolium* Jur., nom. illeg. (*Pottia cavifolia* Fürnr., nom. illeg., *Gymnostomum ovatum* Hedw.,
P. ovatum (Hedw.) Dixon).

(=) *Pharomitrium* Schimp., Syn. Musc.
Eur.: 120. Mar-Apr 1860.
Typus: *P. subsessile* (Brid.) Schimp.
(*Gymnostomum subsessile* Brid.).

Ptychomitrium Fürnr. in Flora 12(2,
Ergänzungsbl.): 19. Jul-Oct 1829
(*'Pthychomitrium'*) (orth. cons.).
Typus: *P. polyphyllum* (Sw.) Bruch
& Schimp. (in Bruch & al., Bryol.
Europ. [3: 82]. Dec 1837) (*Dicran-
um polyphyllum* Sw.).

(=) *Brachysteleum* Rchb., Consp. Regni
Veg.: 34. Dec 1828 - Mar 1829.
Typus: *B. crispatum* (Hedw.) Hornsch.
(in Martius, Fl. Bras. 1(2): 20. 1840)
(*Encalypta crispata* Hedw.).

Pylaisia Schimp. in Bruch & al.,
Bryol. Europ. [5: 87]. 1851 (*'Py-
laisaea'*) (orth. cons.).
Typus: *P. polyantha* (Hedw.) Schimp.
(*Leskea polyantha* Hedw.) (typ. cons.).

(H) *Pilaisaea* Desv. in J. Bot. Agric. 4:
75. Sep 1814 [*Musci*].
Typus: *P. radicans* Bach. Pyl. (in J.
Bot. Agric. 4: 75. Sep 1814).

Rhodobryum (Schimp.) Limpr. in
Rabenh. Krypt.-Fl., ed. 2, 4(2): 444.
Dec 1892 (*Bryum* subg. *Rhodobryum*
Schimp., Syn. Musc. Eur.: 381. Mar-
Apr 1860).
Typus: *Bryum roseum* (Hedw.) Cro-
me (*Mnium roseum* Hedw., *R. ros-
eum* (Hedw.) Limpr.).

(H) *Rhodo-bryum* Hampe in Linnaea 38:
663. Dec 1874 [*Musci*].
Typus: *R. leucocanthum* Hampe

Rhynchostegiella (Schimp.) Limpr.
in Rabenh. Krypt.-Fl., ed. 2, 4(3):
207. Dec 1896 (*Rhynchostegium*
subg. *Rhynchostegiella* Schimp. in
Bruch & al., Bryol. Europ. [5: 201].
1852).
Typus: *Rhynchostegium tenellum*
(Dicks.) Schimp. (*Hypnum tenellum*
Dicks., *Rhynchostegiella tenella*
(Dicks.) Limpr.).

(=) *Remyella* Müll. Hal. in Flora 82:
477. 28 Oct 1896.
Typus: *R. hawaiica* Müll. Hal.

Schistidium Bruch & Schimp. in
Bruch & al., Bryol. Europ. [3: 93].
Aug 1845.
Typus: *S. maritimum* (Turn.) Bruch
& Schimp. (*Grimmia maritima* Turn.)
(typ. cons.).

(H) *Schistidium* Brid., Muscol. Recent.
Suppl. 4: 20. 18 Dec 1818 [*Musci*].
Typus (vide Mårtensson in Kungl.
Svenska Vetenskapsakad. Avh. Na-
turskyddsärenden 14: 106. 1956): *S.
pulvinatum* (Hedw.) Brid. (*Gymnostomum pulvinatum* Hedw.).

Timmia Hedw., Sp. Musc. Frond.:
176. 1 Jan 1801.
Typus: *T. megapolitana* Hedw.

(H) *Timmia* J. F. Gmel., Syst. Nat. 2:
524, 538. Sep (sero) - Nov 1791
[*Monocot.: Amaryllid.*].
Typus: non designatus.

Tortella (Müll. Hal.) Limpr. in
Rabenh. Krypt.-Fl., ed. 2, 4(1): 599.
Oct 1888 (*Barbula* sect. *Tortella*
Müll. Hal., Syn. Musc. Frond. 1:
599. Mar 1849).
Typus: *Barbula caespitosa* Schwägr.
(*T. caespitosa* (Schwägr.) Limpr.) [=
T. humilis (Hedw.) Jenn. (*B. humilis*
Hedw.)].

(=) *Pleurochaete* Lindb. in Öfvers. Förh.
Kongl. Svenska Vetensk.-Akad. 21:
253. 21 Aug 1864.
Typus: *P. squarrosa* (Brid.) Lindb.
(*Barbula squarrosa* Brid.).

Tortula Hedw., Sp. Musc. Frond.:
122. 1 Jan 1801.
Typus: *T. subulata* Hedw. (typ. cons.).

(H) *Tortula* Roxb. ex Willd., Sp. Pl. 3:
6, 359. 1800 [*Dicot.: Verben.*].
Typus: *T. aspera* Roxb. ex Willd.

Trichostomum Bruch in Flora 12: 396. 7 Jul 1829.
Typus: *T. brachydontium* Bruch (typ. cons.).

(H) *Trichostomum* Hedw., Sp. Musc. Frond.: 107. 1 Jan 1801 [*Musci*].
Typus: non designatus.

(=) *Plaubelia* Brid., Bryol. Univ. 1: 522. Jan-Mar 1826.
Typus: *P. tortuosa* Brid.

D. PTERIDOPHYTA

Anemia Sw., Syn. Fil.: 6, 155. Mar-Apr 1806.
Typus: *A. phyllitidis* (L.) Sw. (*Osmunda phyllitidis* L.).

(=) *Ornithopteris* Bernh. in Neues J. Bot. 1(2): 40. Oct-Nov 1805.
Typus (vide Reed in Bol. Soc. Brot., ser. 2, 21: 153. 1947): *O. adiantifolia* (L.) Bernh. (*Osmunda adiantifolia* L.).

Angiopteris Hoffm. in Commentat. Soc. Regiae Sci. Gott. 12: 29. 1796.
Typus: *A. evecta* (G. Forst.) Hoffm. (*Polypodium evectum* G. Forst.).

(H) *Angiopteris* Adans., Fam. Pl. 2: 21, 518. Jul-Aug 1763 [*Pteridoph.*].
≡ *Onoclea* L. 1753.

Araiostegia Copel. in Philipp. J. Sci. 34: 240. 1927.
Typus: *A. hymenophylloides* (Blume) Copel. (*Aspidium hymenophylloides* Blume).

(=) *Gymnogrammitis* Griff., Ic. Pl. Asiat. 2: t. 129, f. 1. 1849; Not. Pl. Asiat. 2: 608. 1849.
Typus: *G. dareiformis* (Hook.) Ching ex Tardieu & C. Chr. (in Notul. Syst. (Paris) 6: 2. 1937) (*Polypodium dareiforme* Hook.).

Ceterach Willd., Anleit. Selbststud. Bot.: 578. 1804.
Typus: *C. officinarum* Willd.

(H) *Ceterac* Adans., Fam. Pl. 2: 20, 536. Jul-Aug 1763 [*Pteridoph.*].
Typus: non designatus.

Cheilanthes Sw., Syn. Fil.: 5, 126. Mar-Apr 1806.
Typus: *C. micropteris* Sw.

(=) *Allosorus* Bernh. in Neues J. Bot. 1(2): 5, 36. Oct-Nov 1805.
Typus (vide Pichi Sermolli in Webbia 9: 394. 1953): *A. pusillus* (Willd. ex Bernh.) Bernh. (*Adiantum pusillum* Willd. ex Bernh.).

Coniogramme Fée [Mém. Foug. 5] in Mém. Soc. Mus. Hist. Nat. Strasbourg 5: 167. 1852.
Typus: *C. javanica* (Blume) Fée (*Gymnogramma javanica* Blume).

(=) *Dictyogramme* Fée [Mém. Foug. 5] in Mém. Soc. Mus. Hist. Nat. Strasbourg 4(1): 206. 1850.
Typus: *D. japonica* (Thunb.) Fée ([Mém. Foug. 5] in Mém. Soc. Mus. Hist. Nat. Strasbourg 5: 375. 1852) (*Hemionitis japonica* Thunb.).

Cystodium J. Sm. in Hooker, Gen. Fil.: ad t. 96. 1841.
Typus: *C. sorbifolium* (Sm.) J. Sm. (*Dicksonia sorbifolia* Sm.).

(H) *Cystodium* Fée, Essai Crypt. Ecorc. 2: 13. 1837 [*Fungi*].
≡ *Gassicurtia* Fée 1824.

Cystopteris Bernh. in Neues J. Bot. 1(2): 5, 26. Oct-Nov 1805.
Typus: *C. fragilis* (L.) Bernh. (*Polypodium fragile* L.).

Danaea Sm. in Mém. Acad. Roy. Sci. (Turin) 5: 420. 1793.
Typus: *D. nodosa* (L.) Sm. (*Acrostichum nodosum* L.).

(H) *Danaa* All., Fl. Pedem. 2: 34. Apr-Jul 1785 [*Dicot.: Umbell.*].
Typus: *D. aquilegiifolia* (All.) All. (*Coriandrum aquilegiifolium* All.).

Doryopteris J. Sm. in J. Bot. (Hooker) 3: 404. Mai 1841.
Typus: *D. palmata* (Willd.) J. Sm. (*Pteris palmata* Willd.).

(=) *Cassebeera* Kaulf., Enum. Filic.: 216. 1824.
Typus (vide Fée [Mém. Foug. 5] in Mém. Soc. Mus. Hist. Nat. Strasbourg 5: 119. 1852): *C. triphylla* (Lam.) Kaulf. (*Adiantum triphyllum* Lam.).

Drymoglossum C. Presl [Tent. Pterid.] in Abh. Königl. Böhm. Ges. Wiss., ser. 4, 5: 227. 1836 (ante 2 Dec).
Typus: *D. piloselloides* (L.) C. Presl (*Pteris piloselloides* L.).

(=) *Pteropsis* Desv. in Mém. Soc. Linn. Paris 6(3): 218. Jul 1827.
Typus (vide Pichi Sermolli in Webbia 9: 403. 1953): *Acrostichum heterophyllum* L.

Drynaria (Bory) J. Sm. in J. Bot. (Hooker) 4: 60. Jul 1841 (*Polypodium* subg. *Drynaria* Bory in Ann. Sci. Nat. (Paris) 5: 463. 1825).
Typus: *Polypodium linnaei* Bory, nom. illeg. (*P. quercifolium* L., *D. quercifolia* (L.) J. Sm.).

Dryopteris Adans., Fam. Pl. 2: 20, 551. Jul-Aug 1763.
Typus: *D. filix-mas* (L.) Schott (Gen. Fil.: ad t. 9. 1834) (*Polypodium filix-mas* L.).

(≡) *Filix* Ség., Pl. Veron. 3: 53. Jul-Aug 1754.

Elaphoglossum Schott ex J. Sm. in J. Bot. (Hooker) 4: 148. Aug 1841.
Typus: *E. conforme* (Sw.) J. Sm. (*Acrostichum conforme* Sw.) (typ. cons.).

(=) *Aconiopteris* C. Presl [Tent. Pterid.] in Abh. Königl. Böhm. Ges. Wiss., ser. 4, 5: 236. 1836 (ante 2 Dec).
Typus: *A. subdiaphana* (Hook. & Grev.) C. Presl (*Acrostichum subdiaphanum* Hook. & Grev.).

Gleichenia Sm. in Mém. Acad. Roy. Sci. (Turin) 5: 419. 1793.
Typus: *G. polypodioides* (L.) Sm. (*Onoclea polypodioides* L.).

Lygodium Sw. in J. Bot. (Schrader) 1800(2): 7, 106. Oct-Dec 1801.
Typus: *L. scandens* (L.) Sw. (*Ophioglossum scandens* L.).

(=) *Ugena* Cav., Icon. 6: 73. Jan-Mai 1801.
Typus (vide Pichi Sermolli in Webbia 9: 418. 1953): *U. semihastata* Cav., nom. illeg. (*Ophioglossum flexuosum* L.).

Marsilea L., Sp. Pl.: 1099. 1 Mai 1753.
Typus: *M. quadrifolia* L. (typ. cons.).

Matteuccia Tod. in Giorn. Sci. Nat. Econ. Palermo 1: 235. 1866.
Typus: *M. struthiopteris* (L.) Tod. (*Osmunda struthiopteris* L.).

(≡) *Pteretis* Raf. in Amer. Monthly Mag. & Crit. Rev. 2: 268. Feb 1818.

Pellaea Link, Fil. Spec.: 59. 3-10 Sep 1841.
Typus: *P. atropurpurea* (L.) Link (*Pteris atropurpurea* L.).

Polystichum Roth, Tent. Fl. Germ. 3: 31, 69. Jun-Sep 1799.
Typus: *P. lonchitis* (L.) Roth (*Polypodium lonchitis* L.).

Pteridium Gled. ex Scop., Fl. Car-
niol.: 169. 15 Jun - 21 Jul 1760.
Typus: *P. aquilinum* (L.) Kuhn (in
Ascherson & al., Bot. Ost-Afrika: 11.
Aug-Sep 1879) (*Pteris aquilina* L.).

Schizaea Sm. in Mém. Acad. Roy. (=) *Lophidium* Rich. in Actes Soc. Hist.
Sci. (Turin) 5: 419. 1793. Nat. Paris 1: 114. 1792.
Typus: *S. dichotoma* (L.) Sm. (*Acro-* Typus: *L. latifolium* Rich.
stichum dichotomum L.).

Selaginella P. Beauv., Prodr. Aethéo- (≡) *Selaginoides* Ség., Pl. Veron. 3: 51.
gam.: 101. 10 Jan 1805. Jul-Aug 1754.
Typus: *S. spinosa* P. Beauv., nom. (=) *Lycopodioides* Boehm. in Ludwig,
illeg. (*Lycopodium selaginoides* L., Def. Gen. Pl. Ed. 3: 485. 1760.
S. selaginoides (L.) Link). Typus (vide Rothmaler in Feddes
Repert. Spec. Nov. Regni Veg. 54:
69. 1944): *L. denticulata* (L.) Kuntze
(Revis. Gen. Pl. 1-2: 824. 5 Nov
1891) (*Lycopodium denticulatum* L.).
(=) *Stachygynandrum* P. Beauv. ex Mirb.
in Lamarck & Mirbel, Hist. Nat.
Vég. 3: 477. 1802, 4: 312. 1802.
Typus (vide Pichi Sermolli in Web-
bia 26: 164. 1971): *S. flabellatum*
(L.) P. Beauv. (Prodr. Aethéogam.:
113. 10 Jan 1805) (*Lycopodium fla-*
bellatum L.).

Sphenomeris Maxon in J. Wash. (≡) *Stenoloma* Fée [Mém. Foug. 5] in
Acad. Sci. 3: 144. 1913. Mém. Soc. Mus. Hist. Nat. Stras-
Typus: *S. clavata* (L.) Maxon (*Adi-* bourg 5: 330. 1852 (typ. des.: Mor-
antum clavatum L.). ton in Taxon 8: 29. 1959).

Thelypteris Schmidel, Icon. Pl., ed. (H) *Thelypteris* Adans., Fam. Pl. 2: 20,
Keller: 3, 45. 18 Oct 1763. 610. Jul-Aug 1763 [*Pteridoph.*].
Typus: *T. palustris* Schott (Gen. Fil.: ≡ *Pteris* L. 1753.
ad t. 10. 1834) (*Acrostichum thelyp-*
teris L.).

Trichomanes L., Sp. Pl.: 1097. 1 Mai
1753.
Typus: *T. crispum* L. (typ. cons.).

E. SPERMATOPHYTA

E1. GYMNOSPERMAE

Agathis Salisb. in Trans. Linn. Soc. London 8: 311. 9 Mar 1807 [*Pin.*]. Typus: *A. loranthifolia* Salisb., nom. illeg. (*Pinus dammara* Lamb., *A. dammara* (Lamb.) Rich.).

Cedrus Trew, Cedr. Lib. Hist. 1: 6. 12 Mai - 13 Oct 1757 [*Pin.*]. Typus: *C. libani* A. Rich. (in Bory, Dict. Class. Hist. Nat. 3: 299. 6 Sep 1823) (*Pinus cedrus* L.).

(H) *Cedrus* Duhamel, Traité Arbr. Arbust. 1: xxviii, 139. 1755 [*Gymnosp.: Cupress.*]. Typus: non designatus.

Cunninghamia R. Br. in Richard, Comm. Bot. Conif. Cycad.: 80, 149. Sep-Nov 1826 [*Pin.*]. Typus: *C. sinensis* R. Br., nom. illeg. (*Pinus lanceolata* Lamb., *C. lanceolata* (Lamb.) Hook.).

(H) *Cunninghamia* Schreb., Gen. Pl.: 789. Mai 1791 [*Dicot.: Rub.*]. ≡ *Malanea* Aubl. 1775.

(≡) *Belis* Salisb. in Trans. Linn. Soc. London 8: 315. 9 Mar 1807.

Dioon Lindl. in Edwards's Bot. Reg. 29 (Misc.): 59. Aug 1843 (*'Dion'*) (orth. cons.) [*Cycad.*]. Typus: *D. edule* Lindl.

Fitzroya Hook. ex Lindl., J. Hort. Soc. London 6: 264. 1 Oct 1851 (*'Fitz-Roy'*) (orth. cons.) [*Pin.*]. Typus: *F. patagonica* Hook. f. ex Lindl.

Metasequoia Hu & W. C. Cheng in Bull. Fan Mem. Inst. Biol., Bot., ser. 2, 1: 154. 15 Mai 1948 [*Pin.*]. Typus: *M. glyptostroboides* Hu & W. C. Cheng (typ. cons.).

(H) *Metasequoia* Miki in Jap. J. Bot. 11: 261. 1941 (post Mar) [*Foss.*]. Typus: *M. disticha* (Heer) Miki (*Sequoia disticha* Heer).

249

Phyllocladus Rich. & Mirb. in Mém. Mus. Hist. Nat. 13: 48. 1825 [*Tax.*]. Typus: *P. billardierei* Mirb., nom. illeg. (*Podocarpus aspleniifolius* Labill., *Phyllocladus aspleniifolius* (Labill.) Hook. f.) (etiam vide *Podocarpus* [*Gymnosp.*]).

Podocarpus Pers., Syn. Pl. 2: 580. Sep 1807 [*Tax.*]. Typus: *P. elongatus* (Aiton) L'Hér. ex Pers. (*Taxus elongata* Aiton) (typ. cons.).

(H)   *Podocarpus* Labill., Nov. Holl. Pl. 2: 71. Aug 1806 [*Gymnosp.: Podocarp.*].
≡ *Phyllocladus* Rich. & Mirb. 1825 (nom. cons.).

(=)   *Nageia* Gaertn., Fruct. Sem. Pl. 1: 191. Dec 1788.
Typus: *N. japonica* Gaertn., nom. illeg. (*Myrica nagi* Thunb.).

Pseudolarix Gordon, Pinetum: 292. Jun-Dec 1858 [*Pin.*]. Typus: [specimen cult. in Anglia] ex Herb. George Gordon (K No. 3455) (typ. cons.) [= *P. amabilis* (J. Nelson) Rehder (*Larix amabilis* J. Nelson)] .

Saxegothaea Lindl., J. Hort. Soc. London 6: 258. 1 Oct 1751 (*'Saxe-Gothaea'*) (orth. cons.) [*Tax.*]. Typus: *S. conspicua* Lindl.

Sequoia Endl., Syn. Conif.: 197. Mai-Jun 1847 [*Pin.*]. Typus: *S. sempervirens* (D. Don) Endl. (*Taxodium sempervirens* D. Don).

Thujopsis Siebold & Zucc. ex Endl., Gen. Pl., Suppl. 2: 24. Mar Jun 1842 [*Cupress.*]. Typus: *T. dolabrata* (L. f.) Siebold & Zucc. (Fl. Jap. 2: 34. Apr-Aug 1844) (*Thuja dolabrata* L. f.).

(≡)   *Dolophyllum* Salisb. in J. Sci. Arts (London) 2: 313. 1817.

Torreya Arn. in Ann. Nat. Hist. 1: 130. Apr 1838 [*Tax.*].
Typus: *T. taxifolia* Arn.

(H)   *Torreya* Raf. in Amer. Monthly Mag. & Crit. Rev. 3: 356. Sep 1818 [*Dicot.: Lab.*].
Typus: *T. grandiflora* Raf.

Welwitschia Hook. f. in Gard. Chron. 1862: 71. 25 Jan 1862 [*Gnet.*].
Typus: *W. mirabilis* Hook. f.

(H)   *Welwitschia* Rchb., Handb. Nat. Pfl.-Syst.: 194. 1-7 Oct 1837 [*Dicot.: Polemon.*].
  ≡ *Eriastrum* Wooton & Standl. 1913.

(≡)   *Tumboa* Welw. in Gard. Chron. 1861: 75. Jan 1861.

Zamia L., Sp. Pl., ed. 2: 1659. Jul-Aug 1763 [*Cycad.*].
Typus: *Z. pumila* L.

(≡)   *Palma-filix* Adans., Fam. Pl. 2: 21, 587. Jul-Aug 1763 (typ. des.: Florin in Taxon 5: 189. 1956).

E2. MONOCOTYLEDONES

Acampe Lindl., Fol. Orchid. 4, *Acampe:* 1. 20 Aug 1853 [*Orchid.*].
Typus: *A. multiflora* (Lindl.) Lindl. (*Vanda multiflora* Lindl.).

(=)   *Sarcanthus* Lindl. in Bot. Reg. 10: ad t. 817. 1 Aug 1824.
Typus: *Epidendrum praemorsum* Roxb.

Aechmea Ruiz & Pav., Fl. Peruv. Prodr.: 47. Oct (prim.) 1794 [*Bromel.*].
Typus: *A. paniculata* Ruiz & Pav. (Fl. Peruv. 3: 37. Aug 1802).

(=)   *Hoiriri* Adans., Fam. Pl. 2: 67, 584. Jul-Aug 1763.
Typus: *Bromelia nudicaulis* L.

Aegilops L., Sp. Pl.: 1050. 1 Mai 1753 [*Gram.*].
Typus: *A. truncialis* L. (typ. cons.: [specimen] Herb. Linnaeus No. 1218.8 (LINN)).

Agapanthus L'Hér., Sert. Angl.: 17. Jan (prim.) 1789 [*Lil.*].
Typus: *A. umbellatus* L'Hér., nom. illeg. (*Crinum africanum* L., *A. africanus* (L.) Hoffmanns.) (etiam vide *Tulbaghia* [*Monocot.*]).

(≡)   *Abumon* Adans., Fam. Pl. 2: 54, 511. Jul-Aug 1763.

Agrostis L., Sp. Pl.: 61. 1 Mai 1753
[*Gram.*].
Typus: *A. canina* L. (typ. cons.: [spe-
cimen] Herb. Burser 1: 3, UPS).

Aira L., Sp. Pl.: 63. 1 Mai 1753
[*Gram.*].
Typus: *A. praecox* L. (typ. cons.: [spe-
cimen] Herb. Linnaeus No. 85.21
(LINN)).

Alocasia (Schott) G. Don in Sweet,
Hort. Brit., ed. 3: 631. 1839 (sero)
(*Colocasia* sect. *Alocasia* Schott in
Schott & Endlicher, Melet. Bot.: 18.
1832) [*Ar.*].
Typus: *Colocasia cucullata* (Lour.)
Schott (*Arum cucullatum* Lour., *Alo-
casia cucullata* (Lour.) G. Don) (typ.
cons.).

(H) *Alocasia* Neck. ex Raf., Fl. Tellur.
3: 64. Nov-Dec 1837 [*Monocot.:
Ar.*].
Typus: non designatus.

Alpinia Roxb. in Asiat. Res. 11:
350. 1810 [*Zingiber.*].
Typus: *A. galanga* (L.) Willd. (Sp.
Pl. 1: 12. Jun 1797) (*Maranta galanga* L.) (typ. cons.).

(H) *Alpinia* L., Sp. Pl.: 2. 1 Mai 1753
[*Monocot.: Zingiber.*].
Typus: *A. racemosa* L.

(=) *Albina* Giseke, Prael. Ord. Nat. Pl.:
207, 227, 248. Apr 1792.
Typus: non designatus.

(=) *Buekia* Giseke, Prael. Ord. Nat. Pl.:
204, 216, 239. Apr 1792.
Typus: *B. malaccensis* (J. König)
Raeusch. (Nomencl. Bot. 1. 1797)
(*Costus malaccensis* J. König).

(=) *Zerumbet* J. C. Wendl., Sert. Han-
nov. 4: 3. Apr-Mai 1798.
Typus: *Z. speciosum* J. C. Wendl.

Amaryllis L., Sp. Pl.: 292. 1 Mai
1753 [*Amaryllid.*].
Typus: *A. belladonna* L. (typ. cons.:
[specimen] Herb. Clifford: 135, *Amaryllis* 2 (BM)).

Amianthium A. Gray in Ann. Lyceum Nat. Hist. New York 4: 121. Nov 1837 [*Lil.*].
Typus: *A. muscitoxicum* (Walter) A. Gray (*Melanthium muscitoxicum* Walter) (typ. cons.).

(=) *Chrosperma* Raf., Neogenyton: 3. 1825.
Typus: *Melanthium laetum* Aiton

Amomum Roxb., Pl. Coromandel 3: 75. 18 Feb 1820 [*Zingiber.*].
Typus: *A. subulatum* Roxb. (typ. cons.).

(H) *Amomum* L., Sp. Pl.: 1. 1 Mai 1753 (typ. des.: Burtt & Smith in Taxon 17: 730. 1968) [*Monocot.: Zingiber.*].
≡ *Zingiber* Mill. 1754 (nom. cons.).

(=) *Etlingera* Giseke, Prael. Ord. Nat. Pl.: 209. Apr 1792.
Typus: *E. littoralis* (J. König) Raeusch. (Nomencl. Bot.: 1. 1797) (*Amomum littorale* J. König).

(=) *Meistera* Giseke, Prael. Ord. Nat. Pl.: 205. Apr 1792.
Typus: *Amomum koenigii* J. F. Gmel.

(=) *Paludana* Giseke, Prael. Ord. Nat. Pl.: 207. Apr 1792.
Typus: *Amomum globba* J. F. Gmel.

(=) *Wurfbainia* Giseke, Prael. Ord. Nat. Pl.: 206. Apr 1792.
Typus: *W. uliginosa* (K. D. Koenig) Giseke (*Amomum uliginosum* K. D. Koenig).

Amorphophallus Blume ex Decne. in Nouv. Ann. Mus. Hist. Nat. 3: 366. 1834 [*Ar.*].
Typus: *A. campanulatus* Decne.

(=) *Thomsonia* Wall., Pl. Asiat. Rar. 1: 83. 1 Sep 1830.
Typus: *T. napalensis* Wall.

(=) *Pythion* Mart. in Flora 14: 458. 1831 (med.).
Typus: *Arum campanulatum* Roxb., nom. illeg. (*Dracontium paeoniifolium* Dennst., *Amorphophallus paeoniifolius* (Dennst.) Nicolson).

Andropogon L., Sp. Pl.: 63. 1 Mai 1753 [*Gram.*].
Typus: *A. distachyos* L. ('*distachyon*') (typ. cons.: [specimen] Herb. Burser 1: 120, UPS).

Anguillaria R. Br., Prodr.: 273. 27 Mar 1810 [*Lil.*].
Typus: *A. dioica* R. Br. (typ. cons.).

(H) *Anguillaria* Gaertn., Fruct. Sem. Pl. 1: 372. Dec 1788 (typ. des.: Rickett & Stafleu in Taxon 8: 234. 1959) [*Dicot.: Myrsin.*].
≡ *Heberdenia* Banks ex DC. 1841 (nom. cons.).

Anoectochilus Blume, Bijdr.: 411. 20 Sep - 7 Dec 1825 (*'Anecochilus'*) (orth. cons.) [*Orchid.*].
Typus: *A. setaceus* Blume

Aponogeton L. f., Suppl. Pl.: 32, 214. Apr 1782 [*Aponogeton.*].
Typus: *A. monostachyos* L. f., nom. illeg. (*Saururus natans* L., *A. natans* (L.) Engl. & Krause).

(H) *Aponogeton* Hill, Brit. Herb.: 480. Dec 1756 [*Monocot.: Potamogeton.*].
≡ *Zannichellia* L. 1753.

Arachnitis Phil. in Bot. Zeitung (Berlin) 22: 217. 15 Jul 1864 [*Burmann.*].
Typus: *A. uniflora* Phil.

(H) *Arachnites* F. W. Schmidt, Fl. Boëm. 1: 74. 9 Apr - 7 Oct 1793 [*Monocot.: Orchid.*].
Typus: non designatus.

Arenga Labill. in Bull. Sci. Soc. Philom. Paris 2: 162. Nov (sero) 1800 [*Palm.*].
Typus: *A. saccharifera* Labill. [= *A. pinnata* (Wurmb) Merr. (*Saguerus pinnatus* Wurmb)].

(=) *Saguerus* Steck, Sagu: 15. 21 Sep 1757.
Typus: *S. pinnatus* Wurmb (in Verh. Batav. Genootsch. Kunsten 1: 351. 1781).

Ascolepis Nees ex Steud., Syn. Pl. Glumac. 2: 105. 10-11 Apr 1855 [*Cyper.*].
Typus: *A. eriocauloides* (Steud.) Nees ex Steud. (*Kyllinga eriocauloides* Steud.) (typ. cons.).

Asplundia Harling in Acta Horti Berg. 17: 41. 1954 (post 3 Nov) [*Cyclanth.*].
Typus: *A. latifolia* (Ruiz & Pav.) Harling (*Carludovica latifolia* Ruiz & Pav.).

(=) *Sarcinanthus* Oersted in Vidensk. Meddel. Dansk Naturhist. Foren. Kjøbenhavn 1857: 196. 1857.
Typus: *S. utilis* Oersted

Astelia Banks & Sol. ex R. Br., (=) *Funckia* Willd. in Ges. Naturf. Freun-
Prodr.: 291. 27 Mar 1810 [*Lil.*].   de Berlin Mag. Neuesten Entdeck.
Typus: *A. alpina* R. Br.   Gesammten Naturk. 2: 19. 1808.
                                           Typus: *F. magellanica* Willd., nom. il-
                                           leg. (*Melanthium pumilum* G. Forst.).

Astrocaryum G. Mey., Prim. Fl. Es- (=) *Avoira* Giseke, Prael. Ord. Nat. Pl.:
seq.: 265. Nov 1818 [*Palm.*].   38, 53. Apr 1792.
Typus: *A. aculeatum* G. Mey.   Typus: *A. vulgaris* Giseke

Babiana Ker Gawl. ex Sims in Bot. (=) *Beverna* Adans., Fam. Pl. 2: (20).
Mag.: ad t. 539. 1 Nov 1801 [*Irid.*].   Jul-Aug 1763.
Typus: *B. plicata* Ker Gawl. (in Bot.   Typus: non designatus.
Mag.: ad t. 576. 1 Aug 1802), nom.
illeg. (*Gladiolus fragrans* Jacq., *B.
fragrans* (Jacq.) Eckl.).

Bambusa Schreb., Gen. Pl.: 236. (≡) *Bambos* Retz., Observ. Bot. 5: 24.
Apr 1789 [*Gram.*].   Sep 1788.
Typus: *B. arundinacea* (Retz.) Willd.
(Sp. Pl. 2: 245. Mar 1799) (*Bambos
arundinacea* Retz.).

Baxteria R. Br. in London J. Bot. 2: (H) *Baxtera* Rchb., Consp. Regni Veg.:
494. 1843 [*Lil.*].   131. Dec 1828 - Mar 1829 [*Dicot.:
Typus: *B. australis* R. Br.   Asclepiad.*].
                                           ≡ *Harrisonia* Hook. 1826, non R.
                                           Br. ex Juss. 1825 (nom. cons.).

Belamcanda Adans., Fam. Pl. 2: 60
(*'Belam-canda'*), 524 (*'Belamkan-
da'*). Jul-Aug 1763 (orth. cons.)
[*Irid.*].
Typus: *B. chinensis* (L.) DC. (in Re-
douté, Liliac. 3: ad t. 121. Jul 1805)
(*Ixia chinensis* L.) (typ. cons.).

Bellevalia Lapeyr. in J. Phys. Chim. (H) *Bellevalia* Scop., Intr. Hist. Nat.: 198.
Hist. Nat. Arts 67: 425. Dec 1808   Jan-Apr 1777 [*Dicot.: Verben.*].
[*Lil.*].   ≡ *Marurang* Rumph. ex Adans.
Typus: *B. operculata* Lapeyr.   1763.

Bessera Schult. f. in Linnaea 4: 121. (H) *Bessera* Schult., Observ. Bot.: 27.
Jan 1829 [*Lil.*].   1809 [*Dicot.: Boragin.*].
Typus: *B. elegans* Schult. f.   Typus: *B. azurea* Schult.

Biarum Schott in Schott & End-licher, Melet. Bot.: 17. 1832 [*Ar.*].
Typus: *B. tenuifolium* (L.) Schott (*Arum tenuifolium* L.).

(≡) *Homaid* Adans., Fam. Pl. 2: 470, 584. Jul-Aug 1763.

Blandfordia Sm., Exot. Bot. 1: 5. 1 Dec 1804 [*Lil.*].
Typus: *B. nobilis* Sm.

(H) *Blandfordia* Andrews in Bot. Re-pos.: ad t. 343. 9 Feb 1804 [*Dicot.: Diapens.*].
Typus: *B. cordata* Andrews

Bletilla Rchb. f. in Fl. Serres Jard. Eur. 8: 246. 5 Oct 1853 [*Orchid.*].
Typus: *B. gebinae* (Lindl.) Rchb. f. (*Bletia gebinae* Lindl.) (typ. cons.).

(=) *Jimensia* Raf., Fl. Tellur. 4: 38. 1838 (med.).
Typus: *J. nervosa* Raf., nom. illeg. (*Limodorum striatum* Thunb., *J. stri-ata* (Thunb.) Garay & R. E. Schult.).

Blysmus Panz. ex Schult., Mant. 2: 41. Jan-Apr 1824 [*Cyper.*].
Typus: *B. compressus* (L.) Panz. ex Link (Hort. Berol. 1: 278. 1 Oct - 27 Nov 1827) (*Schoenus compres-sus* L.).

(≡) *Nomochloa* P. Beauv. ex T. Lestib., Essai Cypér.: 37. 29 Mar 1819.

Bobartia L., Sp. Pl.: 54. 1 Mai 1753 [*Irid.*].
Typus: *B. indica* L. (typ. cons.: [spe-cimen sup. laev.] Herb. Hermann 4: 80 (BM)).

Boophone Herb., Appendix.: 18. 1821 ('*Boophane*') (orth. cons.) [*Am-aryllid.*].
Typus: *B. toxicaria* (Aiton) Herb, nom. illeg. (*Haemanthus toxicarius* Aiton, nom. illeg.); *B. disticha* (L.f.) Herb. (*Amaryllis disticha* L.f.).

Bouteloua Lag. in Varied. Ci. 2(4): 134. 1805 *('Botelua')* (orth. cons.) [*Gram.*].
Typus: *B. racemosa* Lag.

Bowiea Harv. ex Hook. f. in Bot. Mag.: ad t. 5619. 1 Jan 1867 [*Lil.*].
Typus: *B. volubilis* Harv. ex Hook. f.

(H) *Bowiea* Haw. in Philos. Mag. J. 64: 299. 1824 [*Monocot.: Lil.*].
Typus: *B. africana* Haw.

Brachtia Rchb. f. in Linnaea 22: 853. Mai 1850 [*Orchid.*].
Typus: *B. glumacea* Rchb. f.

(H) *Brachtia* Trevis., Sagg. Algh. Coccot.: 57. 1848 [*Chloroph.*].
Typus: *B. crassa* (Naccari) Trevis. (*Palmella crassa* Naccari).

Brassavola R. Br. in Aiton, Hort. Kew., ed. 2, 5: 216. Nov 1813 [*Orchid.*].
Typus: *B. cucullata* (L.) R. Br. (*Epidendrum cucullatum* L.).

(H) *Brassavola* Adans., Fam. Pl. 2: 127, 527. Jul-Aug 1763 [*Dicot.: Comp.*].
≡ *Helenium* L. 1753.

Brodiaea Sm. in Trans. Linn. Soc. London 10: 2. Feb 1810 [*Lil.*].
Typus: *B. grandiflora* Sm., nom. illeg. (*Hookera coronaria* Salisb., *B. coronaria* (Salisb.) Jeps.) (typ. cons.) (etiam vide *Hookeria* [*Musci*]).

Bromus L., Sp. Pl.: 76. 1 Mai 1753 [*Gram.*].
Typus: *B. secalinus* L. (typ. cons.: [specimen] Herb. Linnaeus No. 93.1 (LINN)).

Buchloë Engelm. in Trans. Acad. Sci. St. Louis 1: 432. Jan-Apr 1859 [*Gram.*].
Typus: *B. dactyloides* (Nutt.) Engelm. (*Sesleria dactyloides* Nutt.).

Bulbine Wolf, Gen. Pl.: 84. 1776 [*Lil.*].
Typus: *B. frutescens* (L.) Willd. (Enum. Hort. Berol.: 372. Apr 1809) (*Anthericum frutescens* L.).

Bulbophyllum Thouars, Hist. Orchid., Tabl. Esp.: 3. 1822 [*Orchid.*].
Typus: B. nutans Thouars (typ. cons.).

(≡) *Phyllorkis* Thouars in Nouv. Bull. Sci. Soc. Philom. Paris 1: 319. Apr 1809.

Bulbostylis Kunth, Enum. Pl. 2: 205. 6 Mai 1837 [*Cyper.*].
Typus: *B. capillaris* (L.) Kunth ex C. B. Clarke (in Hooker, Fl. Brit. India 6: 652. Sep 1893) (*Scirpus capillaris* L.) (typ. cons.).

(H) *Bulbostylis* Steven in Mém. Soc. Imp. Naturalistes Moscou 5: 355. 1817 [*Monocot.: Cyper.*].
Typus: non designatus.

(=) *Stenophyllus* Raf., Neogenyton: 4. 1825.
Typus: *S. cespitosus* Raf. (*Scirpus stenophyllus* Elliott).

Burchardia R. Br., Prodr.: 272. 27 Mar 1810 [*Lil.*].
Typus: *B. umbellata* R. Br.

(H) *Burcardia* Heist. ex Duhamel, Traité Arbr. Arbust. 1: xxx, 11. 1755 [*Dicot.: Verben.*].
≡ *Callicarpa* L. 1753.

Calanthe R. Br. in Bot. Reg.: ad t. 573 ('578'). 1 Oct 1821 [*Orchid.*].
Typus: *C. veratrifolia* Ker Gawl. (in Bot. Reg.: ad t. 720. 1 Jul 1823), nom. illeg. (*Orchis triplicata* Willemet, *C. triplicata* (Willemet) Ames).

(=) *Alismorkis* Thouars in Nouv. Bull. Sci. Soc. Philom. Paris 1: 318. Apr 1809.
Typus: non designatus.

Calopogon R. Br. in Aiton, Hort. Kew., ed. 2, 5: 204. Nov 1813 [*Orchid.*].
Typus: *C. pulchellus* R. Br., nom. illeg. (*Limodorum pulchellum* Salisb., nom. illeg., *Limodorum tuberosum* L., *C. tuberosus* (L.) Britton & al.) (etiam vide *Limodorum* [*Monocot.*]).

Calypso Salisb., Parad. Lond.: ad t. 89. 1 Dec 1807 [*Orchid.*].
Typus: *C. borealis* Salisb., nom. illeg. (*Cypripedium bulbosum* L., *Calypso bulbosa* (L.) Oakes).

(H) *Calypso* Thouars, Hist. Vég. Iles France: 29. 1804 (ante 22 Sep) [*Dicot.: Celastr.*].
Typus (vide Stafleu in Index Nom. Gen.: No. 30238. 1970): *C. oppositifolia* J. St.-Hil.

Camassia Lindl. in Edwards's Bot. Reg.: ad t. 1486. 1 Apr 1832 [*Lil.*].
Typus: *C. esculenta* Lindl., nom. illeg. (*Phalangium quamash* Pursh, *C. quamash* (Pursh) Greene).

(=) *Cyanotris* Raf. in Amer. Monthly Mag. & Crit. Rev. 3: 356. Sep 1818.
Typus: *C. scilloides* Raf.

Centotheca Desv. in Nouv. Bull.
Sci. Soc. Philom. Paris 2: 189. Dec
1810 *('Centosteca')* (orth. cons.)
[*Gram.*].
Typus: *C. lappacea* (L.) Desv. (*Cenchrus lappaceus* L.).

Chamaedorea Willd., Sp. Pl. 4: 638,
800. Apr 1806 [*Palm.*].
Typus: *C. gracilis* Willd., nom. illeg. (*Borassus pinnatifrons* Jacq., *C. pinnatifrons* (Jacq.) Oerst.).

(=) *Morenia* Ruiz & Pav., Fl. Peruv.
Prodr.: 150. Oct (prim.) 1794.
Typus: *M. fragrans* Ruiz & Pav.
(Syst. Veg. Fl. Peruv. Chil.: 299.
Dec 1798).

(=) *Nunnezharia* Ruiz & Pav., Fl. Peruv. Prodr.: 147. Oct (prim.) 1794.
Typus: *N. fragrans* Ruiz & Pav.
(Syst. Veg. Fl. Peruv. Chil.: 294.
Dec 1798).

Chionographis Maxim. in Bull. Acad. Imp. Sci. Saint-Pétersbourg 11:
435. 31 Mai 1867 [*Lil.*].
Typus: *C. japonica* (Willd.) Maxim.
(*Melanthium japonicum* Willd.).

(=) *Siraitos* Raf., Fl. Tellur. 4: 26. 1838
(med.).
Typus: *S. aquaticus* Raf.

Chlorogalum Kunth, Enum. Pl. 4:
681. 17-19 Jul 1843 [*Lil.*].
Typus: *C. pomeridianum* (DC.) Kunth
(*Scilla pomeridiana* DC.) (typ. cons.).

(≡) *Laothoë* Raf., Fl. Tellur. 3: 53. Nov-Dec 1837.

Chrysopogon Trin., Fund. Agrost.:
187. Jan 1820 [*Gram.*].
Typus: *C. gryllus* (L.) Trin. (*Andropogon gryllus* L.).

(≡) *Pollinia* Spreng., Pl. Min. Cogn.
Pug. 2: 10. 1815.

(=) *Rhaphis* Lour., Fl. Cochinch.: 538,
552. Sep 1790.
Typus: *R. trivialis* Lour.

(=) *Centrophorum* Trin., Fund. Agrost.:
106. Jan 1820.
Typus: *C. chinense* Trin.

Cirrhopetalum Lindl., Gen. Sp. Orchid. Pl.: 45, 58. Mai 1830 [*Orchid.*].
Typus: *C. thouarsii* Lindl., nom. illeg. (*Epidendrum umbellatum* G.
Forst., *C. umbellatum* (G. Forst.)
Frappier ex Cordemoy).

(=) *Ephippium* Blume, Bijdr.: 308. 20
Sep - 7 Dec 1825.
Typus: non designatus.

(=) *Zygoglossum* Reinw. in Syll. Pl. Nov.
2: 4. 1825.
Typus: *Z. umbellatum* Reinw.

Claderia Hook. f., Fl. Brit. India 5: 810. Apr 1890 [*Orchid.*].
Typus: *C. viridiflora* Hook. f.

(H)   *Claderia* Raf., Sylva Tellur.: 12. Oct-Dec 1838 [*Dicot.: Rut.*].
Typus: *C. parviflora* Raf.

Coleanthus Seidl ['Seidel'] in Roemer & Schultes, Syst. Veg. 2: 11, 276. Nov 1817 [*Gram.*].
Typus: *C. subtilis* (Tratt.) Seidl (*Schmidtia subtilis* Tratt.).

Colocasia Schott in Schott & Endlicher, Melet. Bot.: 18. 1832 [*Ar.*].
Typus: *C. antiquorum* Schott (*Arum colocasia* L.) (typ. cons.).

(H)   *Colocasia* Link, Diss. Bot.: 77. 1795 [*Monocot.: Ar.*].
≡ *Zantedeschia* Spreng. 1826 (nom. cons.).

Corallorhiza Gagnebin in Acta Helv. Phys.-Math. 2: 61. Feb 1755 (*'Corallorrhiza'*) (orth. cons.) [*Orchid.*].
Typus: *C. trifida* Châtel. (Specim. Inaug. Corallorhiza: 8. 13 Mai 1760) (*Ophrys corallorhiza* L.).

Cordyline Comm. ex R. Br., Prodr.: 280. 27 Mar 1810 [*Lil.*].
Typus: *C. cannifolia* R. Br. (typ. cons.).

(H)   *Cordyline* Adans., Fam. Pl. 2: 54, 543. Jul-Aug 1763 [*Monocot.: Lil.*].
≡ *Sansevieria* Thunb. 1794 (nom. cons.).

(=)   *Taetsia* Medik., Theodora: 82. 1786.
Typus: *T. ferrea* Medik., nom. illeg. (*Dracaena ferrea* L., nom. illeg., *Convallaria fruticosa* L.).

Cortaderia Stapf in Gard. Chron., ser. 3, 22: 378. 27 Nov 1897 [*Gram.*].
Typus: *C. argentea* Stapf, nom. illeg. (*Gynerium argenteum* Nees, nom. illeg., *Arundo dioica* Spreng. 1825, non Lour. 1790, *Arundo selloana* Schult. & Schult. f., *C. selloana* (Schult. & Schult. f.) Asch. & Graebn.).

(≡)   *Moorea* Lem. in Ill. Hort. 2: 14. 2 Feb 1855.

Corynephorus P. Beauv., Ess. Agrostogr.: 90, 159. Dec 1812 [*Gram.*].
Typus: *C. canescens* (L.) P. Beauv. (*Aira canescens* L.).

(≡)   *Weingaertneria* Bernh., Syst. Verz.: 23, 51. 1800.

Crypsis Aiton, Hort. Kew. 1: 48. 7 Aug - 1 Oct 1789 [*Gram.*]. Typus: *C. aculeata* (L.) Aiton (*Schoenus aculeatus* L.).

Cryptanthus Otto & A. Dietr. in Allg. Gartenzeitung 4: 297. 1836 [*Bromel.*]. Typus: *C. bromelioides* Otto & A. Dietr.

(H)   *Cryptanthus* Osbeck, Dagb. Ostind. Resa: 215. 1757 [*Spermatoph.*]. Typus: *C. chinensis* Osbeck

Ctenium Panz., Ideen Rev. Gräser: 38, 61. 1813 [*Gram.*]. Typus: *C. carolinianum* Panz., nom. illeg. (*Chloris monostachya* Michx.) [= *Ctenium aromaticum* (Walter) Wood (*Aegilops aromatica* Walter)].

(≡)   *Campulosus* Desv. in Nouv. Bull. Sci. Soc. Philom. Paris 2: 189. Dec 1810.

Culcasia P. Beauv., Fl. Oware, ed. 4°: 4. 2 Oct 1803 [*Ar.*]. Typus: *C. scandens* P. Beauv. (typ. cons.).

Curcuma L., Sp. Pl.: 2. 1 Mai 1753 [*Zingiber.*]. Typus: *C. longa* L. (typ. cons.).

Cyanotis D. Don, Prodr. Fl. Nepal.: 45. 26 Jan - 1 Feb 1825 [*Commelin.*]. Typus: *C. barbata* D. Don

Cymodocea K. D. Koenig in Ann. Bot. (König & Sims) 2: 96. 1 Jun 1805 [*Potamogeton.*]. Typus: *C. aequorea* K. D. Koenig

(=)   *Phucagrostis* Cavolini, Phucagr. Theophr. Anth.: xiii. 1792. Typus: *P. major* Cavolini

Cynodon Rich. in Persoon, Syn. Pl. 1: 85. 1 Apr - 15 Jun 1805 [*Gram.*]. Typus: *C. dactylon* (L.) Pers. (*Panicum dactylon* L.).

(=)   *Dactilon* Vill., Hist. Pl. Dauphiné 2: 69. Feb 1787.

(=)   *Capriola* Adans., Fam. Pl. 2: 31, 532. Jul-Aug 1763. Typus: non designatus.

Cyrtanthus Aiton, Hort. Kew. 1: 414. 7 Aug - 1 Oct 1789 [*Amaryllid.*].
Typus: *C. angustifolius* (L. f.) Aiton (*Crinum angustifolium* L. f.).

(H) *Cyrtanthus* Schreb., Gen. Pl.: 122. Apr 1789 [*Dicot.: Rub.*].
≡ *Posqueria* Aubl. 1775.

Dactylorhiza Necker ex Nevski in Fl. URSS 4: 697, 713. 1935 [*Orchid.*].
Typus: *D. umbrosa* (Kar. & Kir.) Nevski (*Orchis umbrosa* Kar. & Kir.).

(=) *Coeloglossum* Hartm., Handb. Skand. Fl.: 329. 1820.
Typus: *D. viridis* (L.) R. M. Bateman & al. (*Satyrium viride* L.).

Danthonia DC. in Lamarck & Candolle, Fl. Franç., ed. 3, 3: 32. 17 Sep 1805 [*Gram.*].
Typus: *D. spicata* (L.) Roem. & Schult. (Syst. Veg. 2: 690. Nov 1817) (*Avena spicata* L.) (typ. cons.).

(=) *Sieglingia* Bernh., Syst. Verz.: 44. 1800.
Typus: *S. decumbens* (L.) Bernh. (*Festuca decumbens* L.).

Dendrobium Sw. in Nova Acta Regiae Soc. Sci. Upsal., ser. 2, 6: 82. 1799 [*Orchid.*].
Typus: *D. moniliforme* (L.) Sw. (*Epidendrum moniliforme* L.) (typ. cons.).

(=) *Callista* Lour., Fl. Cochinch.: 516, 519. Sep 1790.
Typus: *C. amabilis* Lour.

(=) *Ceraia* Lour., Fl. Cochinch.: 518. Sep 1790.
Typus: *C. simplicissima* Lour.

Desmoncus Mart., Palm. Fam.: 20. 13 Apr 1824 [*Palm.*].
Typus: *D. polyacanthos* Mart. (Hist. Nat. Palm. 2: 85. 1824, serius) (typ. cons.).

Diarrhena P. Beauv., Ess. Agrostogr.: 142, 160, 162. Dec 1812 [*Gram.*].
Typus: *D. americana* P. Beauv. (*Festuca diandra* Michx. 1803, non Moench 1794).

Dichorisandra J. C. Mikan, Del. Fl. Faun. Bras.: ad t. 3. 1820 (sero) [*Commelin.*].
Typus: *D. thyrsiflora* J. C. Mikan

Diectomis Kunth in Mém. Mus. Hist. Nat. 2: 69.1815 [*Gram.*].
Typus: *D. fastigiata* (Sw.) P. Beauv. (Ess. Agrostogr.: 132, 160. Dec 1812) (*Andropogon fastigiatus* Sw.) (typ. cons.).

(H) *Diectomis* P. Beauv., Ess. Agrostogr.: 132. Dec 1812 [*Monocot.: Gram.*].
Typus: *D. fasciculata* P. Beauv.

Dietes Salisb. ex Klatt in Linnaea 34: 583. Feb 1866 [*Irid.*].
Typus: *D. compressa* (L. f.) Klatt (*Iris compressa* L. f.) [= *D. iridioides* (L.) Klatt (*Moraea iridioides* L.)].

(=) *Naron* Medik. in Hist. & Commentat. Acad. Elect. Sci. Theod.-Palat. 6: 419. Apr-Jun 1790.
Typus: *N. orientale* Medik. nom. illeg. (*Moraea iridioides* L., *N. iridioides* (L.) Moench).

Digitaria Haller, Hist. Stirp. Helv. 2: 244. 25 Mar 1768 [*Gram.*].
Typus: *D. sanguinalis* (L.) Scop. (Fl. Carniol., ed. 2, 1: 52. 1771) (*Panicum sanguinale* L.) (typ. cons.).

(H) *Digitaria* Heist. ex Fabr., Enum.: 207. 1759 [*Monocot.: Gram.*].
Typus: non designatus.

Dioscorea L., Sp. Pl.: 1032. 1 Mai 1753 [*Dioscor.*].
Typus: *D. sativa* L. (typ. cons.: [icon in] Linnaeus, Hort. Cliff. t. 28 [stem & lvs]. 1738).

Echinaria Desf., Fl. Atlant. 2: 385. Feb-Jul 1799 [*Gram.*].
Typus: *E. capitata* (L.) Desf. (*Cenchrus capitatus* L.).

(H) *Echinaria* Fabr., Enum.: 206. 1759 [*Monocot.: Gram.*].
≡ *Cenchrus* L. 1753.

(≡) *Panicastrella* Moench, Methodus: 205. 4 Mai 1794.

Echinochloa P. Beauv., Ess. Agrostogr.: 53. Dec 1812 [*Gram.*].
Typus: *E. crusgalli* (L.) P. Beauv. (*Panicum crusgalli* L.).

(≡) *Tema* Adans., Fam. Pl. 2: 496, 610. Jul-Aug 1763.

Ehrharta Thunb. in Kongl. Vetensk. Acad. Handl. 40: 217. Jul-Dec 1779 [*Gram.*].
Typus: *E. capensis* Thunb.

(=) *Trochera* Rich. in Observ. Mém. Phys. 13: 225. Mar 1779.
Typus: *T. striata* Rich.

Eichhornia Kunth, Eichhornia: 3. 1842 [*Ponteder.*].
Typus: *E. azurea* (Sw.) Kunth (*Pontederia azurea* Sw.) (typ. cons.).

(≡) *Piaropus* Raf., Fl. Tellur. 2: 81. Jan-Mar 1837 (typ. des.: Britton, Fl. Bermuda: 64. 1918).

Eleutherine Herb. in Edwards's Bot. Reg. 29: ad t. 57. 1 Nov 1843 [*Irid.*]. Typus: *Marica plicata* Ker Gawl., nom. illeg. (*Moraea plicata* Sw., nom. illeg., *Sisyrinchium latifolium* Sw.) [= *E. bulbosa* (Mill.) Urb. (*Sisyrinchium bulbosum* Mill.)].

Epidendrum L., Sp. Pl., ed. 2: 1347. Jul-Aug 1763 [*Orchid.*]. Typus: *E. nocturnum* Jacq. (Enum. Syst. Pl.: 29. Aug-Sep 1760) (typ. cons.).

(H) *Epidendrum* L., Sp. Pl.: 952. 1 Mai 1753 [*Monocot.: Orchid.*]. Typus (vide Britton & Wilson in Sci. Surv. Porto Rico & Virgin Islands 5: 203. 1924): *E. nodosum* L.

Epipactis Zinn, Cat. Pl. Hort. Gott.: 85. 20 Apr - 21 Mai 1757 [*Orchid.*]. Typus: *E. helleborine* (L.) Crantz (Stirp. Austr. Fasc., ed. 2: 467. Jan-Jul 1769) (*Serapias helleborine* L.) (typ. cons.).

(H) *Epipactis* Ség., Pl. Veron. 3: 253. Jul-Aug 1754 [*Monocot.: Orchid.*]. Typus: *Satyrium repens* L.

(≡) *Helleborine* Mill., Gard. Dict. Abr., ed. 4: [622]. 28 Jan 1754.

Eria Lindl. in Bot. Reg.: ad t. 904. 1 Aug 1825 [*Orchid.*]. Typus: *E. stellata* Lindl.

Eucharis Planch. & Linden in Linden, Cat. Pl. Exot. 8: 3. 1853 [*Amaryllid.*]. Typus: *E. candida* Planch. & Linden

(=) *Caliphruria* Herb. in Edwards's Bot. Reg. 30 (Misc.): 87. Dec 1844. Typus: *C. hartwegiana* Herb.

Eucomis L'Hér., Sert. Angl.: 17. Jan (prim.) 1789 [*Lil.*]. Typus: *E. regia* (L.) L'Hér. (*Fritillaria regia* L.) (typ. cons.).

(≡) *Basilaea* Juss. ex Lam., Encycl. 1: 382. 1 Aug 1785.

Eulophia R. Br. in Bot. Reg.: ad t. 573 ('578'). 1 Oct 1821 (*'Eulophus'*) (orth. cons.) [*Orchid.*]. Typus: *E. guineensis* Ker Gawl. (typ. cons.).

(=) *Graphorkis* Thouars in Nouv. Bull. Sci. Soc. Philom. Paris 1: 318. Apr 1809 (nom. cons.).

(=) *Lissochilus* R. Br. in Bot. Reg.: ad t. 573 ('578'). 1 Oct 1821. Typus: *L. speciosus* R. Br.

Euterpe Mart., Hist. Nat. Palm. 2: 28. Nov 1823 [*Palm.*].
Typus: *E. oleracea* Mart.

(H)   *Euterpe* Gaertn., Fruct. Sem. Pl. 1: 24. Dec 1788 [*Monocot.: Palm.*].
Typus (vide Pfeiffer, Nomencl. Bot. 1: 1317. 1874): *E. globosa* Gaertn.

(=)   *Martinezia* Ruiz & Pav., Fl. Peruv. Prodr.: 148. Oct (prim.) 1794.
Typus: *M. ensiformis* Ruiz & Pav. (Syst. Veg. Fl. Peruv. Chil.: 297. Dec 1798).

(=)   *Oreodoxa* Willd. in Deutsch. Abh. Königl. Akad. Wiss. Berlin 1801: 251. 1803.
Typus: *O. acuminata* Willd.

Ficinia Schrad. in Commentat. Soc. Regiae Sci. Gott. Recent. 7: 143. 1832 [*Cyper.*].
Typus: *F. filiformis* (Lam.) Schrad. (*Schoenus filiformis* Lam.) (typ. cons.).

(=)   *Melancranis* Vahl, Enum. Pl. 2: 239. Oct-Dec 1805.
Typus: non designatus.

(=)   *Hemichlaena* Schrad. in Gött. Gel. Anz. 1821: 2066. 29 Dec 1821.
Typus (vide Goetghebeur & Arnold in Taxon 33: 114. 1984): *H. capillifolia* Schrad.

Fimbristylis Vahl, Enum. Pl. 2: 285. Oct-Dec 1805 [*Cyper.*].
Typus: *F. dichotoma* (L.) Vahl (*Scirpus dichotomus* L.) (typ. cons.).

(=)   *Iria* (Rich.) R. Hedw., Gen. Pl.: 360. Jul 1806 (*Cyperus* subg. *Iria* Rich. in Pers., Syn. Pl. 1: 65. 1 Apr - 15 Jun 1805).
Typus: *Cyperus monostachyos* L.

Freesia Ecklon ex Klatt in Linnaea 34: 672. Dec 1866 [*Irid.*].
Typus: *F. refracta* (Jacq.) Klatt (*Gladiolus refractus* Jacq.) (typ. cons.).

(=)   *Anomatheca* Ker Gawl. in Ann. Bot. (König & Sims) 1: 227. 1 Sep 1804.
Typus: *A. juncea* (L. f.) Ker Gawl. (*Gladiolus junceus* L. f.).

Genyorchis Schltr., Westafr. Kautschuk-Exped.: 280. Dec 1900 [*Orchid.*].
Typus: Cameroon, *Schlechter 12737* (BR) (typ. cons.) [= *G. apetala* (Lindl.) J. J. Verm. (*Bulbophyllum apetalum* Lindl.)].

Glyceria R. Br., Prodr.: 179. 27 Mar 1810 [*Gram.*].
Typus: *G. fluitans* (L.) R. Br. (*Festuca fluitans* L.).

Graphorkis Thouars in Nouv. Bull.
Sci. Soc. Philom. Paris 1: 318. Apr
1809 [*Orchid.*].
Typus: [specimen] Réunion or Madagascar, *Thouars* (P) (typ. cons.) [=
G. concolor (Thouars) Kuntze (*Limodorum concolor* Thouars)] (etiam
vide *Eulophia* [*Monocot.*]).

Hapaline Schott, Gen. Aroid.: 44. (≡) *Hapale* Schott in Oesterr. Bot.
1858 [*Ar.*].   Wochenbl. 7: 85. 12 Mar 1857.
Typus: *H. benthamiana* (Schott)
Schott (*Hapale benthamiana* Schott).

Haworthia Duval, Pl. Succ. Horto (=) *Catevala* Medik., Theodora: 67. 1786.
Alencon.: 7. 1809 [*Lil.*].   Typus: non designatus.
Typus: *H. arachnoidea* (L.) Duval
(*Aloë pumila* var. *arachnoidea* L.).

Helicodiceros Schott in Klotzsch, (≡) *Megotigea* Raf., Fl. Tellur. 3: 64.
App. Gen. Sp. Nov. 1855: 2. Dec   Nov-Dec 1837.
1855 - 1856 (prim.) [*Ar.*].
Typus: *H. muscivorus* (L. f.) Engl.
(in Candolle & Candolle, Monogr.
Phan. 2: 605. Sep 1879) (*Arum mus-
civorum* L. f.).

Heliconia L., Mant. Pl.: 147, 211. (≡) *Bihai* Mill., Gard. Dict. Abr., ed. 4:
Oct 1771 [*Mus.*].   [194]. 28 Jan 1754.
Typus: *H. bihai* (L.) L. (*Musa bihai*
L.).

Heloniopsis A. Gray in Mem. Amer. (=) *Hexonix* Raf., Fl. Tellur. 2: 13. Jan-
Acad. Arts, ser. 2, 6: 416. 1858 [*Lil.*].   Mar 1837.
Typus: *H. pauciflora* A. Gray   Typus: *H. japonica* (Thunb.) Raf.
                                        (*Scilla japonica* Thunb.).
                                    (=) *Kozola* Raf., Fl. Tellur. 2: 25. Jan-
                                        Mar 1837.
                                        Typus: *K. japonica* (Thunb.) Raf.
                                        (*Scilla japonica* Thunb.).

Hessea Herb., Amaryllidaceae: 289. (H) *Hessea* P. J. Bergius ex Schltdl. in
Apr (sero) 1837 [*Amaryllid.*].   Linnaea 1: 252. Apr 1826 [*Mono-
Typus: *H. stellaris* (Jacq.) Herb.   cot.: Amaryllid.*].
(*Amaryllis stellaris* Jacq.).   ≡ *Carpolyza* Salisb. 1807.

266

Hetaeria Blume, Bijdr.: 409. 20 Sep
- 7 Dec 1825 *('Etaeria')* (orth. cons.)
[*Orchid.*].
Typus: *H. oblongifolia* Blume (typ.
cons.).

Heteranthera Ruiz & Pav., Fl. Peruv. Prodr.: 9. Oct (prim.) 1794 [*Ponteder.*].
Typus: *H. reniformis* Ruiz & Pav.
(Fl. Peruv. 1: 43. 1798, med.).

Hierochloë R. Br., Prodr.: 208. 27   (=) *Savastana* Schrank, Baier. Fl. 1:
Mar 1810 [*Gram.*].   100, 337. Jun-Dec 1789.
Typus: *H. odorata* (L.) P. Beauv.   Typus: *S. hirta* Schrank
(Ess. Agrostogr.: 62, 164. Dec 1812)   (=) *Torresia* Ruiz & Pav., Fl. Peruv.
(*Holcus odoratus* L.) (typ. cons.).   Prodr.: 125. Oct (prim.) 1794.
                                                      Typus: *T. utriculata* Ruiz & Pav.
                                                      (Syst. Veg. Fl. Peruv. Chil.: 251.
                                                      Dec (sero) 1798).
                                                (=) *Disarrenum* Labill., Nov. Holl. Pl.
                                                      2: 82. Mar 1807.
                                                      Typus: *D. antarcticum* (G. Forst.)
                                                      Labill. (*Aira antarctica* G. Forst.).

Himantoglossum Spreng., Syst. Veg.
3: 675, 694. Jan-Mar 1826 [Orchid.].
Typus: *H. hircinum* (L.) Spreng.
(*Satyrium hircinum* L.) (typ. cons.).

Hippeastrum Herb., Appendix: 31.   (=) *Leopoldia* Herb. in Trans. Hort. Soc.
Dec 1821 [*Amaryllid.*].   London 4: 181. Jan-Feb 1821.
Typus: *H. reginae* (L.) Herb. (*Ama-   Typus: non designatus.
ryllis reginae* L.) (typ. cons.).

Holcus L., Sp. Pl.: 1047. 1 Mai 1753
[*Gram.*].
Typus: *H. lanatus* L. (typ. cons.).

Holothrix Rich. ex Lindl., Gen. Sp. Orchid. Pl.: 257. Aug 1835 [*Orchid.*]. Typus: *H. parvifolia* Lindl., nom. illeg. (*Orchis hispidula* L. f., *H. hispidula* (L. f.) Durand & Schinz) (typ. cons.).

(=) *Monotris* Lindl. in Edwards's Bot. Reg.: ad t. 1701. 1 Sep 1834. Typus: *M. secunda* Lindl.

(=) *Scopularia* Lindl. in Edwards's Bot. Reg. 20: ad t. 1701. 1 Sep 1834. Typus: *S. burchellii* Lindl.

(=) *Saccidium* Lindl., Gen. Sp. Orchid. Pl.: 258. Aug 1835. Typus: *S. pilosum* Lindl.

(=) *Tryphia* Lindl., Gen. Sp. Orchid. Pl.: 258. Aug 1835. Typus: *T. secunda* (Thunb.) Lindl. (*Orchis secunda* Thunb.).

Hosta Tratt., Arch. Gewächsk. 1: 55. 1812 [*Lil.*]. Typus: *H. japonica* Tratt.

(H) *Hosta* Jacq., Pl. Hort. Schoenbr. 1: 60. 1797 [*Dicot.: Verben.*]. Typus: non designatus.

Hypodiscus Nees in Lindley, Intr. Nat. Syst. Bot., ed. 2: 450. Jul 1836 [*Restion.*]. Typus: *H. aristatus* (Thunb.) Nees ex Mast. (in J. Linn. Soc., Bot. 10: 252. 23 Mai 1868) (*Restio aristatus* Thunb.).

(=) *Lepidanthus* Nees in Linnaea 5: 665. Oct 1830. Typus: *L. willdenowia* Nees, nom. illeg. (*Willdenowia striata* Thunb.).

Hypolaena R. Br., Prodr.: 251. 27 Mar 1810 [*Restion.*]. Typus: *H. fastigiata* R. Br. (typ. cons.).

(=) *Calorophus* Labill., Nov. Holl. Pl. 2: 78. Aug 1806. Typus: *C. elongata* Labill.

Iphigenia Kunth, Enum. Pl. 4: 212. 17-19 Jul 1843 [*Lil.*]. Typus: *I. indica* (L.) Kunth (*Melanthium indicum* L.).

(=) *Aphoma* Raf., Fl. Tellur. 2: 31. Jan-Mar 1837. Typus: *A. angustiflora* Raf.

Ixia L., Sp. Pl., ed. 2: 51. Sep 1762 [*Irid.*]. Typus: *I. polystachya* L. (typ. cons.).

(H) *Ixia* L., Sp. Pl.: 36. 1 Mai 1753 [*Monocot.: Irid.*]. Typus (vide Regnum Veg. 46: 272. 1966): *I. africana* L.

Johnsonia R. Br., Prodr.: 287. 27 Mar 1810 [*Lil.*]. Typus: *J. lupulina* R. Br.

(H) *Johnsonia* Mill., Gard. Dict. Abr., ed. 4: [693]. 28 Jan 1754 [*Dicot.: Verben.*]. Typus: non designatus.

Kniphofia Moench, Methodus: 631. 4 Mai 1794 [*Lil.*].
Typus: *K. alooides* Moench, nom. illeg. (*Aloë uvaria* L., *K. uvaria* (L.) Hook.).

(H) *Kniphofia* Scop., Intr. Hist. Nat.: 327. Jan-Apr 1777 [*Dicot.: Combret.*].
Typus: non designatus.

Kyllinga Rottb., Descr. Icon. Rar. Pl.: 12. Jan-Jul 1773 [*Cyper.*].
Typus: *K. nemoralis* (J. R. Forst. & G. Forst.) Dandy ex Hutch. & Dalziel (Fl. W. Trop. Afr. 2: 487. Feb 1936) (*Thyrocephalon nemorale* J. R. Forst. & G. Forst.) (typ. cons.).

(H) *Killinga* Adans., Fam. Pl. 2: 498, 539. Jul-Aug 1763 [*Dicot.: Umbell.*].
≡ *Athamantha* L. 1753.

Lachnanthes Elliott, Sketch Bot. S.-Carolina 1: 47. 26 Sep 1816 [*Haemodor.*].
Typus: *L. tinctoria* Elliott

Laelia Lindl., Gen. Sp. Orchid. Pl.: 96, 115. Jul 1831 [*Orchid.*].
Typus: *L. grandiflora* (La Llave & Lex.) Lindl. (*Bletia grandiflora* La Llave & Lex.) (typ. cons.).

(H) *Laelia* Adans., Fam. Pl. 2: 423, 567. Jul-Aug 1763 [*Dicot.: Cruc.*].
Typus: *L. orientalis* (L.) Desv. (in J. Bot. Agric. 3: 160. 1815 (prim.)) (*Bunias orientalis* L.).

Lamarckia Moench, Methodus: 201. 4 Mai 1794 (*'Lamarkia'*) (orth. cons.) [*Gram.*].
Typus: *L. aurea* (L.) Moench (*Cynosurus aureus* L.).

(H) *Lamarckia* Olivi, Zool. Adriat.: 258. Sep-Dec 1792 [*Chloroph.*].
Typus: non designatus.

(≡) *Achyrodes* Boehm. in Ludwig, Def. Gen. Pl., ed. 3: 420. 1760.

Lanaria Aiton, Hort. Kew. 1: 462. 7 Aug - 1 Oct 1789 [*Amaryllid.*].
Typus: *L. plumosa* Aiton, nom. illeg. (*Hyacinthus lanatus* L., *L. lanata* (L.) Druce).

(H) *Lanaria* Adans., Fam. Pl. 2: 255, 568. Jul-Aug 1763 [*Dicot.: Caryophyll.*].
≡ *Gypsophila* L. 1753.

(=) *Argolasia* Juss., Gen. Pl.: 60. 4 Aug 1789.
Typus: non designatus.

Laxmannia R. Br., Prodr.: 285. 27 Mar 1810 [*Lil.*].
Typus: *L. gracilis* R. Br. (typ. cons.).

(H) *Laxmannia* J. R. Forst. & G. Forst., Char. Gen. Pl.: 47. 29 Nov 1775 [*Dicot.: Comp.*].
≡ *Petrobium* R. Br. 1817 (nom. cons.).

Leersia Sw., Prodr.: 1, 21. 20 Jun - 29 Jul 1788 [*Gram.*].
Typus: *L. oryzoides* (L.) Sw. (*Phalaris oryzoides* L.) (typ. cons.).

(≡) *Homalocenchrus* Mieg in Acta Helv. Phys.-Math. 4: 307. 1760.

Leopoldia Parl., Fl. Palerm. 1: 435. 1845 [*Lil.*].
Typus: *L. comosa* (L.) Parl. (*Hyacinthus comosus* L.).

(H) *Leopoldia* Herb. in Trans. Hort. Soc. London 4: 181. Jan-Feb 1821 [*Monocot.: Amaryllid.*].
Typus: non designatus.

Leptocarpus R. Br., Prodr.: 250. 27 Mar 1810 [*Restion.*].
Typus: *L. aristatus* R. Br. (typ. cons.).

(=) *Schoenodum* Labill., Nov. Holl. Pl. 2: 79. Aug 1806.
Typus (vide Kunth, Enum. Pl. 3: 445. 1841): *S. tenax* Labill.

Libertia Spreng., Syst. Veg. 1: 127. 1824 (sero) [*Irid.*].
Typus: *L. ixioides* (G. Forst.) Spreng. (*Sisyrinchium ixioides* G. Forst.).

(H) *Libertia* Dumort., Comment. Bot.: 9. Nov (sero) - Dec (prim.) 1822 [*Monocot.: Lil.*].
Typus: *L. recta* Dumort., nom. illeg. (*Hemerocallis caerulea* Andrews).

(=) *Tekel* Adans., Fam. Pl. 2: 497, 610. Jul-Aug 1763.
Typus: non designatus.

Limodorum Boehm. in Ludwig, Def. Gen. Pl., ed. 3: 358. 1760 [*Orchid.*].
Typus: *L. abortivum* (L.) Sw. (in Nova Acta Regiae Soc. Sci. Upsal. 6: 80. 1799) (*Orchis abortiva* L.).

(H) *Limodorum* L., Sp. Pl.: 950. 1 Mai 1753 [*Monocot.: Orchid.*].
≡ *Calopogon* R. Br. 1813 (nom. cons.).

Liparis Rich., De Orchid. Eur.: 21, 30, 38. Aug-Sep 1817 [*Orchid.*].
Typus: *L. loeselii* (L.) Rich. (*Ophrys loeselii* L.).

(=) *Leptorkis* Thouars in Nouv. Bull. Sci. Soc. Philom. Paris 1: 317. Apr 1809.
Typus: non designatus.

Lipocarpha R. Br. in Tuckey, Narr. Exped. Zaire: 459. 5 Mar 1818 [*Cyper.*].
Typus: *L. argentea* R. Br., nom. illeg. (*Hypaelyptum argenteum* Vahl, nom. illeg., *Scirpus senegalensis* Lam., *L. senegalensis* (Lam.) T. Durand & H. Durand).

(=) *Hypaelyptum* Vahl, Enum. Pl. 2: 283. Oct-Dec 1805.
Typus (vide Panigrahi in Taxon 34: 511. 1985): *H. filiforme* Vahl

Listera R. Br. in Aiton, Hort. Kew., ed. 2, 5: 201. Nov 1813 [*Orchid.*]. Typus: *L. ovata* (L.) R. Br. (*Ophrys ovata* L.) (typ. cons.).

(H) *Listera* Adans., Fam. Pl. 2: 321, 572. Jul-Aug 1763 [*Dicot.: Legum.*]. Typus: non designatus.

(=) *Diphryllum* Raf. in Med. Repos., ser. 2, 5: 357. Feb-Apr 1808. Typus: *D. bifolium* Raf.

Lloydia Salisb. ex Rchb., Fl. Germ. Excurs.: 102. Mar-Apr 1830 [*Lil.*]. Typus: *L. serotina* (L.) Rchb. (*Anthericum serotinum* L.).

Loudetia Hochst. ex Steud., Syn. Pl. Glumac. 1: 238. 12-13 Apr 1854 [*Gram.*]. Typus: *L. elegans* Hochst. ex A. Braun (in Flora 24: 713. 7 Dec 1841).

(H) *Loudetia* Hochst. ex A. Braun in Flora 24: 713. 7 Dec 1841 [*Monocot.: Gram.*]. ≡ *Tristachya* Nees 1829.

Ludovia Brongn. in Ann. Sci. Nat., Bot., ser. 4, 15: 361. Jun 1861 [*Cyclanth.*]. Typus: *L. lancifolia* Brongn.

(H) *Ludovia* Pers., Syn. Pl. 2: 576. Sep 1807 [*Monocot.: Cyclanth.*]. Typus: non designatus.

Luzula DC. in Lamarck & Candolle, Fl. Franç., ed. 3, 3: 158. 17 Sep 1805 [*Junc.*]. Typus: *L. campestris* (L.) DC. (*Juncus campestris* L.) (typ. cons.).

(≡) *Juncoides* Ség., Pl. Veron. 3: 88. Jul-Aug 1754.

Luzuriaga Ruiz & Pav., Fl. Peruv. 3: 65. Aug 1802 [*Lil.*]. Typus: *L. radicans* Ruiz & Pav.

(=) *Enargea* Banks ex Gaertn., Fruct. Sem. Pl. 1: 283. Dec 1788. Typus: *E. marginata* Gaertn.

(=) *Callixene* Comm. ex Juss., Gen. Pl.: 41. 4 Aug 1789. Typus: non designatus.

Lyginia R. Br., Prodr.: 248. 27 Mar 1810 [*Restion.*]. Typus: *L. barbata* R. Br. (typ. cons.).

Maianthemum F. H. Wigg., Prim. Fl. Holsat.: 14. 29 Mar 1780 [*Lil.*]. Typus: *M. convallaria* F. H. Wigg., nom. illeg. (*Convallaria bifolia* L., *M. bifolium* (L.) F. W. Schmidt).

Mariscus Vahl, Enum. Pl. 2: 372. Oct-Dec 1805 [*Cyper.*].
Typus: *M. capillaris* (Sw.) Vahl (*Schoenus capillaris* Sw.) (typ. cons.).

(H) *Mariscus* Scop., Meth. Pl.: 22. 25 Mar 1754 [*Monocot.: Cyper.*].
Typus: *Schoenus mariscus* L.

Maximiliana Mart., Palm. Fam.: 20. 13 Apr 1824 [*Palm.*].
Typus: *M. regia* Mart. (Hist. Nat. Palm. 2: 131. Jan-Mar 1826), non *Maximilianea regia* Mart. 1819 (*M. martiana* H. Karst.) (typ. cons.).

(H) *Maximilianea* Mart. in Flora 2: 452. 7 Aug 1819 [*Dicot.: Cochlosperm.*].
Typus: *M. regia* Mart.

Metroxylon Rottb. in Nye Saml. Kongel. Danske Vidensk. Selsk. Skr. 2: 527. 1783 [*Palm.*].
Typus: *M. sagu* Rottb.

(=) *Sagus* Steck, Sagu: 21. 21 Sep 1757. Typus (vide Moore in Taxon 11: 165. 1965): *S. genuina* Giseke (Prael. Ord. Nat. Pl.: 93. Apr 1792).

Micranthus (Pers.) Eckl., Topogr. Verz. Pflanzensamml. Ecklon: 43. Oct 1827 (*Gladiolus* subg. *Micranthus* Pers., Syn. Pl. 1: 46. 1 Apr - 15 Jun 1805) [*Irid.*].
Typus: *Gladiolus alopecuroides* L. (*M. alopecuroides* (L.) Eckl.) (typ. cons.).

(H) *Micranthus* J. C. Wendl., Bot. Beob.: 38. 1798 [*Dicot.: Acanth.*].
≡ *Phaulopsis* Willd. 1800 (nom. cons.).

Microstylis (Nutt.) Eaton, Man. Bot., ed. 3: 115, 347, 353. 23 Mar - 23 Apr 1822 (*Malaxis* sect. *Microstylis* Nutt., Gen. N. Amer. Pl. 2: 196. 14 Jul 1818) [*Orchid.*].
Typus: *Malaxis ophioglossoides* Muhl. ex Willd., nom. illeg. (*Malaxis unifolia* Michx., *Microstylis unifolia* (Michx.) Britton & al.).

(≡) *Achroanthes* Raf. in Amer. Monthly Mag. & Crit. Rev. 4: 195. Jan 1819.

Milligania Hook. f. in Hooker's J. Bot. Kew Gard. Misc. 5: 296. Oct 1853 [*Lil.*].
Typus: *M. longifolia* Hook. f. (typ. cons.).

(H) *Milligania* Hook. f. in Icon. Pl.: ad t. 299. 6 Jan - 6 Feb 1840 [*Dicot.: Gunner.*].
Typus: *M. cordifolia* Hook. f.

Miltonia Lindl. in Edwards's Bot. Reg.: ad t. 1976. 1 Aug 1837 [*Orchid.*].
Typus: *M. spectabilis* Lindl.

Monstera Adans., Fam. Pl. 2: 470, 578. Jul-Aug 1763 [*Ar.*].
Typus: *M. adansonii* Schott (in Wiener Z. Kunst 1830: 1028. 23 Oct 1830) (*Dracontium pertusum* L., non *M. pertusa* (Roxb.) Schott) (typ. cons.).

Montrichardia Crueg. in Bot. Zeitung (Berlin) 12: 25. 13 Jan 1854 [*Ar.*].
Typus: *M. aculeata* (G. Mey.) Schott (Syn. Aroid.: 72. Mar 1856) (*Caladium aculeatum* G. Mey.).

(=) *Pleurospa* Raf., Fl. Tellur. 4: 8. 1838 (med.).
Typus (vide Nicolson in Regnum Veg. 34: 55. 1964): *P. reticulata* Raf., nom. illeg. (*Arum arborescens* L.).

Moraea Mill., Fig. Pl. Gard. Dict.: 159. 27 Jun 1758 (*'Morea'*) (orth. cons.) [*Irid.*].
Typus: *M. vegeta* L. (Sp. Pl., ed. 2: 59. Sep 1762) (typ. cons.).

Munroa Torr., Pac. Railr. Rep. 4 (Pt 5, No. 4): 158. 1857. (*'Monroa'*) (orth. cons.) [*Gram.*].
Typus: *M. squarrosa* (Nutt.) Torr. (*Crypsis squarrosa* Nutt.).

Murdannia Royle, Ill. Bot. Himal. Mts.: 403. Mai-Apr 1840 [*Commelin.*].
Typus: *M. scapiflora* (Roxb.) Royle (*Commelina scapiflora* Roxb.).

(=) *Dilasia* Raf., Fl. Tellur. 4: 122. 1838 (med.).
Typus: *D. vaginata* (L.) Raf. (*Commelina vaginata* L.).

(=) *Streptylis* Raf., Fl. Tellur. 4: 122. 1838 (med.).
Typus: *S. bracteolata* (Lam.) Raf. (*Commelina bracteolata* Lam.).

Narthecium Huds., Fl. Angl.: 127. Jan-Jun 1762 [*Lil.*].
Typus: *N. ossifragum* (L.) Huds. (*Anthericum ossifragum* L.).

(H) *Narthecium* Gérard, Fl. Gallo-Prov.: 142. Mar-Oct 1761 [*Monocot.: Lil.*].
Typus: *Anthericum calyculatum* L.

Neottia Guett. in Hist. Acad. Roy. Sci. Mém. Math. Phys. (Paris, 4°) 1750: 374. 1754 [*Orchid.*].
Typus: *N. nidus-avis* (L.) Rich. (De Orchid. Eur.: 59. Aug-Sep 1817) (*Ophrys nidus-avis* L.).

Nerine Herb. in Bot. Mag.: ad t. 2124. 1 Jan 1820 [*Amaryllid.*].
Typus: *N. sarniensis* (L.) Herb. (*Amaryllis sarniensis* L.) (typ. cons.).

(≡) *Imhofia* Heist., Beschr. Neu. Geschl.: 29. 1755.

Nervilia Comm. ex Gaudich., Voy. Uranie, Bot.: 421. 12 Sep 1829 [*Orchid.*].
Typus: *N. aragoana* Gaudich.

(=) *Stellorkis* Thouars in Nouv. Bull. Sci. Soc. Philom. Paris 1: 317. Apr 1809.
Typus (vide Thouars, Hist. Orchid.: ad t. 24. 1822): *Arethusa simplex* L.

Nicolaia Horan., Prodr. Monogr. Scitam.: 32. 1862 [*Zingiber.*].
Typus: *N. imperialis* Horan. (typ. cons.) [= *N. elatior* (Jack) Horan.].

(=) *Diracodes* Blume, Enum. Pl. Javae 1: 55. Oct-Dec 1827.
Typus: *D. javanica* Blume

Nothoscordum Kunth, Enum. Pl. 4: 457. 17-19 Jul 1843 [*Lil.*].
Typus: *N. striatum* Kunth, nom. illeg. (*Ornithogalum bivalve* L., *N. bivalve* (L.) Britton) (typ. cons.).

Oberonia Lindl., Gen. Sp. Orchid. Pl.: 15. Apr 1830 [*Orchid.*].
Typus: *O. iridifolia* Lindl., nom. illeg. (*Malaxis ensiformis* Sm., *O. ensiformis* (Sm.) Lindl.) (typ. cons.).

(=) *Iridorkis* Thouars in Nouv. Bull. Sci. Soc. Philom. Paris 1: 319. Apr 1809.
Typus (vide Swart in Index Nom. Gen.: No. 30689. 1970): *Epidendrum distichum* Lam.

Oeonia Lindl. in Bot. Reg.: ad t. 817. 1 Aug 1824 *('Aeonia')* (orth. cons.) [*Orchid.*].
Typus: *O. aubertii* Lindl., nom. illeg. (*Epidendrum volucre* Thouars, *O. volucris* (Thouars) Durand & Schinz).

Oncidium Sw. in Kongl. Vetensk. Acad. Nya Handl. 21: 239. Jul-Sep 1800 [*Orchid.*].
Typus: *O. altissimum* (Jacq.) Sw. (*Epidendrum altissimum* Jacq.) (typ. cons.).

Ophiopogon Ker Gawl. in Bot. Mag.: ad t. 1063. 1 Nov 1807 [*Lil.*].
Typus: *O. japonicus* (L. f.) Ker Gawl. (*Convallaria japonica* L. f.).

(=) *Mondo* Adans., Fam. Pl. 2: 496, 578. Jul-Aug 1763.
Typus: non designatus.

Oplismenus P. Beauv., Fl. Oware 2: 14. 6 Aug 1810 [*Gram.*].
Typus: *O. africanus* P. Beauv.

(=) *Orthopogon* R. Br., Prodr.: 194. 27 Mar 1810.
Typus (vide Hitchcock in U.S.D.A. Bull. 772: 238. 1920): *O. compositus* (L.) R. Br. (*Panicum compositum* L.).

Orbignya Mart. ex Endl., Gen. Pl.: 257. Oct 1837 [*Palm.*].
Typus: *O. phalerata* Mart. (Hist. Nat. Palm. 3: 302. 19 Sep 1845).

(H) *Orbignya* Bertero in Mercurio Chileno 16: 737. 15 Jul 1829 [*Dicot.: Euphorb.*].
Typus: *O. trifolia* Bertero

Paepalanthus Mart. in Ann. Sci. Nat., Bot., ser. 2, 2: 28. Jul 1834 [*Eriocaul.*].
Typus: *P. erigeron* Mart. ex Koern. (in Martius, Fl. Bras. 3(1): 390. 1863) (typ. cons.).

(=) *Dupatya* Vell., Fl. Flumin.: 35. 7 Sep - 28 Nov 1829.
Typus: non designatus.

Palisota Rchb. ex Endl., Gen. Pl.: 125. Dec 1836 [*Commelin.*].
Typus: *P. ambigua* (P. Beauv.) C. B. Clarke (in Candolle & Candolle, Monogr. Phan. 3: 131. Jun 1881) (*Commelina ambigua* P. Beauv.).

(=) *Duchekia* Kostel., Allg. Med.-Pharm. Fl. 1: 213. Mai 1831.
Typus: *D. hirsuta* (Thunb.) Kostel. (*Dracaena hirsuta* Thunb.).

Panisea (Lindl.) Lindl., Fol. Orchid. 5, *Panisea*: 1. 20 Jan 1854 (*Coelogyne* sect. *Panisea* Lindl., Gen. Sp. Orchid. Pl.: 44. Mai 1830) [*Orchid.*].
Typus: *Coelogyne parviflora* Lindl. (*P. parviflora* (Lindl.) Lindl.).

(≡) *Androgyne* Griff., Not. Pl. Asiat. 3: 279. 1851.

Paphiopedilum Pfitzer, Morph. Stud. Orchideenbl.: 11. Jan-Jul 1886 [*Orchid.*].
Typus: *P. insigne* (Wall. ex Lindl.) Pfitzer (*Cypripedium insigne* Wall. ex Lindl.) (typ. cons.).

(≡) *Cordula* Raf., Fl. Tellur. 4: 46. 1838 (med.).
(=) *Stimegas* Raf., Fl. Tellur. 4: 45. 1838 (med.).
Typus: *S. venustum* (Wall. ex Sims) Raf. (*Cypripedium venustum* Wall. ex Sims).

Paradisea Mazzuc., Viaggio Bot. Alpi Giulie: 27. 1811 [*Lil.*].
Typus: *P. hemeroanthericoides* Mazzuc., nom. illeg. (*Hemerocallis liliastrum* L., *P. liliastrum* (L.) Bertol.).

(≡) *Liliastrum* Fabr., Enum.: 4. 1759.

Patersonia R. Br. in Bot. Mag.: ad t. 1041. 1 Aug 1807 [*Irid.*].
Typus: *P. sericea* R. Br.

(=)   *Genosiris* Labill., Nov. Holl. Pl. 1: 13. Jan 1805.
Typus: *G. fragilis* Labill.

Pelexia Poit. ex Lindl. in Bot. Reg.: ad t. 985. 1 Jun 1826 [*Orchid.*].
Typus: *P. spiranthoides* Lindl., nom. illeg. (*Satyrium adnatum* Sw., *P. adnata* (Sw.) Spreng.).

(≡)   *Collea* Lindl. in Bot. Reg. 9: ad t. 760. 1 Dec 1823.

Peltandra Raf. in J. Phys. Chim. Hist. Nat. Arts 89: 103. Aug 1819 [*Ar.*].
Typus: *P. undulata* Raf.

Peristylus Blume, Bijdr.: 404. 20 Sep - 7 Dec 1825 [*Orchid.*].
Typus: *P. grandis* Blume

(=)   *Glossula* Lindl. in Bot. Reg. 10: ad t. 862. Feb 1825.
Typus: *G. tentaculata* Lindl.

Petermannia F. Muell., Fragm. 2: 92. Aug 1860 [*Dioscor.*].
Typus: *P. cirrosa* F. Muell.

(H)   *Petermannia* Rchb., Deut. Bot. Herb.-Buch, Syn.: 236. Jul 1841 [*Dicot.: Chenopod.*].
≡ *Cycloloma* Moq. 1840.

Philodendron Schott in Wiener Z. Kunst 1829: 780. 6 Aug 1829 (*'Philodendrum'*) (orth. cons.) [*Ar.*].
Typus: *P. grandifolium* (Jacq.) Schott (*Arum grandifolium* Jacq.).

Phragmipedium Rolfe in Orchid Rev. 4: 330. Nov 1896 [*Orchid.*].
Typus: *P. caudatum* (Lindl.) Rolfe (*Cypripedium caudatum* Lindl.).

(=)   *Uropedium* Lindl., Orchid. Linden.: 28. Nov-Dec 1846.
Typus: *U. lindenii* Lindl.

Phrynium Willd., Sp. Pl. 1: 1, 17. Jun 1797 [*Marant.*].
Typus: *P. capitatum* Willd., nom. illeg. (*Pontederia ovata* L., *Phrynium rheedei* Suresh & Nicolson).

(=)   *Phyllodes* Lour., Fl. Cochinch.: 1, 13. 1790.
Typus: *P. placentaria* Lour.

Phyllostachys Siebold & Zucc. in Abh. Math.-Phys. Cl. Königl. Bayer. Akad. Wiss. 3: 745. 1843 [*Gram.*].
Typus: *P. bambusoides* Siebold & Zucc.

Pigafetta (Blume) Becc. in Malesia 1: 89. 1877 (*'Pigafettia'*) (*Sagus* sect. *Pigafetta* Blume, Rumphia 2: 154. Jan-Aug 1843) (orth. cons.) [*Palm.*]. Typus: *Sagus filaris* Giseke (*P. filaris* (Giseke) Becc.) (typ. cons.).

(H)   *Pigafetta* Adans., Fam. Pl. 2: 223, 590. Jul-Aug 1763 [*Dicot.: Acanth.*]. ≡ *Eranthemum* L. 1753.

Pinellia Ten. in Atti Reale Accad. Sci. Sez. Soc. Reale Borbon. 4: 69. 1839 [*Ar.*]. Typus: *P. tuberifera* Ten., nom. illeg. (*Arum subulatum* Desf.) [= *P. ternata* (Thunb.) Makino (*Arum ternatum* Thunb.)].

(=)   *Atherurus* Blume, Rumphia 1: 135. Apr-Jun 1837. Typus (vide Nicolson in Taxon 16: 515. 1967): *A. tripartitus* Blume

Piptochaetium J. Presl in Presl, Reliq. Haenk. 1: 222. Jan-Jun 1830 [*Gram.*]. Typus: *P. setifolium* J. Presl

(=)   *Podopogon* Raf., Neogenyton: 4. 1825. Typus (vide Clayton in Taxon 32: 649. 1983): *Stipa avenacea* L.

Pitcairnia L'Hér., Sert. Angl.: 7. Jan (prim.) 1789 [*Bromel.*]. Typus: *P. bromeliifolia* L'Hér.

(=)   *Hepetis* Sw., Prodr.: 4, 56. 20 Jun - 29 Jul 1788. Typus: *H. angustifolia* Sw.

Platanthera Rich., De Orchid. Eur.: 20, 26, 35. Aug-Sep 1817 [*Orchid.*]. Typus: *P. bifolia* (L.) Rich. (*Orchis bifolia* L.).

Platylepis A. Rich. in Mém. Soc. Hist. Nat. Paris 4: 34. Sep 1828 [*Orchid.*]. Typus: *P. goodyeroides* A. Rich., nom. illeg. (*Goodyera occulta* Thouars, *P. occulta* (Thouars) Rchb.).

(=)   *Erporkis* Thouars in Nouv. Bull. Sci. Soc. Philom. Paris 1: 317. Apr 1809. Typus: non designatus.

Polystachya Hook., Exot. Fl. 2: ad t. 103. Mai 1824 [*Orchid.*]. Typus: *P. luteola* Hook., nom. illeg. (*Epidendrum minutum* Aubl.) [= *P. extinctoria* Rchb.].

(=)   *Dendrorkis* Thouars in Nouv. Bull. Sci. Soc. Philom. Paris 1: 318. Apr 1809. Typus: non designatus.

Posidonia K. D. Koenig in Ann. Bot. (König & Sims) 2: 95. 1 Jun 1805 [*Potamogeton*.].
Typus: *P. caulinii* K. D. Koenig, nom. illeg. (*Zostera oceanica* L., *P. oceanica* (L.) Delile).

(=) *Alga* Boehm. in Ludwig, Def. Gen. Pl., ed. 3: 503. 1760.
Typus: non designatus.

Prestoea Hook. f. in Bentham & Hooker, Gen. Pl. 3: 875, 899. 14 Apr 1883 [*Palm*.].
Typus: *P. pubigera* (Griseb. & H. Wendl.) Hook. f. (in Rep. (Annual) Roy. Bot. Gard. Kew 1882: 56. 1884) (*Hyospathe pubigera* Griseb. & H. Wendl.).

(=) *Martinezia* Ruiz & Pav., Fl. Peruv. Prodr.: 148. Oct (prim.) 1794.
Typus: *M. ensiformis* Ruiz & Pav. (Syst. Veg. Fl. Peruv. Chil.: 297. Dec 1798).
(=) *Oreodoxa* Willd. in Deutsch. Abh. Königl. Akad. Wiss. Berlin 1801: 251. 1803.
Typus: *O. acuminata* Willd.

Pritchardia Seem. & H. Wendl. in Bonplandia 10: 197. 1 Jul 1862 [*Palm*.].
Typus: *P. pacifica* Seem. & H. Wendl.

(H) *Pritchardia* Unger ex Endl., Gen. Pl., Suppl. 2: 102. Mar-Jun 1842 [Foss.].
Typus: *P. insignis* Unger ex Endl.

Pterostylis R. Br., Prodr.: 326. 27 Mar 1810 [*Orchid*.].
Typus: *P. curta* R. Br. (typ. cons.).

(=) *Diplodium* Sw. in Ges. Naturf. Freunde Berlin Mag. Neuesten Entdeck. Gesammten Naturk. 4: 84. Jul 1810.
Typus: *Disperis alata* Labill. (Nov. Holl. Pl. 2: 59. 1806).

Puccinellia Parl., Fl. Ital. 1: 366. 1848 [*Gram*.].
Typus: *P. distans* (Jacq.) Parl. (*Poa distans* Jacq.) (typ. cons.).

(≡) *Atropis* (Trin.) Rupr. ex Griseb. in Ledebour, Fl. Ross. 4: 388. Sep 1852 (*Poa* sect. *Atropis* Trin. in Mém. Acad. Imp. Sci. Saint-Pétersbourg, Sér. 6, Sci. Math., Seconde Pt. Sci. Nat. 6: 68. Mar 1836).

Reineckea Kunth in Abh. Königl. Akad. Wiss. Berlin 1842: 29. 1844 [*Lil*.].
Typus: *R. carnea* (Andrews) Kunth (*Sansevieria carnea* Andrews).

Renealmia L. f., Suppl. Pl.: 7, 79. Apr 1782 [*Zingiber.*].
Typus: *R. exaltata* L. f.

(H) *Renealmia* L., Sp. Pl.: 286. 1 Mai 1753 [*Monocot.: Bromel.*].
Typus (vide Smith in Index Nom. Gen.: No. 6039. 1958): *R. paniculata* L.

Restio Rottb., Descr. Pl. Rar.: 9. 1772 [*Restion.*].
Typus: *R. triticeus* Rottb.

(H) *Restio* L., Syst. Nat., ed. 12, 2: 735. 15-31 Oct 1767 [*Monocot.: Restion.*].
Typus: *R. dichotomus* L.

Reussia Endl., Gen. Pl.: 139. Dec 1836 [*Ponteder.*].
Typus: *R. triflora* Seub. (in Mart., Fl. Bras. 3(1): 96. 1 Jun 1847).

Rhynchospora Vahl, Enum. Pl. 2: 229. Oct-Dec 1805 (*'Rynchospora'*) (orth. cons.) [*Cyper.*].
Typus: *R. alba* (L.) Vahl (*Schoenus albus* L.) (typ. cons.).

(=) *Dichromena* Michx., Fl. Bor.-Amer. 1: 37. 19 Mar 1803.
Typus: *D. leucocephala* Michx.

Riedelia Oliv. in Hooker's Icon. Pl. 15: 15. Mar 1883 [*Zingiber.*].
Typus: *R. curviflora* Oliv.

(H) *Riedelia* Cham. in Linnaea 7: 240 ('224'). 1832 [*Dicot.: Verben.*].
Typus: *R. lippioides* Cham.

(=) *Nyctophylax* Zipp. in Alg. Konst-Lett.-Bode 1829(1): 298. 8 Mai 1829.
Typus: *N. alba* Zipp.

Romulea Maratti, Pl. Romul. Saturn.: 13. 1772 [*Irid.*].
Typus: *R. bulbocodium* (L.) Sebast. & Mauri (Fl. Roman. Prodr.: 17. 1818) (*Crocus bulbocodium* L.) (typ. cons.).

(≡) *Ilmu* Adans., Fam. Pl. 2: 497, 566. Jul-Aug 1763.

Rottboellia L. f., Suppl. Pl.: 13, 114. Apr 1782 [*Gram.*].
Typus: *R. exaltata* L. f. 1782, non (L.) Naezen 1779 (typ. cons.) [= *R. cochinchinensis* (Lour.) Clayton (*Stegosia cochinchinensis* Lour.)].

(H) *Rottboelia* Scop., Intr. Hist. Nat.: 233. Jan-Apr 1777 [*Dicot.: Olac.*].
≡ *Heymassoli* Aubl. 1775.

(=) *Manisuris* L., Mant. Pl.: 164, 300. Oct 1771.
Typus: *M. myurus* L.

Saccolabium Blume, Bijdr.: 292. 20 Sep - 7 Dec 1825 [*Orchid.*].
Typus: *S. pusillum* Blume

(=) *Gastrochilus* D. Don, Prodr. Fl. Nepal.: 32. 26. Jan - 1 Feb 1825.
Typus: *G. calceolaris* D. Don

Sansevieria Thunb., Prodr. Pl. Cap.
1: [xii], 65. 1794 [*Lil.*].
Typus: *S. thyrsiflora* Thunb., nom.
illeg. (*Aloë hyacinthoides* L., *S. hyacinthoides* (L.) Druce) (etiam vide
Cordyline [*Monocot.*]).

(≡) *Acyntha* Medik., Theodora: 76. 1786.
(=) *Sanseverinia* Petagna, Inst. Bot. 3:
643. 1787.
Typus: *S. thyrsiflora* Petagna

Satyrium Sw. in Kongl. Vetensk.
Acad. Nya Handl. 21: 214. Jul-Sep
1800 [*Orchid.*].
Typus: *S. bicorne* (L.) Thunb. (Prodr.
Pl. Cap.: 6. 1794) (*Orchis bicornis*
L.) (typ. cons.).

(H) *Satyrium* L., Sp. Pl.: 944. 1 Mai 1753
[*Monocot.: Orchid.*].
Typus (vide Green in Sprague & al.,
Nom. Prop. Brit. Bot.: 185. 1929):
S. viride L.

Scaphyglottis Poepp. & Endl., Nov.
Gen. Sp. Pl. 1: 58. 22-28 Mai 1836
[*Orchid.*].
Typus: *S. graminifolia* (Ruiz & Pav.)
Poepp. & Endl. (*Fernandezia gra-
minifolia* Ruiz & Pav.).

(=) *Hexisea* Lindl. in J. Bot. (Hooker) 1:
7. Mar 1834.
Typus: *H. bidentata* Lindl.

Schelhammera R. Br., Prodr.: 273.
27 Mar 1810 [*Lil.*].
Typus: *S. undulata* R. Br. (typ. cons.).

(H) *Schelhameria* Heist. ex Fabr., Enum.:
161. 1759 [*Dicot.: Cruc.*].
Typus: non designatus.

Schmidtia Steud. ex J. A. Schmidt,
Beitr. Fl. Cap Verd. Ins.: 144. 1 Jan
- 13 Feb 1852 [*Gram.*].
Typus: *S. pappophoroides* Steud. ex
J. A. Schmidt

(H) *Schmidtia* Moench, Suppl. Meth.:
217. 2 Mai 1802 [*Dicot.: Comp.*].
Typus: *S. fruticosa* Moench

Schoenolirion Torr. in E. M. Durand, Pl. Pratten. Calif.: 103. Aug
1855 [*Lil.*].
Typus: *S. croceum* (Michx.) A. Gray
(in Amer. Naturalist 10: 427. 1876)
(*Phalangium croceum* Michx.).

(=) *Amblostima* Raf., Fl. Tellur. 2: 26.
Jan-Mar 1837.
Typus: non designatus.
(=) *Oxytria* Raf., Fl. Tellur. 2: 26. Jan-
Mar 1837.
Typus: *O. crocea* Raf. (*Phalangium
croceum* Nutt. 1818, non Michx.
1803).

Schoenoplectus (Rchb.) Palla in Verh. K.K. Zool.-Bot. Ges. Wien 38 (Sitzungsber.): 49. 1888 (*Scirpus* subg. *Schoenoplectus* Rchb., Icon. Fl. Germ. Helv.: 8: 40. 1846) [*Cyper.*].
Typus: *Scirpus lacustris* L. (*Schoenoplectus lacustris* (L.) Palla).

(=) *Heleophylax* P. Beauv. ex T. Lestib., Essai Cypér.: 41. 29 Mar 1819.
Typus: non designatus.

(=) *Elytrospermum* C. A. Mey. in Mém. Acad. Imp. Sci. St.-Pétersbourg Divers Savans 1: 200. Oct 1831.
Typus: *E. californicum* C. A. Mey.

Scirpus L., Sp. Pl.: 47. 1 Mai 1753 [*Cyper.*].
Typus: *S. sylvaticus* L. (typ. cons.).

Scolochloa Link, Hort. Berol. 1: 136. 1 Oct - 27 Nov 1827 [*Gram.*].
Typus: *S. festucacea* (Willd.) Link (*Arundo festucacea* Willd.).

(H) *Scolochloa* Mert. & W. D. J. Koch, Deutschl. Fl., ed. 3, 1: 374, 528. Jan-Mai 1823 [*Monocot.: Gram.*].
Typus: *S. arundinacea* (P. Beauv.) Mert. & W. D. J. Koch (*Donax arundinaceus* P. Beauv.).

Serapias L., Sp. Pl.: 949. 1 Mai 1753 [*Orchid.*].
Typus: *S. lingua* L. (typ. cons.).

Setaria P. Beauv., Ess. Agrostogr.: 51, 178. Dec 1812 [*Gram.*].
Typus: *S. viridis* (L.) P. Beauv. (*Panicum viride* L.) (typ. cons.).

(H) *Setaria* Ach. ex Michx., Fl. Bor.-Amer. 2: 331. 19 Mar 1803 [*Fungi*].
Typus: *S. trichodes* Michx.

Simethis Kunth, Enum. Pl. 4: 618. 17-19 Jul 1843 [*Lil.*].
Typus: *S. bicolor* Kunth, nom. illeg. (*Anthericum planifolium* Vand. ex L., *S. planifolia* (L.) Gren. & Godr.).

Smilacina Desf. in Ann. Mus. Natl. Hist. Nat. 9: 51. 1807 [*Lil.*].
Typus: *S. stellata* (L.) Desf. (*Convallaria stellata* L.) (typ. cons.).

(=) *Vagnera* Adans., Fam. Pl. 2: 496, 617. Jul-Aug 1763.
Typus: non designatus.

(=) *Polygonastrum* Moench, Methodus: 637. 4 Mai 1794.
Typus: *P. racemosum* (L.) Moench (*Convallaria racemosa* L.).

Sorghum Moench, Methodus: 207. 4 Mai 1794 [*Gram.*].
Typus: *S. bicolor* (L.) Moench (*Holcus bicolor* L.).

(H) *Sorgum* Adans., Fam. Pl. 2: 38, 606. Jul-Aug 1763 [*Monocot.: Gram.*].
≡ *Holcus* L. 1753 (nom. cons.).

Spiranthes Rich., De Orchid. Eur.:
20, 28, 36. Aug-Sep 1817 [*Orchid.*].
Typus: *S. autumnalis* Rich., nom. illeg. (*Ophrys spiralis* L., *S. spiralis*
(L.) Chevall.) (typ. cons.).

(≡) *Orchiastrum* Ség., Pl. Veron. 3: 252.
Jul-Dec 1754.

Stelis Sw. in J. Bot. (Schrader)
1799(2): 239. Apr 1800 [*Orchid.*].
Typus: *S. ophioglossoides* (Jacq.)
Sw. (*Epidendrum ophioglossoides*
Jacq.) (typ. cons.).

Stenanthium (A. Gray) Kunth, Enum. Pl. 4: 189. 17-19 Jul 1843 (*Veratrum* subg. *Stenanthium* A. Gray in
Ann. Lyceum Nat. Hist. New York 4:
119. Nov 1837) [*Lil.*].
Typus: *Veratrum angustifolium* Pursh
(*S. angustifolium* (Pursh) Kunth).

(≡) *Anepsa* Raf., Fl. Tellur. 2: 31. Jan-
Mar 1837.

Symphyglossum Schltr. in Orchis 13:
8. 15 Feb 1919 [*Orchid.*].
Typus: *S. sanguineum* (Rchb. f.)
Schltr. (*Mesospinidium sanguineum*
Rchb. f.).

(H) *Symphyoglossum* Turcz. in Bull. Soc.
Imp. Naturalistes Moscou 21: 255.
1848 [*Dicot.: Asclepiad.*].
Typus: *S. hastatum* (Bunge) Turcz.
(*Asclepias hastata* Bunge).

Symplocarpus Salisb. ex W. P. C.
Barton, Veg. Mater. Med. U.S. 1:
124. 1817 [*Ar.*].
Typus: *S. foetidus* (L.) Salisb. ex W.
P. C. Barton (*'foetida'*) (*Dracontium
foetidum* L.).

Syringodea Hook. f. in Bot. Mag.:
ad t. 6072. 1 Dec 1873 [*Irid.*].
Typus: *S. pulchella* Hook. f.

(H) *Syringodea* D. Don in Edinburgh
New Philos. J. 17: 155. Jul 1834 [*Di-
cot.: Eric.*].
Typus: *S. vestita* (Thunb.) D. Don
(*Erica vestita* Thunb.).

Tacca J. R. Forst. & G. Forst., Char.
Gen. Pl.: 35. 29 Nov 1775 [*Tacc.*].
Typus: *T. pinnatifida* J. R. Forst. &
G. Forst.

(=) *Leontopetaloides* Boehm. in Ludwig,
Def. Gen. Pl., ed. 3: 512. 1760.
Typus: *Leontice leontopetaloides* L.

Tapeinochilos Miq. in Ann. Mus. Lugduno-Batavum 4: 101. 21 Feb 1869 (*'Tapeinocheilos'*) (orth. cons.) [*Zingiber.*].
Typus: *T. pungens* (Teijsm. & Binn.) Miq. (*Costus pungens* Teijsm. & Binn.).

Thysanotus R. Br., Prodr.: 282. 27 Mar 1810 [*Lil.*].
Typus: *T. junceus* R. Br., nom. illeg. (*Chlamysporum juncifolium* Salisb., *T. juncifolius* (Salisb.) Willis & Court).

(≡)   *Chlamysporum* Salisb., Parad. Lond.: ad t. 103. 1 Apr 1808.

Tinantia Scheidw. in Allg. Gartenzeitung 7: 365. 16 Nov 1839 [*Commelin.*].
Typus: *T. fugax* Scheidw.

(H)   *Tinantia* Dumort., Anal. Fam. Pl.: 58. 1829 [*Monocot.: Irid.*].
Typus: non designatus.

(=)   *Pogomesia* Raf., Fl. Tellur. 3: 67. Nov-Dec 1837.
Typus: *P. undata* (Humb. & Bonpl. ex Willd.) Raf. (*Tradescantia undata* Humb. & Bonpl. ex Willd.).

Trachyandra Kunth, Enum. Pl. 4: 573. 17-19 Jul 1843 [*Lil.*].
Typus: *T. hispida* (L.) Kunth (*Anthericum hispidum* L.).

(≡)   *Obsitila* Raf., Fl. Tellur. 2: 27. Jan-Mar 1837 (typ. des.: Merrill, Ind. Raf.: 92. 1949).

(=)   *Lepicaulon* Raf., Fl. Tellur. 2: 27. Jan-Mar 1837.
Typus: *L. squameum* (L. f.) Raf. (*Anthericum squameum* L. f.).

Tragus Haller, Hist. Stirp. Helv. 2: 203. 25 Mar 1768 [*Gram.*].
Typus: *T. racemosus* (L.) All. (Fl. Pedem. 2: 241. Apr-Jun 1785) (*Cenchrus racemosus* L.).

(≡)   *Nazia* Adans., Fam. Pl. 2: 31, 581. Jul-Aug 1763.

Trichophorum Pers., Syn. Pl. 1: 69. 1 Apr - 15 Jun 1805 [*Cyper.*].
Typus: *T. alpinum* (L.) Pers. (*Eriophorum alpinum* L.) (typ. cons.).

Tricyrtis Wall., Tent. Fl. Napal.: 61. Sep-Dec 1826 [*Lil.*].
Typus: *T. pilosa* Wall.

(=)   *Compsoa* D. Don, Prodr. Fl. Nepal.: 50. 26 Jan - 1 Feb 1825.
Typus: *C. maculata* D. Don

Tulbaghia L., Mant. Pl.: 148, 223. Oct 1771 *('Tulbagia')* (orth. cons.) [*Lil.*].
Typus: *T. capensis* L.

(H) *Tulbaghia* Heist., Beschr. Neu. Geschl.: 15. 1755 [*Monocot.: Lil.*].
≡ *Agapanthus* L'Hér. 1789 (nom. cons.).

Urceolina Rchb., Consp. Regni Veg.: 61. Dec 1828 - Mar 1829 [*Amaryllid.*].
Typus: *Urceolaria pendula* Herb., nom. illeg. (*Crinum urceolatum* Ruiz & Pav., *U. urceolata* (Ruiz & Pav.) M. L. Green).

(=) *Leperiza* Herb., Appendix: 41. Dec 1821.
Typus: *L. latifolia* (Ruiz & Pav.) Herb. (*Pancratium latifolium* Ruiz & Pav.).

Vallota Salisb. ex Herb., Appendix: 29. Dec 1821 [*Amaryllid.*].
Typus: *V. purpurea* Herb., nom. illeg. (*Crinum speciosum* L. f., *V. speciosa* (L. f.) Voss).

(H) *Valota* Adans., Fam. Pl. 2: 495, 617. Jul-Aug 1763 [*Monocot.: Gram.*].
Typus: *V. insularis* (L.) Chase (in Proc. Biol. Soc. Wash. 19: 188. 1906) (*Andropogon insulare* L.).

Veitchia H. Wendl. in Seemann, Fl. Vit.: 270. 31 Jul 1868 [*Palm.*].
Typus: *V. joannis* H. Wendl. (typ. cons.).

(H) *Veitchia* Lindl. in Gard. Chron. 1861: 265. Mar 1861 [*Gymnosp.: Pin.*].
Typus: *V. japonica* Lindl.

Ventenata Koeler, Descr. Gram.: 272. 1802 [*Gram.*].
Typus: *V. avenacea* Koeler, nom. illeg. (*Avena dubia* Leers, *V. dubia* (Leers) Coss.) (typ. cons.).

(H) *Vintenatia* Cav., Icon. 4: 28. Sep-Dec 1797 [*Dicot.: Epacrid.*].
Typus: non designatus.

(=) *Heteranthus* Borkh. [Fl. Grafsch. Catznelnb. 2] in Andre, Botaniker Compend. Biblioth. 16-18: 71. 1796.
Typus: *H. bromoides* Borkh., nom. illeg. (*Bromus triflorus* L.).

Vossia Wall. & Griff. in J. Asiat. Soc. Bengal 5: 572. Sep 1836 [*Gram.*].
Typus: *V. procera* Wall. & Griff., nom. illeg. (*Ischaemum cuspidatum* Roxb., *V. cuspidata* (Roxb.) Griff.).

(H) *Vossia* Adans., Fam. Pl. 2: 243, 619. Jul-Aug 1763 [*Dicot.: Aiz.*].
Typus: non designatus.

Vriesea Lindl. in Edwards's Bot. Reg. 29: ad t. 10. 7 Feb 1843 *('Vriesia')* (orth. cons.) [*Bromel.*].
Typus: *V. psittacina* (Hook.) Lindl. (*Tillandsia psittacina* Hook.).

(H) *Vriesea* Hassk. in Flora 25(2), Beibl.: 27. 21-28 Jul 1842 [*Dicot.: Scrophular.*].
Typus: non designatus.

(≡) *Hexalepis* Raf., Fl. Tellur. 4: 24. 1838 (med.).

Warmingia Rchb. f., Otia Bot. Hamburg.: 87. 8 Aug 1881 [*Orchid.*].
Typus: *W. eugenii* Rchb. f.

(H)   *Warmingia* Engl. in Martius, Fl. Bras. 12(2): 281. 1 Sep 1874 [*Dicot.: Anacard.*].
Typus: *W. pauciflora* Engl.

Washingtonia H. Wendl. in Bot. Zeitung (Berlin) 37: 68. 31 Jan 1879 [*Palm.*].
Typus: *W. filifera* (Linden ex André) H. Wendl. (in Bot. Zeitung (Berlin) 37: lxi. 1880 (prim.)) (*Pritchardia filifera* Linden ex André).

Watsonia Mill., Fig. Pl. Gard. Dict.: 184. 22 Dec 1758 [*Irid.*].
Typus: *W. meriana* (L.) Mill. (Gard. Dict., ed. 8: *Watsonia* No. 1. 16 Apr 1768) (*Antholyza meriana* L.) (etiam vide *Meriania* [*Dicot.*]).

Wolffia Horkel ex Schleid., Beitr. Bot. 1: 233. 11-13 Jul 1844 [*Lemn.*].
Typus: *W. michelii* Schleid.

(H)   *Wolfia* Schreb., Gen. Pl.: 801. Mai 1791 [*Dicot.: Flacourt.*].
Typus: non designatus.

Zantedeschia Spreng., Syst. Veg. 3: 756, 765. Jan-Mar 1826 [*Ar.*].
Typus: *Z. aethiopica* (L.) Spreng. (*Calla aethiopica* L.) (etiam vide *Colocasia* [*Monocot.*]).

Zephyranthes Herb., Appendix: 36. Dec 1821 [*Amaryllid.*].
Typus: *Z. atamasca* (L.) Herb. (*Amaryllis atamasca* L.) (typ. cons.).

($\equiv$)   *Atamosco* Adans., Fam. Pl. 2: 57, 522. Jul-Aug 1763.

Zeugites P. Browne, Civ. Nat. Hist. Jamaica: 341. 10 Mar 1756 [*Gram.*].
Typus: *Z. americanus* Willd. (Sp. Pl. 4: 204. 1805) (*Apluda zeugites* L.).

Zeuxine Lindl., Orchid. Scelet.: 9. Jan 1826 (*'Zeuxina'*) (orth. cons.) [*Orchid.*].
Typus: *Z. sulcata* (Roxb.) Lindl. (Gen. Sp. Orchid. Pl.: 485. Sep 1840) (*Pterygodium sulcatum* Roxb.).

Zingiber Mill., Gard. Dict. Abr., ed.
4: [1545]. 28 Jan 1754 *('Zinziber')*
(orth. cons.) [*Zingiber.*].
Typus: *Z. officinale* Roscoe (in
Trans. Linn. Soc. London 8: 358. 9
Mar 1807) (*Amomum zingiber* L.)
(etiam vide *Amomum* [*Monocot.*]).

Zoysia Willd. in Ges. Naturf. Freunde Berlin Neue Schriften 3: 440.
1801 (post 21 Apr) [*Gram.*].
Typus: *Z. pungens* Willd.

E3. DICOTYLEDONES

Acacia Mill., Gard. Dict. Abr. ed. 4:
[25]: 28 Jan 1754. [*Legum.*].
Typus: *A. penninervis* Sieber ex DC.
(typ. cons.).

Acantholimon Boiss., Diagn. Pl. Ori-
ent. 7: 69. Jul-Oct 1846 [*Plumbagin.*].
Typus: *A. glumaceum* (Jaub. & Spach)
Boiss. (*Statice glumacea* Jaub. &
Spach).

(≡) *Armeriastrum* (Jaub. & Spach) Lindl.,
Veg. Kingd.: 641. Jan-Mai 1846 (*Statice* subg. *Armeriastrum* Jaub. &
Spach in Ann. Sci. Nat., Bot., ser. 2,
20: 248. Oct 1843).

Acanthonema Hook. f. in Bot. Mag.:
ad t. 5339. 1 Oct 1862 [*Gesner.*].
Typus: *A. strigosum* Hook. f.

(H) *Acanthonema* J. Agardh in Öfvers.
Förh. Kongl. Svenska Vetensk.-Akad.
3: 104. 1846 [*Rhodoph.*].
≡ *Camontagnea* Pujals 1981.

Acanthospermum Schrank, Pl. Rar.
Hort. Monac.: ad t. 53. Apr-Mai
1820. [*Comp.*].
Typus: *A brasilum* Schrank

(=) *Centrospermum* Kunth in Humboldt
& al., Nov. Gen. Sp. 4, ed. f°: 212.
26 Oct 1818.
Typus: *C. xanthioides* Kunth

Achimenes Pers., Syn. Pl. 2: 164.
Nov 1806 [*Gesner.*].
Typus: *A. coccinea* (Scop.) Pers.
(*Buchnera coccinea* Scop.) (typ.
cons.).

(H) *Achimenes* P. Browne, Civ. Nat.
Hist. Jamaica: 270. 10 Mar 1756 [*Di-
cot.: Gesner.*].
Typus: non designatus.

286

Achyranthes L., Sp. Pl.: 204. 1 Mai 1753 [*Amaranth.*].
Typus: *A. aspera* L. (typ. cons.: [specimen] Herb. Hermann 2: 69, No. 105 (BM)).

Acidoton Sw., Prodr.: 6, 83. 20 Jun - 29 Jul 1788 [*Euphorb.*].
Typus: *A. urens* Sw.

(H) *Acidoton* P. Browne, Civ. Nat. Hist. Jamaica: 355. 10 Mar 1756 [*Dicot.: Euphorb.*].
Typus: *Adelia acidoton* L.

Acranthera Arn. ex Meisn., Pl. Vasc. Gen. 1: 162; 2: 115. 16-22 Sep 1838 [*Rub.*].
Typus: *A. ceylanica* Arn. ex Meisn.

(=) *Psilobium* Jack in Malayan Misc. 2(7): 84. 1822.
Typus: *P. nutans* Jack

Acridocarpus Guill. & Perr., Fl. Seneg. Tent.: 123, t. 29. Sept 1831. [*Malpigh.*].
Typus: *A. plagiopterus* Guill. & Perr.

(=) *Anomalopteris* (DC.) G. Don, Gen. Hist. 1: 634, 647. prim. Aug 1831 (*Heteropteris* sect. *Anomalopteris* DC., Prodr. 1: 592. 1824.
Typus: *A. smeathmannii* (DC.) Guill. & Perr. (*Heteropteris smeathmannii* DC.).

Acronychia J. R. Forst. & G. Forst., Char. Gen. Pl.: 27. 29 Nov 1775 [*Rut.*].
Typus: *A. laevis* J. R. Forst. & G. Forst.

(=) *Jambolifera* L., Sp. Pl.: 349. 1 Mai 1753.
Typus: *J. pedunculata* L.

(=) *Cunto* Adans., Fam. Pl. 2: 446, 547. Jul-Aug 1763.
Typus: non designatus.

Actinomeris Nutt., Gen. N. Amer. Pl. 2: 181. 14 Jul 1818 [*Comp.*].
Typus: *A. squarrosa* Nutt., nom. illeg. (*Coreopsis alternifolia* L., *A. alternifolia* (L.) DC.) (typ. cons.).

(≡) *Ridan* Adans., Fam. Pl. 2: 130, 598. Jul-Aug 1763.

Adelia L., Syst. Nat., ed. 10: 1285, 1298. 7 Jun 1759 [*Euphorb.*].
Typus: *A. ricinella* L. (typ. cons.).

(H) *Adelia* P. Browne, Civ. Nat. Hist. Jamaica: 361. 10 Mar 1756 [*Dicot.: Ol.*].
≡ *Forestiera* Poir. 1810 (nom. cons.).

(=) *Bernardia* Mill., Gard. Dict. Abr., ed. 4: [185]. 28 Jan 1754.
Typus (vide Buchheim in Willdenowia 3: 217. 1962): *B. carpinifolia* Griseb. (Fl. Brit. W.I.: 45. Dec 1859) (*Adelia bernardia* L.).

Adenandra Willd., Enum. Pl.: 256. Apr 1809 [*Rut.*].
Typus: *A. uniflora* (L.) Willd. (*Diosma uniflora* L.).

(=) *Haenkea* F. W. Schmidt, Neue Selt. Pfl.: 19. 1793 (ante 17 Jun).
Typus: non designatus.

(=) *Glandulifolia* J. C. Wendl., Coll. Pl. 1: 35. 1805.
Typus: *G. umbellata* J. C. Wendl.

Adenocalymma Mart. ex Meisn., Pl. Vasc. Gen. 1: 300; 2: 208. 25-31 Oct 1840. (*'Adenocalymna'*) (orth. cons.) [*Bignon.*].
Lectotypus: *A. comosum* (Chamisso) Candolle (vide Sandwith, Kew Bull. 15: 453. 1962).

Adesmia DC. in Ann. Sci. Nat. (Paris) 4: 94. Jan 1825 [*Legum.*].
Typus: *A. muricata* (Jacq.) DC. (*Hedysarum muricatum* Jacq.) (typ. cons.).

(≡) *Patagonium* Schrank in Denkschr. Königl. Akad. Wiss. München 1808: 93. 1809.

Adlumia Raf. ex DC., Syst. Nat. 2: 111. Mai (sero) 1821 [*Papaver.*].
Typus: *A. cirrhosa* Raf. ex DC., nom. illeg. (*Fumaria fungosa* Aiton, *A. fungosa* (Aiton) Greene ex Britton & al.).

Adonis L., Sp. Pl.: 547. 1 Mai 1753 [*Ranunc.*].
Typus: *A. annua* L. (typ. cons.: [specimen] Herb. Linnaeus No. 714.3 (LINN)).

Aegle Corrêa in Trans. Linn. Soc. London 5: 222. 1800 [*Rut.*].
Typus: *A. marmelos* (L.) Corrêa (*Crateva marmelos* L.).

(≡) *Belou* Adans., Fam. Pl. 2: 408, 525. Jul-Aug 1763.

Aerva Forssk., Fl. Aegypt.-Arab.: 170. 1 Oct 1775 [*Amaranth.*].
Typus: *A. tomentosa* Forssk. (typ. cons.).

(=) *Ouret* Adans., Fam. Pl. 2: 268, 596. Jul-Aug 1763.
Typus (vide Rickett & Stafleu in Taxon 8: 268. 1959): *Achyranthes lanata* L.

Aeschynanthus Jack in Trans. Linn. Soc. London 14: 42. 28 Mai - 12 Jun 1823 [*Gesner.*]. Typus: *A. volubilis* Jack (typ. cons.).

(=) *Trichosporum* D. Don in Edinburgh Philos. J. 7: 84. 1822. Typus: non designatus.

Afzelia Sm. in Trans. Linn. Soc. London 4: 221. 24 Mai 1798 [*Legum.*]. Typus: *A. africana* Sm. ex Pers. (Syn. Pl. 1: 455. 1 Apr - 15 Jun 1805).

(H) *Afzelia* J. F. Gmel., Syst. Nat. 2: 927. Apr (sero) - Oct 1792 [*Dicot.: Scrophular.*]. ≡ *Seymeria* Pursh 1814 (nom. cons.).

Agalinis Raf., New Fl. 2: 61. Jul-Dec 1837 [*Scrophular.*]. Typus: *A. palustris* Raf., nom. illeg. (*Gerardia purpurea* L., *A. purpurea* (L.) Pennell) (typ. cons.).

(=) *Virgularia* Ruiz & Pav., Fl. Peruv. Prodr.: 92. Oct (prim.) 1794. Typus (vide D'Arcy in Taxon 28: 419-420. 1979): *V. lanceolata* Ruiz & Pav. (Syst. Veg. Fl. Peruv. Chil.: 161. Dec (sero) 1798).

(=) *Chytra* C. F. Gaertn., Suppl. Carp.: 184. 1807. Typus: *C. anomala* C. F. Gaertn.

(=) *Tomanthera* Raf., New Fl. 2: 65. Jul-Dec 1837. Typus (vide D'Arcy in Taxon 28: 419-420. 1979): *T. lanceolata* Raf.

Agathosma Willd., Enum. Pl.: 259. Apr 1809 [*Rut.*]. Typus: *A. villosa* (Willd.) Willd. (*Diosma villosa* Willd.).

(≡) *Bucco* J. C. Wendl., Coll. Pl. 1: 13. 1805.

(=) *Hartogia* L., Syst. Nat., ed. 10: 939, 1365. 7 Jun 1759. Typus: *H. capensis* L.

Aglaia Lour., Fl. Cochinch.: 98, 173. Sep 1790 [*Mel.*]. Typus: *A. odorata* Lour.

(H) *Aglaia* F. Allam. in Nova Acta Phys.-Med. Acad. Caes. Leop.-Carol. Nat. Cur. 4: 93. 1770 [*Monocot.: Cyper.*]. Typus: non designatus.

(=) *Nialel* Adans., Fam. Pl. 2: 446, 582. Jul-Aug 1763. Typus (vide Nicolson & Suresh in Taxon 35: 388. 1986): *Nyalel racemosa* Dennst. ex Kostel. (Allg. Med.-Pharm. Fl.: 2005. Jan-Sep 1836).

Agonis (DC.) Sweet, Hort. Brit., ed. 2: 209. Oct-Dec 1830 (*Leptospermum* sect. *Agonis* DC., Prodr. 3: 226. Mar (med.) 1828) [*Myrt.*].
Typus: *Leptospermum flexuosum* (Willd.) Spreng. (*Metrosideros flexuosa* Willd., *A. flexuosa* (Willd.) Sweet) (typ. cons.).

Ailanthus Desf. in Mém. Acad. Sci. (Paris) 1786: 265. 1788 [*Simaroub.*]. Typus: *A. glandulosa* Desf.

(=) *Pongelion* Adans., Fam. Pl. 2: 319, 593. Jul-Aug 1763.
Typus: *Ailanthus triphysa* (Dennst.) Alston (*Adenanthera triphysa* Dennst.).

Alangium Lam., Encycl. 1: 174. 2 Dec 1783 [*Corn.*].
Typus: *A. decapetalum* Lam. (typ. cons.).

(≡) *Angolam* Adans., Fam. Pl. 2: 85, 518. Jul-Aug 1763.
(=) *Kara-angolam* Adans., Fam. Pl. 2: 84, 532. Jul-Aug 1763.
Typus: *Alangium hexapetalum* Lam.

Aldina Endl., Gen. Pl.: 1322. Oct 1840 [*Legum.*].
Typus: *A. insignis* (Benth.) Endl. (in Walpers, Repert. Bot. Syst. 1: 843. 26-29 Jan 1843) (*Allania insignis* Benth.).

(H) *Aldina* Adans., Fam. Pl. 2: 328, 514. Jul-Aug 1763 [*Dicot.: Legum.*].
≡ *Brya* P. Browne 1756.

Alkanna Tausch in Flora 7: 234. 21 Apr 1824 [*Boragin.*].
Typus: *A. tinctoria* Tausch (typ. cons.).

(H) *Alkanna* Adans., Fam. Pl. 2: 444, 514. Jul-Aug 1763 [*Dicot.: Lythr.*].
≡ *Lawsonia* L. 1753.

Allionia L., Syst. Nat., ed. 10: 883, 890, 1361. 7 Jun 1759 [*Nyctagin.*].
Typus: *A. incarnata* L. (typ. cons.) (etiam vide *Wedelia* [*Dicot.*]).

(H) *Allionia* Loefl., Iter Hispan.: 181. Dec 1758 [*Dicot.: Nyctagin.*].
Typus: *A. violacea* L. (Syst. Nat., ed. 10: 890. 7 Jun 1759).

Alloplectus Mart., Nov. Gen. Sp. Pl. 3: 53. Jan-Jun 1829 [*Gesner.*].
Typus: *A. hispidus* (Kunth) Mart. (*Besleria hispida* Kunth) (typ. cons.).

(=) *Crantzia* Scop., Intr. Hist. Nat.: 173. Jan-Apr 1777.
Typus: *Besleria cristata* L.
(=) *Vireya* Raf. in Specchio Sci. 1: 194. 1 Jun 1814.
Typus: *V. sanguinolenta* Raf.

Alstonia R. Br., Asclepiadeae: 64. 3 Apr 1810 [*Apocyn.*].
Typus: *A. scholaris* (L.) R. Br. (*Echites scholaris* L.) (typ. cons.).

(H) *Alstonia* Scop., Intr. Hist. Nat.: 198. Jan-Apr 1777 [*Dicot.: Apocyn.*].
≡ *Pacouria* Aubl. 1775 (nom. rej. sub *Landolphia*).

Alvesia Welw. in Trans. Linn. Soc. London 27: 55. 24 Dec 1869 [*Lab.*].
Typus: *A. rosmarinifolia* Welw.

(H) *Alvesia* Welw. in Ann. Cons. Ultramarino, ser. 1: 587. Dec 1859 [*Dicot.: Legum.*].
Typus: *A. bauhinioides* Welw.

Alysicarpus Desv. in J. Bot. Agric. 1: 120. Feb 1813 [*Legum.*].
Typus: *A. bupleurifolius* (L.) DC. (Prodr. 2: 352. Nov (med.) 1825) (*Hedysarum bupleurifolium* L.) (typ. cons.).

Alyxia Banks ex R. Br., Prodr.: 469. 27 Mar 1810 [*Apocyn.*].
Typus: *A. spicata* R. Br. (typ. cons.).

(=) *Gynopogon* J. R. Forst. & G. Forst., Char. Gen. Pl.: 18. 29 Nov 1775.
Typus (vide Grant & al. in Smithsonian Contr. Bot. 17: 46. 1974): *G. stellatus* J. R. Forst. & G. Forst..

Amaracus Gled., Syst. Pl. Stamin. Situ: 189. 1764 (ante 13 Sep) [*Lab.*].
Typus: *A. dictamnus* (L.) Benth. (Labiat. Gen. Sp.: 323. Mai 1834) (*Origanum dictamnus* L.) (typ. cons.).

(H) *Amaracus* Hill, Brit. Herb.: 381. 13 Oct 1756 [*Dicot.: Lab.*].
≡ *Majorana* Mill. 1754 (nom. cons.).

(=) *Hofmannia* Heist. ex Fabr., Enum.: 61. 1759.
Typus: *Origanum sipyleum* L.

Amasonia L. f., Suppl. Pl.: 48, 294. Apr 1782 [*Verben.*].
Typus: *A. erecta* L. f.

(=) *Taligalea* Aubl., Hist. Pl. Guiane: 625. Jun-Dec 1775.
Typus: *T. campestris* Aubl.

Amberboa Vaill. in Königl. Akad. Wiss. Paris Phys. Abh. 5: 182. Jan-Apr 1754 (*'Amberboi'*) (orth. cons.) [*Comp.*].
Typus: *Centaurea moschata* L. (*A. moschata* (L.) DC.).

Amellus L., Syst. Nat., ed. 10: 1189, 1225, 1377. 7 Jun 1759 [*Comp.*].
Typus: *A. lychnitis* L. (typ. cons.).

(H) *Amellus* P. Browne, Civ. Nat. Hist. Jamaica: 317. 10 Mar 1756 [*Dicot.: Comp.*].
Typus: *Santolina amellus* L.

Ampelocissus Planch. in Vigne Amér. Vitic. Eur. 8: 371. Dec 1884 [*Vit.*]. Typus: *A. latifolia* (Roxb.) Planch. (*Vitis latifolia* Roxb.) (typ. cons.).

(=) *Botria* Lour., Fl. Cochinch.: 96, 153. Sep 1790. Typus: *B. africana* Lour.

Amphicarpaea Elliott ex Nutt., Gen. N. Amer. Pl. 2: 113. 14 Jul 1818 (*'Amphicarpa'*) (orth. cons.) [*Legum.*]. Typus: *A. monoica* Elliott ex Nutt., nom. illeg. (*Glycine monoica* L., nom. illeg., *Glycine bracteata* L., *A. bracteata* (L.) Fernald).

(=) *Falcata* J. F. Gmel., Syst. Nat. 2: 1131. Apr (sero) - Oct 1791. Typus: *F. caroliniana* J. F. Gmel.

Amphirrhox Spreng., Syst. Veg. 4(2): 51, 99. Jan-Jun 1827 [*Viol.*]. Typus: *A. longifolia* (A. St.-Hil.) Spreng. (*Spathularia longifolia* A. St.-Hil.).

Amsinckia Lehm., Sem. Hort. Bot. Hamburg. 1831: 3, 7. 1831 [*Boragin.*]. Typus: *A. lycopsoides* Lehm.

Anacampseros L., Opera Var.: 232. 1758 [*Portulac.*]. Typus: *A. telephiastrum* DC. (Cat. Pl. Horti Monsp.: 77. Feb-Mar 1813) (*Portulaca anacampseros* L.).

(H) *Anacampseros* Mill., Gard. Dict. Abr., ed. 4: [73]. 28 Jan 1754 [*Dicot.: Crassul.*]. Typus: non designatus.

Anarrhinum Desf., Fl. Atlant. 2: 51. Oct 1798 [*Scrophular.*]. Typus: *A. pedatum* Desf. (typ. cons.).

(=) *Simbuleta* Forssk., Fl. Aegypt.-Arab.: 115. 1 Oct 1775. Typus: *S. forskaohlii* J. F. Gmel. (Syst. Nat. 2: 242. Sep (sero) - Nov 1791).

Ancistrocarpus Oliv. in J. Linn. Soc., Bot. 9: 173. 12 Oct 1865 [*Til.*]. Typus: *A. brevispinosus* Oliv. (typ. cons.).

(H) *Ancistrocarpus* Kunth in Humboldt & al., Nov. Gen. Sp. 2, ed. 4°: 186; ed. f°: 149. 8 Dec 1817 [*Dicot.: Phytolacc.*]. Typus: *A. maypurensis* Kunth

Ancistrocladus Wall., Numer. List: No. 1052. 1829 [*Ancistroclad.*]. Typus: *A. hamatus* (Vahl) Gilg (in Engler & Prantl, Nat. Pflanzenfam. 3(6): 276. 19 Feb 1895) (*Wormia hamata* Vahl).

(=) *Bembix* Lour., Fl. Cochinch.: 259, 282. Sep 1790. Typus: *B. tectoria* Lour.

Andira Lam., Encycl. 1: 171. 1783 [*Legum.*]. Typus: *A. inermis* (W. Wright) DC. (*Geoffrea inermis* W. Wright) (typ. cons.).

Androstachys Prain in Bull. Misc. Inform. Kew 1908: 438. Dec 1908 [*Euphorb.*]. Typus: *A. johnsonii* Prain

(H) *Androstachys* Grand'Eury in Mém. Divers Savants Acad. Roy. Sci. Inst. Roy. France, Sci. Math. 24(1): 190. 1877 [Foss.]. Typus: *A. frondosa* Grand'Eury

Anemone L., Sp. Pl.: 538. 1 Mai 1753 [*Ranunc.*]. Typus: *A. coronaria* L. (typ. cons.: [specimen] Herb. Linnaeus No. 710.9 (LINN)).

Anemopaegma Mart. ex Meisn., Pl. Vasc. Gen. 1: 300; 2: 208. 25-31 Oct 1840 (*'Anemopaegmia'*) (orth. cons.) [*Bignon.*]. Typus: *A. mirandum* (Cham.) DC. (*Bignonia miranda* Cham.).

(=) *Cupulissa* Raf., Fl. Tellur. 2: 57. Jan-Mar 1837. Typus: *C. grandifolia* (Jacq.) Raf. (*Bignonia grandifolia* Jacq.).

(=) *Platolaria* Raf., Sylva Tellur.: 78. Oct-Dec 1838. Typus: *P. flavescens* Raf., nom. illeg. (*Bignonia orbiculata* Jacq.).

Angianthus J. C. Wendl., Coll. Pl. 2: 31. 1808 [*Comp.*]. Typus: *A. tomentosus* J. C. Wendl.

(=) *Siloxerus* Labill, Nov. Holl. Pl. 2: 57. Jun 1806. Typus: *S. humifusus* Labill.

Angostylis Benth. in Hooker's J. Bot. Kew Gard. Misc. 6: 328. Nov 1854 (*'Angostyles'*) (orth. cons.). [*Euphorb.*]. Typus: *A. longifolia* Benth.

Anisotes Nees in Candolle, Prodr. 11: 424. 25 Nov 1847 [*Acanth.*]. Typus: *A. trisulcus* (Forssk.) Nees (*Dianthera trisulca* Forssk.).

(H) *Anisotes* Lindl. ex Meisn., Pl. Vasc. Gen. 1: 117, 2: 84. 8-14 Apr 1838 [*Dicot.: Lythr.*]. Typus: *A. hilariana* Meisn., nom. illeg. (*Lythrum anomalum* A. St.-Hil.).

(≡) *Calasias* Raf., Fl. Tellur. 4: 64. 1838 (med.).

Anneslea Wall., Pl. Asiat. Rar. 1: 5. Sep 1829 [*The.*]. Typus: *A. fragrans* Wall.

(H) *Anneslia* Salisb., Parad. Lond.: ad t. 64. 1 Mar 1807 [*Dicot.: Legum.*]. Typus: *A. falcifolia* Salisb., nom. illeg. (*Gleditsia inermis* L.).

Antennaria Gaertn., Fruct. Sem. Pl. 2: 410. Sep-Dec 1791 [*Comp.*]. Typus: *A. dioica* (L.) Gaertn. (*Gnaphalium dioicum* L.).

(H) *Antennaria* Link in Neues J. Bot. 3(1,2): 16. Apr 1809 : Fr., Syst. Mycol. 1: xlvii. 1 Jan 1821 [*Fungi*]. Typus: *A. ericophila* Link:Fr.

Anthriscus Pers., Syn. Pl. 1: 320. 1 Apr - 15 Jun 1805 [*Umbell.*]. Typus: *A. vulgaris* Pers. 1805, non Bernh. 1800 (*Scandix anthriscus* L., *A. caucalis* M. Bieb.) (typ. cons.).

(H) *Anthriscus* Bernh., Syst. Verz.: 113. 1800 [*Dicot.: Umbell.*]. Typus: *A. vulgaris* Bernh. (*Tordylium anthriscus* L.).

(=) *Cerefolium* Fabr., Enum.: 36. 1759. Typus: *Scandix cerefolium* L.

Antiaris Lesch. in Ann. Mus. Natl. Hist. Nat. 16: 478. 1810 [*Mor.*]. Typus: *A. toxicaria* Lesch.

(=) *Ipo* Pers., Syn. Pl. 2: 566. Sep 1807. Typus: *I. toxicaria* Pers.

Aphananthe Planch. in Ann. Sci. Nat., Bot., ser. 3, 10: 265. Nov 1848 [*Ulm.*]. Typus: *A. philippinensis* Planch. in Ann. Sci. Nat., Bot., ser. 3, 10: 337. Dec 1848.

(H) *Aphananthe* Link, Enum. Hort. Berol. Alt. 1: 383. 16 Mar - 30 Jun 1821 [*Dicot.: Phytolacc.*]. Typus: *A. celosioides* (Spreng.) Link (*Galenia celosioides* Spreng.).

Apios Fabr., Enum.: 176. 1759 [*Legum.*]. Typus: *A. americana* Medik. (in Vorles. Churpfälz. Phys.-Ökon. Ges. 2: 354. 1787) (*Glycine apios* L.).

Aptosimum Burch. ex Benth. in Edwards's Bot. Reg.: ad t. 1882. 1 Aug 1836 [*Scrophular.*].
Typus: *A. depressum* Burch. ex Benth., nom. illeg. (*Ohlendorffia procumbens* Lehm., *A. procumbens* (Lehm.) Steud.).

(≡) *Ohlendorffia* Lehm., Sem. Hort. Bot. Hamburg. 1835: 7. 1835.

Apuleia Mart. in Flora 20(2, Beibl.): 123. 21 Nov 1837 (*'Apuleja'*) (orth. cons.) [*Legum.*].
Typus: *A. praecox* Mart.

(H) *Apuleja* Gaertn., Fruct. Sem. Pl. 2: 439. Sep-Dec 1791 [*Dicot.: Comp.*].
≡ *Berkheya* Ehrh. 1788 (nom. cons.).

Aquilaria Lam., Encycl. 1: 49. 2 Dec 1783 [*Thymel.*].
Typus: *A. malaccensis* Lam.

(=) *Agallochum* Lam., Encycl. 1: 48. Dec 1783.
Typus: non designatus.

Arabidopsis Heynh. in Holl & Heynhold, Fl. Sachsen 1(2): 538. 1842 [*Cruc.*].
Typus: *A. thaliana* (L.) Heynh. (*Arabis thaliana* L.) (typ. cons.).

Araliopsis Engl. in Engler & Prantl, Nat. Pflanzenfam. 3(4): 175. Mar 1896 [*Rut.*].
Typus: *A. soyauxii* Engl.

Arceuthobium M. Bieb., Fl. Taur.-Caucas. 3: 629. 1819 (sero) - 1820 (prim.) [*Loranth.*].
Typus: *A. oxycedri* (DC.) M. Bieb. (*Viscum oxycedri* DC.).

(=) *Razoumofskya* Hoffm., Hort. Mosq.: 1. Jun-Dec 1808.
Typus: *R. caucasica* Hoffm.

Arctostaphylos Adans., Fam. Pl. 2: 165, 520. Jul-Aug 1763 [*Eric.*].
Typus: *A. uva-ursi* (L.) Spreng. (*Arbutus uva-ursi* L.).

(≡) *Uva-ursi* Duhamel, Traité Arbr. Arbust. 2: 371. 1755.

Ardisia Sw., Prodr.: 3, 48. 20 Jun - 29 Jul 1788 [*Myrsin.*].
Typus: *A. tinifolia* Sw. (typ. cons.).

(=) *Katoutheka* Adans., Fam. Pl. 2: 159, 534. Jul-Aug 1763.
Typus (vide Ridsdale in Manilal, Bot. Hist. Hort. Malab.: 136. 1980): *Psychotria dalzellii* Hook. f.

(=) *Vedela* Adans., Fam. Pl. 2: 502, 617. Jul-Aug 1763.
Typus: non designatus.

(=) *Icacorea* Aubl., Hist. Pl. Guiane: Suppl. 1. Jun-Dec 1775.
Typus: *I. guianensis* Aubl.

(=) *Bladhia* Thunb., Nova Gen. Pl.: 6. 24 Nov 1781.
Typus: *B. japonica* Thunb.

Aremonia Neck. ex Nestl., Monogr. Potentilla: iv, 17. Jun 1816.
Typus: *A. agrimonoides* (L.) DC. (Prodr. 2: 588. Nov (med.) 1825) (*Agrimonia agrimonoides* L.) [*Ros.*].

(≡) *Agrimonoides* Mill., Gard. Dict. Abr., ed. 4: [42]. 28 Jan 1754.

Argania Roem. & Schult., Syst. Veg. 4: xlvi, 502. Mar-Jun 1819 [*Sapot.*].
Typus: *A. sideroxylon* Roem. & Schult., nom. illeg. (*Sideroxylon spinosum* L., *A. spinosa* (L.) Skeels).

Argyrolobium Eckl. & Zeyh., Enum. Pl. Afric. Austral.: 184. Jan 1836 [*Legum.*].
Typus: *A. argenteum* (Jacq.) Eckl. & Zeyh. (*Crotalaria argentea* Jacq.) (typ. cons.).

(=) *Lotophyllus* Link, Handbuch 2: 156. Jan-Aug 1831.
Typus: *L. argenteus* Link

Aristotelia L'Hér., Stirp. Nov.: 31. Dec 1785 (sero) - Jan 1786 [*Elaeocarp.*].
Typus: *A. macqui* L'Hér.

(H) *Aristotela* Adans., Fam. Pl. 2: 125, 520. Jul-Aug 1763 [*Dicot.: Comp.*].
≡ *Othonna* L. 1753.

Armeria Willd., Enum. Pl.: 333. Apr 1809 [*Plumbagin.*].
Typus: *A. vulgaris* Willd. (*Statice armeria* L.).

(≡) *Statice* L., Sp. Pl.: 274. 1 Mai 1753 (typ. des.: Hitchcock in Sprague & al., Nom. Prop. Brit. Bot.: 143. 1929).

Armoracia P. Gaertn. & al., Oekon. Fl. Wetterau 2: 426. Mai-Jul 1800 [*Cruc.*].
Typus: *A. rusticana* P. Gaertn. & al. (*Cochlearia armoracia* L.).

(≡) *Raphanis* Moench, Methodus: 267. 4 Mai 1794.

Aronia Medik., Philos. Bot. 1: 155. Apr 1789 [*Ros.*].
Typus: *A. arbutifolia* (L.) Pers. (Syn. Pl. 2: 39. Nov 1806) (*Mespilus arbutifolia* L.).

(H) *Aronia* Mitch., Diss. Princ. Bot.: 28. 1769 [*Monocot.: Ar.*].
≡ *Orontium* L. 1753.

Artanema D. Don in Sweet, Brit. Fl. Gard. 6: ad t. 234. 1 Apr 1834 [*Scrophular.*].
Typus: *A. fimbriatum* (Hook. ex Graham) D. Don (*Torenia fimbriata* Hook. ex Graham).

(=) *Bahel* Adans., Fam. Pl. 2: 210, 523. Jul-Aug 1763.
Typus: *Columnea longifolia* L.

Arthrocereus A. Berger, Kakteen: 337. Jul-Aug 1929 [*Cact.*].
Typus: [icon] *'Cereus damazoi'* in Monatsschr. Kakteenk. 28: [63]. 1918 (typ. cons.) [= *A. glaziovii* (K. Schum.) N. P. Taylor & Zappi (*Cereus glaziovii* K. Schum.)].

Artocarpus J. R. Forst. & G. Forst., Char. Gen. Pl.: 51. 29 Nov 1775 [*Mor.*].
Typus: *A. communis* J. R. Forst. & G. Forst.

(=) *Sitodium* Parkinson, J. Voy. South Seas: 45. Jul 1773.
Typus: *S. altile* Parkinson

Asperula L., Sp. Pl.: 103. 1 Mai 1753 [*Rub.*].
Typus: *A. arvensis* L. (typ. cons.).

Aspidosperma Mart. & Zucc. in Flora 7 (1, Beil.): 135. Mai-Jun 1824 [*Apocyn.*].
Typus: *A. tomentosum* Mart. & Zucc. (typ. cons.).

(=) *Coutinia* Vell., Quinogr. Port.: 166. 1799.
Typus: *C. illustris* Vell.
(=) *Macaglia* Rich. ex Vahl in Skr. Naturhist.-Selsk. 6: 107. 1810.
Typus (vide Woodson in Ann. Missouri Bot. Gard. 38: 136. 1951): *M. alba* Vahl

Astronidium A. Gray, U.S. Expl. Exped., Phan.: 581. Jun 1854 [*Melastomat.*].
Typus: *A. parviflorum* A. Gray

(=) *Lomanodia* Raf., Sylva Tellur.: 97. Oct-Dec 1838.
Typus (vide Veldkamp in Taxon 32: 134. 1983): *L. glabra* (G. Forst.) Raf. (*Melastoma glabrum* G. Forst.).

Atalantia Corrêa in Ann. Mus. Natl. Hist. Nat. 6: 383, 385, 386. 1805 [*Rut.*].
Typus: *A. monophylla* (L.) DC. (Prodr. 1: 535. Jan (med.) 1824) (*Limonia monophylla* L.) (typ. cons.: [specimen] India, Kerala State, Idukki District, Chinnar, c. 400 m, 10 Oct 1994, *Santhosh Kumar 17590* (BM)).

(=) *Malnaregam* Adans., Fam. Pl. 2: 345, 574. Jul-Aug 1763.
Typus: *M. malabarica* Raf. (Sylva Tellur.: 143. Oct-Dec 1838).

Athenaea Sendtn. in Martius, Fl. Bras. 10: 133. 1 Jul 1846 [*Solan.*].
Typus: *A. picta* (Mart.) Sendtn. (*Witheringia picta* Mart.) (typ. cons.).

(H) *Athenaea* Adans., Fam. Pl. 2: 121, 522. Jul-Aug 1763 [*Dicot.: Comp.*]. ≡ *Struchium* P. Browne 1756.

(=) *Deprea* Raf., Sylva Tellur.: 57. Oct-Dec 1838.
Typus (vide D'Arcy in Ann. Missouri Bot. Gard. 60: 624. 1974): *D. orinocensis* (Kunth) Raf. (*Physalis orinocensis* Kunth).

Atriplex L., Sp. Pl.: 1052. 1 Mai 1753 [*Chenopod.*].
Typus: *A. hortensis* L. (typ. cons.: [specimen] Herb. Clifford: 469, *Atriplex* 1, BM).

Augea Thunb., Prodr. Pl. Cap. 1: [viii], 80. 1794 [*Zygophyll.*].
Typus: *A. capensis* Thunb.

(H) *Augia* Lour., Fl. Cochinch.: 327, 337. Sep 1790 [*Dicot.: Anacard.*].
Typus: *A. sinensis* Lour.

Augusta Pohl, Pl. Bras. Icon. Descr. 2: 1. 1828 (sero) - Feb 1829 [*Rub.*].
Typus: *A. lanceolata* Pohl (typ. cons.) [= *A. longifolia* (Spreng.) Rehder (*Ucriana longifolia* Spreng.)].

(H) *Augusta* Leandro in Denkschr. Königl. Akad. Wiss. München 7: 235. Jul-Dec 1821 [*Dicot.: Comp.*].
Typus: non designatus.

Baccharis L., Sp. Pl.: 860. 1 Mai 1753 [*Comp.*].
Typus: *B. halimifolia* L. (typ. cons.).

Bacopa Aubl., Hist. Pl. Guiane: 128. Jun-Dec 1775 [*Scrophular.*]. Typus: *B. aquatica* Aubl.

(=)   *Moniera* P. Browne, Civ. Nat. Hist. Jamaica: 269. 10 Mar 1756. Typus: non designatus.

(=)   *Brami* Adans., Fam. Pl. 2: 208, 527. Jul-Aug 1763. Typus: *B. indica* Lam. (Encycl. 1: 456. 1 Aug 1785).

Balanites Delile, Descr. Egypte, Hist. Nat. 2: 221. 1813 (sero) - 1814 (prim.) [*Zygophyll.*]. Typus: *B. aegyptiacus* (L.) Delile (*Ximenia aegyptiaca* L.).

(≡)   *Agialid* Adans., Fam. Pl. 2: 508, 514. Jul-Aug 1763.

Balbisia Cav. in Anales Ci. Nat. 7: 61. Feb 1804 [*Geran.*]. Typus: *B. verticillata* Cav.

(H)   *Balbisia* Willd., Sp. Pl. 3: 1486, 2214. Apr-Dec 1803 [*Dicot.: Comp.*]. Typus: *B. elongata* Willd.

Balboa Planchon & Triana in Ann. Sci. Nat., Bot., ser. 4, 13: 315. Mai 1860 [*Guttif.*]. Typus: *B. membranaceum* Planchon & Triana

(H)   *Balboa* Liebm. ex Didr. in Vidensk. Meddel. Dansk Naturhist. Foren. Kjøbenhavn 1853: 106. 1853 [*Dicot.: Legum.*]. Typus: *B. diversifolia* Liebm. ex Didr.

Balduina Nutt., Gen. N. Amer. Pl. 2: 175. 14 Jul 1818 [*Comp.*]. Typus: *B. uniflora* Nutt. (typ. cons.).

(=)   *Mnesiteon* Raf., Fl. Ludov.: 67. Oct - Dec (prim.) 1817. Typus: non designatus.

Baltimora L., Mant. Pl.: 158, 288. Oct 1771 [*Comp.*] Typus: *B. recta* L. (typ. cons.: [specimen] Vera Cruz, s.d., *Houstoun* (BM)).

Banksia L. f., Suppl. Pl.: 15, 126. Apr 1782 [*Prot.*]. Typus: *B. serrata* L. f.

(H)   *Banksia* J. R. Forst. & G. Forst., Char. Gen. Pl.: 4. 29 Nov 1775 [*Dicot.: Thymel.*]. Typus: non designatus.

Barbarea W. T. Aiton, Hort. Kew., ed. 2, 4: 109. Dec 1812 [*Cruc.*]. Typus: *B. vulgaris* W. T. Aiton (*Erysimum barbarea* L.).

(H)   *Barbarea* Scop., Fl. Carniol.: 522. 15 Jun - 21 Jul 1760 [*Dicot.: Cruc.*]. Typus: *Dentaria bulbifera* L.

Barclaya Wall. in Trans. Linn. Soc. London 15: 442. 11-20 Dec 1827 [*Nymph.*].
Typus: *B. longifolia* Wall.

(≡) *Hydrostemma* Wall. in Philos. Mag. Ann. Chem. 1: 454. Jun 1827.

Barosma Willd., Enum. Pl.: 257. Apr 1809 [*Rut.*].
Typus: *B. serratifolia* (Curt.) Willd. (*Diosma serratifolia* Curt.).

(=) *Parapetalifera* J. C. Wendl., Coll. Pl. 1: 49. 1806.
Typus: *P. odorata* J. C. Wendl.

Barringtonia J. R. Forst. & G. Forst., Char. Gen. Pl.: 38. 29 Nov 1775 [*Lecythid.*].
Typus: *B. speciosa* J. R. Forst. & G. Forst.

(=) *Huttum* Adans., Fam. Pl. 2: 88, 616. Jul-Aug 1763.
Typus: non designatus.

Bartonia Muhl. ex Willd. in Ges. Naturf. Freunde Berlin Neue Schriften 3: 444. 1801 (post 21 Apr) [*Gentian.*].
Typus: *B. tenella* Willd. [= *B. virginica* (L.) Britton & al. (*Sagina virginica* L.)].

Bartsia L., Sp. Pl.: 602. 1 Mai 1753 [*Scrophular.*].
Typus: *B. alpina* L. (typ. cons.).

Bejaria Mutis in L., Mant. Pl.: 152, 242. Oct 1771 (*'Befaria'*) (orth. cons.) [*Eric.*]
Typus: *B. aestuans* Mutis (*'stuans'*) (typ. cons.: [specimen] Colombia, Santander, road above Río Chicamocha, 60 km NNE of Barbosa, 1700 m, 9 Mai 1979, *Luteyn & al. 7616* (NY; isotype: COL)).

Bellucia Neck. ex Raf., Sylva Tellur.: 92. Oct-Dec 1838 [*Melastomat.*].
Typus: *B. nervosa* Raf., nom. illeg. (*Blakea quinquenervia* Aubl., *Bellucia quinquenervia* (Aubl.) H. Karst.) [= *Bellucia grossularioides* (L.) Triana (*Melastoma grossularioides* L.)].

(H) *Belluccia* Adans., Fam. Pl. 2: 344, 525. Jul-Aug 1763 [*Dicot.: Rut.*].
≡ *Ptelea* L. 1753.

(≡) *Apatitia* Desv. ex Ham., Prodr. Pl. Ind. Occid.: 42. 1825.

Belmontia E. Mey., Comment. Pl. Afr. Austr.: 183. 1-8 Jan 1838 [*Gentian.*].
Typus: *B. cordata* E. Mey., nom. illeg. (*Sebaea cordata* Roem. & Schult., nom. illeg., *Gentiana exacoides* L., *B. exacoides* (L.) Druce) (typ. cons.).

(≡) *Parrasia* Raf., Fl. Tellur. 3: 78. Nov-Dec 1837.

Berchemia Neck. ex DC., Prodr. 2: 22. Nov (med.) 1825 [*Rhamn.*].
Typus: *B. volubilis* (L. f.) DC. (*Rhamnus volubilis* L. f.) [= *B. scandens* (Hill) K. Koch (*Rhamnus scandens* Hill)].

(≡) *Oenoplea* Michx. ex R. Hedw., Gen. Pl.: 151. Jul 1806.

Bergenia Moench, Methodus: 664. 4 Mai 1794 [*Saxifrag.*].
Typus: *B. bifolia* Moench, nom. illeg. (*Saxifraga crassifolia* L., *B. crassifolia* (L.) Fritsch).

(H) *Bergena* Adans., Fam. Pl. 2: 345, 525. Jul-Aug 1763 [*Dicot.: Lecythid.*].
≡ *Lecythis* Loefl. 1758.

Berkheya Ehrh. in Neues Mag. Aerzte 6: 303. 12 Mai - 7 Sep 1784 [*Comp.*].
Typus: *B. fruticosa* (L.) Ehrh. (*Atractylis fruticosa* L.).

(≡) *Crocodilodes* Adans., Fam. Pl. 2: 127, 545. Jul-Aug 1763.

Berlinia Sol. ex Hook. f. in Hooker, Niger Fl.: 326. Nov-Dec 1849 [*Legum.*].
Typus: *B. acuminata* Sol. ex Hook. f.

(=) *Westia* Vahl in Skr. Naturhist.-Selsk. 6: 117. 1810.
Typus: non designatus.

Bernieria Baill. in Bull. Mens. Soc. Linn. Paris: 434. 1884 [*Laur.*].
Typus: *B. madagascariensis* Baill.

(H) *Berniera* DC., Prodr. 7: 18. Apr (sero) 1838 [*Dicot.: Comp.*].
Typus: *B. nepalensis* DC., nom. illeg. (*Chaptalia maxima* D. Don).

Bernoullia Oliv. in Hooker's Icon. Pl.: ad t. 1169-1170. Dec 1873 [*Bombac.*].
Typus: *B. flammea* Oliv.

(H) *Bernullia* Neck. ex Raf., Autik. Bot.: 173. 1840 [*Dicot.: Ros.*].
Typus: non designatus.

Berrya Roxb., Pl. Coromandel 3: 60. 18 Feb 1820 (*'Berria'*) (orth. cons.) [*Til.*].
Typus: *B. ammonilla* Roxb.

(=) *Espera* Willd. in Ges. Naturf. Freunde Berlin Neue Schriften 3: 450. 1801 (post 21 Apr).
Typus: *E. cordifolia* Willd.

Bertolonia Raddi, Quar. Piant. Nuov. Bras.: 5. 1820 [*Melastomat.*].
Typus: *B. nymphaeifolia* Raddi

(H)   *Bertolonia* Spin, Jard. St. Sébastien, ed. 1909: 24. 1809 [*Dicot.: Myopor.*].
Typus: *B. glandulosa* Spin

Bifora Hoffm., Gen. Pl. Umbell., ed. 2: xxxiv, 191. 1816 (post 15 Mai) [*Umbell.*].
Typus: *B. dicocca* Hoffm., nom. illeg. (*Coriandrum testiculatum* L., *B. testiculata* (L.) Spreng. ex Roem. & Schult.).

Bigelowia DC., Prodr. 5: 329. 1-10 Oct 1836 [*Comp.*].
Typus: *B. nudata* (Michx.) DC. (*Chrysocoma nudata* Michx.) (typ. cons.).

(H)   *Bigelowia* Raf. in Amer. Monthly Mag. & Crit. Rev. 1: 442. Oct 1817 [*Dicot.: Caryophyll.*].
Typus: *B. montana* Raf.

(≡)   *Pterophora* L., Pl. Rar. Afr.: 17. 20 Dec 1760.

Bignonia L., Sp. Pl.: 622. 1 Mai 1753 [*Bignon.*].
Typus: *B. capreolata* L. (typ. cons.).

Bikkia Reinw. in Syll. Pl. Nov. 2: 8. 1825 [*Rub.*].
Typus: *B. grandiflora* Reinw., nom. illeg. (*Portlandia tetrandra* L. f., *B. tetrandra* (L. f.) A. Gray).

Billia Peyr. in Bot. Zeitung (Berlin) 16: 153. 28 Mai 1858 [*Hippocastan.*].
Typus: *B. hippocastanum* Peyr.

(H)   *Billya* Cass. in Cuvier, Dict. Sci. Nat. 34: 38. Apr 1825 [*Dicot.: Comp.*].
Typus: *B. bergii* Cass.

Biscutella L., Sp. Pl.: 652. 1 Mai 1753 [*Cruc.*].
Typus: *B. didyma* L. (typ. cons.: [specimen] Herb. Clifford: 329, *Biscutella* 2 (BM)).

Bivonaea DC. in Mém. Mus. Hist. Nat. 7: 241. 20 Apr 1821 [*Cruc.*].
Typus: *B. lutea* (Biv.) DC. (Syst. Nat. 2: 255. Mai (sero) 1821) (*Thlaspi luteum* Biv.).

(H)   *Bivonea* Raf. in Specchio Sci. 1: 156. 1 Mai 1814 [*Dicot.: Euphorb.*].
Typus: *B. stimulosa* (Michx.) Raf. (*Jatropha stimulosa* Michx.).

Blachia Baill., Etude Euphorb.: 385. 1858 [*Euphorb.*].
Typus: *B. umbellata* (Willd.) Baill. (*Croton umbellatus* Willd.).

(=) *Bruxanellia* Dennst. ex Kostel., Allg. Med.-Pharm. Fl.: 2002. Jan-Sep 1836.
Typus: *B. indica* Dennst. ex Kostel.

Blumea DC. in Arch. Bot. (Paris) 2: 514. 23 Dec 1833 [*Comp.*].
Typus: *B. balsamifera* (L.) DC. (Prodr. 5: 447. 1-10 Oct 1836) (*Conyza balsamifera* L.) (typ. cons.).

(H) *Blumia* Nees in Flora 8: 152. 14 Mar 1825 [*Dicot.: Magnol.*].
Typus: *B. candollei* (Blume) Nees (*Talauma candollei* Blume).

(=) *Placus* Lour., Fl. Cochinch.: 475, 496. Sep 1790.
Typus (vide Merrill in Trans. Amer. Philos. Soc., ser. 2, 24: 387. 1935): *P. tomentosus* Lour.

Blumenbachia Schrad. in Gött. Gel. Anz. 1825: 1705. 24 Oct 1825 [*Loas.*].
Typus: *B. insignis* Schrad.

(H) *Blumenbachia* Koeler, Descr. Gram.: 28. 1802 [*Monocot.: Gram.*].
Typus: *B. halepensis* (L.) Koeler (*Holcus halepensis* L.).

Boenninghausenia Rchb. ex Meisn., Pl. Vasc. Gen. 1: 60; 2: 44. 21-27 Mai 1837 [*Rut.*].
Typus: *B. albiflora* (Hook.) Meisn. (*Ruta albiflora* Hook.).

(H) *Boenninghausia* Spreng., Syst. Veg. 3: 153, 245. Jan-Mar 1826 [*Dicot.: Legum.*].
Typus: *B. vincentina* (Ker-Gawl.) Spreng. (*Glycine vincentina* Ker-Gawl.).

Bombacopsis Pittier in Contr. U.S. Natl. Herb. 18: 162. 3 Mar 1916 [*Bombac.*].
Typus: *B. sessilis* (Benth.) Pittier (*Pachira sessilis* Benth.).

(=) *Pochota* Ram. Goyena, Fl. Nicarag. 1: 198. 1909.
Typus: *P. vulgaris* Ram. Goyena

Bombax L., Sp. Pl.: 511. 1 Mai 1753 [*Bombac.*].
Typus: *B. ceiba* L. (typ. cons.: [icon in] Rheede, Hort. Malab. 3: t. 52. 1682).

Bonamia Thouars, Hist. Vég. Iles France: 33. 1804 (ante 22 Sep) [*Convolvul.*].
Typus: *B. alternifolia* J. St.-Hil. (Expos. Fam. Nat. 2: 349. Feb-Aug 1805).

Bonannia Guss., Fl. Sicul. Syn. 1: 355. Feb 1843 [*Umbell.*].
Typus: *B. resinifera* Guss., nom. illeg. (*Ferula nudicaulis* Spreng., *B. nudicaulis* (Spreng.) Rickett & Stafleu).

(H) *Bonannia* Raf. in Specchio Sci. 1: 115. 1 Apr 1814 [*Dicot.: Sapind.*].
Typus: *B. nitida* Raf.

Bonnetia Mart., Nov. Gen. Sp. Pl. 1: 114. Jan-Mar 1826 [*The.*].
Typus: *B. anceps* Mart. (typ. cons.).

(H) *Bonnetia* Schreb., Gen. Pl.: 363. Apr 1789 [*Dicot.: The.*].
≡ *Mahurea* Aubl. 1775.

(=) *Kieseria* Nees in Wied-Neuwied, Reise Bras. 2: 338. Jan-Jun 1821.
Typus: *K. stricta* Nees

Borreria G. Mey., Prim. Fl. Esseq.: 79. Nov 1818 [*Rub.*].
Typus: *B. suaveolens* G. Mey. (typ. cons.).

(H) *Borrera* Ach., Lichenogr. Universalis: 93, 496. Apr-Mai 1810 [*Fungi*].
Typus: non designatus.

(=) *Tardavel* Adans., Fam. Pl. 2: 145, 609. Jul-Aug 1763.
Typus: non designatus.

Boscia Lam. ex J. St.-Hil., Expos. Fam. Nat. 2: 3. Feb-Apr 1805 [*Cappar.*].
Typus: *B. senegalensis* (Pers.) Lam. ex Poir. (in Lamarck, Encycl., Suppl. 1: 680. 2 Mai 1811) (*Podoria senegalensis* Pers.) (typ. cons.).

(H) *Boscia* Thunb., Prodr. Pl. Cap.: [x], 32. 1794 [*Dicot.: Rut.*].
Typus: *B. undulata* Thunb.

Bossiaea Vent., Descr. Pl. Nouv.: ad t. 7. 1 Mai 1753 [*Legum.*].
Typus: *B. heterophylla* Vent.

(=) *Platylobium* Sm., Spec. Bot. New Holland 1: 17. 15 Oct 1793.
Typus: *P. formosum* Sm.

Botryophora Hook. f., Fl. Brit. India 5: 476. Dec 1888 [*Euphorb.*].
Typus: *B. kingii* Hook. f.

(H) *Botryophora* Bompard in Hedwigia 6: 129. Sep 1867 [*Chloroph.*].
Typus: *B. dichotoma* Bompard

Bouchea Cham. in Linnaea 7: 252. 1832 [*Verben.*].
Typus: *B. pseudogervao* (A. St.-Hil.) Cham. (*Verbena pseudogervao* A. St.-Hil.) (typ. cons.).

Bougainvillea Comm. ex Juss., Gen.
Pl.: 91. 4 Aug 1789 *('Buginvillaea')*
(orth. cons.) [*Nyctagin.*].
Typus: *B. spectabilis* Willd. (Sp. Pl.
2: 348. 1799) (typ. cons.).

Bourreria P. Browne, Civ. Nat. Hist.
Jamaica: 168. 10 Mar 1756 [*Borag-in.*].
Typus: *B. baccata* Raf. (Sylva Tellur.:
42. Oct-Dec 1838) (*Cordia bourreria*
L.).

(H) *Beureria* Ehret, Pl. Papil. Rar.: ad t.
13. 1755 [*Dicot.: Calycanth.*].
Typus: non designatus.

Boykinia Nutt. in J. Acad. Nat. Sci.
Philadelphia 7: 113. 28 Oct 1834
[*Saxifrag.*].
Typus: *B. aconitifolia* Nutt.

(H) *Boykiana* Raf., Neogenyton: 2. 1825
[*Dicot.: Lythr.*].
Typus: *B. humilis* (Michx.) Raf. (*Am-mannia humilis* Michx.).

Brachyandra Phil., Fl. Atacam.: 34.
1860 [*Comp.*].
Typus: *B. macrogyne* Phil.

(H) *Brachyandra* Naudin in Ann. Sci.
Nat., Bot., ser. 3, 2: 143. Sep 1844
[*Dicot.: Melastomat.*].
Typus: *B. perpusilla* Naudin

Brachynema Benth. in Trans. Linn.
Soc. London 22: 126. 21 Nov 1857
[*Eben.*].
Typus: *B. ramiflorum* Benth.

(H) *Brachynema* Griff., Not. Pl. Asiat.
4: 176. 1854 [*Dicot.: Verben.*].
Typus: *B. ferrugineum* Griff.

Brachystelma R. Br. in Bot. Mag.:
ad t. 2343. 2 Sep 1822 [*Asclepiad.*].
Typus: *B. tuberosum* (Meerb.) R. Br.
(*Stapelia tuberosa* Meerb.).

(=) *Microstemma* R. Br., Prodr.: 459. 27
Mar 1810.
Typus: *M. tuberosum* R. Br.

Bradburia Torr. & A. Gray, Fl. N.
Amer. 2: 250. Apr 1842 [*Comp.*].
Typus: *B. hirtella* Torr. & A. Gray

(H) *Bradburya* Raf., Fl. Ludov.: 104. Oct-Dec (prim.) 1817 [*Dicot.: Legum.*].
Typus (vide Cowan in Regnum Veg.
100: 232. 1979): *B. scandens* Raf.

Brexia Noronha ex Thouars, Gen.
Nov. Madagasc.: 20. 17 Nov 1806
[*Saxifrag.*].
Typus: *B. madagascariensis* (Lam.)
Ker Gawl. (in Bot. Reg.: ad t. 730.
1 Aug 1823) (*Venana madagascar-iensis* Lam.).

(≡) *Venana* Lam., Tabl. Encycl. 2: 99. 6
Nov 1797.

Breynia J. R. Forst. & G. Forst., Char. Gen. Pl.: 73. 29 Nov 1775 [*Euphorb.*].
Typus: *B. disticha* J. R. Forst. & G. Forst.

(H)   *Breynia* L., Sp. Pl.: 503. 1 Mai 1753 [*Dicot.: Cappar.*].
≡ *Linnaeobreynia* Hutch. 1967.

Brickellia Elliott, Sketch Bot. S.-Carolina 2: 290. 1823 [*Comp.*].
Typus: *B. cordifolia* Elliott

(H)   *Brickellia* Raf. in Med. Repos., ser. 2, 5: 353. Feb-Apr 1808 [*Dicot.: Polemon.*].
≡ *Ipomopsis* Michx. 1803.

(=)   *Kuhnia* L., Sp. Pl., ed. 2: 1662. Jul-Aug 1763.
Typus: *K. eupatorioides* L.

(=)   *Coleosanthus* Cass. in Bull. Sci. Soc. Philom. Paris 1817: 67. Apr 1817.
Typus: *C. cavanillesii* Cass.

Bridelia Willd., Sp. Pl. 4: 978. Apr 1806 (*'Briedelia'*) (orth. cons.) [*Euphorb.*]
Typus: *B. scandens* (Roxb.) Willd. (*Clutia scandens* Roxb.).

Bridgesia Bertero ex Cambess. in Nouv. Ann. Mus. Hist. Nat. 3: 234. 1834 [*Sapind.*].
Typus: *B. incisifolia* Bertero ex Cambess.

(H)   *Bridgesia* Hook. in Bot. Misc. 2: 222. 1831 (ante 11 Jun) [*Dicot.: Comp.*].
Typus: *B. echinopsoides* Hook.

Brosimum Sw., Prodr. 1: 12. 20 Jun - 29 Jul 1788 [*Mor.*].
Typus: *B. alicastrum* Sw. (typ. cons.).

(≡)   *Alicastrum* P. Browne, Civ. Nat. Hist. Jamaica: 372. 10 Mar 1756.

(=)   *Piratinera* Aubl., Hist. Pl. Guiane: 888. Jun-Dec 1775.
Typus: *P. guianensis* Aubl.

(=)   *Ferolia* Aubl., Hist. Pl. Guiane: Suppl. 7. Jun-Dec 1775.
Typus: *F. guianensis* Aubl.

Broussonetia L'Hér. ex Vent., Tabl. Règne Vég. 3: 547. 5 Mai 1799 [*Mor.*].
Typus: *B. papyrifera* (L.) Vent. (*Morus papyrifera* L.).

(H)   *Broussonetia* Ortega, Nov. Pl. Descr. Dec.: 61. 1798 (post 15 Mai) [*Dicot.: Legum.*].
Typus: *B. secundiflora* Ortega

Brownea Jacq., Enum. Syst. Pl.: 6, 26. Aug-Sep 1760 *('Brownaea')* (orth. cons.) [*Legum.*].
Typus: *B. coccinea* Jacq.

(=) *Hermesias* Loefl., Iter Hispan.: 278. Dec 1758.
Typus: non designatus.

Brownlowia Roxb., Pl. Coromandel 3: 61. 18 Feb 1820 [*Til.*].
Typus: *B. elata* Roxb.

(=) *Glabraria* L., Mant. Pl. 156, 276. Oct 1771.
Typus: *G. tersa* L.

Brucea J. F. Mill. [Icon. Anim. Pl.]: t. 25. 1779-1780 [*Simaroub.*].
Typus: *B. antidysenterica* J. F. Mill.

Brunfelsia L., Sp. Pl.: 191. 1 Mai 1753 *('Brunsfelsia')* (orth. cons.) [*Solan.*].
Typus: *B. americana* L.

Brunia Lam., Encycl. 1: 474. 1 Aug 1785 [*Brun.*].
Typus: *B. paleacea* Bergius (Descr. Pl. Cap.: 56. Sep 1767).

(H) *Brunia* L., Sp. Pl.: 199. 1 Mai 1753 [*Dicot.: Brun.*].
Typus: *B. lanuginosa* L.

Buchenavia Eichler in Flora 49: 164. 17 Apr 1866 [*Combret.*].
Typus: *B. capitata* (Vahl) Eichler (*Bucida capitata* Vahl).

(=) *Pamea* Aubl., Hist. Pl. Guiane: 946. Jun-Dec 1775.
Typus: *P. guianensis* Aubl.

Bucida L., Syst. Nat., ed. 10: 1012, 1025, 1368. 7 Jun 1759 [*Combret.*].
Typus: *B. buceras* L.

(≡) *Buceras* P. Browne, Civ. Nat. Hist. Jamaica: 221. 10 Mar 1756.

Buckleya Torr. in Amer. J. Sci. Arts 45: 170. Jun 1843 [*Santal.*].
Typus: *B. distichophylla* (Nutt.) Torr. (*Borya distichophylla* Nutt.).

(=) *Nestronia* Raf., New Fl. 3: 12. Jan-Mar 1838.
Typus: *N. umbellula* Raf.

Bumelia Sw., Prodr.: 3, 49. 20 Jun - 29 Jul 1788 [*Sapot.*].
Typus: *B. retusa* Sw. (typ. cons.).

(=) *Robertia* Scop., Intr. Hist. Nat.: 154. Jan-Apr 1777.
Typus: *Sideroxylon decandrum* L.

Buraeavia Baill. in Adansonia 11: 83. 15 Nov 1873 [*Euphorb.*].
Typus: *B. carunculata* (Baill.) Baill. (*Baloghia carunculata* Baill.).

(H) *Bureava* Baill. in Adansonia 1: 71. 1 Nov 1860 [*Dicot.: Combret.*].
Typus: *B. crotonoides* Baill.

Bursera Jacq. ex L., Sp. Pl., ed. 2: 471. Sep 1762 [*Burser.*].
Typus: *B. gummifera* L., nom. illeg. (*Pistacia simaruba* L., *B. simaruba* (L.) Sarg.) (etiam vide *Simarouba* [*Dicot.*]).

(=) *Elaphrium* Jacq., Enum. Syst. Pl.: 3, 19. Aug-Sep 1760.
Typus (vide Rose in N. Amer. Fl. 25: 241. 1911): *E. tomentosum* Jacq.

Burtonia R. Br. in Aiton, Hort. Kew., ed. 2, 3: 12. Oct-Nov 1811 [*Legum.*].
Typus: *B. scabra* (Sm.) R. Br. (*Gompholobium scabrum* Sm.).

(H) *Burtonia* Salisb., Parad. Lond.: ad t. 73. 1 Jun 1807 [*Dicot.: Dillen.*].
Typus: *B. grossulariifolia* Salisb.

Butea Roxb. ex Willd., Sp. Pl. 3: 857, 917. 1-10 Nov 1802 [*Legum.*].
Typus: *B. frondosa* Roxb. ex Willd., nom. illeg. (*Erythrina monosperma* Lam., *B. monosperma* (Lam.) Taub.) (typ. cons.).

(≡) *Plaso* Adans., Fam. Pl. 2: 325, 592. Jul-Aug 1763 (typ. des.: Panigrahi & Mishra in Taxon 33: 119. 1984).

Byrsanthus Guill. in Delessert, Icon. Sel. Pl. 3: 30. Feb 1838 [*Flacourt.*].
Typus: *B. brownii* Guill.

(H) *Byrsanthes* C. Presl, Prodr. Monogr. Lobel.: 41. Jul-Aug 1836 [*Dicot.: Campanul.*].
Typus (vide Rickett & Stafleu in Taxon 8: 314. 1959): *B. humboldtiana* C. Presl, nom. illeg. (*Lobelia nivea* Willd.).

Bystropogon L'Hér., Sert. Angl.: 19. Jan (prim.) 1789 [*Lab.*].
Typus: *B. plumosus* L'Hér. (typ. cons.).

Byttneria Loefl., Iter Hispan.: 313. Dec 1758 [*Stercul.*].
Typus: *B. scabra* L. (Syst. Nat., ed. 10: 939. 7 Jun 1759).

(H) *Butneria* Duhamel, Traité Arbr. Arbust. 1: 113. 1755 [*Dicot.: Calycanth.*].
Typus: non designatus.

Cajanus Adans., Fam. Pl. 2: 326, 529. Jul-Aug 1763 (*'Cajan'*) (orth. cons.) [*Legum.*].
Typus: *C. cajan* (L.) Huth (in Helios 11: 133. 1893) (*Cytisus cajan* L.).

Calandrinia Kunth in Humboldt & al., Nov. Gen. Sp. 6, ed. f°: 62. 14 Apr 1823 [*Portulac.*].
Typus: *C. caulescens* Kunth (typ. cons.).

(=) *Baitaria* Ruiz & Pav., Fl. Peruv. Prodr.: 63. Oct (prim.) 1794.
Typus: *B. acaulis* Ruiz & Pav. (Syst. Veg. Fl. Peruv. Chil.: 111. Dec (sero) 1798).

Calceolaria L. in Kongl. Vetensk. Acad. Handl. 31: 286. Oct-Dec 1770 [*Scrophular.*].
Typus: *C. pinnata* L.

(H) *Calceolaria* Loefl., Iter Hispan.: 183, 185. Dec 1758 [*Dicot.: Viol.*].
Typus: non designatus.

Calliandra Benth. in J. Bot. (Hooker) 2: 138. Apr 1840 [*Legum.*].
Typus: *C. houstonii* Benth., nom. illeg. (*Mimosa houstoniana* Mill., *C. houstoniana* (Mill.) Standl.) (typ. cons.).

Callistephus Cass. in Cuvier, Dict. Sci. Nat. 37: 491. Dec 1825 [*Comp.*].
Typus: *C. chinensis* (L.) Nees (Gen. Sp. Aster.: 222. Jul-Dec 1832) (*Aster chinensis* L.).

(≡) *Callistemma* Cass. in Bull. Sci. Soc. Philom. Paris 1817: 32. Feb 1817.

Calodendrum Thunb., Nov. Gen. Pl.: 41. 10 Jul 1782 [*Rut.*].
Typus: *C. capense* Thunb.

(=) *Pallassia* Houtt., Nat. Hist. 2(4): 382. 4 Aug 1775.
Typus: *P. capensis* Christm. (Vollst. Pflanzensyst. 3: 318. 1778).

Calycanthus L., Syst. Nat., ed. 10: 1053, 1066, 1371. 7 Jun 1759 [*Calycanth.*].
Typus: *C. floridus* L.

(=) *Basteria* Mill., Fig. Pl. Gard. Dict.: 40. 30 Dec 1755.
Typus: non designatus.

Calycera Cav., Icon. 4: 34. Sep-Dec 1797 (*'Calicera'*) (orth. cons.) [*Calycer.*].
Typus: *C. herbacea* Cav.

Calyptranthes Sw., Prodr.: 5, 79. 20 Jun - 29 Jul 1788 [*Myrt.*].
Typus: *C. chytraculia* (L.) Sw. (*Myrtus chytraculia* L.) (typ. cons.).

(≡) *Chytraculia* P. Browne, Civ. Nat. Hist. Jamaica: 239. 10 Mar 1756.

Calystegia R. Br., Prodr.: 483. 27 Mar 1810 [*Convolvul.*].
Typus: *C. sepium* (L.) R. Br. (*Convolvulus sepium* L.) (typ. cons.).

(≡) *Volvulus* Medik., Philos. Bot. 2: 42. Mai 1791.

Cambessedesia DC., Prodr. 3: 110. Mar (med.) 1828 [*Melastomat.*].
Typus: *C. hilariana* (Kunth) DC. (*Rhexia hilariana* Kunth) (typ. cons.).

(H) *Cambessedea* Kunth in Ann. Sci. Nat. (Paris) 2: 336. 1824 [*Dicot.: Anacard.*].
Typus: *Mangifera axillaris* Desr.

Camoënsia Welw. ex Benth. & Hook. f., Gen. Pl. 1: 456, 557. 19 Oct 1865 [*Legum.*].
Typus: *C. maxima* Welw. ex Benth. (in Trans. Linn. Soc. London 25: 301. 30 Nov 1865) (typ. cons.).

(=) *Giganthemum* Welw. in Ann. Cons. Ultramarino, ser. 1, 1858: 585. Dec 1859.
Typus: *G. scandens* Welw.

Campnosperma Thwaites in Hooker's J. Bot. Kew Gard. Misc. 6: 65. Mar 1854 [*Anacard.*].
Typus: *C. zeylanicum* Thwaites

(=) *Coelopyrum* Jack in Malayan Misc. 2(7): 65. 1822.
Typus: *C. coriaceum* Jack

Campsis Lour., Fl. Cochinch.: 358, 377. Sep 1790 [*Bignon.*].
Typus: *C. adrepens* Lour.

(=) *Notjo* Adans., Fam. Pl. 2: 226, 582. Jul-Aug 1763.
Typus: non designatus.

Camptocarpus Decne. in Candolle, Prodr. 8: 493. Mar (med.) 1844 [*Asclepiad.*].
Typus: *C. mauritianus* (Lam.) Decne. (*Cynanchum mauritianum* Lam.) (typ. cons.).

(H) *Camptocarpus* K. Koch in Linnaea 17: 304. Jan 1844 [*Dicot.: Boragin.*].
≡ *Oskampia* Moench 1794.

Cananga (DC.) Hook. f. & Thomson, Fl. Ind.: 129. 1-19 Jul 1855 (*Unona* subsect. *Cananga* DC., Syst. Nat. 1: 485. 1-15 Nov 1817) [*Annon.*].
Typus: *Unona odorata* (Lam.) Dunal (*Uvaria odorata* Lam., *C. odorata* (Lam.) Hook. f. & Thomson).

(H) *Cananga* Aubl., Hist. Pl. Guiane: 607. Jun-Dec 1775 [*Dicot.: Annon.*].
Typus: *C. ouregou* Aubl.

Canarina L., Mant. Pl.: 148, 225, 588. Oct 1771 [*Campanul.*].
Typus: *C. campanula* L., nom. illeg. (*Campanula canariensis* L., *Canarina canariensis* (L.) Vatke).

(≡) *Mindium* Adans., Fam. Pl. 2: 134, 578. Jul-Aug 1763 (per typ. des.).

Canavalia Adans., Fam. Pl. 2: 325, 531. Jul-Aug 1763 *('Canavali')* (orth. cons.) [*Legum.*].
Typus: *C. ensiformis* (L.) DC. (Prodr. 2: 404. Nov. (med.) 1825) (*Dolichos ensiformis* L.).

Canella P. Browne, Civ. Nat. Hist. Jamaica: 275. 10 Mar 1756 [*Canell.*].
Typus: *C. winterana* (L.) Gaertn. (Fruct. Sem. Pl. 1: 373. Dec 1788) (*Laurus winterana* L.).

Cansjera Juss., Gen. Pl.: 448. 4 Aug 1789 [*Opil.*].
Typus: *C. rheedei* J. F. Gmel. (Syst. Nat. 2: 4, 20. Sep-Nov 1791).

(≡) *Tsjeru-caniram* Adans., Fam. Pl. 2: 80, 614. Jul-Aug 1763.

Capsella Medik., Pfl.-Gatt.: 85, 99. 22 Apr 1792 [*Cruc.*].
Typus: *C. bursa-pastoris* (L.) Medik. (*Thlaspi bursa-pastoris* L.) (typ. cons.).

(≡) *Bursa-pastoris* Ség., Pl. Veron. 3: 166. Jul-Aug 1754.

Carallia Roxb., Pl. Coromandel 3: 8. Jul 1811 [*Rhizophor.*].
Typus: *C. lucida* Roxb.

(=) *Karekandel* Wolf, Gen. Pl.: 73. 1776. Typus (vide Ross in Acta Bot. Neerl. 15: 158. 1966): *Karkandela malabarica* Raf.

(=) *Barraldeia* Thouars, Gen. Nov. Madagasc.: 24. 17 Nov 1806. Typus: non designatus.

Careya Roxb., Pl. Coromandel 3: 13. Jul 1811 [*Lecythid.*].
Typus: *C. herbacea* Roxb. (typ. cons.).

Carissa L., Syst. Nat., ed. 12, 2: 135, 189; Mant. Pl.: 7, 52. 15-31 Oct 1767 [*Apocyn.*].
Typus: *C. carandas* L.

(≡) *Carandas* Adans., Fam. Pl. 2: 171, 532. Jul-Aug 1763.

Carlowrightia A. Gray in Proc. Amer. Acad. Arts 13: 364. 5 Apr 1878 [*Acanth.*].
Typus: *C. linearifolia* (Torr.) A. Gray (*Schaueria linearifolia* Torr.).

(=) *Cardiacanthus* Nees & Schauer in Candolle, Prodr. 11: 331. 25 Nov 1847.
Typus: *C. neesianus* Schauer ex Nees

Carrichtera DC. in Mém. Mus. Hist. Nat. 7: 244. 20 Apr 1821 [*Cruc.*]. Typus: *C. annua* (L.) DC. (*Vella annua* L.) (typ. cons.).

(H)   *Carrichtera* Adans., Fam. Pl. 2: 421, 533. Jul-Aug 1763 [*Dicot.: Cruc.*]. Typus: *Vella pseudocytisus* L.

Carya Nutt., Gen. N. Amer. Pl. 2: 220. 14 Jul 1818 [*Jugland.*]. Typus: *C. tomentosa* (Poir.) Nutt. (*Juglans tomentosa* Poir.) (typ. cons.).

(=)   *Hicorius* Raf., Fl. Ludov.: 109. Oct-Dec (prim.) 1817. Typus: non designatus.

Casselia Nees & Mart. in Nova Acta Phys.-Med. Acad. Caes. Leop.-Carol. Nat. Cur. 11: 73. 1823 [*Verben.*]. Typus: *C. serrata* Nees & Mart. (typ. cons.).

(H)   *Casselia* Dumort., Comment. Bot.: 21. Nov (sero) - Dec (prim.) 1822 [*Dicot.: Boragin.*]. ≡ *Mertensia* Roth 1797 (nom. cons.).

Cassia L., Sp. Pl.: 376. 1 Mai 1753 [*Legum.*]. Typus: *C. fistula* L. (typ. cons.).

Cassine L., Sp. Pl.: 268. 1 Mai 1753 [*Celastr.*]. Typus: *C. peragua* L. (typ. cons.: [icon in] Dillenius, Hort. Eltham. t. 236, f. 305. 1732).

Cassinia R. Br., Observ. Compos.: 126. 1817 (ante Sep) [*Comp.*]. Typus: *C. aculeata* (Labill.) R. Br. (*Calea aculeata* Labill.) (typ. cons.).

(H)   *Cassinia* R. Br. in Aiton, Hort. Kew., ed. 2, 5: 184. Nov 1813 [*Dicot.: Comp.*]. Typus: *C. aurea* R. Br.

Castanopsis (D. Don) Spach, Hist. Nat. Vég. 11: 142, 185. 25 Dec 1841 (*Quercus* [unranked] *Castanopsis* D. Don, Prodr. Fl. Nepal.: 56. 26 Jan - 1 Feb 1825) [*Fag.*]. Typus: *Quercus armata* Roxb. (*C. armata* (Roxb.) Spach).

(=)   *Balanoplis* Raf., Alsogr. Amer.: 29. 1838. Typus (vide Pichi Sermolli in Taxon 3: 113. 1954): *B. tribuloides* (Sm.) Raf. (*Quercus tribuloides* Sm.).

Castela Turpin in Ann. Mus. Natl. Hist. Nat. 7: 78. 1806 [*Simaroub.*]. Typus: *C. depressa* Turpin (typ. cons.).

(H)   *Castelia* Cav. in Anales Ci. Nat. 3: 134. 1801 [*Dicot.: Verben.*]. Typus: *C. cuneato-ovata* Cav.

Cavendishia Lindl. in Edwards's Bot. Reg.: ad t. 1791. 1 Sep 1835 [*Eric.*]. Typus: *C. nobilis* Lindl.

(H) *Cavendishia* Gray, Nat. Arr. Brit. Pl. 1: 678, 689. 1 Nov 1821 [*Hepat.*]. ≡ *Antoiria* Raddi 1818.

(=) *Chupalon* Adans., Fam. Pl. 2: 164, 538. Jul-Aug 1763. Typus: non designatus.

Cayaponia Silva Manso, Enum. Subst. Braz.: 31. 1836 [*Cucurbit.*]. Typus: *C. diffusa* Silva Manso (typ. cons.).

Caylusea A. St.-Hil., Deux. Mém. Réséd.: 29. 1837 (sero) - Jan (prim.) 1838 [*Resed.*]. Typus: *C. canescens* Webb (in Hooker, Niger Fl.: 101. Nov-Dec 1849), non (L.) Walp. 1843 [= *C. hexagyna* (Forssk.) M. L. Green (*Reseda hexagyna* Forssk.)].

Cayratia Juss. in Cuvier, Dict. Sci. Nat. 10: 103. 23 Mai 1818 [*Vit.*]. Typus: *Columella pedata* Lour. [= *Cayratia pedata* (Lam.) Gagnep. (*Cissus pedata* Lam.)] (etiam vide *Columellia* [*Dicot.*]).

(=) *Lagenula* Lour., Fl. Cochinch.: 65, 88. Sep 1790. Typus: *L. pedata* Lour.

Cecropia Loefl., Iter Hispan.: 272. Dec 1758 [*Mor.*]. Typus: *C. peltata* L. (Syst. Nat., ed. 10: 1286. 7 Jun 1759).

(≡) *Coilotapalus* P. Browne, Civ. Nat. Hist. Jamaica: 111. 10 Mar 1756.

Celastrus L., Sp. Pl.: 196. 1 Mai 1753 (gend. masc. cons.) [*Celastr.*]. Typus: *C. scandens* L.

Celmisia Cass. in Cuvier, Dict. Sci. Nat. 37: 259. Dec 1825 [*Comp.*]. Typus: *C. longifolia* Cass. (typ. cons.).

(H) *Celmisia* Cass. in Bull. Sci. Soc. Philom. Paris 1817: 32. Feb 1817 [*Dicot.: Comp.*]. Typus: *C. rotundifolia* Cass. (in Cuvier, Dict. Sci. Nat. 7: 356. 24 Mai 1817).

Centaurea L., Sp. Pl.: 990. 1 Mai 1753 [*Comp.*].
Typus: *C. paniculata* L. (typ. cons.).

Centrosema (DC.) Benth., Comm. Legum. Gen.: 53. Jun 1837 (*Clitoria* sect. *Centrosema* DC., Prodr. 2: 234. Nov (med.) 1825) [*Legum.*].
Typus: *Clitoria brasiliana* L. (*Centrosema brasilianum* (L.) Benth.) (typ. cons.).

(=) *Steganotropis* Lehm., Sem. Hort. Bot. Hamburg. 1826: 18. 1826.
Typus: *S. conjugata* Lehm.

Cephaëlis Sw., Prodr.: 3, 45. 20 Jun - 29 Jul 1788 [*Rub.*].
Typus: *C. muscosa* (Jacq.) Sw. (*Morinda muscosa* Jacq.) (typ. cons.).

(=) *Evea* Aubl., Hist. Pl. Guiane: 100. Jun-Dec 1775.
Typus: *E. guianensis* Aubl.
(=) *Carapichea* Aubl., Hist. Pl. Guiane: 167. Jun-Dec 1775.
Typus: *C. guianensis* Aubl.
(=) *Tapogomea* Aubl., Hist. Pl. Guiane: 157. Jun-Dec 1775.
Typus: *T. violacea* Aubl.

Cephalaria Schrad. in Roemer & Schultes, Syst. Veg. 3: 1, 43. Apr-Jul 1818 [*Dipsac.*].
Typus: *C. alpina* (L.) Roem. & Schult. (*Scabiosa alpina* L.) (typ. cons.).

(=) *Lepicephalus* Lag., Gen. Sp. Pl.: 7. Jun-Dec 1816.
Typus: non designatus.

Cephalotus Labill., Nov. Holl. Pl. 2: 6. Feb 1806 [*Cephalot.*].
Typus: *C. follicularis* Labill.

(H) *Cephalotos* Adans., Fam. Pl. 2: 189, 534. Jul-Aug 1763 [*Dicot.: Lab.*].
Typus: *Thymus cephalotos* L.

Chaenomeles Lindl. in Trans. Linn. Soc. London 13: 96, 97. 23 Mai - 21 Jun 1821 (*'Choenomeles'*) (orth. cons.) [*Ros.*].
Typus: *C. japonica* (Thunb.) Lindl. ex Spach (Hist. Nat. Vég. 2: 159. 12 Jul 1834) (*Pyrus japonica* Thunb.).

Chaenostoma Benth. in Companion Bot. Mag. 1: 374. 1 Jul 1836 [*Scrophular.*].
Typus: *C. aethiopicum* (L.) Benth. (*Buchnera aethiopica* L.) (typ. cons.).

(=) *Palmstruckia* Retz., Obs. Bot. Pugill.: 15. 14 Nov 1810.
Typus: *Manulea foetida* (Andrews) Pers. (*Buchnera foetida* Andrews).

Chaetocarpus Thwaites in Hooker's J. Bot. Kew Gard. Misc. 6: 300. Oct 1854 [*Euphorb.*].
Typus: *C. castanicarpus* (Roxb.) Thwaites (Enum. Pl. Zeyl.: 275. 1861) (*Adelia castanicarpa* Roxb.).

(H) *Chaetocarpus* Schreb., Gen. Pl.: 75. Apr 1789 [*Dicot.: Sapot.*].
≡ *Pouteria* Aubl. 1775.

Chamaedaphne Moench, Methodus: 457. 4 Mai 1794 [*Eric.*].
Typus: *C. calyculata* (L.) Moench (*Andromeda calyculata* L.).

(H) *Chamaedaphne* Mitch., Diss. Princ. Bot.: 44. 1769 [*Dicot.: Rub.*].
≡ *Mitchella* L. 1753.

Chamissoa Kunth in Humboldt & al., Nov. Gen. Sp. 2, ed. 4°: 196; ed. f°: 158. Feb 1818 [*Amaranth.*].
Typus: *C. altissima* (Jacq.) Kunth (*Achyranthes altissima* Jacq.) (typ. cons.).

(=) *Kokera* Adans., Fam. Pl. 2: 269, 541. Jul-Aug 1763.
Typus: non designatus.

Chaptalia Vent., Descr. Pl. Nouv.: ad t. 61. 22 Mar 1802 [*Comp.*].
Typus: *C. tomentosa* Vent.

Chascanum E. Mey., Comment. Pl. Afr. Austr.: 275. 1-8 Jan 1838 [*Verben.*].
Typus: *C. cernuum* (L.) E. Mey. (*Buchnera cernua* L.) (typ. cons.).

(=) *Plexipus* Raf., Fl. Tellur. 2: 104. Jan-Mar 1837.
Typus: *P. cuneifolius* (L. f.) Raf. (*Buchnera cuneifolia* L. f.).

Chiliophyllum Phil. in Linnaea 33: 132. Aug 1864 [*Comp.*].
Typus: *C. densifolium* Phil.

(H) *Chiliophyllum* DC., Prodr. 5: 554. 1-10 Oct 1836 [*Dicot.: Comp.*].
≡ *Hybridella* Cass. 1817.

Chimonanthus Lindl. in Bot. Reg.: ad t. 404. 1 Oct 1819 [*Calycanth.*].
Typus: *C. fragrans* Lindl., nom. illeg. (*Calycanthus praecox* L., *Chimonanthus praecox* (L.) Link).

(≡) *Meratia* Loisel., Herb. Gén. Amat. 3: ad t. 173. Jul 1818.

Chloroxylon DC., Prodr. 1: 625. Jan (med.) 1824 [*Rut.*].
Typus: *C. swietenia* DC. (*Swietenia chloroxylon* Roxb.).

(H) *Chloroxylum* P. Browne, Civ. Nat. Hist. Jamaica: 187. 10 Mar 1756 [*Dicot.: Rhamn.*].
Typus: *Ziziphus chloroxylon* (L.) Oliv. (*Laurus chloroxylon* L.).

Chomelia Jacq., Enum. Syst. Pl.: 1, 12. Aug-Sep 1760 [*Rub.*].
Typus: *C. spinosa* Jacq.

(H) *Chomelia* L., Opera Var.: 210. 1758 [*Dicot.: Rub.*].
Typus (vide Dandy in Taxon 18: 470. 1969): *Rondeletia asiatica* L.

Chonemorpha G. Don, Gen. Hist. 4: 69, 76. 1837 [*Apocyn.*].
Typus: *C. macrophylla* G. Don, nom. illeg. (*Echites fragrans* Moon, *C. fragrans* (Moon) Alston).

(≡) *Belutta-kaka* Adans., Fam. Pl. 2: 172, 525. Jul-Aug 1763.

Chorispora R. Br. ex DC. in Mém. Mus. Hist. Nat. 7: 237. 20 Apr 1821 [*Cruc.*].
Typus: *C. tenella* (Pall.) DC. (Syst. Nat. 2: 435. Mai (sero) 1821) (*Raphanus tenellus* Pall.).

(≡) *Chorispermum* W. T. Aiton, Hort. Kew., ed. 2, 4: 129. Dec 1812.

Chrozophora A. Juss., Euphorb. Gen.: 27. 21 Feb 1824 (*'Crozophora'*) (orth. cons.) [*Euphorb.*].
Typus: *C. tinctoria* (L.) A. Juss. (*Croton tinctorius* L.).

(≡) *Tournesol* Adans., Fam. Pl. 2: 356, 612. Jul-Aug 1763.

Chrysanthemum L., Sp. Pl.: 887. 1 Mai 1753 [*Comp.*].
Typus: *C. indicum* L. (typ. cons.).

Chrysopsis (Nutt.) Elliott, Sketch Bot. S. Carolina 2: 333. 1823 (*Inula* subg. *Chrysopsis* Nutt., Gen. N. Amer. Pl. 2: 150. 14 Jul 1818) [*Comp.*].
Typus: *Inula mariana* L. (*C. mariana* (L.) Elliott) (typ. cons.).

(≡) *Diplogon* Raf. in Amer. Monthly Mag. & Crit. Rev. 4: 195. Jan 1819.

Chrysothamnus Nutt. in Trans. Amer. Philos. Soc., ser. 2, 7: 323. Oct-Dec 1840 [*Comp.*].
Typus: *C. pumilus* Nutt. (typ. cons.).

Cinnamomum Schaeff., Bot. Exped.: 74. Oct-Dec 1760 [*Laur.*].
Typus: *C. verum* J. Presl (in Berchtold & Presl, Přir. Rostlin 2(2): 36. 1825) (*Laurus cinnamomum* L.).

(=) *Camphora* Fabr., Enum.: 218. 1759.
Typus: *Laurus camphora* L.

Citrullus Schrad. ex Eckl. & Zeyh., Enum. Pl. Afric. Austral.: 279. Jan 1836 [*Cucurbit.*].
Typus: *C. vulgaris* Schrad. ex Eckl. & Zeyh. (*Cucurbita citrullus* L.) (typ. cons.).

(≡) *Anguria* Mill., Gard. Dict. Abr., ed. 4: [93]. 28 Jan 1754.
(=) *Colocynthis* Mill., Gard. Dict. Abr., ed. 4: [357]. 28 Jan 1754.
Typus: non designatus.

Clarisia Ruiz & Pav., Fl. Peruv. Prodr.: 128. Oct (prim.) 1794 [*Mor.*].
Typus: *C. racemosa* Ruiz & Pav. (Syst. Veg. Fl. Peruv. Chil.: 255. Dec (sero) 1798) (typ. cons.).

(H) *Clarisia* Abat in Mem. Acad. Soc. Med. Sevilla 10: 418. 1792 [*Dicot.: Basell.*].
≡ *Anredera* Juss. 1789.

Cleyera Thunb., Nov. Gen. Pl.: 68. 18 Jun 1783 [*The.*].
Typus: *C. japonica* Thunb.

(H) *Cleyera* Adans., Fam. Pl. 2: 224, 540. Jul-Aug 1763 [*Dicot.: Logan.*].
≡ *Polypremum* L. 1753.

Clianthus Sol. ex Lindl. in Edwards's Bot. Reg.: ad t. 1775. 1 Jul 1835 [*Legum.*].
Typus: *C. puniceus* (G. Don) Sol. ex Lindl. (*Donia punicea* G. Don).

(=) *Sarcodum* Lour., Fl. Cochinch.: 425, 461. Sep 1790.
Typus: *S. scandens* Lour.

Cnicus L., Sp. Pl.: 826. 1 Mai 1753 [*Comp.*].
Typus: *C. benedictus* L. (typ. cons.).

Coccocypselum P. Browne, Civ. Nat. Hist. Jamaica: 144. 10 Mar 1756 (*'Coccocipsilum'*) (orth. cons.) [*Rub.*].
Typus: *C. repens* Sw. (Prodr.: 31. 20 Jun - 29 Jul 1788) (typ. cons.).

(=) *Sicelium* P. Browne, Civ. Nat. Hist. Jamaica: 144. 10 Mar 1756.
Typus: non designatus.

Coccoloba P. Browne, Civ. Nat. Hist. Jamaica: 209. 10 Mar 1756 (*'Coccolobis'*) (orth. cons.) [*Polygon.*].
Typus: *C. uvifera* (L.) L. (*Polygonum uvifera* L.) (typ. cons.).

(=) *Guaiabara* Mill., Gard. Dict. Abr., ed. 4: [590]. 28 Jan 1754.
Typus: non designatus.

Cocculus DC., Syst. Nat. 1: 515. 1-15 Nov 1817 [*Menisperm.*].
Typus: *C. hirsutus* (L.) W. Theob. (in Mason, Burmah 2: 657. 1883) (*Menispermum hirsutum* L.) (typ. cons.).

(=) *Cebatha* Forssk., Fl. Aegypt.-Arab.: 171. 1 Oct 1775.
Typus: *Cocculus cebatha* DC.

(=) *Leaeba* Forssk., Fl. Aegypt.-Arab.: 172. 1 Oct 1775.
Typus: *L. dubia* J. F. Gmel. (Syst. Nat. 2: 567. Sep-Nov 1791).

(=) *Epibaterium* J. R. Forst. & G. Forst., Char. Gen. Pl.: 54. 29 Nov 1775.
Typus: *E. pendulum* J. R. Forst. & G. Forst.

(=) *Nephroia* Lour., Fl. Cochinch.: 539, 565. Sep 1790.
Typus: *N. sarmentosa* Lour.

(=) *Baumgartia* Moench, Methodus: 650. 4 Mai 1794.
Typus: *B. scandens* Moench

(=) *Androphylax* J. C. Wendl., Bot. Beob.: 37, 38. 1798.
Typus: *A. scandens* J. C. Wendl.

Cochlospermum Kunth in Humboldt & al., Nov. Gen. Sp. 5, ed. 4°: 297; ed. f°: 231. Jun 1822 [*Cochlosperm.*].
Typus: *Bombax gossypium* L., nom. illeg. (*Bombax religiosum* L., *C. religiosum* (L.) Alston).

Codiaeum A. Juss., Euphorb. Gen.: 33. 21 Feb 1824 [*Euphorb.*].
Typus: *C. variegatum* (L.) A. Juss. (*Croton variegatus* L.) (typ. cons.).

(≡) *Phyllaurea* Lour., Fl. Cochinch.: 540, 575. Sep 1790.

Codonanthe (Mart.) Hanst. in Linnaea 26: 209. Apr 1854 (*Hypocyrta* sect. *Codonanthe* Mart., Nov. Gen. Sp. Pl. 3: 50. Jan-Jun 1829) [*Gesner.*].
Typus: *Hypocyrta gracilis* Mart. (*C. gracilis* (Mart.) Hanst.) (typ. cons.).

(H) *Codonanthus* G. Don, Gen. Hist. 4: 164, 166. 1837 [*Dicot.: Logan.*].
Typus: *C. africanus* G. Don

Cola Schott & Endl., Melet. Bot.: 33. 1832 [*Stercul.*].
Typus: *C. acuminata* (P. Beauv.) Schott & Endl. (*Sterculia acuminata* P. Beauv.) (typ. cons.).

(=) *Bichea* Stokes, Bot. Mater. Med. 2: 564. 1812.
Typus: *B. solitaria* Stokes

Colea Bojer ex Meisn., Pl. Vasc. Gen. 1: 301; 2: 210. 25-31 Oct 1840 [*Bignon.*].
Typus: *C. colei* (Bojer ex Hook.) M. L. Green (in Sprague & al., Nom. Prop. Brit. Bot.: 107. Aug 1929) (*Bignonia colei* Bojer ex Hook.) (typ. cons.).

(≡) *Odisca* Raf., Sylva Tellur.: 80. Oct-Dec 1838.

(=) *Uloma* Raf., Fl. Tellur. 2: 62. Jan-Mar 1837.
Typus: *U. telfairiae* (Bojer) Raf. (*Bignonia telfairiae* Bojer).

Colletia Comm. ex Juss., Gen. Pl.: 380. 4 Aug 1789 [*Rhamn.*].
Typus: *C. spinosa* Lam. (Tabl. Encycl. 2: 91. 6 Nov 1797).

(H) *Colletia* Scop., Intr. Hist. Nat.: 207. Jan-Apr 1777 [*Dicot.: Ulm.*].
Typus: *Rhamnus iguanaeus* Jacq.

Colophospermum J. Kirk ex J. Léonard in Bull. Jard. Bot. Etat 19: 390. Dec 1949. [*Legum.*].
Typus: *C. mopane* (J. Kirk ex Benth.) J. Léonard (*Copaifera mopane* J. Kirk. ex Benth.).

(=) *Hardwickia* Roxb., Pl. Coromandel 3: 6, t. 209. Jul 1811.
Typus: *H. binata* Roxb.

Colubrina Rich. ex Brongn., Mém. Fam. Rhamnées: 61. Jul 1826 [*Rhamn.*].
Typus: *C. ferruginosa* Brongn. (*Rhamnus colubrinus* Jacq.) (typ. cons.).

Columellia Ruiz & Pav., Fl. Peruv. Prodr.: 3. Oct (prim.) 1794 [*Columell.*].
Typus: *C. oblonga* Ruiz & Pav. (Fl. Peruv. 1: 28. 1798 (med.)) (typ. cons.).

(H) *Columella* Lour., Fl. Cochinch.: 64, 85. Sep 1790 [*Dicot.: Vit.*].
≡ *Cayratia* Juss. 1818 (nom. cons.).

Combretum Loefl., Iter Hispan.: 308. Dec 1758 [*Combret.*].
Typus: *C. fruticosum* (Loefl.) Stuntz (U.S.D.A. Bur. Pl. Industr. Invent. Seeds 31: 86. 1914) (*Gaura fruticosa* Loefl.).

(=) *Grislea* L., Sp. Pl.: 348. 1 Mai 1753.
Typus: *G. secunda* L.

Commiphora Jacq., Pl. Hort. Schoenbr. 2: 66. 1797 [*Burser.*].
Typus: *C. madagascarensis* Jacq.

(=) *Balsamea* Gled. in Schriften Berlin. Ges. Naturf. Freunde 3: 127. 1782.
Typus: *B. meccanensis* Gled.

Condalia Cav. in Anales Hist. Nat. 1: 39. Oct 1799 [*Rhamn.*]. Typus: *C. microphylla* Cav.

(H) *Condalia* Ruiz & Pav., Fl. Peruv. Prodr.: 11. Oct (prim.) 1794 [*Dicot.: Rub.*]. Typus: *C. repens* Ruiz & Pav. (Fl. Peruv. 1: 54. 1798 (med.)).

Conopodium W. D. J. Koch in Nova Acta Phys.-Med. Acad. Caes. Leop.-Carol. Nat. Cur. 12: 118. 1824 (ante 28 Oct) [*Umbell.*]. Typus: *C. denudatum* W. D. J. Koch, nom. illeg. (*Bunium denudatum* DC., nom. illeg., *B. majus* Gouan, *C. majus* (Gouan) Loret) (typ. cons.).

Conyza Less., Syn. Gen. Compos.: 203. Jul-Aug 1832 [*Comp.*]. Typus: *C. chilensis* Spreng. (Novi Provent.: 14. Dec 1818) (typ. cons.).

(H) *Conyza* L., Sp. Pl.: 861. 1 Mai 1753 [*Dicot.: Comp.*]. Typus (vide Green in Sprague & al., Nom. Prop. Brit. Bot.: 181. 1929): *C. squarrosa* L.

(=) *Eschenbachia* Moench, Methodus: 573. 4 Mai 1794. Typus: *E. globosa* Moench, nom. illeg. (*Erigeron aegyptiacus* L.).

(=) *Dimorphanthes* Cass. in Bull. Sci. Soc. Philom. Paris 1818: 30. Feb 1818. Typus: non designatus.

(=) *Laënnecia* Cass. in Cuvier, Dict. Sci. Nat. 25: 91. 1822. Typus: *L. gnaphalioides* (Kunth) Cass. (*Conyza gnaphalioides* Kunth).

Copaifera L., Sp. Pl., ed. 2: 557. Sep 1762 [*Legum.*]. Typus: *C. officinalis* (Jacq.) L. (*Copaiva officinalis* Jacq.).

(≡) *Copaiva* Jacq., Enum. Syst. Pl.: 4, 21. Sep-Nov 1760.

(=) *Copaiba* Mill., Gard. Dict. Abr., ed. 4: [371]. 28 Jan 1754. Typus: non designatus.

Coptophyllum Korth. in Ned. Kruidk. Arch. 2(2): 161. 1851 [*Rub.*]. Typus: *C. bracteatum* Korth.

(H) *Coptophyllum* Gardner in London J. Bot. 1: 133. 1 Mar 1842 [*Pteridoph.*]. Typus: *C. buniifolium* Gardner

Cordia L., Sp. Pl.: 190. 1 Mai 1753
[*Borag.*].
Typus: *C. myxa* L. (typ. cons.: [spe-
cimen] Linnaean Herb. fiche 94.5,
S).

Cordylanthus Nutt. ex Benth. in Can-   (≡)   *Adenostegia* Benth. in Lindley, Intr.
dolle, Prodr. 10: 597. 8 Apr 1846   Nat. Syst. Bot., ed. 2: 445. Jul 1836.
[*Scrophular.*].
Typus: *C. filifolius* Nutt. ex Benth.,
nom. illeg. (*Adenostegia rigida* Benth.,
C. rigidus (Benth.) Jeps.).

Coronilla L., Sp. Pl.: 742. 1 Mai
1753 [*Legum.*].
Typus: *C. valentina* L. (typ. cons.:
[specimen] Herb. Linnaeus No.
917.4 (LINN)).

Coronopus Zinn, Cat. Pl. Hort. Gott.:   (H)   *Coronopus* Mill., Gard. Dict. Abr., ed.
325. 20 Apr - 21 Mai 1757 [*Cruc.*].   4: [387]. 28 Jan 1754 [*Dicot.: Plan-*
Typus: *C. ruellii* All. (Fl. Pedem. 1:   *tagin.*].
256. Apr-Jul 1785) (*Cochlearia co-*   Typus: non designatus.
ronopus L.).

Correa Andrews in Bot. Repos.: ad   (H)   *Correia* Vand., Fl. Lusit. Bras.
t. 18. 1 Apr 1798 [*Rut.*].   Spec.: 28. 1788 [*Dicot.: Ochn.*].
Typus: *C. alba* Andrews   Typus: non designatus.

Corydalis DC. in Lamarck & Candolle, Fl. Franç., ed. 3, 4: 637. 17 Sep 1805 [*Papaver.*].
Typus: *C. bulbosa* (L.) DC., *comb. rej.* (*Fumaria bulbosa* L., nom. utique rej., *Fumaria bulbosa* var. *solida* L., *C. solida* (L.) Clairv.) (typ. cons.: [specimen] Herb. Linnaeus No. 881.5 (LINN)).

(H)   *Corydalis* Medik., Philos. Bot. 1: 96. Apr 1789 [*Dicot.: Papaver.*].
≡ *Cysticapnos* Mill. 1754 (nom. rej. sub *Corydalis*).

(≡)   *Pistolochia* Bernh., Syst. Verz.: 57, 74. 1800.

(=)   *Capnoides* Mill., Gard. Dict. Abr., ed. 4: [249]. 28 Jan 1754.
Typus: *C. sempervirens* (L.) Borkh. (in Arch. Bot. (Leipzig) 1(2): 44. Mai-Dec 1797) (*Fumaria sempervirens* L.).

(=)   *Cysticapnos* Mill., Gard. Dict. Abr., ed. 4: [427]. 28 Jan 1754.
Typus: *C. vesicaria* (L.) Fedde (in Repert. Spec. Nov. Regni Veg. 19: 287. 20 Feb 1924) (*Fumaria vesicaria* L.).

(=)   *Pseudo-fumaria* Medik., Philos. Bot. 1: 110. Apr 1789.
Typus: *P. lutea* (L.) Borkh. (in Arch. Bot. (Leipzig) 1(2): 45. Mai-Dec 1797) (*Fumaria lutea* L.).

Coryphantha (Engelm.) Lem., Cactées: 32. Aug 1868 (*Mammillaria* subg. *Coryphanta* Engelm. in Emory, Rep. U.S. Mex. Bound. 2: 10. 1858) [*Cact.*].
Typus: *Mammillaria sulcata* Engelm. (*C. sulcata* (Engelm.) Britton & Rose).

(=)   *Aulacothele* Monv., Cat. Pl. Exot.: 21. 1846.
Typus: *Mammillaria aulacothele* Lem.

Cosmibuena Ruiz & Pav., Fl. Peruv. 3: 2. Aug 1802 [*Rub.*].
Typus: *C. obtusifolia* Ruiz & Pav., nom. illeg. (*Cinchona grandiflora* Ruiz & Pav., *Cosmibuena grandiflora* (Ruiz & Pav.) Rusby) (typ. cons.).

(H)   *Cosmibuena* Ruiz & Pav., Fl. Peruv. Prodr.: 10. Oct (prim.) 1794 [*Dicot.: Ros.*].
Typus (vide Pichi Sermolli in Taxon 3: 121. 1954): *Hirtella cosmibuena* Lam.

Cotyledon L., Sp. Pl.: 429. 1 Mai 1753 [*Crassul.*].
Typus: *C. orbiculata* L. (typ. cons.: [icon in] Hermann, Hort. Lugd.-Bat. t. 551. 1687).

Crabbea Harv. in London J. Bot. 1: 27. 1 Jan 1842 [*Acanth.*].
Typus: *C. hirsuta* Harv.

(H)   *Crabbea* Harv., Gen. S. Afr. Pl.: 276. Aug-Dec 1838 [*Dicot.: Acanth.*].
Typus: *C. pungens* Harv.

Cracca Benth. in Vidensk. Meddel. Dansk Naturhist. Foren. Kjøbenhavn 1853: 8. 1853 [*Legum.*]. Typus: *C. glandulifera* Benth.

(H) *Cracca* L., Sp. Pl.: 752. 1 Mai 1753 [*Dicot.: Legum.*]. ≡ *Tephrosia* Pers. 1807 (nom. cons.).

Cristaria Cav., Icon. 5: 10. Jun-Sep 1799 [*Malv.*]. Typus: *C. glaucophylla* Cav.

(H) *Cristaria* Sonn., Voy. Indes Orient., ed. 4°, 2: 247; ed. 8°, 3: 284. 1782 [*Dicot.: Combret.*]. Typus: *C. coccinea* Sonn.

Crotalaria L., Sp. Pl.: 268. 1 Mai 1753 [*Legum.*]. Typus: *C. lotifolia* L. (typ. cons.: [icon in] Sloane, Voy. Mad. Jam. 2: t. 176. 1725).

Cruckshanksia Hook. & Arn. in Bot. Misc. 3: 361. 1 Aug 1833 [*Rub.*]. Typus: *C. hymenodon* Hook. & Arn.

(H) *Cruckshanksia* Hook. in Bot. Misc. 2: 211. 1831 (ante 11 Jun) [*Dicot.: Geran.*]. Typus: *C. cistiflora* Hook.

Crudia Schreb., Gen. Pl.: 282. Apr 1789 [*Legum.*]. Typus: *C. spicata* (Aubl.) Willd. (Sp. Pl. 2: 539. Mar 1799) (*Apalatoa spicata* Aubl.) (typ. cons.).

(≡) *Apalatoa* Aubl., Hist. Pl. Guiane: 382. Jun-Dec 1775.

(=) *Touchiroa* Aubl., Hist. Pl. Guiane: 384. 1775. Typus: *T. aromatica* Aubl.

Cryptocarya R. Br., Prodr.: 402. 27 Mar 1810 [*Laur.*]. Typus: *C. glaucescens* R. Br.

(=) *Ravensara* Sonn., Voy. Indes Orient. 2: 226. 1782. Typus: *R. aromatica* Sonn.

Cryptogyne Hook. f. in Bentham & Hooker, Gen. Pl. 2: 652, 656. Mai 1876 [*Sapot.*]. Typus: *C. gerrardiana* Hook f.

(H) *Cryptogyne* Cass. in Cuvier, Dict. Sci. Nat. 50: 491, 493, 498. Nov 1827 [*Dicot.: Comp.*]. Typus: *C. absinthioides* Cass.

Cryptotaenia DC., Coll. Mém. 5: 42. 12 Sep 1829 [*Umbell.*]. Typus: *C. canadensis* (L.) DC. (*Sison canadense* L.).

(≡) *Deringa* Adans., Fam. Pl. 2: 498, 549. Jul-Aug 1763.

Cudrania Trécul in Ann. Sci. Nat., Bot., ser. 3, 8: 122. Jul-Dec 1847 [*Mor.*]. Typus: *C. javanensis* Trécul (typ. cons.).

(=) *Vanieria* Lour., Fl. Cochinch.: 539, 564. Sep 1790. Typus (vide Merrill in Trans. Amer. Philos. Soc., ser. 2, 24: 134. 1935): *V. cochinchinensis* Lour.

Cumingia Vidal, Phan. Cuming. Phi-   (H)   *Cummingia* D. Don in Sweet, Brit.
lipp.: 211. Nov 1885 [*Bombac.*].   Fl. Gard. 3: ad t. 257. Apr 1828 [*Mo-*
Typus: *C. philippinensis* Vidal   *nocot.: Haemodor.*].
                                Typus: *C. campanulata* (Lindl.) D.
                                Don (*Conanthera campanulata* Lindl.).

Cunila L., Syst. Nat., ed. 10: 1359.   (H)   *Cunila* L. ex Mill., Gard. Dict. Abr.,
7 Jun 1759 [*Lab.*].   ed. 4: [414]. 28 Jan 1754 [*Dicot.:*
Typus: *C. mariana* L., nom. illeg.   *Lab.*].
(*Satureja origanoides* L., *C. origan-*   Typus (vide Reveal & Strachan in
oides (L.) Britton).   Taxon 29: 333. 1980): *Sideritis rom-*
                                ana L.

Cunonia L., Syst. Nat., ed. 10: 1013,   (H)   *Cunonia* Mill., Fig. Pl. Gard. Dict. 1:
1025, 1368. 7 Jun 1759 [*Cunon.*].   75. 28 Sep 1756 [*Monocot.: Irid.*].
Typus: *C. capensis* L.   Typus: *C. antholyza* Mill. (Gard.
                                Dict., ed. 8: *Cunonia* No. [1]. 16 Apr
                                1768) (*Antholyza cunonia* L.).

Curtisia Aiton, Hort. Kew. 1: 162. 7   (H)   *Curtisia* Schreb., Gen. Pl.: 199. Apr
Aug - 1 Oct 1789 [*Corn.*].   1789 [*Dicot.: Rut.*].
Typus: *C. faginea* Aiton, nom. illeg.   Typus: *C. schreberi* J. F. Gmel.
(*Sideroxylon dentatum* Burm. f., *C.*   (Syst. Nat. 2: 498. Sep (sero) - Nov
dentata (Burm. f.) C. A. Sm.).   1791).

Cuspidaria DC. in Biblioth. Univer-   (H)   *Cuspidaria* (DC.) Besser, Enum. Pl.:
selle Genève, ser. 2, 17: 125. Sep   104. 1822 (post 25 Mai) (*Erysimum*
1838 [*Bignon.*].   sect. *Cuspidaria* DC., Syst. Nat. 2:
Typus: *C. pterocarpa* (Cham.) DC.   493. Mai 1821) [*Dicot.: Cruc.*].
(Prodr. 9: 178. 1 Jan 1845) (*Bignon-*   ≡ *Acachmena* H. P. Fuchs 1960.
ia pterocarpa Cham.).

Cuviera DC. in Ann. Mus. Natl. Hist.   (H)   *Cuviera* Koeler, Descr. Gram.: 328
Nat. 9: 222. 30 Apr 1807 [*Rub.*].   ('382'). 1802 [*Monocot.: Gram.*].
Typus: *C. acutiflora* DC.   ≡ *Hordelymus* (Jess.) Harz 1885 (*Hor-*
                                deum [unranked] *Hordelymus* Jess.
                                1863).

Cyananthus Wall. ex Benth. in   (H)   *Cyananthus* Raf., Anal. Nat.: 192.
Royle, Ill. Bot. Himal. Mts.: 309.   Apr-Jul 1815 [*Dicot.: Comp.*].
Mai 1836 [*Campanul.*].   ≡ *Cyanus* P. Mill. 1754.
Typus: *C. lobatus* Wall. ex Benth.

Cyathula Blume, Bijdr.: 548. 24 Jan 1826 [*Amaranth.*].
Typus: *C. prostrata* (L.) Blume (*Achyranthes prostrata* L.).

(H) *Cyathula* Lour., Fl. Cochinch.: 93, 101. Sep 1790 [*Dicot.: Amaranth.*].
Typus: *C. geniculata* Lour.

Cybianthus Mart., Nov. Gen. Sp. Pl. 3: 87. Jan-Mar 1831 [*Myrsin.*].
Typus: *C. penduliflorus* Mart.

(=) *Peckia* Vell., Fl. Flumin.: 51. 7 Sep - 28 Nov 1829.
Typus: non designatus.

Cyclobalanopsis Oerst. in Vidensk. Meddel. Dansk Naturhist. Foren. Kjøbenhavn 1866: 77. 5 Jul 1867 [*Fag.*].
Typus: *C. velutina* Oerst. (*Quercus velutina* Lindl. ex Wall., non Lam.) (typ. cons.).

(=) *Perytis* Raf., Alsogr. Amer.: 29. 1838.
Typus (vide Pichi Sermolli in Taxon 3: 114. 1954): *P. lamellosa* (Sm.) Raf. (*Quercus lamellosa* Sm.).

Cyclospermum Lag., Amen. Nat. Españ. 1(2): 101. 1821 (*'Ciclospermum'*) (orth. cons.) [*Umbell.*].
Typus: *C. leptophyllum* (Pers.) Sprague ex Britton & P. Wilson (Bot. Porto Rico 6: 52. 14 Jan 1925) (*Pimpinella leptophylla* Pers.) (typ. cons.).

Cytinus L., Gen. Pl., ed. 6: 576 ('566'). Jun 1764 [*Raffles.*].
Typus: *C. hypocistis* (L.) L. (*Asarum hypocistis* L.).

(≡) *Hypocistis* Mill., Gard. Dict. Abr., ed. 4: [662]. 28 Jan 1754.

Cytisus Desf., Fl. Atlant. 2: 139. Nov 1798 [*Legum.*].
Typus: *C. triflorus* L'Hér. 1791, non Lam. 1786 (typ. cons.) [= *C. villosus* Pourr.].

(H) *Cytisus* L., Sp. Pl.: 739. 1 Mai 1753 [*Dicot.: Legum.*].
Typus (vide Green in Sprague & al., Nom. Prop. Brit. Bot.: 175. 1929): *C. sessilifolius* L.

Daboecia D. Don in Edinburgh New Philos. J. 17: 160. Jul 1834 [*Eric.*].
Typus: *D. polifolia* D. Don, nom. illeg. (*Vaccinium cantabricum* Huds., *D. cantabrica* (Huds.) K. Koch).

325

Dalbergia L. f., Suppl. Pl.: 52, 316. Apr 1782 [*Legum.*].
Typus: *D. lanceolaria* L. f.

(=) *Amerimnon* P. Browne, Civ. Nat. Hist. Jamaica: 288. 10 Mar 1756.
Typus: *A. brownei* Jacq. (Enum. Syst. Pl.: 27. Aug-Sep 1760).

(=) *Ecastaphyllum* P. Browne, Civ. Nat. Hist. Jamaica: 299. 10 Mar 1756.
Typus: *E. brownei* Pers. (Syn. Pl. 2: 277. Sep 1807) (*Hedysarum ecasta-phyllum* L.).

(=) *Acouroa* Aubl., Hist. Pl. Guiane: 753. Jun-Dec 1775.
Typus: *A. violacea* Aubl.

Dalea L., Opera Var.: 244. 1758 [*Legum.*].
Typus: *D. cliffortiana* Willd. (*Psoralea dalea* L.).

(H) *Dalea* Mill., Gard. Dict. Abr., ed. 4: [433]. 28 Jan 1754 [*Dicot.: Solan.*].
≡ *Browallia* L. 1753.

Darlingtonia Torr. in Smithsonian Contr. Knowl. 6(4): 4. Apr 1853 [*Sarracen.*].
Typus: *D. californica* Torr.

(H) *Darlingtonia* DC. in Ann. Sci. Nat. (Paris) 4: 97. Jan 1825 [*Dicot.: Legum.*].
Typus (vide Pichi Sermolli in Taxon 3: 115. 1954): *D. brachyloba* (Willd.) DC. (*Acacia brachyloba* Willd.).

Decaisnea Hook. f. & Thomson in Proc. Linn. Soc. Lond. 2: 350. 1 Mai 1855 [*Lardizabal.*].
Typus: *D. insignis* (Griff.) Hook. f. & Thomson (*Slackia insignis* Griff.).

(H) *Decaisnea* Brongn. in Duperrey, Voy. Monde, Phan.: 192. Jan 1834 [*Monocot.: Orchid.*].
Typus: *D. densiflora* Brongn.

Denhamia Meisn., Pl. Vasc. Gen. 1: 18; 2: 16. 26 Mar - 1 Apr 1837 [*Celastr.*].
Typus: *D. obscura* (A. Rich.) Meisn. ex Walp. (Repert. Bot. Syst. 1: 203. 18-20 Sep 1842) (*Leucocarpum obscurum* A. Rich.).

(H) *Denhamia* Schott in Schott & Endlicher, Melet. Bot.: 19. 1832 [*Monocot.: Ar.*].
≡ *Culcasia* P. Beauv. 1803 (nom. cons.).

(≡) *Leucocarpum* A. Rich. in Urville, Voy. Astrolabe 2: 46. 1834.

Derris Lour., Fl. Cochinch.: 423, 432. Sep 1790 [*Legum.*].
Typus: *D. trifoliata* Lour. (typ. cons.).

(=) *Salken* Adans., Fam. Pl. 2: 322, 600. Jul-Aug 1763.
Typus: non designatus.

(=) *Solori* Adans., Fam. Pl. 2: 327, 606. Jul-Aug 1763.
Typus: non designatus.

(=) *Deguelia* Aubl., Hist. Pl. Guiane: 750. Jun-Dec 1775.
Typus: *D. scandens* Aubl.

Descurainia Webb & Berthel., Hist. Nat. Iles Canaries 3(2,1): 72. Nov 1836 [*Cruc.*].
Typus: *D. sophia* (L.) Webb ex Prantl (in Engler & Prantl, Nat. Pflanzenfam. 3(2): 192. Mar 1891) (*Sisymbrium sophia* L.) (typ. cons.).

(≡) *Sophia* Adans., Fam. Pl. 2: 417, 606. Jul-Aug 1763.

(=) *Hugueninia* Rchb., Fl. Germ. Excurs.: 691. 1832.
Typus: *H. tanacetifolia* (L.) Rchb. (*Sisymbrium tanacetifolium* L.).

Desmanthus Willd., Sp. Pl. 4: 888, 1044. Apr 1806 [*Legum.*].
Typus: *D. virgatus* (L.) Willd. (*Mimosa virgata* L.).

(≡) *Acuan* Medik., Theodora: 62. 1786.

Desmodium Desv. in J. Bot. Agric. 1: 122. Feb 1813 [*Legum.*].
Typus: *D. scorpiurus* (Sw.) Desv. (*Hedysarum scorpiurus* Sw.) (typ. cons.).

(=) *Meibomia* Heist. ex Fabr., Enum.: 168. 1759.
Typus: *Hedysarum canadense* L.

(=) *Grona* Lour., Fl. Cochinch.: 424, 459. Sep 1790.
Typus: *G. repens* Lour.

(=) *Pleurolobus* J. St.-Hil. in Nouv. Bull. Sci. Soc. Philom. Paris 3: 192. Dec 1812.
Typus: non designatus.

Diamorpha Nutt., Gen. N. Amer. Pl. 1: 293. 14 Jul 1818 [*Crassul.*].
Typus: [specimen] N. of Camden, S. Carolina, winter 1816, *Nuttall* (PH) (typ. cons.) [= *D. smallii* Britton].

Dicentra Bernh. in Linnaea 8: 457, 468. 1833 (post Jul) [*Papaver.*].
Typus: *D. cucullaria* (L.) Bernh. (*Fumaria cucullaria* L.).

(≡) *Diclytra* Borkh. in Arch. Bot. (Leipzig) 1(2): 46. Mai-Dec 1797.

(=) *Capnorchis* Mill., Gard. Dict. Abr., ed. 4: [250]. 28 Jan 1754.
Typus: non designatus.

(=) *Bikukulla* Adans., Fam. Pl. 2: (23). Jul-Aug 1763.
Typus: non designatus.

(=) *Dactylicapnos* Wall., Tent. Fl. Napal.: 51. Sep-Dec 1826.
Typus: *D. thalictrifolia* Wall.

Dichrostachys (DC.) Wight & Arn., Prodr. Fl. Ind. Orient.: 271. Oct (prim.) 1834 (*Desmanthus* sect. *Dichrostachys* DC., Prodr. 2: 445. Nov (med.) 1825) [*Legum.*].
Typus: *Desmanthus cinereus* (L.) Willd. (*Mimosa cinerea* L., *Dichrostachys cinerea* (L.) Wight & Arn.).

(=) *Cailliea* Guill. & Perr. in Guillemin & al., Fl. Seneg. Tent.: 239. 2 Jul 1832.
Typus: *C. dicrostachys* Guill. & Perr., nom. illeg. (*Mimosa nutans* Pers.).

Dicliptera Juss. in Ann. Mus. Natl. Hist. Nat. 9: 267. Jul 1807 [*Acanth.*].
Typus: *D. chinensis* (L.) Juss. (*Justicia chinensis* L.) (typ. cons.).

(≡) *Diapedium* K. D. Koenig in Ann. Bot. (König & Sims) 2: 189. 1 Jun 1805.

Dicrastylis Drumm. ex Harv., Hooker's J. Bot. Kew Gard. Misc. 7: 56. 1855 [*Verb.*].
Typus: *D. fulva* Drumm. ex Harv.

(=) *Mallophora* Endl., Ann. Wien Mus. 2: 206. 1839.
Typus (vide Munir, Brunoniana 1: 567. 1978): *M. globifera* Endl.

(=) *Lachnocephalus* Turcz., Bull. Soc. Nat. Mosc. 22(2): 36. 1849.
Typus: *L. lepidotus* Turcz.

Dictyoloma A. Juss. in Mém. Mus. Hist. Nat. 12: 499. 1825 [*Rut.*].
Typus: *D. vandellianum* A. Juss.

Didelta L'Hér., Stirp. Nov.: 55. Mar 1786 [*Comp.*].
Typus: *D. tetragoniifolia* L'Hér.

(=) *Breteuillia* Buc'hoz, Grand Jard.: t. 62. 1785.
Typus: *B. trianensis* Buc'hoz

Didissandra C. B. Clarke in Candolle & Candolle, Monogr. Phan. 5: 65. Jul 1883 [*Gesner.*].
Typus: *D. elongata* (Jack) C. B. Clarke (*Didymocarpus elongata* Jack) (typ. cons.).

(=)   *Ellobum* Blume, Bijdr.: 746. 1826.
Typus: *E. montanum* Blume

Didymocarpus Wall. in Edinburgh Philos. J. 1: 378. 1819 [*Gesner.*].
Typus: *D. primulifolius* D. Don (Prodr. Fl. Nepal.: 123. 26 Jan - 1 Feb 1825) (typ. cons.).

(=)   *Henckelia* Spreng., Anleit. Kenntn. Gew., ed. 2, 2: 402. 20 Apr 1817.
Typus: *H. incana* (Vahl) Spreng. (*Roettlera incana* Vahl).

Dimorphotheca Vaill., Königl. Akad. Wiss. Paris Phys. Abh. 5: 547. Jan-Apr 1754. [*Comp.*].
Typus: *D. pluvialis* (L.) Moench (*Calendula pluvialis* L.) (typ. cons.).

Dipholis A. DC., Prodr. 8: 188. Mar (med.) 1844 [*Sapot.*].
Typus: *D. salicifolia* (L.) A. DC. (*Achras salicifolia* L.).

(=)   *Spondogona* Raf., Sylva Tellur.: 35. Oct-Dec 1838.
Typus: *S. nitida* Raf., nom. illeg. (*Bumelia pentagona* Sw.).

Diplandra Hook. & Arn., Bot. Beechey Voy.: 291. Dec 1838 [*Onagr.*].
Typus: *D. lopezioides* Hook. & Arn.

(H)   *Diplandra* Bertero in Mercurio Chileno 13: 612. 15 Apr 1829 [*Monocot.: Hydrocharit.*].
Typus: *D. potamogeton* Bertero

Dipteryx Schreb., Gen. Pl.: 485. Mai 1791 [*Legum.*].
Typus: *D. odorata* (Aubl.) Willd. (Sp. Pl. 3: 910. 1-10 Nov 1802) (*Coumarouna odorata* Aubl.) (typ. cons.).

(≡)   *Coumarouna* Aubl., Hist. Pl. Guiane: 740. Jun-Dec 1775.

(=)   *Taralea* Aubl., Hist. Pl. Guiane: 745. Jun-Dec 1775.
Typus: *T. oppositifolia* Aubl.

Disparago Gaertn., Fruct. Sem. Pl. 2: 463. Sep-Dec 1791 [*Comp.*].
Typus: *D. ericoides* (P. J. Bergius) Gaertn. (*Stoebe ericoides* P. J. Bergius).

Dissotis Benth. in Hooker, Niger Fl.: 346. Nov-Dec 1849 [*Melastomat.*].
Typus: *D. grandiflora* (Sm.) Benth. (*Osbeckia grandiflora* Sm.).

(≡)   *Hedusa* Raf., Sylva Tellur.: 101. Oct-Dec 1838.

(=)   *Kadalia* Raf., Sylva Tellur.: 101. Oct-Dec 1838.
Typus: non designatus.

Dolichandrone (Fenzl) Seem. in Ann. Mag. Nat. Hist., ser. 3, 10: 31. 1862 (*Dolichandra* [unranked] *Dolichandrone* Fenzl in Denkschr. Königl.-Baier. Bot. Ges. Regensburg 3: 265. 1841) [*Bignon.*].
Typus: *Dolichandrone spathacea* (L. f.) K. Schum. (*Bignonia spathacea* L. f.).

(≡) *Pongelia* Raf., Sylva Tellur.: 78. Oct-Dec 1838.

Dolichos L., Sp. Pl.: 725. 1 Mai 1753 [*Legum.*].
Typus: *D. trilobus* L. (typ. cons.).

Dombeya Cav., Diss. 2, App.: [4]. Jan-Apr 1786 [*Stercul.*].
Typus: *D. palmata* Cav. (Diss. 3: 122. Feb. 1787)

(H) *Dombeya* L'Hér., Stirp. Nov.: 33. Dec (sero) 1785 - Jan 1786 [*Dicot.: Bignon.*].
≡ *Tourrettia* Foug. 1787 (nom. cons.).

(=) *Assonia* Cav., Diss. 2. App.: [5]. Jan-Apr 1786.
Typus: *A. populnea* Cav.

Donatia J. R. Forst. & G. Forst., Char. Gen. Pl.: 5. 29 Nov 1775 [*Saxifrag.*].
Typus: *D. fascicularis* J. R. Forst. & G. Forst.

Dontostemon Andrz. ex C. A. Mey. in Ledeb., Fl. Altaic. 3: 4, 118. Jul-Dec 1831 [*Cruc.*].
Typus: *D. integrifolius* (L.) C. A. Mey. (*Sisymbrium integrifolium* L.).

Doona Thwaites in Hooker's J. Bot. Kew Gard. Misc. 3: t. 12. 1851 (sero) [*Dipterocarp.*].
Typus: *D. zeylanica* Thwaites

(=) *Caryolobis* Gaertn., Fruct. Sem. Pl. 1: 215. Dec 1788.
Typus: *C. indica* Gaertn.

Douglasia Lindl. in Quart. J. Sci. Lit. Arts 1827: 385. Oct-Dec 1827 [*Primul.*].
Typus: *D. nivalis* Lindl.

(H) *Douglassia* Mill., Gard. Dict. Abr., ed. 4: [452]. 28 Jan 1754 [*Dicot.: Verben.*].
≡ *Volkameria* L. 1753.

(=) *Vitaliana* Sesl. in Donati, Essai Hist. Nat. Mer Adriat.: 69. Jan-Mar 1758. Typus: *V. primuliflora* Bertol. (Fl. Ital. 2: 368. Mai 1835 - Mar 1836) (*Primula vitaliana* L.).

Downingia Torr. in Rep. Explor. Railroad Pacif. Ocean 4(1,4): 116. Aug-Sep 1857 [*Campanul.*].
Typus: *D. elegans* (Douglas ex Lindl.) Torr. (*Clintonia elegans* Douglas ex Lindl.).

(=) *Bolelia* Raf. in Atlantic J. 1: 120. 1832 (aut.).

Dracocephalum L., Sp. Pl.: 594. 1 Mai 1753 [*Lab.*].
Typus: *D. moldavica* L. (typ. cons.).

Dregea E. Mey., Comment. Pl. Afr. Austr.: 199. 1-8 Jan 1838 [*Asclepiad.*].
Typus: *D. floribunda* E. Mey.

(H) *Dregea* Eckl. & Zeyh., Enum. Pl. Afric. Austral.: 350. Apr 1837 [*Dicot.: Umbell.*].
Typus: non designatus.

Drimys J. R. Forst. & G. Forst., Char. Gen. Pl.: 42. 29 Nov 1775 [*Magnol.*].
Typus: *D. winteri* J. R. Forst. & G. Forst. (typ. cons.).

Dryandra R. Br. in Trans. Linn. Soc. London 10: 211. Feb 1810 [*Prot.*].
Typus: *D. formosa* R. Br. (typ. cons.).

(H) *Dryandra* Thunb., Nov. Gen. Pl.: 60. 18 Jun 1783 [*Dicot.: Euphorb.*].
Typus: *D. cordata* Thunb.

(=) *Josephia* R. Br. ex Knight, Cult. Prot.: 110. Dec 1809.
Typus: non designatus.

Duguetia A. St.-Hil., Fl. Bras. Merid. 1, ed. 4°: 35; ed. f°: 28. 23 Feb 1824 [*Annon.*].
Typus: *D. lanceolata* A. St.-Hil.

Dunalia Kunth in Humboldt & al., Nov. Gen. Sp. 3, ed. 4°: 55; ed. f°: 43. Sep (sero) 1818 [*Solan.*]. Typus: *D. solanacea* Kunth

(H) *Dunalia* Spreng., Pl. Min. Cogn. Pug. 2: 25. 1815 [*Dicot.: Rub.*]. ≡ *Lucya* DC. 1830 (nom. cons.).

Durandea Planch. in London J. Bot. 6: 594. 1847 [*Lin.*]. Typus: *D. serrata* Planch. (in London J. Bot. 7: 528. 1848).

(H) *Durandea* Delarbre, Fl. Auvergne, ed. 2, 365. Aug 1800 [*Dicot.: Cruc.*]. Typus: *D. unilocularis* Delarbre, nom. illeg. (*Raphanus raphanistrum* L.).

Duroia L. f., Suppl. Pl.: 30, 209. Apr 1782 [*Rub.*]. Typus: *D. eriopila* L. f.

(=) *Pubeta* L., Pl. Surin.: 16. 23 Jun 1775. Typus: non designatus.

Ecballium A. Rich. in Bory, Dict. Class. Hist. Nat. 6: 19. 9 Oct 1824 [*Cucurbit.*]. Typus: *E. elaterium* (L.) A. Rich. (*Momordica elaterium* L.).

(≡) *Elaterium* Mill., Gard. Dict. Abr., ed. 4: [459]. 28 Jan 1754.

Echinocystis Torr. & A. Gray, Fl. N. Amer. 1: 542. Jun 1840 [*Cucurbit.*]. Typus: *E. lobata* (Michx.) Torr. & A. Gray (*Sicyos lobata* Michx.).

(=) *Micrampelis* Raf. in Med. Repos., ser. 2, 5: 350. Feb-Apr 1808. Typus: *M. echinata* Raf.

Eclipta L., Mant. Pl.: 157, 286. Oct 1771 [*Comp.*]. Typus: *E. erecta* L., nom. illeg. (*Verbesina alba* L., *E. alba* (L.) Hassk.) (typ. cons.).

(≡) *Eupatoriophalacron* Mill., Gard. Dict. Abr., ed. 4: [479]. 28 Jan 1754.

Edraianthus A. DC. in Meisner, Pl. Vasc. Gen. 2: 149. 18-24 Aug 1839 [*Campanul.*]. Typus: *E. graminifolius* (L.) A. DC. (*Campanula graminifolia* L.).

(≡) *Pilorea* Raf., Fl. Tellur. 2: 80. Jan-Mar 1837.

Elatostema J. R. Forst. & G. Forst., Char. Gen. Pl.: 53. 29 Nov 1775 [*Urtic.*]. Typus: *E. sessile* J. R. Forst. & G. Forst. (typ. cons.).

Elionurus Humb. & Bonpl. ex Willd., Sp. Pl. 4: 941. Apr 1806 *('Elyonurus')* [*Gram.*].
Typus: *E. tripsacoides* Humb. & Bonpl. ex Willd.

Ellisia L., Sp. Pl., ed. 2: 1662. Jul-Aug 1763 [*Hydrophyll.*].
Typus: *E. nyctelea* (L.) L. (*Ipomoea nyctelea* L.).

(H) *Ellisia* P. Browne, Civ. Nat. Hist. Jamaica: 262. 10 Mar 1756 [*Dicot.: Verben.*].
Typus: *E. acuta* L. (Syst. Nat., ed. 10, 2: 1121. 7 Jun 1759).

Elytraria Michx., Fl. Bor.-Amer. 1: 8. 19 Mar 1803 [*Acanth.*].
Typus: *E. virgata* Michx., nom. illeg. (*Tubiflora caroliniensis* J. F. Gmel., *E. caroliniensis* (J. F. Gmel.) Pers.).

(≡) *Tubiflora* J. F. Gmel., Syst. Nat. 2: 27. Sep (sero) - Nov 1791.

Embelia Burm. f., Fl. Indica: 62. 1 Mar - 6 Apr 1768 [*Myrsin.*].
Typus: *E. ribes* Burm. f.

(≡) *Ghesaembilla* Adans., Fam. Pl. 2: 449, 561. Jul-Aug 1763.
(=) *Pattara* Adans., Fam. Pl. 2: 447, 588. Jul-Aug 1763.
Typus: *Ardisia tserim-cottam* Roem. & Schult.

Emex Campd., Monogr. Rumex: 56. 1819 [*Polygon.*].
Typus: *E. spinosa* (L.) Campd. (*Rumex spinosus* L.) (typ. cons.).

(≡) *Vibo* Medik., Philos. Bot. 1: 178. Apr 1789.

Enallagma (Miers) Baill., Hist. Pl. 10: 54. Nov-Dec 1888 (*Crescentia* sect. *Enallagma* Miers in Trans. Linn. Soc. London 26: 174. 1868) [*Bignon.*].
Typus: *Crescentia cucurbitina* L. (*E. cucurbitinum* (L.) Baill. ex K. Schum.).

(≡) *Dendrosicus* Raf., Sylva Tellur.: 80. Oct-Dec 1838.

Endlicheria Nees in Linnaea 8: 37. 1833 [*Laur.*].
Typus: *E. hirsuta* (Schott) Nees (*Cryptocarya hirsuta* Schott) (typ. cons.).

(H) *Endlichera* C. Presl, Symb. Bot. 1: 73. Jan-Feb 1832 [*Dicot.: Rub.*].
Typus: *E. brasiliensis* C. Presl

Endospermum Benth., Fl. Hongk.:   (H) *Endespermum* Blume, Catalogus: 24.
304. Feb 1861 [*Euphorb.*].   Feb-Sep 1823 [*Dicot.: Legum.*].
Typus: *E. chinense* Benth.   Typus: *E. scandens* Blume

Enicostema Blume, Bijdr.: 848. Jul-
Dec 1826 [*Gentian.*].
Typus: *E. littorale* Blume

Enneastemon Exell in J. Bot. 70,   (=) *Clathrospermum* Planch. ex Hook. f.
Suppl. 1: 209. Feb 1932 [*Annon.*].   in Bentham & Hooker, Gen. Pl. 1:
Typus: *E. angolensis* Exell   29. 7 Aug 1862.
                             Typus: *C. vogelii* (Hook. f.) Benth.
                             (*Uvaria vogelii* Hook. f.).

Entada Adans., Fam. Pl. 2: 318, 554.   (=) *Gigalobium* P. Browne, Civ. Nat.
Jul-Aug 1763 [*Legum.*].   Hist. Jamaica: 362. 10 Mar 1756.
Typus: *E. rheedei* Spreng. (Syst. Veg.   Typus (vide Panigrahi in Taxon 34:
2: 325. Jan-Mai 1825) (*Mimosa en-*   714. 1985): *Entada gigas* (L.) Fawc.
tada L.).   & Rendle (*Mimosa gigas* L.).

Epacris Cav., Icon. 4: 25. Sep-Dec   (H) *Epacris* J. R. Forst. & G. Forst.,
1797 [*Epacrid.*].   Char. Gen. Pl.: 10. 29 Nov 1775 [*Di-*
Typus: *E. longiflora* Cav. (typ. cons.).   *cot.: Epacrid.*].
                             Typus (vide Pichi Sermolli in Taxon
                             3: 119. 1954): *E. longifolia* J. R.
                             Forst. & G. Forst.

Epifagus Nutt., Gen. N. Amer. Pl.
2: 60. 14 Jul 1818 [*Orobanch.*].
Typus: *E. americana* Nutt., nom.
illeg. (*Orobanche virginiana* L., *E.
virginianus* (L.) Barton).

Eranthis Salisb. in Trans. Linn. Soc.   (≡) *Cammarum* Hill, Brit. Herb.: 47. 23
London 8: 303. 9 Mar 1807 [*Ranun-*   Feb 1756.
cul.].
Typus: *E. hyemalis* (L.) Salisb. (*Helleborus hyemalis* L.).

Eriosema (DC.) Desv. in Ann. Sci.
Nat. (Paris) 9: 421. Dec 1826 (*'Eur-
iosma'*) (*Rhynchosia* sect. *Eriosema*
DC., Prodr. 2: 388. Nov (med.)
1825) [*Legum.*].
Typus: *Rhynchosia rufa* (Kunth) DC.
(*Glycine rufa* Kunth, *E. rufum*
(Kunth) G. Don).

Erophila DC. in Mém. Mus. Hist. Nat. 7: 234. 20 Apr 1821 [*Cruc.*]. Typus: *E. verna* (L.) Chevall. (Fl. Gén. Env. Paris 2: 898. 5 Jan 1828) (*Draba verna* L.) (typ. cons.).

(≡)   *Gansblum* Adans., Fam. Pl. 2: 420, 561. Jul-Aug 1763.

Erythrospermum Thouars, Hist. Vég. Isles Austral. Afriq.: 65. Jan 1808 [*Flacourt.*]. Typus: *E. pyrifolium* Poir. (in Lamarck, Encycl., Suppl. 2: 584. 3 Jul 1812) (typ. cons.).

(=)   *Pectinea* Gaertn., Fruct. Sem. Pl. 2: 136. Sep (sero) - Nov 1790. Typus: *P. zeylanica* Gaertn.

Euclidium W. T. Aiton, Hort. Kew., ed. 2, 4: 74. Dec 1812 [*Cruc.*]. Typus: *E. syriacum* (L.) W. T. Aiton (*Anastatica syriaca* L.).

(≡)   *Hierochontis* Medik., Pfl.-Gatt.: 51. 22 Apr 1792.

(=)   *Soria* Adans., Fam. Pl. 2: 421, 606. Jul-Aug 1763. Typus: non designatus.

Eucnide Zucc., Delect. Sem. Hort. Monac. 1844: [4]. 28 Dec 1844 [*Loas.*]. Typus: *E. bartonioides* Zucc.

(=)   *Microsperma* Hook. in Icon. Pl.: ad t. 234. Jan-Feb 1839. Typus: *M. lobatum* Hook.

Euonymus L., Sp. Pl.: 197. 1 Mai 1753 (*'Evonymus'*) (orth. cons.) [*Celastr.*]. Typus: *E. europaeus* L.

Euscaphis Siebold & Zucc., Fl. Jap. 1: 122. 1840 [*Staphyl.*]. Typus: *E. staphyleoides* Siebold & Zucc., nom. illeg. (*Sambucus japonica* Thunb., *E. japonica* (Thunb.) Kanitz).

(≡)   *Hebokia* Raf., Alsogr. Amer.: 47. 1838.

Eusideroxylon Teijsm. & Binn. in Natuurk. Tijdschr. Ned. Indië 25: 292. 1863 [*Laur.*]. Typus: *E. zwageri* Teijsm. & Binn.

Exocarpos Labill., Voy. Rech. Pérouse 1: 155. 22 Feb - 4 Mar 1800 [*Santal.*]. Typus: *E. cupressiformis* Labill.

(=)   *Xylophylla* L., Mant. Pl.: 147, 221. Oct 1771. Typus: *X. longifolia* L.

Eysenhardtia Kunth in Humboldt & al., Nov. Gen. Sp. 6, ed. 4°: 489; ed. f°: 382. Sep 1824 [*Legum.*].
Typus: *E. amorphoides* Kunth

(=)   *Viborquia* Ortega, Nov. Pl. Descr. Dec.: 66. 1798.
Typus: *V. polystachya* Ortega

Fagara L., Syst. Nat., ed. 10: 885, 897, 1362. 7 Jun 1759 [*Rut.*].
Typus: *F. pterota* L.

(H)   *Fagara* Duhamel, Traité Arbr. Arbust. 1: 229. 1755 [*Dicot.: Rut.*].
Typus: non designatus.

(≡)   *Pterota* P. Browne, Civ. Nat. Hist. Jamaica: 146. 10 Mar 1756.

Fagopyrum Mill., Gard. Dict. Abr., ed. 4: [495]. 28 Jan 1754 [*Polygon.*].
Typus: *F. esculentum* Moench (Methodus: 290. 4 Mai 1794) (*Polygonum fagopyrum* L.) (typ. cons.).

Falcaria Fabr., Enum.: 34. 1759 [*Umbell.*].
Typus: *F. vulgaris* Bernh. (Syst. Verz.: 176. 1800) (*Sium falcaria* L.).

Falkia Thunb., Nov. Gen. Pl.: 17. 24 Nov 1781 (*'Falckia'*) (orth. cons.) [*Convolvul.*].
Typus: *F. repens* Thunb.

Fedia Gaertn., Fruct. Sem. Pl. 2: 36. Sep (sero) - Nov 1790 [*Valerian.*].
Typus: *F. cornucopiae* (L.) Gaertn. (*Valeriana cornucopiae* L.) (typ. cons.).

(H)   *Fedia* Adans., Fam. Pl. 2: 152, 557. Jul-Aug 1763 [*Dicot.: Valerian.*].
Typus: *Valeriana ruthenica* Willd.

Felicia Cass. in Bull. Sci. Soc. Philom. Paris 1818: 165. Nov 1818 [*Comp.*].
Typus: *F. tenella* (L.) Nees (Gen. Sp. Aster.: 208. Jul-Dec 1832) (*Aster tenellus* L.).

(=)   *Detris* Adans., Fam. Pl. 2: 131, 549. Jul-Aug 1763.
Typus: non designatus.

Filago L., Sp. Pl.: 927, 1199, [add. post indicem]. 1 Mai 1753 [*Comp.*].
Typus: *F. pyramidata* L. (typ. cons.).

Fittonia Coem. in Fl. Serres Jard. Eur. 15: 185. 1865 [*Acanth.*].
Typus: *F. verschaffeltii* (Lem.) Van Houtte (*Gymnostachyum verschaffeltii* Lem.).

(=) *Adelaster* Lindl. ex Veitch in Gard. Chron. 1861: 499. 1 Jun 1861.
Typus: *A. albivenis* Lindl. ex Veitch

Flemingia Roxb. ex W. T. Aiton, Hort. Kew., ed. 2, 4: 349. Dec 1812 [*Legum.*].
Typus: *F. strobilifera* (L.) W. T. Aiton (*Hedysarum strobiliferum* L.).

(H) *Flemingia* Roxb. ex Rottler in Ges. Naturf. Freunde Berlin Neue Schriften 4: 202. 1803 [*Dicot.: Acanth.*].
Typus: *F. grandiflora* Roxb. ex Rottler

(≡) *Luorea* Neck. ex J. St.-Hil. in Nouv. Bull. Sci. Soc. Philom. Paris 3: 193. Dec 1812.

Forestiera Poir. in Lamarck, Encycl., Suppl. 1: 132. 3 Sep 1810 [*Ol.*].
Typus: *F. cassinoides* (Willd.) Poir. (in Lamarck, Encycl., Suppl. 2: 665. 3 Jul 1812) (*Borya cassinoides* Willd.) (etiam vide *Adelia* [*Dicot.*]).

Forsythia Vahl, Enum. Pl. 1: 39. Jul-Dec 1804 [*Ol.*].
Typus: *F. suspensa* (Thunb.) Vahl (*Ligustrum suspensum* Thunb.).

(H) *Forsythia* Walter, Fl. Carol.: 153. Apr-Jun 1788 [*Dicot.: Saxifrag.*].
Typus: *F. scandens* Walter

Franseria Cav., Icon. 2: 78. Dec 1793 -Jan 1794 [*Comp.*].
Typus: *F. ambrosioides* Cav., nom. illeg. (*Ambrosia arborescens* Mill., non *Franseria arborescens* Brandegee 1903) (etiam vide *Gaertnera* [*Dicot.*]) [nom. legit. sub *Franseria* deest].

Freziera Willd., Sp. Pl. 2(2): 1122, 1179. Dec 1799 [*The.*].
Typus: *F. undulata* (Sw.) Willd. (*Eroteum undulatum* Sw.) (typ. cons.).

(≡) *Eroteum* Sw., Prodr.: 5, 85. 20 Jun - 29 Jul 1788.

(=) *Lettsomia* Ruiz & Pav., Fl. Peruv. Prodr.: 77. Oct (prim.) 1794.
Typus: non designatus.

Gaertnera Lam., Tabl. Encycl. 1: 379. 30 Jul 1792 [*Rub.*].
Typus: *G. vaginata* Lam. (Tabl. Encycl. 2: 273. 31 Oct 1819).

(H) *Gaertnera* Schreb., Gen. Pl.: 290. Apr 1789 [*Dicot.: Malpigh.*].
≡ *Hiptage* Gaertn. 1790 (nom. cons.).

(H) *Gaertneria* Medik., Philos. Bot. 1: 45. Apr 1789 [*Dicot.: Comp.*].
≡ *Franseria* Cav. 1793 (nom. cons.).

Galactites Moench, Methodus: 558. 4 Mai 1794 [*Comp.*].
Typus: *G. tomentosus* Moench (*Centaurea galactites* L.) (typ. cons.).

Galax Sims in Bot. Mag.: ad t. 754. Jun 1804 [*Diapens.*].
Typus: *G. urceolata* (Poir.) Brummitt (*Pyrola urceolata* Poir.).

(H) *Galax* L., Sp. Pl.: 200. 1 Mai 1753 [*Dicot.: Hydrophyll.*].
Typus: *G. aphylla* L.

Galearia Zoll. & Moritzi in Moritzi, Syst. Verz.: 19. Mai-Jun 1846 [*Euphorb.*].
Typus: *G. pedicellata* Zoll. & Moritzi

(H) *Galearia* C. Presl, Symb. Bot. 1: 49. Sep-Dec 1831 [*Dicot.: Legum.*].
Typus (vide Hossain in Notes Roy. Bot. Gard. Edinburgh 23: 446. 1961): *G. fragifera* (L.) C. Presl (*Trifolium fragiferum* L.).

Gardenia J. Ellis in Philos. Trans. 51: 935. 1761 [*Rub.*].
Typus: *G. jasminoides* J. Ellis

(H) *Gardenia* Colden in Essays Observ. Phys. Lit. Soc. Edinburgh 2: 2. 1756 [*Dicot.: Guttif.*].
Typus: non designatus.

Gaylussacia Kunth in Humboldt & al., Nov. Gen. Sp. 3, ed. 4°: 275; ed. f°: 215. 9 Jul 1819 [*Eric.*].
Typus: *G. buxifolia* Kunth

Gazania Gaertn., Fruct. Sem. Pl. 2: 451. Sep-Dec 1791 [*Comp.*].
Typus: *G. rigens* (L.) Gaertn. (*Othonna rigens* L.).

(=) *Meridiana* Hill, Veg. Syst. 2: 121**. Oct 1761.
Typus: *M. tesselata* Hill (Hort. Kew.: 26. 1768).

Gentianella Moench, Methodus: 482. 4 Mai 1794 [*Gentian.*].
Typus: *G. tetrandra* Moench, nom. illeg. (*Gentiana campestris* L., *Gentianella campestris* (L.) Börner).

(=) *Amarella* Gilib., Fl. Lit. Inch. 1: 36. 1782.
Typus (vide Rauschert in Taxon 25: 192. 1976): *Gentiana amarella* L.

Geophila D. Don, Prodr. Fl. Nepal.:
136. 26 Jan - 1 Feb 1825 [*Rub.*].
Typus: *G. reniformis* D. Don, nom.
illeg. (*Psychotria herbacea* Jacq., *G. herbacea* (Jacq.) K. Schum.).

(H)   *Geophila* Bergeret, Fl. Basses-Pyré-
nées 2: 184. 1803 [*Monocot.: Lil.*].
Typus: *G. pyrenaica* Bergeret

Gerbera L., Opera Var.: 247. 1758
[*Comp.*].
Typus: *G. linnaei* Cass. (in Cuvier,
Dict. Sci. Nat. 18: 460. 6 Apr 1821)
(*Arnica gerbera* L.) (typ. cons.).

Glechoma L., Sp. Pl.: 578. 1 Mai
1753 (*'Glecoma'*) (orth. cons.) [*Lab.*].
Typus: *G. hederacea* L.

Glochidion J. R. Forst. & G. Forst.,
Char. Gen. Pl.: 57. 29 Nov 1775 [*Eu-
phorb.*].
Typus: *G. ramiflorum* J. R. Forst. &
G. Forst.

(=)   *Agyneia* L., Mant. Pl.: 161, 296, 576.
Oct 1771.
Typus (vide Regnum Veg. 46: 307.
1966): *A. pubera* L.

Glossostigma Wight & Arn. in Nova
Acta Phys.-Med. Acad. Caes. Leop.-
Carol. Nat. Cur. 18: 355. 1836
[*Scrophular.*].
Typus: *G. spathulatum* Arn., nom. il-
leg. (*Limosella diandra* L., *F. dian-
drum* (L.) Kuntze).

(≡)   *Peltimela* Raf. in Atlantic J. 1: 199.
1833 (aut.).

Glycine Willd., Sp. Pl. 3: 854, 1053.
1-10 Nov 1802 [*Legum.*].
Typus: *G. clandestina* J. C. Wendl.
(Bot. Beob.: 54. 1798).

(H)   *Glycine* L., Sp. Pl.: 753. 1 Mai 1753
[*Dicot.: Legum.*].
Typus (vide Green in Sprague & al.,
Nom. Prop. Brit. Bot.: 176. 1929):
G. javanica L.

(=)   *Soja* Moench, Methodus: 153, index.
4 Mai 1794.
Typus: *S. hispida* Moench

Glycosmis Corrêa in Ann. Mus.
Natl. Hist. Nat. 6: 384. 1805 [*Rut.*].
Typus: *G. arborea* (Roxb.) DC.
(Prodr. 1: 538. Jan (med.) 1824) (*Li-
monia arborea* Roxb.).

(=)   *Panel* Adans., Fam. Pl. 2: 447, 587.
Jul-Aug 1763.
Typus: *Limonia winterlia* Steud.

Goetzea Wydler in Linnaea 5: 423. Jul 1830 [*Solan.*].
Typus: *G. elegans* Wydler

(H) *Goetzea* Rchb., Consp. Regni Veg.: 150. Dec 1828 - Mar 1829 [*Dicot.: Legum.*].
≡ *Rothia* Pers. 1807 (nom. cons.).

Goldbachia DC. in Mém. Mus. Hist. Nat. 7: 242. 20 Apr 1821 [*Cruc.*].
Typus: *G. laevigata* (M. Bieb.) DC. (Syst. Nat. 2: 577. Mai (sero) 1821) (*Raphanus laevigatus* M. Bieb.).

(H) *Goldbachia* Trin. in Sprengel, Neue Entd. 2: 42. Jan 1821 [*Monocot.: Gram.*].
Typus: *G. mikanii* Trin.

Gordonia J. Ellis in Philos. Trans. 60: 520. 1771 [*The.*].
Typus: *G. lasianthus* (L.) J. Ellis (*Hypericum lasianthus* L.) (etiam vide *Lasianthus* [*Dicot.*]).

Grevillea R. Br. ex Knight, Cult. Prot.: xvii, 120. Dec 1809 (*'Grevillia'*) (orth. cons.) [*Prot.*].
Typus: *G. aspleniifolia* R. Br. ex Knight.

(=) *Lysanthe* Salisb. ex Knight, Cult. Prot.: 116. Dec 1809.
Typus: non designatus.

(=) *Stylurus* Salisb. ex Knight, Cult. Prot.: 115. Dec 1809.
Typus: non designatus.

Guaiacum L., Sp. Pl.: 321. 1 Mai 1753 (*'Guajacu'*) (orth. cons.) [*Zygophyll.*].
Typus: *G. officinale* L.

Guarea F. Allam. in L., Mant. Pl.: 150, 228. Oct 1771 [*Mel.*].
Typus: *G. trichilioides* L., nom. illeg. (*Melia guara* Jacq., *Guarea guara* (Jacq.) P. Wilson).

(=) *Elutheria* P. Browne, Civ. Nat. Hist. Jamaica: 369. 10 Mar 1756.
Typus: *E. microphylla* (Hook.) M. Roem. (Fam. Nat. Syn. Monogr. 1: 122. 14 Sep - 15 Oct 1846) (*Guarea microphylla* Hook.).

Guatteria Ruiz & Pav., Fl. Peruv. Prodr.: 85. Oct (prim.) 1794 [*Annon.*].
Typus: *G. glauca* Ruiz & Pav. (Syst. Veg. Fl. Peruv. Chil.: 145. Dec (sero) 1798).

(=) *Aberemoa* Aubl., Hist. Pl. Guiane 1: 610. Jun-Dec 1775.
Typus: *A. guianensis* Aubl.

Guizotia Cass. in Cuvier, Dict. Sci. Nat. 59: 237, 247, 248. Jun 1829 [*Comp.*].
Typus: *G. abyssinica* (L. f.) Cass. (*Polymnia abyssinica* L. f.).

Gustavia L., Pl. Surin.: 12, 17, 18. 23 Jun 1775 [*Lecythid.*].
Typus: *G. augusta* L.

(=) *Japarandiba* Adans., Fam. Pl. 2: 448, 564. Jul-Aug 1763.
Typus (vide Miers in Trans. Linn. Soc. London 30: 183. 1874): *Gustavia marcgraviana* Miers

Gymnocladus Lam., Encycl.: 733. 1 Aug 1785 (gend. masc. cons.) [*Legum.*].
Typus: *G. canadensis* Lam., nom. illeg. (*Guilandina dioica* L., *Gymnocladus dioicus* (L.) K. Koch) (typ. cons.).

Gymnosporia (Wight & Arn.) Benth. & Hook. f., Gen. Pl. 1: 359, 365. 7 Aug 1862 (*Celastrus* sect. *Gymnosporia* Wight & Arn., Prodr. Fl. Ind. Orient.: 159. 10 Oct 1834) [*Celastr.*].
Typus: *Celastrus montanus* Roth ex Roem. & Schult. (*G. montana* (Roth ex Roem. & Schult.) Benth.) (typ. cons.).

(=) *Catha* Forssk. ex Scop., Intr. Hist. Nat.: 228. Jan-Apr 1777.
Typus (vide Friis in Taxon 33: 662. 1984): *C. edulis* (Vahl) Endl. (Ench. Bot.: 575. 15-21 Aug 1841) (*Celastrus edulis* Vahl).

(=) *Scytophyllum* Eckl. & Zeyh, Enum. Pl. Afric. Austral.: 124. Dec 1834 - Mar 1835.
Typus: non designatus.

(=) *Encentrus* C. Presl in Abh. Königl. Böhm. Ges. Wiss., ser 5, 3: 463. Jul-Dec 1845.
Typus: *E. linearis* (L. f.) C. Presl (*Celastrus linearis* L. f.).

(=) *Polyacanthus* C. Presl in Abh. Königl. Böhm. Ges. Wiss., ser. 5, 3: 463. Jul-Dec 1845.
Typus: *P. stenophyllus* (Eckl. & Zeyh.) C. Presl (*Celastrus stenophyllus* Eckl. & Zeyh.).

Gynandropsis DC., Prodr. 1: 237. Jan (med.) 1824 [*Cappar.*]. Typus: *G. pentaphylla* DC., nom. illeg. (*Cleome pentaphylla* L., nom. illeg., *Cleome gynandra* L., *G. gynandra* (L.) Briq.).

(≡) *Pedicellaria* Schrank in Bot. Mag. (Römer & Usteri) 3(8): 10. Apr 1790.

Gynura Cass. in Cuvier, Dict. Sci. Nat. 34: 391. Apr 1825 [*Comp.*]. Typus: *G. auriculata* Cass. (Opusc. Phytol. 3: 100. 19 Apr 1834) (typ. cons.).

(=) *Crassocephalum* Moench, Methodus: 516. 4 Mai 1794. Typus: *C. cernuum* Moench, nom. illeg. (*Senecio rubens* B. Juss. ex Jacq., *C. rubens* (B. Juss. ex Jacq.) S. Moore).

Halenia Borkh. in Arch. Bot. (Leipzig) 1(1): 25. 1796 [*Gentian.*]. Typus: *H. sibirica* Borkh., nom. illeg. (*Swertia corniculata* L., *H. corniculata* (L.) Cornaz).

Halesia J. Ellis ex L., Syst. Nat., ed. 10: 1041, 1044, 1369. 7 Jun 1759 [*Styrac.*]. Typus: *H. carolina* L.

(H) *Halesia* P. Browne, Civ. Nat. Hist. Jamaica: 205. 10 Mar 1756 [*Dicot.: Rub.*]. Typus: *Guettarda argentea* Lam.

Haplolophium Cham. in Linnaea 7: 556. 1832 (*'Aplolophium'*) (orth. cons.) [*Bignon.*]. Typus: *H. bracteatum* Cham.

Haplopappus Cass. in Cuvier, Dict. Sci. Nat. 56: 168. Sep 1828 (*'Aplopappus'*) (orth. cons.) [*Comp.*]. Typus: *H. glutinosus* Cass.

Haplophyllum A. Juss. in Mém. Mus. Hist. Nat. 12: 464. 1825 (*'Aplophyllum'*) (orth. cons.) [*Rut.*]. Typus: *H. tuberculatum* (Forssk.) A. Juss. (*Ruta tuberculata* Forssk.).

(H) *Aplophyllum* Cass. in Cuvier, Dict. Sci. Nat. 33: 463. Dec. 1824 [*Dicot.: Comp.*]. Typus: non designatus.

Harrisonia R. Br. ex A. Juss. in Mém. Mus. Hist. Nat. 12: 517. 1825 [*Simaroub.*]. Typus: *H. brownii* A. Juss.

(H) *Harissona* Adans. ex Léman in Cuvier, Dict. Sci. Nat. 20: 290. 29 Jun 1821 [*Musci*]. Typus: non designatus.

Hebecladus Miers in London J. Bot. 4: 321. 1845 [*Solan.*].
Typus: *H. umbellatus* (Ruiz & Pav.) Miers (*Atropa umbellata* Ruiz & Pav.) (typ. cons.).

(≡) *Kokabus* Raf., Sylva Tellur.: 55. Oct-Dec 1838.

(=) *Ulticona* Raf., Sylva Tellur.: 55. Oct-Dec 1838.
Typus (vide D'Arcy in Solanaceae Newslett. 2(4): 20. 1986): *U. biflora* (Ruiz & Pav.) Raf. (*Atropa biflora* Ruiz & Pav.).

(=) *Kukolis* Raf., Sylva Tellur.: 55. Oct-Dec 1838.
Typus: *K. bicolor* Raf.

Heberdenia Banks ex A. DC. in Ann. Sci. Nat., Bot., ser. 2, 16: 79. Aug 1841 [*Myrsin.*].
Typus: *H. excelsa* Banks ex A. DC. (Prodr. 8: 106. Mar (med.) 1844), nom. illeg. (*Anguillaria bahamensis* Gaertn., *H. bahamensis* (Gaertn.) Sprague) (etiam vide *Anguillaria* [*Monocot.*]).

Hedyotis L., Sp. Pl.: 101. 1 Mai 1753 [*Rub.*].
Typus: *H. fruticosa* L. (typ. cons.: [specimen] Herb. Hermann 1: 18, No. 63 (BM)).

Hedysarum L., Sp. Pl.: 745. 1 Mai 1753 [*Legum.*].
Typus: *H. alpinum* L. (typ. cons.).

Heisteria Jacq., Enum. Syst. Pl.: 4, 20. Aug-Sep 1760 [*Olac.*].
Typus: *H. coccinea* Jacq.

(H) *Heisteria* L., Opera Var.: 242. 1758 [*Dicot.: Polygal.*].
≡ *Muraltia* DC. 1824 (nom. cons.).

Helichrysum Mill., Gard. Dict. Abr., ed. 4: [462]. 28 Jan 1754 (*'Elichrysum'*) (orth. cons.) [*Comp.*].
Typus: *H. orientale* (L.) Gaertn. (Fruct. Sem. Pl. 2: 404. Sep-Dec 1791) (*Gnaphalium orientale* L.) (typ. cons.).

Helinus E. Mey. ex Endl., Gen. Pl.: 1102. Apr 1840 [*Rhamn.*]. Typus: *H. mystacinus* (Aiton) E. Mey. ex Steud. (Nomencl. Bot., ed. 2, 1: 742. Nov 1840) (*Rhamnus mystacinus* Aiton).

(≡) *Mystacinus* Raf., Sylva Tellur.: 30. Oct-Dec 1838.

Heliopsis Pers., Syn. Pl. 2: 473. Sep 1807 [*Comp.*]. Typus: *H. helianthoides* (L.) Sweet (Hort. Brit. 2: 487. Sep-Oct 1826) (*Buphthalmum helianthoides* L.) (typ. cons.).

Helosis Rich. in Mém. Mus. Hist. Nat. 8: 416, 432. 1822 [*Balano-phor.*]. Typus: *H. guyannensis* Rich., nom. illeg. (*Cynomorium cayanense* Sw., *H. cayanensis* (Sw.) Spreng.).

Helwingia Willd., Sp. Pl. 4: 634, 716. Apr 1806 [*Corn.*]. Typus: *H. rusciflora* Willd., nom. illeg. (*Osyris japonica* Thunb., *H. japonica* (Thunb.) F. G. Dietr.).

(H) *Helvingia* Adans., Fam. Pl. 2: 345, 553. Jul-Aug 1763 [*Dicot.: Flacourt.*]. ≡ *Thamnia* P. Browne 1756 (nom. rej. sub *Thamnea*).

Hemimeris L. f., Suppl. Pl.: 45, 280. Apr 1782 [*Scrophular.*]. Typus: *H. montana* L. f. (typ. cons.).

(H) *Hemimeris* L., Pl. Rar. Afr.: 8. 20 Dec 1760 [*Dicot.: Scrophular.*]. Typus: *H. bonae-spei* L.

Hesperochiron S. Watson, Botany [Fortieth Parallel]: 281. Sep-Dec 1871 [*Hydrophyll.*]. Typus: *H. californicus* (Benth.) S. Watson (*Ourisia californica* Benth.).

(=) *Capnorea* Raf., Fl. Tellur. 3: 74. Nov-Dec 1837. Typus: *C. nana* (Lindl.) Raf. (*Nicotiana nana* Lindl.).

Heterolepis Cass. in Bull. Sci. Soc. Philom. Paris 1820: 26. Feb 1820 [*Comp.*]. Typus: *H. decipiens* Cass., nom. illeg. (*Arnica inuloides* Vahl) [= *H. aliena* (L. f.) Druce (*Oedera aliena* L. f.)] (etiam vide *Heteromorpha* [*Dicot.*]).

Heteromorpha Cham. & Schltdl. in Linnaea 1: 385. Aug-Oct 1826 [*Umbell.*].
Typus: *H. arborescens* Cham. & Schltdl. (*Bupleurum arborescens* Thunb. 1794, non Jacq. 1788).

(H)   *Heteromorpha* Cass. in Bull. Sci. Soc. Philom. Paris 1817: 12. Jan 1817 [*Dicot.: Comp.*].
≡ *Heterolepis* Cass. 1820 (nom. cons.).

Heteropterys Kunth in Humboldt & al., Nov. Gen. Sp. 5, ed. 4°: 163; ed. f°: 126] 25 Feb 1822 (*'Heteropteris'*) (orth. cons.) [*Malpigh.*].
Typus: *H. purpurea* (L.) Kunth (*Banisteria purpurea* L.) (typ. cons.).

(=)   *Banisteria* L., Sp. Pl.: 427. 1 Mai 1753.
Typus (vide Sprague in Gard. Chron., ser. 3, 75: 104. 1924): *B. brachiata* L.

Heteropyxis Harv., Thes. Cap. 2: 18. 1863 [*Heteropyxid.*].
Typus: *H. natalensis* Harv.

(H)   *Heteropyxis* Griff., Not. Pl. Asiat. 4: 524. 1854 [*Dicot.: Bombac.*].
Typus: *Boschia griffithii* Mast.

Hibiscus L., Sp. Pl.: 693. 1 Mai 1753 [*Malv.*].
Typus: *H. syriacus* L. (typ. cons.).

Hieronyma Allemão, Hyeronima Alchorneoides: [1] (*'Hyeronima'*), t. [1] (*'Hieronima'*). 1848 (orth. cons.) [*Euphorb.*].
Typus: *H. alchorneoides* Allemão.

Hiptage Gaertn., Fruct. Sem. Pl. 2: 169. Sep (sero) - Nov 1790 [*Malpigh.*].
Typus: *H. madablota* Gaertn., nom. illeg. (*Banisteria tetraptera* Sonn.) [= *H. benghalensis* (L.) Kurz (*Banisteria benghalensis* L.)] (etiam vide *Gaertnera* [*Dicot.*]).

Hoffmannseggia Cav., Icon. 4: 63. 14 Mai 1798 (*'Hoffmanseggia'*) (orth. cons.) [*Legum.*].
Typus: *H. falcaria* Cav., nom. illeg. (*Larrea glauca* Ortega, *H. glauca* (Ortega) Eifert) (etiam vide *Larrea* [*Dicot.*]).

Holigarna Buch.-Ham. ex Roxb., Pl. Coromandel 3: 79. 18 Feb 1820 [*Anacard.*].
Typus: *H. longifolia* Buch.-Ham. ex Roxb.

(=) *Katou-tsjeroë* Adans., Fam. Pl. 2: 84, 534. Jul-Aug 1763.
Typus: non designatus.

Holodiscus (C. Koch) Maxim. in Trudy Imp. S.-Peterburgsk. Bot. Sada 6: 253. Jul-Dec 1879 (*Spiraea* [unranked] *Holodiscus* K. Koch, Dendrologie 1: 309. Jan 1869) [*Ros.*].
Typus: *Spiraea discolor* Pursh (*H. discolor* (Pursh) Maxim.).

(=) *Sericotheca* Raf., Sylva Tellur.: 152. Oct-Dec 1838.
Typus: *S. argentea* (L. f.) B. D. Jacks. (*Spiraea argentea* L. f.).

Homalanthus A. Juss., Euphorb. Gen.: 50. 21 Feb 1824 (*'Omalanthus'*) (orth. cons.). [*Euphorb.*].
Typus: *H. leschenaultianus* A. Juss.

Hopea Roxb., Pl. Coromandel 3: 7. Jul 1811 [*Dipterocarp.*].
Typus: *H. odorata* Roxb.

(H) *Hopea* Garden ex L., Syst. Nat., ed. 12, 2: 509; Mant. Pl.: 14, 105. 15-31 Oct 1767 [*Dicot.: Symploc.*].
Typus: *H. tinctoria* L.

Houttuynia Thunb. in Kongl. Vetensk. Acad. Nya Handl. 4: 149. Apr-Jun 1783 (*'Houtuynia'*) (orth. cons.) [*Saurur.*].
Typus: *H. cordata* Thunb.

(H) *Houttuynia* Houtt., Nat. Hist. 2(12): 448. 5 Jul 1780 [*Monocot.: Irid.*].
Typus: *H. capensis* Houtt.

Humboldtia Vahl, Symb. Bot. 3: 106. 1794 [*Legum.*].
Typus: *H. laurifolia* Vahl

(H) *Humboltia* Ruiz & Pav., Fl. Peruv. Prodr.: 121. Oct (prim.) 1794 [*Monocot.: Orchid.*].
Typus (vide Garay & Sweet in J. Arnold Arbor. 53: 522. 1972): *H. purpurea* Ruiz & Pav. (Syst. Veg. Fl. Peruv. Chil.: 235. Dec (sero) 1798).

Humiria Aubl., Hist. Pl. Guiane: 564. Jun-Dec 1775 (*'Houmiri'*) (orth. cons.) [*Humir.*].
Typus: *H. balsamifera* Aubl.

Hybanthus Jacq., Enum. Syst. Pl.: 2, 17. Aug-Sep 1760 [*Viol.*]. Typus: *H. havanensis* Jacq.

Hydrocera Bl. ex Wight & Arn., Prodr. Fl. Ind. Orient.: 140. 10 Oct 1834. [*Balsamin.*]. Typus: *Impatiens natans* Willd. [= *Hydrocera triflora* (L.) Wight & Arn. (*Impatiens triflora* L.)].

(≡) *Tytonia* G. Don., Gen. Hist. 1: 749. Aug. (prim.) 1831. Lectotypus (vide Raju & al. in Taxon 51: 383. 2002): *T. natans* (Willd.) G. Don (*Impatiens natans* Willd.).

Hydrolea L., Sp. Pl., ed. 2: 328. Sep 1762 [*Hydrophyll.*]. Typus: *H. spinosa* L.

Hymenocarpos Savi, Fl. Pis. 2: 205. 1798 [*Legum.*]. Typus: *H. circinnatus* (L.) Savi (*Medicago circinnata* L.).

(≡) *Circinnus* Medik. in Vorles. Churpfälz. Phys.-Öcon. Ges. 2: 384. 1787.

Hymenodictyon Wall. in Roxburgh, Fl. Ind. 2: 148. Mar-Jun 1824 [*Rub.*]. Typus: *H. excelsum* (Roxb.) DC. (Prodr. 4: 358. Sep (sero) 1824) (*Cinchona excelsa* Roxb.).

(=) *Benteca* Adans., Fam. Pl. 2: 166, 525. Jul-Aug 1763. Typus: *B. rheedei* Roem. & Schult. (Syst. Veg. 4: 706. Mar-Jun 1819).

Hyperbaena Miers ex Benth. in J. Proc. Linn. Soc., Bot. 5, Suppl. 2: 47. 1861 [*Menisperm.*]. Typus: *H. domingensis* (DC.) Benth. (*Cocculus domingensis* DC.) (typ. cons.).

(=) *Alina* Adans., Fam. Pl. 2: 84, 512. Jul-Aug 1763. Typus: non designatus.

Hyptis Jacq. in Collect. Bot. Spectentia (Vienna) 1: 101, 103. Jan-Sep 1787 [*Lab.*]. Typus: *H. capitata* Jacq. (typ. cons.).

(=) *Mesosphaerum* P. Browne, Civ. Nat. Hist. Jamaica: 257. 10 Mar 1756. Typus: *M. suaveolens* (L.) Kuntze (Revis. Gen. Pl. 2: 525. 5 Nov 1891) (*Ballota suaveolens* L.).

(=) *Condea* Adans., Fam. Pl. 2: 504, 542. Jul-Aug 1763. Typus (vide Kuntze, Revis. Gen. Pl. 2: 525. 1891): *Satureja americana* Poir.

Ichnocarpus R. Br., Asclepiadeae:
50. 3 Apr 1810 [*Apocyn.*].
Typus: *I. frutescens* (L.) W. T. Aiton (Hort. Kew., ed. 2, 2: 69. Feb-Mai 1811) (*Apocynum frutescens* L.).

Idesia Maxim. in Bull. Acad. Imp. Sci. Saint-Pétersbourg, ser. 3, 10: 485. 8 Sep 1866 [*Flacourt.*].
Typus: *I. polycarpa* Maxim.

(H) *Idesia* Scop., Intr. Hist. Nat.: 199. Jan-Apr 1777 [*Dicot.: Verben.*].
≡ *Ropourea* Aubl. 1775.

Inocarpus J. R. Forst. & G. Forst., Char. Gen. Pl.: 33. 29 Nov 1775 [*Legum.*].
Typus: *I. edulis* J. R. Forst. & G. Forst.

(=) *Aniotum* Parkinson, J. Voy. South Seas: 39. Jul 1773.
Typus: *A. fagiferum* Parkinson

Iochroma Benth. in Edwards's Bot. Reg. 31: ad t. 20. 1 Apr 1845 [*Solan.*].
Typus: *I. tubulosum* Benth., nom. illeg. (*Habrothamnus cyaneus* Lindl., *I. cyaneum* (Lindl.) M. L. Green) (typ. cons.).

(=) *Diplukion* Raf., Sylva Tellur.: 53. Oct-Dec 1838.
Typus (vide D'Arcy, Solanaceae Newslett. 2(4): 21. 1986): *D. cornifolium* (Kunth) Raf. (*Lycium cornifolium* Kunth).

(=) *Trozelia* Raf., Sylva Tellur.: 54. Oct-Dec 1838.
Typus: *T. umbellata* (Ruiz & Pav.) Raf. (*Lycium umbellatum* Ruiz & Pav.).

(=) *Valteta* Raf., Sylva Tellur.: 53. Oct-Dec 1838.
Typus (vide D'Arcy in Solanaceae Newslett. 2(4): 21. 1986): *V. gesnerioides* (Kunth) Raf. (*Lycium gesnerioides* Kunth).

Iphiona Cass. in Bull. Sci. Soc. Philom. Paris 1817: 153. Oct 1817 [*Comp.*].
Typus: *I. dubia* Cass., nom. illeg. (*Conyza pungens* Lam.) (typ. cons.) [= *I. mucronata* (Forssk.) Asch. & Schweinf. (*Chrysocoma mucronata* Forssk.)].

Ipomoea L., Sp. Pl.: 159. 1 Mai
1753 [*Convolvul.*].
Typus: *I. pes-tigridis* L. (typ. cons.).

Iresine P. Browne, Civ. Nat. Hist.
Jamaica: 358. 10 Mar 1756 [*Amaranth.*].
Typus: [specimen] *P. Browne*, Herb.
Linnaeus No. 288.5 (LINN) (typ.
cons.) [= *I. diffusa* L.].

Isoglossa Oerst. in Vidensk. Meddel. Dansk Naturhist. Foren. Kjøbenhavn 1854: 155. 1854 [*Acanth.*].
Typus: *I. origanoides* (Nees) S. Moore
(in Trans. Linn. Soc. London, Bot. 4:
34. 1894) (*Rhytiglossa origanoides*
Nees).

(≡) *Rhytiglossa* Nees in Lindley, Intr.
Nat. Syst. Bot., ed. 2: 444. Jul 1836.

Isopogon R. Br. ex Knight, Cult.
Prot.: 93. Dec 1809 [*Prot.*].
Typus: *I. anemonifolius* (Salisb.)
Knight (*Protea anemonifolia* Salisb.)
(typ. cons.).

(=) *Atylus* Salisb., Parad. Lond.: ad t.
67. 1 Jun 1806 -1 Mai 1807.
Typus: non designatus.

Isopyrum L., Sp. Pl.: 557. 1 Mai 1753
[*Ranuncul.*].
Typus: *I. thalictroides* L. (typ. cons.).

Jacobinia Nees ex Moricand, Pl.
Nouv. Amér.: 156. Jan-Jun 1847 [*Acanth.*].
Typus: *J. lepida* Nees ex Moricand

Jacquinia L., Fl. Jamaica: 27. 22
Dec 1759 *('Jaquinia')* (orth. cons.)
[*Theophrast.*].
Typus: *J. ruscifolia* Jacq. (Enum.
Syst. Pl.: 15. Aug-Sep 1760).

Jambosa Adans., Fam. Pl. 2: 88,
564. Jul-Aug 1763 *('Jambos')* (orth.
cons.) [*Myrt.*].
Typus: *J. vulgaris* DC., nom. illeg.
(*Eugenia jambos* L., *J. jambos* (L.)
Millsp.) (typ. cons.) (etiam vide *Syzygium* [*Dicot.*]).

Jatropha L., Sp. Pl.: 1006. 1 Mai
1753 [*Euphorb.*].
Typus: *B. gossypiifolia* L. (typ. cons.:
[specimen] Herb. Linnaeus No.
1141.1 (LINN)).

Jamesia Torr. & A. Gray, Fl. N.
Amer. 1: 593. Jun 1840 [*Saxifrag.*].
Typus: *J. americana* Torr. & A. Gray

(H) *Jamesia* Raf. in Atlantic J. 1: 145.
1832 (sero) [*Dicot.: Legum.*].
Typus: *J. obovata* Raf., nom. illeg.
(*Psoralea jamesii* Torr.).

Julocroton Mart. in Flora 20 (2,
Beibl.): 119. 21 Nov 1837 [*Eu-
phorb.*].
Typus: *J. phagedaenicus* Mart.

(=) *Cieca* Adans., Fam. Pl. 2: 356, 612.
Jul-Aug 1763.
Typus: *Croton argenteus* L.

Junellia Moldenke in Lilloa 5: 392.
3 Dec 1940 [*Verben.*].
Typus: *J. serpyllifolia* (Speg.) Moldenke (*Monopyrena serpyllifolia*
Speg.).

(≡) *Monopyrena* Speg. in Revista Fac.
Agron. Univ. Nac. La Plata 3: 559.
Jun-Jul 1897.

(=) *Thryothamnus* Phil. in Anales Univ.
Chile 90: 618. Mai 1895.
Typus: *T. junciformis* Phil.

Jungia L. f., Suppl. Pl.: 58. 390. Apr
1782 (*'Iungia'*) (orth. cons.) [*Comp.*].
Typus: *J. ferruginea* L. f.

(H) *Jungia* Heist. ex Fabr., Enum.: 47.
1759 (*'Jvngia'*) [*Dicot.: Lab.*].
Typus: *Salvia mexicana* L.

Kennedia Vent., Jard. Malmaison:
ad t. 104. Jul 1805 [*Legum.*].
Typus: *K. rubicunda* (Schneev.) Vent.
(*Glycine rubicunda* Schneev.).

Kernera Medik., Pfl.-Gatt.: 77, 95.
22 Apr 1792 [*Cruc.*].
Typus: *K. myagrodes* Medik., nom.
illeg. (*Cochlearia saxatilis* L., *K.
saxatilis* (L.) Rchb.).

(H) *Kernera* Schrank, Baier. Reise: 50.
post 5 Apr 1786 [*Dicot.: Scrophular.*].
Typus: *K. bavarica* Schrank

Knightia R. Br. in Trans. Linn. Soc.
London 10: 193. Feb 1810 [*Prot.*].
Typus: *K. excelsa* R. Br.

(=) *Rymandra* Salisb. ex Knight, Cult.
Prot.: 124. Dec 1809.
Typus: *R. excelsa* Knight

Kohautia Cham. & Schltdl. in Lin-
naea 4: 156. Apr 1829 [*Rub.*].
Typus: *K. senegalensis* Cham. &
Schltdl.

(=) *Duvaucellia* Bowdich in Bowdich &
Bowdich, Exc. Madeira: 259. 1825.
Typus: *D. tenuis* Bowdich

Kopsia Blume, Catalogus: 12. Feb-Sep 1823 [*Apocyn.*].
Typus: *K. arborea* Blume

(H)   *Kopsia* Dumort., Comment. Bot.: 16. Nov (sero) - Dec (prim.) 1822 [*Dicot.: Orobanch.*].
Typus (vide Bullock in Taxon 5: 197. 1956): *K. ramosa* (L.) Dumort. (*Orobanche ramosa* L.).

Kosteletzkya C. Presl, Reliq. Haenk. 2: 130. Jun-Jul 1835 [*Malv.*].
Typus: *K. hastata* C. Presl

(=)   *Thorntonia* Rchb., Consp. Regni Veg.: 202. Dec 1828 - Mar 1829.
Typus: *Hibiscus pentaspermus* Bertero ex DC.

Krigia Schreb., Gen. Pl.: 532. Mai 1791 [*Comp.*].
Typus: *K. virginica* (L.) Willd. (Sp. Pl. 3: 1618. Apr-Dec 1803) (*Tragopogon virginicus* L.) (typ. cons.).

Kundmannia Scop., Intr. Hist. Nat.: 116. Jan-Apr 1777 [*Umbell.*].
Typus: *K. sicula* (L.) DC. (Prodr. 4: 143. Sep (sero) 1830) (*Sium siculum* L.).

(≡)   *Arduina* Adans., Fam. Pl. 2: 499, 520. Jul-Aug 1763.

Kunzea Rchb., Consp. Regni Veg.: 175. Dec 1828 - Mar 1829 [*Myrt.*].
Typus: *K. capitata* (Sm.) Rchb. ex Heynh. (Nomencl. Bot. Hort.: 337. 27 Sep - 3 Oct 1840) (*Metrosideros capitata* Sm.).

(=)   *Tillospermum* Salisb. in Monthly Rev. 75: 74. 1814.
Typus: *Leptospermum ambiguum* Sm.

Labatia Sw., Prodr.: 2, 32. 20 Jun - 29 Jul 1788 [*Sapot.*].
Typus: *L. sessiliflora* Sw.

(H)   *Labatia* Scop., Intr. Hist. Nat.: 197. Jan-Apr 1777 [*Dicot.: Aquifol.*].
≡ *Macoucoua* Aubl. 1775.

Labisia Lindl. in Edwards's Bot. Reg. 31: ad t. 48. Sep 1845 [*Myrsin.*].
Typus: *L. pothoina* Lindl.

(=)   *Angiopetalum* Reinw. in Syll. Pl. Nov. 2: 7. 1825.
Typus: *A. punctatum* Reinw.

Laetia Loefl. ex L., Syst. Nat., ed. 10: 1068, 1074, 1373. 7 Jun 1759 [*Flacourt.*].
Typus: *L. americana* L.

Lagascea Cav. in Anales Ci. Nat. 6: 331. Jun 1803 (*'Lagasca'*) (orth. cons.) [*Comp.*].
Typus: *L. mollis* Cav.

(=) *Nocca* Cav., Icon. 3: 12. Apr 1795.
Typus: *N. rigida* Cav.

Lagenophora Cass. in Bull. Sci. Soc. Philom. Paris 1816: 199. Dec 1816 (*'Lagenifera'*) (orth. cons.) [*Comp.*].
Typus: *Calendula magellanica* Willd., nom. illeg. (*Aster nudicaulis* Lam., *L. nudicaulis* (Lam.) Dusén).

Lamourouxia Kunth in Humboldt & al., Nov. Gen. Sp. 2, ed. 4°: 335; ed. f°: 269. 8 Jun 1818 [*Scrophular.*].
Typus: *L. multifida* Kunth

(H) *Lamourouxia* C. Agardh, Syn. Alg. Scand.: xiv. Mai-Dec 1817 [*Rhodoph.*].
≡ *Claudea* J. V. Lamour. 1813.

Lampranthus N. E. Br. in Gard. Chron., ser. 3, 87: 71. 25 Jan 1930 [*Aiz.*].
Typus: *L. multiradiatus* (Jacq.) N. E. Br. (*Mesembryanthemum multiradiatum* Jacq.).

(=) *Oscularia* Schwantes in Möller's Deutsche Gärtn.-Zeitung 42: 187. 21 Mai 1927.
Typus: *O. deltoides* (L.) Schwantes (*Mesembryanthemum deltoides* L.).

Landolphia P. Beauv., Fl. Oware 1: 54. Mai 1806 [*Apocyn.*].
Typus: *L. owariensis* P. Beauv.

(=) *Pacouria* Aubl., Hist. Pl. Guiane: 268. Jun-Dec 1775.
Typus: *P. guianensis* Aubl.

Lannea A. Rich. in Guillemin & al., Fl. Seneg. Tent.: 153. Sep 1831 [*Anacard.*].
Typus: *L. velutina* A. Rich. (typ. cons.).

(=) *Calesiam* Adans., Fam. Pl. 2: 446, 530. Jul-Aug 1763.
Typus: *Calesiam malabarica* Raf. (Sylva Tellur.: 12. Oct-Dec 1838).

Lantana L., Sp. Pl.: 626. 1 Mai 1753 [*Verben.*].
Typus: *L. camara* L. (typ. cons.: [specimen] Herb. Linnaeus No. 783.4 (LINN)).

Laplacea Kunth in Humboldt & al., Nov. Gen. Sp. 5, ed. 4°: 207; ed. f.°: 161. 25 Feb 1822 [*The.*].
Typus: *L. speciosa* Kunth

Laportea Gaudich., Voy. Uranie, Bot.: 498. 6 Mar 1830 [*Urtic.*].
Typus: *L. canadensis* (L.) Wedd. (in Ann. Sci. Nat., Bot., ser. 4, 1: 181: Mar 1854) (*Urtica canadensis* L.).

(≡) *Urticastrum* Heist. ex Fabr., Enum.: 204. 1759.

Larrea Cav. in Anales Hist. Nat. 2: 119. Jun 1800 [*Zygophyll.*].
Typus: *L. nitida* Cav. (typ. cons.).

(H) *Larrea* Ortega, Nov. Pl. Descr. Dec.: 15. 1797 [*Dicot.: Legum.*].
≡ *Hoffmannseggia* Cav. 1798 (nom. cons.).

Lasianthus Jack in Trans. Linn. Soc. London 14: 125. 28 Mai - 12 Jun 1823 [*Rub.*].
Typus: *L. cyanocarpus* Jack (typ. cons.).

(H) *Lasianthus* Adans., Fam. Pl. 2: 398, 568. Jul-Aug 1763 [*Dicot.: The.*].
≡ *Gordonia* J. Ellis 1771 (nom. cons.).
(=) *Dasus* Lour., Fl. Cochinch.: 96, 141. Sep 1790.
Typus: *D. verticillata* Lour.

Laurelia Juss. in Ann. Mus. Natl. Hist. Nat. 14: 134. 1809 [*Monim.*].
Typus: *L. sempervirens* (Ruiz & Pav.) Tul. (in Arch. Mus. Hist. Nat. 8: 416. 1855) (*Pavonia sempervirens* Ruiz & Pav.) (typ. cons.).

Layia Hook. & Arn. ex DC., Prodr. 7: 294. Apr (sero) 1838 [*Comp.*].
Typus: *L. gaillardioides* (Hook. & Arn.) DC. (*Tridax gaillardioides* Hook. & Arn.).

(H) *Layia* Hook. & Arn., Bot. Beechey Voy.: 182. Oct 1833 [*Dicot.: Legum.*].
Typus: *L. emarginata* Hook. & Arn.
(=) *Blepharipappus* Hook., Fl. Bor.-Amer. 1: 316. 1833 (sero).
Typus: *B. scaber* Hook.

Lebetanthus Endl., Gen. Pl.: 1411. Feb-Mar 1841 (*'Lebethanthus'*) (orth. cons.) [*Epacrid.*].
Typus: *L. americanus* (Hook.) Endl. ex Hook. f. (Fl. Antarct.: 327. 1846 (ante 1 Jun)) (*Prionitis americana* Hook.) (typ. cons.).

(≡) *Allodape* Endl., Gen. Pl.: 749. Mar 1839.

Leea D. Royen ex L., Syst. Nat., ed. 12, 2: 608, 627; Mant. Pl.: 17, 124. 15-31 Oct 1767 [*Vit.*].
Typus: *L. aequata* L. (typ. cons.).

(=) *Nalagu* Adans., Fam. Pl. 2: 445, 581. Jul-Aug 1763.
Typus: *Leea asiatica* (L.) Ridsdale (*Phytolacca asiatica* L.).

Lens Mill., Gard. Dict. Abr., ed. 4: [765]. 28 Jan 1754 [*Legum.*].
Typus: *L. culinaris* Medik. (in Vorles. Churpfälz. Phys.-Öcon. Ges. 2: 361. 1787) (*Ervum lens* L.).

Leontodon L., Sp. Pl.: 798. 1 Mai 1753 [*Comp.*].
Typus: *L. hispidus* L. (typ. cons.).

Lepidostemon Hook. f. & Thomson in J. Proc. Linn. Soc., Bot. 5: 131, 156. 27 Mar 1861 [*Cruc.*].
Typus: *L. pedunculosus* Hook. f. & Thomson (typ. cons.).

Leptospermum J. R. Forst. & G. Forst., Char. Gen. Pl.: 36. 29 Nov 1775 [*Myrt.*].
Typus: *L. scoparium* J. R. Forst. & G. Forst. (typ. cons.).

Lerchea L., Mant. Pl.: 155, 256. Oct 1771 [*Rub.*].
Typus: *L. longicauda* L.

(H) *Lerchia* Haller ex Zinn, Cat. Pl. Hort. Gott.: 30. 20 Apr - 21 Mai 1757 [*Dicot.: Chenopod.*].
Typus: non designatus.

Lessertia DC., Astragalogia, ed. 4°: 5, 19, 47; ed. f°: 4, 15, 37. 15 Nov 1802 [*Legum.*].
Typus: *L. perennans* (Jacq.) DC. (*Colutea perennans* Jacq.).

(≡) *Sulitra* Medik. in Vorles. Churpfälz. Phys.-Öcon. Ges. 2: 366. 1787 (typ. des.: Brummitt in Regnum Veg. 40: 24. 1965).

(=) *Coluteastrum* Fabr., Enum., ed. 2: 317. Sep-Dec 1763.
Typus: *C. herbaceum* (L.) Kuntze (Revis. Gen. Pl. 1: 171. 5 Nov 1891) (*Colutea herbacea* L.).

Leucadendron R. Br. in Trans. Linn. Soc. London 10: 50. Feb 1810 [*Prot.*].
Typus: *L. argenteum* (L.) R. Br. (*Protea argentea* L.) (typ. cons.) (etiam vide *Protea* [*Dicot.*]).

(H) *Leucadendron* L., Sp. Pl.: 91. 1 Mai 1753 [*Dicot.: Prot.*].
Typus (vide Hitchcock in Sprague & al., Nom. Prop. Brit. Bot.: 122. 1929): *L. lepidocarpodendron* L.

Leucaena Benth. in J. Bot. (Hooker)
4: 416. Jun 1842. [*Legum.*].
Typus: *L. diversifolia* (Schltdl.) Benth.
(*Acacia diversifolia* Schltdl.) (typ.
cons.).

Leucopogon R. Br., Prodr.: 541. 27   (=) *Perojoa* Cav., Icon. 4: 29. Sep-Dec
Mar 1810 [*Epacrid.*].   1797.
Typus: *L. lanceolatus* R. Br., nom. il-   Typus: *P. microphylla* Cav.
leg. (*Styphelia parviflora* Andrews,
L. parviflorus (Andrews) Lindl.).

Leucospermum R. Br. in Trans.
Linn. Soc. London 10: 95. Feb 1810
[*Prot.*].
Typus: *L. hypophyllum* R. Br., nom.
illeg. (*Leucadendron hypophyllocar-
podendron* L., *Leucospermum hypo-
phyllocarpodendron* (L.) Druce) (typ.
cons.).

Levisticum Hill, Brit. Herb.: 423.   (H) *Levisticum* Hill, Brit. Herb.: 410.
Nov 1756 [*Umbell.*].   Nov 1756 [*Dicot.: Umbell.*].
Typus: *L. officinale* W. D. J. Koch   ≡ *Ligusticum* L. 1753.
(*Ligusticum levisticum* L.).

Liatris Gaertn. ex Schreb., Gen. Pl.: (≡) *Lacinaria* Hill, Veg. Syst. 4: 49.
542. Mai 1791 [*Comp.*].   1762.
Typus: *L. squarrosa* (L.) Michx. (Fl.
Bor.-Amer. 2: 92. 19 Mar 1803)
(*Serratula squarrosa* L.) (typ. cons.).

Libanotis Haller ex Zinn, Cat. Pl.   (H) *Libanotis* Hill, Brit. Herb.: 420. Nov
Hort. Gott.: 226. 20 Apr - 21 Mai   1756 [*Dicot.: Umbell.*].
1757 [*Umbell.*].   Typus (vide Rauschert in Taxon 31:
Typus: *L. montana* Crantz (Stirp.   755. 1982): *Selinum cervaria* L.
Austr. Fasc. 3: 117. 1767) (*Atha-
manta libanotis* L.).

Lichtensteinia Cham. & Schltdl. in   (H) *Lichtensteinia* Willd. in Ges. Naturf.
Linnaea 1: 394. Aug-Oct 1826 [*Um-   Freunde Berlin Mag. Neuesten Ent-
bell.*].   deck. Gesammten Naturk. 2: 19.
Typus: *L. lacera* Cham. & Schltdl.   1808 [*Monocot.: Lil.*].
(typ. cons.).   Typus: non designatus.

Ligularia Cass. in Bull. Sci. Soc. Philom. Paris 1816: 198. Dec 1816 [*Comp.*].
Typus: *L. sibirica* (L.) Cass. (in Cuvier, Dict. Sci. Nat. 26: 402. Mai 1823) (*Othonna sibirica* L.).

(≡) *Jacobaeoides* Vaill., Königl. Akad. Wiss. Paris Phys. Abh. 5: 570. Jan-Apr 1754.
(H) *Ligularia* Duval, Pl. Succ. Horto Alencon.: 11. 1809 [*Dicot.: Saxifrag.*].
≡ *Sekika* Medik. 1791.
(=) *Senecillis* Gaertn., Fruct. Sem. Pl. 2: 453. Sep-Dec 1791.
Typus: non designatus.

Limnanthes R. Br. in London Edinburgh Philos. Mag. & J. Sci. 3: 70. Jul 1833 [*Limnanth.*].
Typus: *L. douglasii* R. Br.

(H) *Limnanthes* Stokes, Bot. Mat. Med. 1: 300. 1812 [*Dicot.: Gentian.*].
≡ *Limnanthemum* S. G. Gmel. 1770.

Limnophila R. Br., Prodr.: 442. 27 Mar 1810 [*Scrophular.*].
Typus: *L. gratioloides* R. Br., nom. illeg. (*Hottonia indica* L., *L. indica* (L.) Druce).

(≡) *Hydropityon* C. F. Gaertn., Suppl. Carp.: 19. 24-26 Jun 1805.
(=) *Ambuli* Adans., Fam. Pl. 2: 208, 516. Jul-Aug 1763.
Typus: *A. aromatica* Lam. (Encycl. 1: 128. 2 Dec 1783).
(=) *Diceros* Lour., Fl. Cochinch.: 358, 381, post 722. Sep 1790.
Typus: *D. cochinchinesis* Lour.

Limonium Mill., Gard. Dict. Abr., ed. 4: [1328]. 28 Jan 1754 [*Plumbagin.*].
Typus: *L. vulgare* Mill. (*Statice limonium* L.) (typ. cons.).

Lindera Thunb., Nov. Gen. Pl.: 64. 18 Jun 1783 [*Laur.*].
Typus: *L. umbellata* Thunb.

(H) *Lindera* Adans., Fam. Pl. 2: 499, 571. Jul-Aug 1763 [*Dicot.: Umbell.*].
Typus: *Chaerophyllum coloratum* L.
(=) *Benzoin* Schaeff., Bot. Exped.: 60. 1 Oct - 24 Dec 1760.
Typus: *B. odoriferum* Nees (in Wallich, Pl. Asiat. Rar. 2: 63. 6 Sep 1831) (*Laurus benzoin* L.).

Lindleya Kunth in Humboldt & al., Nov. Gen. Sp. 6, ed. 4°: 239; ed. f°: 188. 5 Jan 1824 [*Ros.*].
Typus: *L. mespiloides* Kunth

(H) *Lindleya* Nees in Flora 4: 299. 21 Mai 1821 [*Dicot.: The.*].
≡ *Wikstroemia* Schrad., 5 Mai 1821 (nom. rej. sub *Wikstroemia*).

Linociera Sw. ex Schreb., Gen. Pl.: 784. Mai 1791 [*Ol.*].
Typus: *L. ligustrina* (Sw.) Sw. (*Thouinia ligustrina* Sw.).

(=) *Mayepea* Aubl., Hist. Pl. Guiane: 81. Jun-Dec 1775.
Typus: *M. guianensis* Aubl.

(=) *Ceranthus* Schreb., Gen. Pl.: 14. Apr 1789.
Typus: *C. schreberi* J. F. Gmel. (Syst. Nat. 2: 26. Sep (sero) - Nov 1791).

Lithophragma (Nutt.) Torr. & A. Gray, Fl. N. Amer. 1: 583. Jun 1840 (*Tellima* [unranked] *Lithophragma* Nutt. in J. Acad. Nat. Sci. Philadelphia 7: 26. 28 Oct 1834) [*Saxifrag.*].
Typus: *Tellima parviflora* Hook. (*L. parviflorum* (Hook.) Torr. & A. Gray) (typ. cons.).

(≡) *Pleurendotria* Raf., Fl. Tellur. 2: 73. Jan-Mar 1837.

Litsea Lam., Encycl. 3: 574. 13 Feb 1792 [*Laur.*].
Typus: *L. chinensis* Lam.

(=) *Malapoënna* Adans., Fam. Pl. 2: 447, 573. Jul-Aug 1763.
Typus: *Darwinia quinqueflora* Dennst.

Lobularia Desv. in J. Bot. Agric. 3: 162. 1815 (prim.) [*Cruc.*].
Typus: *L. maritima* (L.) Desv. (*Clypeola maritima* L.).

(≡) *Aduseton* Adans., Fam. Pl. 2: (23), 420 (*'Konig'*), 542 (*'Konig'*). Jul-Aug 1763.

Logania R. Br., Prodr.: 454. 27 Mar 1810 [*Logan.*].
Typus: *L. floribunda* R. Br., nom. illeg. (*Euosma albiflora* Andrews, *L. albiflora* (Andrews) Druce) (typ. cons.).

(H) *Loghania* Scop., Intr. Hist. Nat.: 236. Jan-Apr 1777 [*Dicot.: Marcgrav.*].
≡ *Souroubea* Aubl. 1775.

(≡) *Euosma* Andrews in Bot. Repos.: ad t. 520. Mai 1808.

Loiseleuria Desv. in J. Bot. Agric. 1: 35. Jan 1813 [*Eric.*].
Typus: *L. procumbens* (L.) Desv. (*Azalea procumbens* L.).

(≡) *Azalea* L., Sp. Pl.: 150. 1 Mai 1753.

Lomatia R. Br. in Trans. Linn. Soc. London 10: 199. Feb 1810 [*Prot.*].
Typus: *L. silaifolia* (Sm.) R. Br. (*Embothrium silaifolium* Sm.) (typ. cons.).

(=) *Tricondylus* Salisb. ex Knight, Cult. Prot.: 121. Dec 1809.
Typus: non designatus.

Lonchocarpus Kunth in Humboldt & al., Nov. Gen. Sp. 6, ed. f°: 300. Apr 1824 [*Legum.*].
Typus: *L. sericeus* (Poir.) DC. (Prodr. 2: 260. Nov (med.) 1825) (*Robinia sericea* Poir.) (typ. cons.).

(=) *Clompanus* Aubl., Hist. Pl. Guiane: 773. Jun-Dec 1775.
Typus: *C. paniculata* Aubl.
(=) *Coublandia* Aubl., Hist. Pl. Guiane 937. Jun-Dec 1775 (nom. rej. sub *Muellera*).
Typus: *C. frutescens* Aubl.
(=) *Muellera* L. f., Suppl. Pl.: 53, 329. Apr 1782 (nom. cons.).

Lonchostoma Wikstr. in Kongl. Ve-tensk. Acad. Handl. 1818: 350. 1818 [*Brun.*].
Typus: *L. obtusiflorum* Wikstr., nom. illeg. (*Passerina pentandra* Thunb., *L. pentandrum* (Thunb.) Druce).

(=) *Ptyxostoma* Vahl in Skr. Naturhist.-Selsk. 6: 95. 1810.
Typus: non designatus.

Lophanthera A. Juss., Malpigh. Syn.: 53. Mai 1840 [*Malpigh.*].
Typus: *L. kunthiana* A. Juss., nom. illeg. (*Galphimia longifolia* Kunth, *L. longifolia* (Kunth) Griseb.).

(H) *Lophanthera* Raf., New Fl. 2: 58. Jul-Dec 1837 [*Dicot.: Scrophular.*].
Typus: *Gerardia delphiniifolia* L.

Loranthus Jacq., Enum. Stirp. Vin-dob.: 55, 230. Mai 1762 [*Loranth.*].
Typus: *L. europaeus* Jacq.

(H) *Loranthus* L., Sp. Pl.: 331. 1 Mai 1753 [*Dicot.: Loranth.*].
Typus: *L. americanus* L.
(=) *Scurrula* L., Sp. Pl.: 110. 1 Mai 1753.
Typus: *S. parasitica* L.

Lotononis (DC.) Eckl. & Zeyh., En-um. Pl. Afric. Austral.: 176. Jan 1836 (*Ononis* sect. *Lotononis* DC., Prodr. 2: 166. Nov (med.) 1825) [*Legum.*].
Typus: *Crotalaria vexillata* E. Mey. (*L. vexillata* (E. Mey.) Eckl. & Zeyh.) (typ. cons.) [= *L. prostata* (L.) Benth. (*Ononis prostrata* L.)].

(=) *Amphinomia* DC., Prodr. 2: 522. Nov (med.) 1825.
Typus: *A. decumbens* (Thunb.) DC. (*Connarus decumbens* Thunb.).
(=) *Leobordea* Delile in Laborde, Voy. Arabie Pétrée: 82, 86. 1830.
Typus: *L. lotoidea* Delile

Lucya DC., Prodr. 4: 434. Sep (se-ro) 1830 [*Rub.*].
Typus: *L. tuberosa* DC., nom. illeg. (*Peplis tetrandra* L., *L. tetrandra* (L.) K. Schum.) (etiam vide *Dunalia* [*Dicot.*]).

Luehea Willd. in Ges. Naturf. Freunde Berlin Neue Schriften 3: 410. 1801 (post 21 Apr) [*Til.*].
Typus: *L. speciosa* Willd.

(H) *Luehea* F. W. Schmidt, Neue Selt. Pfl.: 23. 1793 (ante 17 Jun) [*Dicot.: Verben.*].
Typus: *L. ericoides* F. W. Schmidt

Lunania Hook. in London J. Bot. 3: 317. 1844 [*Flacourt.*].
Typus: *L. racemosa* Hook.

(H) *Lunanea* DC., Prodr. 2: 92. Nov (med.) 1825 [*Dicot.: Stercul.*].
≡ *Bichea* Stokes 1812 (nom. rej. sub *Cola*).

Lundia DC. in Biblioth. Universelle Genève, ser. 2, 17: 127. Sep 1838 [*Bignon.*].
Typus: *L. glabra* DC. (Prodr. 9: 180. 1 Jan 1845) (typ. cons.).

(H) *Lundia* Schumach., Beskr. Guin. Pl.: 231. 1827 [*Dicot.: Flacourt.*].
Typus: *L. monacantha* Schumach.

Lycianthes (Dunal) Hassl. in Annuaire Conserv. Jard. Bot. Genève 20: 180. 1 Oct 1917 (*Solanum* subsect. *Lycianthes* Dunal in Candolle, Prodr. 13(1): 29. 10 Mai 1852) [*Solan.*].
Typus: *Solanum lycioides* L. (*L. lycioides* (L.) Hassl.).

(≡) *Otilix* Raf., Med. Fl. 2: 87. 1830.
(=) *Parascopolia* Baill., Hist. Pl. 9: 338. Feb-Mar 1888.
Typus: *P. acapulcensis* Baill.

Lyonia Nutt., Gen. N. Amer. Pl. 1: 266. 14 Jul 1818 [*Eric.*].
Typus: *L. ferruginea* (Walter) Nutt. (*Andromeda ferruginea* Walter).

(H) *Lyonia* Raf. in Med. Repos., ser. 2, 5: 353. Feb-Apr 1808 [*Dicot.: Polygon.*].
≡ *Polygonella* Michx. 1803.

Machaerium Pers., Syn. Pl. 2: 276. Sep 1807 [*Legum.*].
Typus: *M. ferrugineum* Pers., nom. illeg. (*Nissolia ferruginea* Willd., nom. illeg., *N. quinata* Aubl., *M. quinatum* (Aubl.) Sandwith).

(=) *Nissolius* Medik. in Vorles. Churpfälz. Phys.-Öcon. Ges. 2: 389. 1787.
Typus: *N. arboreus* (Jacq.) Medik. (*Nissolia arborea* Jacq.).
(=) *Quinata* Medik. in Vorles. Churpfälz. Phys.-Öcon. Ges. 2: 389. 1787.
Typus: *Q. violacea* Medik. (*Nissolia quinata* Aubl.).

Mackaya Harv., Thes. Cap. 1: 8. 1859 [*Acanth.*].
Typus: *M. bella* Harv.

(H) *Mackaia* Gray, Nat. Arr. Brit. Pl. 1: 320, 391. 1 Nov. 1821 [*Phaeoph.*].
Typus: non designatus.

Maclura Nutt., Gen. N. Amer. Pl. 2: 233. 14 Jul 1818 [*Mor.*].
Typus: *M. aurantiaca* Nutt.

(=) *Ioxylon* Raf. in Amer. Monthly Mag. & Crit. Rev. 2: 118. Dec 1817.
Typus: *I. pomiferum* Raf.

Macrolobium Schreb., Gen. Pl.: 30. Apr 1789 [*Legum*.].
Typus: *M. vuapa* J. F. Gmel. (Syst. Nat. 2: 93. Sep (sero) - Nov 1791), nom. illeg. (*Vouapa bifolia* Aubl., *M. bifolium* (Aubl.) Pers.) (typ. cons.).

(≡) *Vouapa* Aubl., Hist. Pl. Guiane: 25. Jun-Dec 1775.

(=) *Outea* Aubl., Hist. Pl. Guiane: 28. Jun-Dec 1775.
Typus: *O. guianensis* Aubl.

Macrosiphonia Müll. Arg. in Martius, Fl. Bras. 6(1): 137. 30 Jul 1860. [*Apocyn*.].
Typus: *M. velame* (A. St.-Hil.) Müll. Arg. (*Echites velame* A. St.-Hil.).

(H) *Macrosyphonia* Duby in Mém. Soc. Phys. Genève 10(2): 426. Apr-Dec 1844 [*Dicot.: Primul*.].
Typus: *M. cespitosa* (Duby) Duby (*Gregoria cespitosa* Duby).

Macrotyloma (Wight & Arn.) Verdc. in Kew Bull. 24: 322. 1 Apr 1970 (*Dolichos* sect. *Macrotyloma* Wight & Arn., Prodr. Fl. Ind. Orient.: 248. Oct 1834) [*Legum*.].
Typus: *Dolichos uniflorus* Lam. (*M. uniflorum* (Lam.) Verdc.).

(=) *Kerstingiella* Harms in Ber. Deutsch. Bot. Ges. 26a: 230. 23 Apr 1908.
Typus: *K. geocarpa* Harms

Mahonia Nutt., Gen. N. Amer. Pl. 1: 211. 14 Jul 1818 [*Berberid*.].
Typus: *M. aquifolium* (Pursh) Nutt. (*Berberis aquifolium* Pursh) (typ. cons.).

Majorana Mill., Gard. Dict. Abr., ed. 4: [829]. 28 Jan 1754 [*Lab*.].
Typus: *M. hortensis* Moench (Methodus: 406. 4 Mai 1794) (*Origanum majorana* L.) (typ. cons.).

Malcolmia W. T. Aiton, Hort. Kew., ed. 2, 4: 121. Dec 1812 (*'Malcomia'*) (orth. cons.) [*Cruc*.].
Typus: *M. maritima* (L.) W. T. Aiton (*Cheiranthus maritimus* L.) (typ. cons.).

(≡) *Wilckia* Scop., Intr. Hist. Nat.: 317. Jan-Apr 1777.

Malvastrum A. Gray in Mem. Amer. Acad. Arts, ser. 2, 4: 21. 10 Feb 1849 [*Malv*.].
Typus: *M. wrightii* A. Gray [= *M. aurantiacum* (Scheele) Walp., *Malva aurantiaca* Scheele].

(=) *Malveopsis* C. Presl. in Abh. Königl. Böhm. Ges. Wiss., ser. 5, 3: 449. Jul-Dec 1845.
Typus: *M. anomala* (Link & Otto) C. Presl (*Malva anomala* Link & Otto).

Mammillaria Haw., Syn. Pl. Succ.: 177. 1812 [*Cact.*].
Typus: *M. simplex* Haw., nom. illeg. (*Cactus mammillaris* L., *M. mammillaris* (L.) H. Karst.) (typ. cons.).

(H)   *Mammillaria* Stackh. in Mém. Soc. Imp. Naturalistes Moscou 2: 55, 74. 1809 [*Rhodoph.*].
Typus: non designatus.

(≡)   *Cactus* L., Sp. Pl.: 466. 1 Mai 1753.

Mancoa Wedd., Chlor. Andina 2: t. 86d. 10 Oct 1859 [*Cruc.*].
Typus: *M. hispida* Wedd.

(H)   *Mancoa* Raf., Fl. Tellur. 3: 56. Nov-Dec 1837 [*Dicot.: Phytolacc.*].
Typus: *M. secunda* (Ruiz & Pav.) Raf. (*Rivina secunda* Ruiz & Pav.).

Manettia Mutis ex L., Mant. Pl.: 553, 558. Oct 1771 [*Rub.*].
Typus: *M. reclinata* L.

(H)   *Manettia* Boehm. in Ludwig, Def. Gen. Pl., ed. 3: 99. 1760 [*Dicot.: Scrophular.*].
≡ *Selago* L. 1753.

(=)   *Lygistum* P. Browne, Civ. Nat. Hist. Jamaica: 142. 10 Mar 1756.
Typus: *Petesia lygistum* L.

Manilkara Adans., Fam. Pl. 2: 166, 574. Jul-Aug 1763 [*Sapot.*].
Typus: *M. kauki* (L.) Dubard (in Ann. Inst. Bot.-Géol. Colon. Marseille, ser. 3, 3: 9. 1915) (*Mimusops kauki* L.) (typ. cons.).

(=)   *Achras* L., Sp. Pl.: 1190. 1 Mai 1753.
Typus: *A. zapota* L.

(=)   *Sapota* Mill., Gard. Dict. Abr., ed. 4: [1249]. 28 Jan 1754.
≡ *Achras* L. (nom. rej.).

Manulea L., Syst. Nat., ed. 12, 2: 385, 419; Mant. Pl.: 12, 88. 15-31 Oct 1767 [*Scrophular.*].
Typus: *M. cheiranthus* (L.) L. (*Lobelia cheiranthus* L.).

(≡)   *Nemia* P. J. Bergius, Descr. Pl. Cap.: 160, 162. Sep 1767.

Mappia Jacq., Pl. Hort. Schoenbr. 1: 22. 1797 [*Icacin.*].
Typus: *M. racemosa* Jacq.

(H)   *Mappia* Heist. ex Fabr., Enum.: 58. 1759 [*Dicot.: Lab.*].
≡ *Cunila* Mill. 1754.

Marsdenia R. Br., Prodr.: 460. 27 Mar 1810 [*Asclepiad.*].
Typus: *M. tinctoria* R. Br.

(=)   *Stephanotis* Thouars, Gen. Nov. Madagasc.: 11. 17 Nov 1806.
Typus (vide Forster in Taxon 39: 364. 1990): *S. thouarsii* Brongn.

Marshallia Schreb., Gen. Pl.: 810.
Mai 1791 [*Comp.*].
Typus: *M. obovata* (Walter) Beadle
& F. E. Boynton (in Biltmore Bot.
Stud. 1: 5. 8 Apr 1901) (*Athanasia
obovata* Walter) (typ. cons.).

Matricaria L., Sp. Pl.: 890. 1 Mai
1753 [*Comp.*].
Typus: *M. recutita* L. (typ. cons.:
[specimen] Czech Rep., in ruderatis
ad urbem Brno, ca. 180 m, 15 Jun
1925, *J. Podpěra* in Fl. Exs. Reip.
Boh.-Slov. 946.II, (K)).

Matthiola W. T. Aiton, Hort. Kew.,
ed. 2, 4: 119. Dec 1812 *('Mathiola')*
(orth. cons.) [*Cruc.*].
Typus: *M. incana* (L.) W. T. Aiton
(*Cheiranthus incanus* L.) (typ. cons.).

(H) *Matthiola* L., Sp. Pl.: 1192. 1 Mai
1753 [*Dicot.: Rub.*].
Typus: *M. scabra* L.

Medicago L., Sp. Pl.: 778. 1 Mai
1753 [*Legum.*].
Typus: *M. sativa* L. (typ. cons.).

Melaleuca L., Syst. Nat., ed. 12, 2:
507, 509 . 15-31 Oct 1767 [*Myrt.*].
Typus: *M. leucadendra* (L.) L. (*Myrtus leucadendra* L.) (typ. cons.).

(≡) *Kajuputi* Adans., Fam. Pl. 2: 84,
530. Jul-Aug 1763.

Melocactus Link & Otto in Verh.
Vereins Beförd. Gartenbaues Königl.
Preuss. Staaten 3: 417. 1827 [*Cact.*].
Typus: *M. communis* (W. T. Aiton)
Link & Otto (*Cactus melocactus* var.
communis W. T. Aiton, Hort. Kew.
ed. 2, 3: 175. 1811) (typ. cons.).

(H) *Melocactus* Boehm. in Ludwig, Def.
Gen. Pl., ed. 3: 79. 1760 [*Dicot.:
Cact.*].
≡ *Cactus* L. 1753 (nom. rej. sub
Mammillaria).

Melochia L., Sp. Pl.: 674. 1 Mai
1753 [*Stercul.*].
Typus: *M. corchorifolia* L. (typ.
cons.: [icon in] Dillenius, Hort. Eltham. t. 176, f. 217. 1732).

Meriania Sw., Fl. Ind. Occid.: 823. (H)   *Meriana* Trew, Pl. Select.: 11. 1754.
Jan-Jun 1798 [*Melastomat.*].   [*Monocot.: Irid.*].
Typus: *M. leucantha* (Sw.) Sw. (*Rhe-*   ≡ *Watsonia* Mill. 1759 (nom. cons.).
xia leucantha Sw.) (typ. cons.).

Merremia Dennst. ex Endl., Gen. Pl.: (=)   *Operculina* Silva Manso, Enum.
1403. Feb-Mar 1841 [*Convolvul.*].   Subst. Braz.: 16. 1836.
Typus: *M. hederacea* (Burm. f.) Hal-   Typus: *O. turpethum* (L.) Silva Man-
lier f. (in Bot. Jahrb. Syst. 18: 118.   so (*Convolvulus turpethum* L.).
22 Dec 1893) (*Evolvulus hederaceus* (=)   *Camonea* Raf., Fl. Tellur. 4: 81.
Burm. f.).   1838 (med.).
                                            Typus: *C. bifida* (Vahl) Raf. (*Con-*
                                            volvulus bifidus Vahl).

Mertensia Roth, Catal. Bot. 1: 34. (=)   *Pneumaria* Hill, Veg. Syst. 7: 40.
Jan-Feb 1797 [*Boragin.*].   1764.
Typus: *M. pulmonarioides* Roth   Typus: non designatus.

Mesembryanthemum L., Sp. Pl.:
480. 1 Mai 1753 [*Aiz.*].
Typus: *M. nodiflorum* L. (typ. cons.).

Mesoptera Hook. f. in Bentham & (H)   *Mesoptera* Raf., Herb. Raf.: 73. 1833
Hooker, Gen. Pl. 2: 25, 130. 7-9 Apr   [*Monocot.: Orchid.*].
1873 [*Rub.*].   ≡ *Liparis* Rich. 1818 (nom. cons.).
Typus: *M. maingayi* Hook. f.

Metrosideros Banks ex Gaertn., (=)   *Nani* Adans., Fam. Pl. 2: 88, 581.
Fruct. Sem. Pl. 1: 170. Dec 1788   Jul-Aug 1763.
[*Myrt.*].   Typus: *Metrosideros vera* Lindl.
Typus: *M. spectabilis* Gaertn. (typ.
cons.).

Michauxia L'Hér., Michauxia: ad t.
1. Mar-Apr 1788 [*Campanul.*].
Typus: *M. campanuloides* L'Hér.
(typ. cons.).

Miconia Ruiz & Pav., Fl. Peruv. (=)   *Leonicenia* Scop., Intr. Hist. Nat.:
Prodr.: 60. Oct (prim.) 1794 [*Mela-*   212. Jan-Apr 1777.
stomat.].   Typus: *Fothergilla mirabilis* Aubl.
Typus: *M. triplinervis* Ruiz & Pav.
(Syst. Veg. Fl. Peruv. Chil.: 104.
Dec (sero) 1798) (typ. cons.).

Micrandra Benth. in Hooker's J. Bot. Kew Gard. Misc. 6: 371. Dec 1854 [*Euphorb.*].
Typus: *M. siphonioides* Benth. (typ. cons.).

(H) *Micrandra* R. Br. in Bennett, Pl. Jav. Rar.: 237. 4 Jun 1844 [*Dicot.: Euphorb.*].
Typus: *M. ternata* R. Br.

Micranthemum Michx., Fl. Bor.-Amer. 1: 10. 19 Mar 1803 [*Scrophu-lar.*].
Typus: *M. orbiculatum* Michx., nom. illeg. (*Globifera umbrosa* J. F. Gmel., *M. umbrosum* (J. F. Gmel.) Blake).

(≡) *Globifera* J. F. Gmel., Syst. Nat. 2: 32. Sep (sero) - Nov 1791.

Microlepis (DC.) Miq., Comm. Phytogr.: 71. 16-21 Mar 1840 (*Osbeckia* sect. *Microlepis* DC., Prodr. 3: 139. Mar (med.) 1828) [*Melastomat.*].
Typus: *Osbeckia oleifolia* DC. (*M. oleifolia* (DC.) Triana) (typ. cons.).

Micromelum Blume, Bijdr.: 137. 20 Aug 1825 [*Rut.*].
Typus: *M. pubescens* Blume

(=) *Aulacia* Lour., Fl. Cochinch.: 258, 273. Sep 1790.
Typus: *A. falcata* Lour.

Micromeria Benth. in Edwards's Bot. Reg.: ad t. 1282. Mar-Dec 1829 [*Lab.*].
Typus: *M. juliana* (L.) Benth. ex Rchb. (Fl. Germ. Excurs.: 311. Jul 1831 - Jul 1832 (*Satureja juliana* L.).

(=) *Xenopoma* Willd. in Ges. Naturf. Freunde Berlin Mag. Neuesten Entdeck. Gesammten Naturk. 5: 399. 1811.
Typus: *X. obovatum* Willd.

(=) *Zygis* Desv. ex Ham., Prodr. Pl. Ind. Occid.: 40. 1825.
Typus: *Z. aromatica* Ham.

Microtropis Wall. ex Meisn., Pl. Vasc. Gen. 1: 68; 2: 49. 27 Aug - 3 Sep 1837 [*Celastr.*].
Typus: *M. discolor* (Wall.) Meisn. (*Cassine discolor* Wall.).

(H) *Microtropis* E. Mey., Comm. Pl. Afr. Austr.: 65. 14 Feb - 5 Jun 1836 [*Legum.*].
≡ *Euchlora* Eckl. & Zeyh. Jan-Feb 1836.

Mikania Willd., Sp. Pl. 3: 1481, 1742. Apr-Dec 1803 [*Comp.*].
Typus: *M. scandens* (L.) Willd. (*Eupatorium scandens* L.) (typ. cons.).

Millettia Wight & Arn., Prodr. Fl. Ind. Orient: 263. Oct (prim.) 1834 [*Legum.*].
Typus: *M. rubiginosa* Wight & Arn.

(=) *Pongamia* Adans., Fam. Pl. 2: 322, 593. Jul-Aug 1763 (nom. cons.).

Miquelia Meisn., Pl. Vasc. Gen. 1: 152; 2: 109. 16-22 Sep 1838 [*Icacin.*].
Typus: *M. kleinii* Meisn.

(H) *Miquelia* Blume in Bull. Sci. Phys. Nat. Néerl. 1: 94. 30 Jun 1838 [*Dicot.: Gesner.*].
Typus: *M. caerulea* Blume

Mischocarpus Blume, Bijdr.: 238. 20 Sep - 7 Dec 1825 [*Sapind.*].
Typus: *M. sundaicus* Blume

(=) *Pedicellia* Lour., Fl. Cochinch.: 641, 655. Sep 1790.
Typus: *P. oppositifolia* Lour.

Mitragyna Korth., Observ. Naucl. Indic.: 19. 1839 [*Rub.*].
Typus: *M. parvifolia* (Roxb.) Korth. (*Nauclea parvifolia* Roxb.).

(H) *Mitragyne* R. Br., Prodr.: 452. 27 Mar 1810 [*Dicot.: Logan./Gentian.*].
≡ *Mitrasacme* Labill. 1804.

(=) *Mamboga* Blanco, Fl. Filip.: 140. 1837.
Typus: *M. capitata* Blanco

Mitraria Cav. in Anales Ci. Nat. 3: 230. Mar 1801 [*Gesner.*].
Typus: *M. coccinea* Cav.

(H) *Mitraria* J. F. Gmel., Syst. Nat. 2: 771, 799. Sep (sero) - Nov 1791 [*Dicot.: Lecythid.*].
≡ *Commersona* Sonn. 1776, non *Commersonia* J. R. Forst. & G. Forst. 1775.

Moenchia Ehrh. in Neues Mag. Aerzte 5: 203. 1783 (post 11 Jun) [*Caryophyll.*].
Typus: *M. quaternella* Ehrh., nom. illeg. (*Sagina erecta* L., *M. erecta* (L.) P. Gaertn. & al.).

Mollia Mart., Nov. Gen. Sp. Pl. 1: 96. Jan-Mar 1826 [*Til.*].
Typus: *M. speciosa* Mart.

(H) *Mollia* J. F. Gmel., Syst. Nat. 2: 303, 420. Sep (sero) - Nov 1791 [*Dicot.: Myrt.*].
Typus: *M. imbricata* (Gaertn.) J. F. Gmel. (*Jungia imbricata* Gaertn.).

Monochaetum (DC.) Naudin in Ann. Sci. Nat., Bot., ser. 3, 4: 48. Jul 1845 (*Arthrostemma* sect. *Monochaetum* DC., Prodr. 3: 135, 138. Mar (med.) 1828) [*Melastomat.*].
Typus: *Arthrostemma calcaratum* DC. (*M. calcaratum* (DC.) Triana) (typ. cons.).

(=) *Ephynes* Raf., Sylva Tellur.: 101. Oct-Dec 1838.
Typus: *E. bonplandii* (Humb. & Bonpl.) Raf. (*Rhexia bonplandii* Humb. & Bonpl.).

Moquinia DC., Prodr. 7: 22. Apr (sero) 1838 [*Comp.*].
Typus: *M. racemosa* (Spreng.) DC. (*Conyza racemosa* Spreng.) (typ. cons.).

(H) *Moquinia* A. Spreng., Tent. Suppl.: 9. 20 Sep 1828 [*Dicot.: Loranth.*].
≡ *Moquiniella* Balle 1954.

Moscharia Ruiz & Pav., Fl. Peruv. Prodr.: 103. Oct (prim.) 1794 [*Comp.*].
Typus: *M. pinnatifida* Ruiz & Pav. (Syst. Veg. Fl. Peruv. Chil.: 186. Dec (sero) 1798).

(H) *Moscharia* Forssk., Fl. Aegypt.-Arab.: 158. 1 Oct 1775 [*Dicot.: Lab.*].
Typus: *M. asperifolia* Forssk.

Mucuna Adans., Fam. Pl. 2: 325, 579. Jul-Aug 1763 [*Legum.*].
Typus: *M. urens* (L.) DC. (Prodr. 2: 405. Nov. (med.) 1825) (*Dolichos urens* L.) (typ. cons.).

(≡) *Zoophthalmum* P. Browne, Civ. Nat. Hist. Jamaica: 295. 10 Mar 1756.
(=) *Stizolobium* P. Browne, Civ. Nat. Hist. Jamaica: 290. 10 Mar 1756.
Typus (vide Piper in U.S.D.A. Bur. Pl. Industr. Bull. 179: 9. 1910): *S. pruriens* (L.) Medik. (*Dolichos pruriens* L.).

Muehlenbeckia Meisn., Pl. Vasc. Gen. 1: 316; 2: 227. 18-24 Jul 1841 [*Polygon.*].
Typus: *M. australis* (G. Forst.) Meisn. (*Coccoloba australis* G. Forst.).

(≡) *Calacinum* Raf., Fl. Tellur. 2: 33. Jan-Mar 1837.
(=) *Karkinetron* Raf., Fl. Tellur. 3: 11. Nov-Dec 1837.
Typus: non designatus.

Muellera L. f., Suppl. Pl.: 53, 329. Apr 1782 [*Legum.*].
Typus: *M. moniliformis* L. f. (etiam vide *Lonchocarpus* [*Dicot.*]).

(=) *Coublandia* Aubl., Hist. Pl. Guiane: 937. Jun-Dec 1775 (nom. rej. sub *Lonchocarpus*).
Typus: *C. frutescens* Aubl.

Muraltia DC., Prodr. 1: 335. Jan (med.) 1824 [*Polygal.*].
Typus: *M. heisteria* (L.) DC. (*Polygala heisteria* L.) (typ. cons.) (etiam vide *Heisteria* [*Dicot.*]).

(H) *Muralta* Adans., Fam. Pl. 2: 460, 580. Jul-Aug 1763 [*Dicot.: Ranuncul.*].
Typus: *Clematis cirrhosa* L.

Murraya J. König ex L., Mant. Pl.: 554, 563. Oct 1771 (*'Murraea'*) (orth. cons.) [*Rut.*].
Typus: *M. exotica* L.

(=) *Camunium* Adanson., Fam. Pl. 2: 166. Jul-Aug 1763.
Typus: *Chalcas paniculata* L.

(=) *Bergera* J. König ex L., Mant. Pl.: 555, 563. Oct 1771.
Typus: *B. koenigii* L.

Myristica Gronov., Fl. Orient.: 141. Apr-Nov 1755 [*Myristic.*].
Typus: *M. fragrans* Houtt. (Nat. Hist. 2(3): 333. Dec (sero) 1774).

Myroxylon L. f., Suppl. Pl.: 34, 233. Apr 1782 [*Legum.*].
Typus: *M. peruiferum* L. f.

(H) *Myroxylon* J. R. Forst. & G. Forst., Char. Gen. Pl.: 63. 29 Nov 1775 [*Dicot.: Flacourt.*].
≡ *Xylosma* G. Forst. 1786 (nom. cons.).

(=) *Toluifera* L., Sp. Pl.: 384. 1 Mai 1753.
Typus: *T. balsamum* L.

Myrteola O. Berg in Linnaea 27: 348. Jan 1856 [*Myrt.*].
Typus: *M. microphylla* (Humb. & Bonpl.) Berg (*Myrtus microphylla* Humb. & Bonpl.) (typ. cons.).

(≡) *Amyrsia* Raf., Sylva Tellur.: 106. Oct-Dec 1838.

(=) *Cluacena* Raf., Sylva Tellur.: 104. Oct-Dec 1838.
Typus (vide McVaugh in Taxon 5: 139. 1956): *C. vaccinioides* (Kunth) Raf. (*Myrtus vaccinioides* Kunth).

Myrtopsis Engl. in Engler & Prantl, Nat. Pflanzenfam. 3(4): 137. Mar 1896 [*Rut.*].
Typus: *M. novae-caledoniae* Engl.

(H) *Myrtopsis* O. Hoffm. in Linnaea 43: 133. Jan 1881 [*Dicot.: Lecythid.*].
Typus: *M. malangensis* O. Hoffm.

Nama L., Syst. Nat., ed. 10: 908, 950. 7 Jun 1759 [*Hydrophyll.*].
Typus: *N. jamaicensis* L. (typ. cons.).

(H) *Nama* L., Sp. Pl.: 226. 1 Mai 1753 [*Dicot.: Hydrophyll.*].
Typus: *N. zeylanica* L.

Naravelia Adans., Fam. Pl. 2: 460, 581. Jul-Aug 1763 *('Naravel')* (orth. cons.) [*Ranuncul.*].
Typus: *N. zeylanica* (L.) DC. (Syst. Nat. 1: 167. 1-15 Nov 1817) (*Atragene zeylanica* L.).

Naregamia Wight & Arn., Prodr. Fl. Ind. Orient.: 116. Oct (prim.) 1834 [*Mel.*].
Typus: *N. alata* Wight & Arn.

(≡) *Nelanaregam* Adans., Fam. Pl. 2: 343, 581. Jul-Aug 1763.

Nasturtium W. T. Aiton, Hort. Kew., ed. 2, 4: 109. Dec 1812 [*Cruc.*].
Typus: *N. officinale* W. T. Aiton (*Sisymbrium nasturtium-aquaticum* L.) (typ. cons.).

(H) *Nasturtium* Mill., Gard. Dict. Abr., ed. 4: [946]. 28 Jan 1754 [*Dicot.: Cruc.*].
Typus: non designatus.

(≡) *Cardaminum* Moench, Methodus: 262. 4 Mai 1794.

Naudinia Planch. & Linden in Ann. Sci. Nat., Bot., ser. 3, 19: 79. Feb 1853 [*Rut.*].
Typus: *N. amabilis* Planch. & Linden

(H) *Naudinia* A. Rich. in Sagra, Hist. Phys. Cuba, Bot. Pl. Vasc.: 561. 1846 [*Dicot.: Melastomat.*].
Typus (vide Mansfeld in Kew Bull. 1935: 438. 1935): *N. argyrophylla* A. Rich.

Nautilocalyx Linden ex Hanst. in Linnaea 26: 181, 206-207. Apr 1854 [*Gesner.*].
Typus: *N. hastatus* Linden ex Hanst., nom. illeg. (*Centrosolenia bractescens* Hook.) [= *N. bracteatus* (Planch.) Sprague (*Centrosolenia bracteata* Planch.)].

(=) *Centrosolenia* Benth. in London J. Bot. 5: 362. 1846.
Typus: *C. hirsuta* Benth.

Nectandra Rol. ex Rottb. in Acta Lit. Univ. Hafn. 1: 279. 1778 [*Laur.*].
Typus: *N. sanguinea* Rol. ex Rottb. (typ. cons.).

(H) *Nectandra* P. J. Bergius, Descr. Pl. Cap.: 131. Sep 1767 [*Dicot.: Thymel.*].
Typus (vide Mansfeld in Bull. Misc. Inform. Kew 1935: 439. 1935): *N. sericea* (L.) P. J. Bergius (*Passerina sericea* L.).

Neesia Blume in Nova Acta Phys.- (H)
Med. Acad. Caes. Leop.-Carol. Nat.
Cur. 17: 83. 1835 [*Bombac.*].
Typus: *N. altissima* (Blume) Blume
(*Esenbeckia altissima* Blume).

(H) *Neesia* Spreng., Anleit. Kenntn. Gew.,
ed. 2, 2: 547. 31 Mar 1818 [*Dicot.:
Comp.*].
Typus: non designatus.

Nematanthus Schrad. in Gött. Gel.
Anz. 1821: 718. 5 Mai 1821 [*Gesner.*].
Typus: *N. corticicola* Schrad.

(=) *Orobanchia* Vand., Fl. Lusit. Bras.
Spec.: 41. 1788.
Typus (vide Chautems in Taxon 36:
656. 1987): *O. radicans* Poir. (in Lamarck, Encycl. Suppl. 4: 203. 29 Jun
1816).

Nemopanthus Raf. in Amer. Monthly Mag. & Crit. Rev. 4: 357. Mar
1819 [*Aquifol.*].
Typus: *N. fascicularis* Raf., nom. illeg. (*Ilex canadensis* Michx., *N. canadensis* (Michx.) DC.).

(=) *Ilicioides* Dum. Cours., Bot. Cult. 4:
27. 1-4 Jul 1802.
Typus: non designatus.

Nemophila Nutt. in J. Acad. Nat.
Sci. Philadelphia 2: 179. 1822 (med.)
[*Hydrophyll.*].
Typus: *N. phacelioides* Nutt.

(=) *Viticella* Mitch., Diss. Princ. Bot.:
42. 1769.
Typus: non designatus.

Neolitsea (Benth. & Hook. f.) Merr.
in Philipp. J. Sci., C, 1, Suppl. 1: 56.
15 Apr 1906 (*Litsea* sect. *Neolitsea*
Benth. & Hook. f., Gen. Pl. 3: 161.
Feb 1880) [*Laur.*].
Typus: *Litsea zeylanica* Nees & T.
Nees (*N. zeylanica* (Nees & T. Nees)
Merr.).

(=) *Bryantea* Raf., Sylva Tellur.: 165.
Oct-Dec 1838.
Typus: *B. dealbata* (R. Br.) Raf. (*Tetranthera dealbata* R. Br.).

Nertera Banks ex Gaertn., Fruct.
Sem. Pl. 1: 124. Dec 1788 [*Rub.*].
Typus: *N. depressa* Gaertn.

(=) *Gomozia* Mutis ex L. f., Suppl. Pl.:
17, 129. Apr 1782.
Typus: *G. granadensis* L. f.

Nesaea Comm. ex Kunth in Humboldt & al., Nov. Gen. Sp. 6, ed. f°:
151. 6 Aug 1823 [*Lythr.*].
Typus: *N. triflora* (L. f.) Kunth (*Lythrum triflorum* L. f.).

(H) *Nesaea* J. V. Lamour. in Nouv. Bull.
Sci. Soc. Philom. Paris 3: 185. Dec
1812 [*Chloroph.*].
Typus: non designatus.

Neslia Desv. in J. Bot. Agric. 3: 162. 1815 (prim.) [*Cruc.*].
Typus: *N. paniculata* (L.) Desv. (*Myagrum paniculatum* L.) (etiam vide *Rapistrum* [*Dicot.*]).

Nicandra Adans., Fam. Pl. 2: 219, 582. Jul-Aug 1763 [*Solan.*].
Typus: *N. physalodes* (L.) Gaertn. (Fruct. Sem. Pl. 2: 237. Apr-Mai 1791) (*Atropa physalodes* L.).

(≡) *Physalodes* Boehm. in Ludwig, Def. Gen. Pl., ed. 3: 41. 1760.

Niemeyera F. Muell., Fragm. 7: 114. Dec 1870 [*Sapot.*].
Typus: *N. prunifera* (F. Muell.) F. Muell. (*Chrysophyllum pruniferum* F. Muell.).

(H) *Niemeyera* F. Muell., Fragm. 6: 96. Dec 1867 [*Monocot.: Orchid.*].
Typus: *N. stylidioides* F. Muell.

Nissolia Jacq., Enum. Syst. Pl.: 7, 27. Aug-Sep 1760 [*Legum.*].
Typus: *N. fruticosa* Jacq. (typ. cons.).

(H) *Nissolia* Mill., Gard. Dict. Abr., ed. 4: [954]. 28 Jan 1754 [*Dicot.: Legum.*].
Typus: non designatus.

Nothofagus Blume, Mus. Bot. 1: 307. 1851 (prim.) [*Fag.*].
Typus: *N. antarctica* (G. Forst.) Oerst. (Bidr. Egefam.: 24. 1871) (*Fagus antarctica* G. Forst.) (typ. cons.).

(=) *Fagaster* Spach, Hist. Nat. Vég. 11: 142. 25 Dec 1841.
Typus: *Fagus dombeyi* Mirbel

(=) *Calucechinus* Hombr. & Jacquinot in Urville, Voy. Pôle Sud, Bot., Atlas (Dicot.): t. 6. Sep-Dec 1843.
Typus: *C. antarctica* Hombr. & Jacquinot

(=) *Calusparassus* Hombr. & Jacquinot in Urville, Voy. Pôle Sud, Bot., Atlas (Dicot.): t. 6. Sep-Dec 1843.
Typus: *C. forsteri* Hombr. & Jacquinot

Nothopegia Blume, Mus. Bot. 1: 203. Oct 1850 [*Anacard.*].
Typus: *N. colebrookiana* (Wight) Blume (*Pegia colebrookiana* Wight).

(=) *Glycycarpus* Dalzell in J. Roy. Asiat. Soc. Bombay 3(1): 69. 1849.
Typus: *G. edulis* Dalzell

Nuphar Sm., Fl. Graec. Prodr. 1: 361. Mai-Nov 1809 [*Nymph.*].
Typus: *N. lutea* (L.) Sm. (*Nymphaea lutea* L.).

(≡) *Nymphozanthus* Rich., Démonstr. Bot.: 63, 68, 103. Mai 1808.

Nymphaea L., Sp. Pl.: 510. 1 Mai 1753 [*Nymph.*].
Typus: *N. alba* L. (typ. cons.).

Odontonema Nees in Linnaea 16: 300. Jul-Aug (prim.) 1842 [*Acanth.*]. Typus: [specimen cult. s. ann., s. coll.] (GZU) (typ. cons.) [= *O. rubrum* (Vahl) Kuntze (*Justicia rubra* Vahl)].

(H) *Odontonema* Nees ex Endl., Gen. Pl., Suppl. 2: 63. Mar-Jun 1842 [*Dicot.: Acanth.*].
Typus (vide Baum & Reveal in Taxon 29: 336. 1980): *Justicia lucida* Andrews

Oedera L., Mant. Pl.: 159, 291. Oct 1771 [*Comp.*].
Typus: *O. prolifera* L., nom. illeg. (*Buphthalmum capense* L., *O. capensis* (L.) Druce).

(H) *Oedera* Crantz, Duab. Drac. Arbor.: 30. 1768 [*Monocot.: Lil.*].
Typus: *O. dragonalis* Crantz

Olearia Moench, Suppl. Meth.: 254. 2 Mai 1802 [*Comp.*].
Typus: *O. dentata* Moench, nom. illeg. (*Aster tomentosus* J. C. Wendl., *O. tomentosa* (J. C. Wendl.) DC.).

(=) *Shawia* J. R. Forst. & G. Forst., Char. Gen. Pl.: 48. 29 Nov 1775.
Typus: *S. paniculata* J. R. Forst. & G. Forst.

Oligomeris Cambess. in Jacquemont, Voy. Inde 4, Bot.: 23. 1839 [*Resed.*].
Typus: *O. glaucescens* Cambess.

(=) *Dipetalia* Raf., Fl. Tellur. 3: 73. Nov-Dec 1837.
Typus: *D. capensis* (Burm. f.) Raf. (*Reseda capensis* Burm. f.).

(=) *Ellimia* Nutt. in Torr. & A. Gray, Fl. N. Amer. 1: 125. Jul 1838.
Typus: *E. ruderalis* Nutt.

Olinia Thunb. in Arch. Bot. (Leipzig) 2(1): 4. Mai-Jul 1800 [*Olin.*].
Typus: *O. cymosa* (L. f.) Thunb. (*Sideroxylon cymosum* L. f.).

(=) *Plectronia* L., Syst. Nat., ed. 12, 2: 138, 183; Mant. Pl.: 6, 52. 15-31 Oct 1767.
Typus: *P. ventosa* L.

Omphalea L., Syst. Nat., ed. 10: 1254, 1264, 1378. 7 Jun 1759 [*Euphorb.*].
Typus: *O. triandra* L. (typ. cons.).

(≡) *Omphalandria* P. Browne, Civ. Nat. Hist. Jamaica: 334. 10 Mar 1756.

Orbaea Haw., Syn. Pl. Succ.: 37. 1812 [*Asclepiad.*].
Typus: *O. variegata* (L.) Haw. (*Stapelia variegata* L.) (typ. cons.).

Oreocharis Benth. in Bentham & Hooker, Gen. Pl. 2: 995, 1021. Mai 1876 [*Gesner.*].
Typus: *O. benthamii* C. B. Clarke (in Candolle & Candolle, Monogr. Phan. 5: 63. Jul 1883) (*Didymocarpus oreocharis* Hance) (typ. cons.).

(H) *Oreocharis* (Decne.) Lindl., Veg. Kingd.: 656. Jan-Mai 1846 (*Lithospermum* subg. *Oreocharis* Decne. in Jacquemont, Voy. Inde 4, Bot.: 122. 1843) [*Dicot.: Boragin.*].
Typus: non designatus.

Ormocarpum P. Beauv., Fl. Oware 1: 95. 23 Feb 1807 [*Legum.*].
Typus: *O. verrucosum* P. Beauv.

(=) *Diphaca* Lour., Fl. Cochinch.: 424, 453. Sep 1790.
Typus: *D. cochinchinensis* Lour.

Ormosia Jacks. in Trans. Linn. Soc. London 10: 360. 7 Sep 1811 [*Legum.*].
Typus: *O. coccinea* (Aubl.) Jacks. (*Robinia coccinea* Aubl.) (typ. cons.).

(=) *Toulichiba* Adans., Fam. Pl. 2: 326, 612. Jul-Aug 1763.
Typus: non designatus.

Orphium E. Mey., Comment. Pl. Afr. Austr.: 181. 1-8 Jan 1838 [*Gentian.*].
Typus: *O. frutescens* (L.) E. Mey. (*Chironia frutescens* L.).

Osmorhiza Raf. in Amer. Monthly Mag. & Crit. Rev. 4: 192. Jan 1819 [*Umbell.*].
Typus: *O. claytonii* (Michx.) C. B. Clarke (in Hooker, Fl. Brit. India 2: 680. Mai 1879) (*Myrrhis claytonii* Michx.).

(≡) *Uraspermum* Nutt., Gen. N. Amer. Pl. 1: 192. 14 Jul 1818.

Osteospermum L., Sp. Pl.: 923. 1 Mai 1753 [*Comp.*].
Typus: *O. spinosum* L. (typ. cons.: [specimen] Herb. Linnaeus No. 1037.1 (LINN)).

Ostrya Scop., Fl. Carniol.: 414. 15 Jun - 21 Jul 1760 [*Betul.*].
Typus: *O. carpinifolia* Scop. (*Carpinus ostrya* L.).

(H) *Ostrya* Hill, Brit. Herb.: 513. Jan 1757 [*Dicot.: Betul.*].
≡ *Carpinus* L. 1753.

Ouratea Aubl., Hist. Pl. Guiane: 397. Jun-Dec 1775 [*Ochn.*].
Typus: *O. guianensis* Aubl.

Ovidia Meisn. in Candolle, Prodr. 14(2): 524. Nov (sero) 1857 [*Thymel.*].
Typus: *O. pillopillo* (Gay) Meisn. (*Daphne pillopillo* Gay) (typ. cons.).

(H)   *Ovidia* Raf., Fl. Tellur. 3: 68. Nov-Dec 1837 [*Monocot.: Commelin.*].
Typus: *O. gracilis* (Ruiz & Pav.) Raf. (*Commelina gracilis* Ruiz & Pav.).

Oxylobium Andrews in Bot. Repos.: ad t. 492. Nov 1807 [*Legum.*].
Typus: *O. cordifolium* Andrews

(=)   *Callistachys* Vent., Jard. Malmaison: ad t. 115. Nov 1805.
Typus: *C. lanceolata* Vent.

Oxypetalum R. Br., Asclepiadeae: 30. 3 Apr 1810 [*Asclepiad.*].
Typus: *O. banksii* Schult. (in Roemer & Schultes, Syst. Veg. 6: 91. Aug-Dec 1820).

(=)   *Gothofreda* Vent., Choix Pl.: ad t. 60. 1808.
Typus: *G. cordifolia* Vent.

Oxytropis DC., Astragalogia, ed. 4°: 66; ed. f°: 53. 15 Nov 1802 [*Legum.*].
Typus: *O. montana* (L.) DC. (*Astragalus montanus* L.) (typ. cons.).

Pachyrhizus Rich. ex DC., Prodr. 2: 402. Nov (med.) 1825 [*Legum.*].
Typus: *P. angulatus* Rich. ex DC., nom. illeg. (*Dolichos erosus* L., *P. erosus* (L.) Urban) (typ. cons.).

(≡)   *Cacara* Thouars in Cuvier, Dict. Sci. Nat. 6: 35. 1806.

Paederia L., Syst. Nat., ed. 12, 2: 135, 189; Mant. Pl.: 7, 52. 15-31 Oct 1767 [*Rub.*].
Typus: *P. foetida* L.

(≡)   *Daun-contu* Adans., Fam. Pl. 2: 146, 549. Jul-Aug 1763.

(=)   *Hondbessen* Adans., Fam. Pl. 2: 158, 584. Jul-Aug 1763.
Typus: *Paederia valli-kara* Juss.

Pallenis Cass. in Bull. Sci. Soc. Philom. Paris 1818: 166. Nov 1818 [*Comp.*].
Typus: *P. spinosa* (L.) Cass. (in Cuvier, Dict. Sci. Nat. 37: 276. Dec 1825) (*Buphthalmum spinosum* L.).

Pancheria Brongn. & Gris in Bull. Soc. Bot. France 9: 74. 1862 [*Cunon.*].
Typus: *P. elegans* Brongn. & Gris

(H)   *Panchezia* Montrouz. in Mém. Acad. Roy. Sci. Lyon, Sect. Sci. 10: 223. 1860 [*Dicot.: Rub.*].
Typus: *P. collina* Montrouz.

Pancovia Willd., Sp. Pl. 2: 280, 285. Mar 1799 [*Sapind.*].
Typus: *P. bijuga* Willd.

(H) *Pancovia* Heist. ex Fabr., Enum.: 64. 1759 [*Dicot.: Ros.*].
≡ *Comarum* L. 1753.

Parahebe W. R. B. Oliv. in Rec. Domin. Mus. 1: 229. 1944 [*Scrophular.*].
Typus: *P. catarractae* (G. Forst.) W. R. B. Oliv. (*Veronica catarractae* G. Forst.).

(=) *Derwentia* Raf., Fl. Tellur. 4: 55. 1838 (med.).
Typus (vide Garnock-Jones & al. in Taxon 39: 536. 1990): *D. suaveolens* Raf., nom. illeg. (*Veronica derwentiana* Andrews).

Parodia Speg. in Anales Soc. Ci. Argent. 96: 70. 1923 [*Cact.*].
Typus: *P. microsperma* (F. A. C. Weber) Speg. (*Echinocactus microspermus* F. A. C. Weber).

(=) *Frailia* Britton & Rose, Cact. 3: 208. 12 Oct 1922.
Typus: *F. cataphracta* (Dams) Britton & Rose (*Echinocactus cataphractus* Dams).

Parsonsia R. Br., Asclepiadeae: 53. 3 Apr 1810 [*Apocyn.*].
Typus: *P. capsularis* (G. Forst.) R. Br. ex Endl. (in Ann. Wiener Mus. Naturgesch. 1: 175. 1836) (*Periploca capsularis* G. Forst.) (typ. cons.).

(H) *Parsonsia* P. Browne, Civ. Nat. Hist. Jamaica: 199. 10 Mar 1756 [*Dicot.: Lythr.*].
Typus: *P. herbacea* J. St.-Hil. (Expos. Fam. Nat. 2: 178. Feb-Apr 1805) (*Lythrum parsonsia* L.).

Parthenocissus Planch. in Candolle & Candolle, Monogr. Phan. 5: 447. Jul 1887 [*Vit.*].
Typus: *P. quinquefolia* (L.) Planch. (*Hedera quinquefolia* L.) (typ. cons.).

Passiflora L., Sp. Pl.: 955. 1 Mai 1753 [*Passiflor.*].
Typus: *P. incarnata* L. (typ. cons.: [specimen] Herb. Linnaeus No. 1070.25 (LINN)).

Patrinia Juss. in Ann. Mus. Natl. Hist. Nat. 10: 311. Oct 1807 [*Valerian.*].
Typus: *P. sibirica* (L.) Juss. (*Valeriana sibirica* L.) (typ. cons.).

Pavonia Cav., Diss. 2, App.: [5]. Jan-Apr 1786 [*Malv.*].
Typus: *P. paniculata* Cav. (Diss. 3: 135. Feb. 1787).

(=) *Lass* Adans., Fam. Pl. 2: 400, 568. Jul-Aug 1763.
Typus: *Hibiscus spinifex* L.

(=) *Malache* B. Vogel in Trew, Pl. Select.: 50. 1772.
Typus: *M. scabra* B. Vogel

Payera Baill. in Bull. Mens. Soc. Linn. Paris: 178. 1878 [*Rub.*].
Typus: *P. conspicua* Baill.

(H) *Payeria* Baill. in Adansonia 1: 50. 1 Oct 1860 [*Dicot.: Mel.*].
Typus: *P. excelsa* Baill.

Pectinaria Haw., Suppl. Pl. Succ.: 14. Mai 1819 [*Asclepiad.*].
Typus: *P. articulata* (Aiton) Haw. (*Stapelia articulata* Aiton).

(H) *Pectinaria* Bernh., Syst. Verz.: 113, 221. 1800 [*Dicot.: Umbell.*].
≡ *Scandix* L. 1753 (typ. des.: Hitchcock in Sprague & al., Nom. Prop. Brit. Bot.: 141. 1929).

Pedilanthus Necker ex Poit. in Ann. Mus. Natl. Hist. Nat. 19: 388. 1812 [*Euphorb.*].
Typus: *P. tithymaloides* (L.) Poit. (*Euphorbia tithymaloides* L.) (etiam vide *Tithymalus* [*Dicot.*]).

(≡) *Tithymaloides* Ortega, Tab. Bot.: 9. 1773.

Pellionia Gaudich., Voy. Uranie, Bot.: 494. 6 Mar 1830 [*Urtic.*].
Typus: *P. elatostemoides* Gaudich. (typ. cons.).

(=) *Polychroa* Lour., Fl. Cochinch.: 538, 559. Sep 1790.
Typus: *P. repens* Lour.

Peltaea (C. Presl) Standl. in Contr. U.S. Natl. Herb. 18: 113. 11 Feb 1916 (*Malachra* sect. *Peltaea* C. Presl, Reliq. Haenk. 2: 125. Jan-Jul 1835) [*Malv.*].
Typus: *Malachra ovata* C. Presl (*P. ovata* (C. Presl) Standl.).

(=) *Peltostegia* Turcz. in Bull. Soc. Imp. Naturalistes Moscou 31(1): 223. 27 Mai 1858.
Typus: *P. parviflora* Turcz.

Peltanthera Benth. in Bentham & Hooker, Gen. Pl. 2: 788, 797. Mai 1876 [*Logan.*].
Typus: *P. floribunda* Benth. & Hook. f.

(H) *Peltanthera* Roth, Nov. Pl. Sp.: 132. Apr 1821 [*Dicot.: Apocyn.*].
Typus: *P. solanacea* Roth

Peltogyne Vogel in Linnaea 11: 410. Apr-Jul 1837 [*Legum.*].
Typus: *P. discolor* Vogel

(=) *Orectospermum* Schott in Schreibers, Nachr. Österr. Naturf. Bras. 2, App.: 54. 1822.
Typus: non designatus.

Peltophorum (Vogel) Benth. in J. Bot. (Hooker) 2: 75. Mar 1840 (*Caesalpinia* sect. *Peltophorum* Vogel in Linnaea 11: 406. Apr-Jul 1837) [*Legum.*].
Typus: *Caesalpinia dubia* Spreng. (*P. dubium* (Spreng.) Taub.).

(=) *Baryxylum* Lour., Fl. Cochinch.: 257, 266. Sep 1790.
Typus: *B. rufum* Lour.

Pentaceras Hook. f. in Bentham & Hooker, Gen. Pl. 1: 298. 7 Aug 1862 [*Rut.*].
Typus: *P. australe* (F. Muell.) Benth. (Fl. Austral. 1: 365. 30 Mai 1863) (*Cookia australis* F. Muell.).

(H) *Pentaceros* G. Mey., Prim. Fl. Esseq.: 136. Nov 1818 [*Dicot.: Byttner.*].
Typus: *P. aculeatum* G. Mey.

Pentagonia Benth., Bot. Voy. Sulphur: t. 39. 25 Oct 1844 [*Rub.*].
Typus: *P. macrophylla* Benth.

(H) *Pentagonia* Heist. ex Fabr., Enum., ed. 2: 336. Sep-Dec 1763 [*Dicot.: Solan.*].
≡ *Nicandra* Adanson, Jul-Aug 1763 (nom. cons.).

Pericampylus Miers in Ann. Mag. Nat. Hist., ser. 2, 7: 36, 40. Jan 1851 [*Menisperm.*].
Typus: *P. incanus* (Colebr.) Hook. f. & Thomson (Fl. Ind.: 194. 1-19 Jul 1855) (*Cocculus incanus* Colebr.).

(=) *Pselium* Lour., Fl. Cochinch.: 600, 621. Sep 1790.
Typus: *P. heterophyllum* Lour.

Pernettya Gaudich. in Ann. Sci. Nat. (Paris) 5: 102. 1825 (*'Pernettia'*) (orth. cons.) [*Eric.*].
Typus: *P. empetrifolia* (Lam.) Gaudich. (*Andromeda empetrifolia* Lam.).

(H) *Pernetya* Scop., Intr. Hist. Nat.: 150. Jan-Apr 1777 [*Dicot.: Campanul.*].
≡ *Canarina* L. 1771 (nom. cons.).

Persea Mill., Gard. Dict. Abr., ed. 4: [1030]. 28 Jan 1754 [*Laur.*].
Typus: *P. americana* Mill. (Gard. Dict., ed. 8: *Persea* No. [1]. 16 Apr 1768) (*Laurus persea* L.).

Persoonia Sm. in Trans. Linn. Soc. London 4: 215. 24 Mai 1798 [*Prot.*]. Typus: *P. lanceolata* Andrews (in Bot. Repos.: ad t. 74. Nov-Dec 1799) (typ. cons.).

(=) *Linkia* Cav., Icon. 4: 61. 14 Mai 1798. Typus: *L. levis* Cav.

Petalostemon Michx., Fl. Bor.-Amer. 2: 48. 19 Mar 1803 (*'Petalostemum'*) (orth. cons.) [*Legum.*]. Typus: *P. candidus* (Willd.) Michx. (*Dalea candida* Willd.).

(=) *Kuhnistera* Lam., Encycl. 3: 370. 13 Feb 1792. Typus: *K. caroliniensis* Lam.

Petrobium R. Br. in Trans. Linn. Soc. London 12(1): 113. ante Sep 1817 [*Comp.*] Typus: *P. arboreum* (J. R. Forst. & G. Forst.) Spreng. (*Laxmannia arborea* J. R. Forst. & G. Forst.).

Petteria C. Presl in Abh. Königl. Böhm. Ges. Wiss., ser. 5, 3: 569. Jul-Dec 1845 [*Legum.*]. Typus: *P. ramentacea* (Sieber) C. Presl (*Cytisus ramentaceus* Sieber).

(H) *Pettera* Rchb., Icon. Fl. Germ. Helv. 5: 33. Mar 1841-Aug 1842 [*Dicot.: Caryophyll.*]. Typus: *P. graminifolia* (Ard.) Rchb. (*Arenaria graminifolia* Ard.).

Petunia Juss. in Ann. Mus. Natl. Hist. Nat. 2: 215. 1803 [*Solan.*]. Typus: *P. nyctaginiflora* Juss. (typ. cons.).

Peumus Molina, Sag. Stor. Nat. Chili: 185, 350. 12-13 Oct 1782 [*Monim.*]. Typus: *P. boldus* Molina (typ. cons.).

(=) *Boldu* Adans., Fam. Pl. 2: 446, 526. Jul-Aug 1763. Typus: non designatus.

Peyrousea DC., Prodr. 6: 76. Jan (prim.) 1838 [*Comp.*]. Typus: *P. oxylepis* DC., nom. illeg. (*Cotula umbellata* L. f., *P. umbellata* (L. f.) Fourc.) (typ. cons.).

(H) *Peyrousia* Poir. in Cuvier, Dict. Sci. Nat. 39: 363. Apr 1826 [*Monocot.: Irid.*]. ≡ *Lapeirousia* Pourr. 1788.

Phaedranthus Miers in Proc. Hort. Soc. London, ser. 2, 3: 182. 1863 [*Bignon.*]. Typus: *P. lindleyanus* Miers

(=) *Sererea* Raf., Sylva Tellur.: 107. Oct-Dec 1838. Typus: *S. heterophylla* Raf., nom. illeg. (*Bignonia heterophylla* Willd., nom. illeg., *Bignonia kerere* Aubl.).

Pharbitis Choisy in Mém. Soc. Phys. Genève 6: 438. 1833 [*Convolvul.*]. Typus: *P. hispida* Choisy, nom. illeg. (*Convolvulus purpureus* L., *P. purpurea* (L.) Voigt) (typ. cons.).

(≡) *Convolvuloides* Moench, Methodus: 451. 4 Mai 1794.

(=) *Diatremis* Raf. in Ann. Gén. Sci. Phys. 8: 271. 1821. Typus: *Convolvulus nil* L.

(=) *Diatrema* Raf., Herb. Raf.: 80. 1833. Typus: *D. trichocarpa* Raf., nom. illeg. (*Convolvulus carolinus* L.).

Phaulopsis Willd., Sp. Pl. 3: 4, 342. 1800 (*'Phaylopsis'*) (orth. cons.) [*Acanth.*]. Typus: *P. parviflora* Willd., nom. illeg. (*Micranthus oppositifolius* J. C. Wendl., *P. oppositifolia* (J. C. Wendl.) Lindau) (etiam vide *Micranthus* [*Monocot.*]).

Physocarpus (Cambess.) Raf., New Fl. 3: 73. Jan-Mar 1838 (*'Physocarpa'*) (*Spiraea* sect. *Physocarpus* Cambess. in Ann. Sci. Nat. (Paris) 1: 239, 385. 1824, *'Physocarpos'*) (orth. cons.) [*Ros.*]. Typus: *Spiraea opulifolia* L. (*P. opulifolius* (L.) Maxim.).

(=) *Epicostorus* Raf. in Atlantic J. 1: 144. 1832 (sero). Typus: *E. montanus* Raf., nom. illeg. (*Spiraea monogyna* Torr.).

Phytocrene Wall., Pl. Asiat. Rar. 3: 11. 10 Dec 1831 [*Icacin.*]. Typus: *P. gigantea* Wall. [= *P. macrophylla* (Blume) Blume (*Gynocephalum macrophyllum* Blume)].

(=) *Gynocephalum* Blume, Bijdr. 483. 1825. Typus: *G. macrophyllum* Blume

Pickeringia Nutt. in Torr. & A. Gray, Fl. N. Amer. 1: 388. Jun 1840 [*Legum.*]. Typus: *P. montana* Nutt.

(H) *Pickeringia* Nutt. in J. Acad. Nat. Sci. Philadelphia 7: 95. 28 Oct 1834 [*Dicot.: Myrsin.*]. Typus: *P. paniculata* (Nutt.) Nutt. (*Cyrilla paniculata* Nutt.).

Picramnia Sw., Prodr.: 2, 27. 20 Jun - 29 Jul 1788 [*Simaroub.*]. Typus: *P. antidesma* Sw.

(=) *Pseudo-brasilium* Adans., Fam. Pl. 2: 341, 595. Jul-Aug 1763. Typus: *Brasiliastrum americanum* Lam.

(=) *Tariri* Aubl., Hist. Pl. Guiane: Suppl. 37. Jun-Dec 1775. Typus: *T. guianensis* Aubl.

Picrodendron Griseb., Fl. Brit. W.
I.: 176. Jun 1860 [*Simaroub.*].
Typus: [specimen] Jamaica, *Macfad-
yen* (K) (typ. cons.) [= *P. baccatum*
(L.) Krug (*Juglans baccata* L.)].

(H) *Picrodendron* Planch. in London J.
Bot. 5: 579. 1846 [*Dicot.: Sapind.*].
Typus: *P. arboreum* (Mill.) Planch.
(*Toxicodendron arboreum* Mill.).

Pierrea F. Heim in Bull. Mens. Soc.
Linn. Paris: 958. 1891 [*Dipterocarp.*].
Typus: *P. pachycarpa* F. Heim

(H) *Pierrea* Hance in J. Bot. 15: 339.
Nov 1877 [*Dicot.: Flacourt.*].
Typus: *P. dictyoneura* Hance

Pilea Lindl., Collect. Bot.: ad t. 4. 1
Apr 1821 [*Urtic.*].
Typus: *P. muscosa* Lindl., nom. illeg. (*Parietaria microphylla* L., *Pilea
microphylla* (L.) Liebm.).

Piliocalyx Brongn. & Gris in Bull.
Soc. Bot. France 12: 185. 1865 (post
Apr) [*Myrt.*].
Typus: *P. robustus* Brongn. & Gris.

(H) *Pileocalyx* Gasp. in Ann. Sci. Nat.,
Bot., ser. 3, 9: 220. Apr 1848 [*Di-
cot.: Cucurbit.*].
≡ *Mellonia* Gasp. 1847.

Piliostigma Hochst. in Flora 29: 598.
14 Oct 1846 [*Legum.*].
Typus: *P. reticulatum* (DC.) Hochst.
(*Bauhinia reticulata* DC.) (typ. cons.).

(=) *Elayuna* Raf., Sylva Tellur.: 145.
Oct-Dec 1838.
Typus: *E. biloba* Raf., nom. illeg.
(*Bauhinia tamarindacea* Delile).

Pimelea Banks ex Gaertn., Fruct.
Sem. Pl. 1: 186. Dec 1788 [*Thy-
mel.*].
Typus: *P. laevigata* Gaertn., nom. il-
leg. (*Banksia prostrata* J. R. Forst. &
G. Forst., *P. prostrata* (J. R. Forst.
& G. Forst.) Willd.).

Piptolepis Sch. Bip. in Jahresber.
Pollichia 20-21: 380. Jul-Dec 1863
[*Comp.*].
Typus: *P. ericoides* Sch. Bip.

(H) *Piptolepis* Benth., Pl. Hartw.: 29.
Feb 1840 [*Dicot.: Ol.*].
Typus: *P. phillyreoides* Benth.

Piscidia L., Syst. Nat., ed. 10: 1151,
1155, 1376. 7 Jun 1759 [*Legum.*].
Typus: *P. erythrina* L., nom. illeg.
(*Erythrina piscipula* L., *P. piscipula*
(L.) Sarg.).

(≡) *Ichthyomethia* P. Browne, Civ. Nat.
Hist. Jamaica: 296. 10 Mar 1756.

Pithecellobium Mart. in Flora 20 (2, Beibl.): 114. 21 Oct 1837 *('Pithecollobium')* (orth. cons.) [*Legum.*]. Typus: *P. unguis-cati* (L.) Benth. (in London J. Bot. 3: 200. 1844) (*Mimosa unguis-cati* L.) (typ. cons.).

(=) *Zygia* P. Browne, Civ. Nat. Hist. Jamaica: 279. 10 Mar 1756. Typus (vide Fawcett & Rendle, Fl. Jamaica 4: 150. 1920): *Z. latifolia* (L.) Fawc. & Rendle (Fl. Jamaica 4: 150. Jul 1920) (*Mimosa latifolia* L.).

Pittosporum Banks ex Gaertn., Fruct. Sem. Pl. 1: 286. Dec 1788 [*Pittospor.*]. Typus: *P. tenuifolium* Gaertn. (typ. cons.).

(=) *Tobira* Adans., Fam. Pl. 2: 449, 611. Jul-Aug 1763. Typus: non designatus.

Planchonella Pierre, Not. Bot. Sapot.: 34. 30 Dec 1890 [*Sapot.*]. Typus: *P. obovata* (R. Br.) Pierre (*Sersalisia obovata* R. Br.) (typ. cons.).

(=) *Hormogyne* A. DC., Prodr. 8: 176. Mar (med.) 1844. Typus: *H. cotinifolia* A. DC.

Plathymenia Benth. in J. Bot. (Hooker) 2: 134. Apr 1840 [*Legum.*]. Typus: *P. foliolosa* Benth.

(=) *Echyrospermum* Schott in Schreibers, Nachr. Österr. Naturf. Bras. 2, App.: 55. 1822. Typus: non designatus.

Platonia Mart., Nov. Gen. Sp. Pl. 3: 168. Sep 1832 [*Guttif.*]. Typus: *P. insignis* Mart.

(H) *Platonia* Raf., Caratt. Nuov. Gen.: 73. Apr-Dec 1810 [*Dicot.: Cist.*]. ≡ *Helianthemum* Mill. 1754.

Platylophus D. Don in Edinburgh New Philos. J. 9: 92. Apr-Jun 1830 [*Cunon.*]. Typus: *P. trifoliatus* (L. f.) D. Don (*Weinmannia trifoliata* L. f.).

(H) *Platylophus* Cass. in Cuvier, Dict. Sci. Nat. 44: 36. Dec 1826 [*Dicot.: Comp.*]. Typus: *Centaurea nigra* L.

Plectranthus L'Hér., Stirp. Nov.: 84. Mar-Apr 1788 [*Lab.*]. Typus: *P. fruticosus* L'Hér. (typ. cons.).

Plenckia Reissek in Martius, Fl. Bras. 11(1): 29. 15 Feb 1861 [*Celastr.*]. Typus: *P. populnea* Reissek

(H) *Plenckia* Raf. in Specchio Sci. 1: 194. 1 Jun 1814 [*Dicot.: Aiz.*]. Typus: *P. setiflora* (Forssk.) Rafin. (*Glinus setiflorus* Forssk.).

Podalyria Willd., Sp. Pl. 2: 492, 501. Mar 1799 [*Legum.*].
Typus: *P. retzii* (J. F. Gmel.) Rickett & Stafleu (*Sophora retzii* J. F. Gmel., *Sophora biflora* Retz. 1799, non L. 1759) (typ. cons.).

Podanthus Lag., Gen. Sp. Pl.: 24. Jun-Dec 1816 [*Comp.*].
Typus: *P. ovatifolius* Lag.

(H) *Podanthes* Haw., Syn. Pl. Succ.: 32. 1812 [*Dicot.: Asclepiad.*].
Typus (vide Mansfeld in Bull. Misc. Inform. Kew 1935: 451. 1935): *P. pulchra* Haw.

Podolepis Labill., Nov. Holl. Pl. 2: 56. Jun 1806 [*Comp.*].
Typus: *P. rugata* Labill.

Podospermum DC. in Lamarck & Candolle, Fl. Franç., ed. 3, 4: 61. 17 Sep 1805 [*Comp.*].
Typus: *P. laciniatum* (L.) DC. (*Scorzonera laciniata* L.) (typ. cons.).

(≡) *Arachnospermum* F. W. Schmidt in Samml. Phys.-Oekon. Aufsätze 1: 274. 1795.

Podotheca Cass. in Cuvier, Dict. Sci. Nat. 23: 561. Nov 1822 [*Comp.*].
Typus: *P. angustifolia* (Labill.) Less. (*Podosperma angustifolium* Labill.).

(≡) *Podosperma* Labill., Nov. Holl. Pl. 2: 35. Apr 1806.

Poiretia Vent. in Mém. Cl. Sci. Math. Inst. Natl. France 1807(1): 4. Jul 1807 [*Legum.*].
Typus: *P. scandens* Vent.

(H) *Poiretia* J. F. Gmel., Syst. Nat. 2: 213, 263. Sep (sero) - Nov 1791 [*Dicot.: Rub.*].
Typus: non designatus.

Polemannia Eckl. & Zeyh., Enum. Pl. Afric. Austral.: 347. Apr-Jun 1837 [*Umbell.*].
Typus: *P. grossulariifolia* Eckl. & Zeyh.

(H) *Polemannia* K. Bergius ex Schltdl. in Linnaea 1: 250. Apr 1826 [*Monocot.: Lil.*].
Typus: *P. hyacinthiflora* K. Bergius ex Schltdl.

Pollichia Aiton, Hort. Kew. 1: 5. 7 Aug - 1 Oct 1789 [*Caryophyll.*].
Typus: *P. campestris* Aiton

(H) *Polichia* Schrank in Acta Acad. Elect. Mogunt. Sci. Util. Erfurti 3: 35. 1781 [*Dicot.: Lab.*].
Typus: *P. galeobdolon* (L.) Willd. (Fl. Berol. Prodr.: 198. 1787) (*Galeopsis galeobdolon* L.).

Polycarpaea Lam. in J. Hist. Nat. 2: 3, 5. 1792 [*Caryophyll.*].
Typus: *P. teneriffae* Lam. (typ. cons.).

(=) *Polia* Lour., Fl Cochinch.: 97, 164. Sep 1790.
Typus: *P. arenaria* Lour.

Polygonum L., Sp. Pl.: 359. 1 Mai 1753 [*Polygon.*].
Typus: *P. aviculare* L. (typ. cons.).

Polypompholyx Lehm. in Bot. Zeitung (Berlin) 2: 109. 9 Feb 1844 [*Lentibular.*].
Typus: *P. tenella* Lehm. (Nov. Stirp. Pug. 8: 48. Apr-Mai (prim.) 1844) (typ. cons.).

(=) *Cosmiza* Raf., Fl. Tellur. 4: 110. 1838 (med.).
Typus: *C. coccinea* Raf., nom. illeg. (*Utricularia multifida* R. Br.).

Pongamia Adans., Fam. Pl. 2: 322, 593. Jul-Aug 1763 *('Pongam')* (orth. cons.) [*Legum.*].
Typus: *P. pinnata* (L.) Pierre (Fl. Forest. Cochinch.: ad t. 385. 15 Apr 1899) (*Cytisus pinnatus* L.) (typ. cons.) (etiam vide *Millettia* [*Dicot.*]).

Premna L., Mant. Pl.: 154, 252. Oct 1771 [*Verben.*].
Typus: *P. serratifolia* L. (typ. cons.).

(=) *Appella* Adans., Fam. Pl. 2: 445, 519. Jul-Aug 1763.
Typus: non designatus.

Prestonia R. Br., Asclepiadeae: 58. 3 Apr 1810 [*Apocyn.*].
Typus: *P. tomentosa* R. Br.

(H) *Prestonia* Scop., Intr. Hist. Nat.: 281. Jan-Apr 1777 [*Dicot.: Malv.*].
≡ *Lass* Adans. 1763.

Printzia Cass. in Cuvier, Dict. Sci. Nat. 37: 488. Dec 1825 [*Comp.*].
Typus: *P. cernua* (P. J. Bergius) Druce (Bot. Soc. Exch. Club Brit. Isles 1916: 642. 1917) (*Inula cernua* P. J. Bergius).

(≡) *Asteropterus* Vaillant, Königl. Akad. Wiss. Paris Phys. Abh. 5: 585. Jan-Apr 1754.

Prismatocarpus L'Hér., Sert. Angl.: 1. Jan (prim.) 1789 [*Campanul.*].
Typus: *P. paniculatus* L'Hér. (typ. cons.).

Prolongoa Boiss., Voy. Bot. Espagne: 320. 23 Sep 1840 [*Comp.*].
Typus: *P. hispanica* G. López & C. E. Jarvis (in Anales Jard. Bot. Madrid 40: 343. Mai 1984) (typ. cons.).

Protea L., Mant. Pl.: 187, 194, 328. Oct 1771 [*Prot.*].
Typus: *P. cynaroides* (L.) L. (*Leucadendron cynaroides* L.) (typ. cons.).

(H) *Protea* L., Sp. Pl.: 94. 1 Mai 1753 (typ. des.: Hitchcock in Sprague & al., Nom. Prop. Brit. Bot.: 113. 1929) [*Dicot.: Prot.*].
≡ *Leucadendron* R. Br. 1810 (nom. cons.).

(=) *Lepidocarpus* Adans., Fam. Pl. 2: 284, 569. Jul-Aug 1763.
≡ *Leucadendron* L. 1753 (nom. rej. sub *Leucadendron*).

Protium Burm. f., Fl. Indica: 88. 1 Mar - 6 Apr 1768 [*Burser.*].
Typus: *P. javanicum* Burm. f. (*Amyris protium* L.).

Pseudelephantopus Rohr in Skr. Naturhist.-Selsk. 2(1). 214. 1792 (*'Pseudo-elephantopus'*) (orth. cons.) [*Comp.*].
Typus: *P. spicatus* (B. Juss. ex Aubl.) C. F. Baker (in Trans. Acad. Sci. St. Louis 12: 45, 55, 56. 20 Mai 1902) (*Elephantopus spicatus* B. Juss. ex Aubl.).

Psilanthus Hook. f. in Bentham & Hooker, Gen. Pl. 2: 23, 115. 7-9 Apr 1873 [*Rub.*].
Typus: *P. mannii* Hook. f. (in Hooker's Icon. Pl. 12: 28. Apr 1873).

(H) *Psilanthus* (DC.) Juss. ex M. Roem., Fam. Nat. Syn. Monogr. 2: 132, 198. Dec 1846 (*Tacsonia* sect. *Psilanthus* DC., Prodr. 3: 335. Mar (med.) 1828) [*Dicot.: Passiflor.*].
≡ *Synactila* Raf. 1838.

Psophocarpus Neck. ex DC., Prodr. 2: 403. Nov (med.) 1825 [*Legum.*].
Typus: *P. tetragonolobus* (L.) DC. (*Dolichos tetragonolobus* L.).

(≡) *Botor* Adans., Fam. Pl. 2: 326, 527. Jul-Aug 1763.

Psychotria L., Syst. Nat., ed. 10: 906, 929, 1364. 7 Jun 1759 [*Rub.*]. Typus: *P. asiatica* L.

(=) *Psychotrophum* P. Browne, Civ. Nat. Hist. Jamaica: 160. 10 Mar 1756. Typus: non designatus.

(=) *Myrstiphyllum* P. Browne, Civ. Nat. Hist. Jamaica: 152. 10 Mar 1756. Typus: *Psychotria myrstiphyllum* Sw.

Pterocarpus Jacq., Sel. Stirp. Amer. Hist.: 283. 5 Jan 1763 [*Legum.*]. Typus: *P. officinalis* Jacq. (typ. cons.).

(H) *Pterocarpus* L., Herb. Amb.: 10. 11 Mai 1754 [*Dicot.: Legum.*]. Typus: non designatus.

Pterococcus Hassk. in Flora 25 (2, Beibl.): 41. 7 Aug 1842 [*Euphorb.*]. Typus: *P. glaberrimus* Hassk., nom. illeg. (*Plukenetia corniculata* Sm., *Pterococcus corniculatus* (Sm.) Pax & K. Hoffm.).

(H) *Pterococcus* Pall., Reise Russ. Reich. 2: 738. 1773 [*Dicot.: Polygon.*]. Typus: *P. aphyllus* Pall.

Pterolepis (DC.) Miq., Comm. Phytogr.: 72. 16-21 Mar 1840 (*Osbeckia* sect. *Pterolepis* DC., Prodr. 3: 140. Mar (med.) 1828) [*Melastomat.*]. Typus: *Osbeckia parnassiifolia* DC. (*P. parnassiifolia* (DC.) Triana) (typ. cons.).

(H) *Pterolepis* Schrad. in Gött. Gel. Anz. 1821: 2071. 29 Dec 1821 [*Monocot.: Cyper.*]. Typus: *P. scirpoides* Schrad.

Pterolobium R. Br. ex Wight & Arn., Prodr. Fl. Ind. Orient.: 283. Oct (prim.) 1834 [*Legum.*]. Typus: *P. lacerans* (Roxb.) Wight & Arn. (*Caesalpinia lacerans* Roxb.).

(H) *Pterolobium* Andrz. ex C. A. Mey., Verz. Pfl. Casp. Meer.: 185. Nov-Dec 1831 [*Dicot.: Cruc.*]. Typus: *P. biebersteinii* Andrz. ex C. A. Mey., nom. illeg. (*Thlaspi latifolium* M. Bieb.).

(=) *Cantuffa* J. F. Gmel., Syst. Nat. 2: 677. Sep (sero) - Nov 1791. Typus: *C. excelsa* J. F. Gmel.

Pteronia L., Sp. Pl., ed. 2: 1176. Jul-Aug 1763 [*Comp.*]. Typus: *P. camphorata* (L.) L. (*Pterophora camphorata* L.).

(≡) *Pterophorus* Vaill., Königl. Akad. Wiss. Paris 5: 375. ca. 14 Apr 1754.

(≡) *Pterophora* L., Pl. Rar. Afr.: 17. 20 Dec 1760 (typ. des.: Greuter & al. in Taxon 54: 162. 2005).

Pterospermum Schreb., Gen. Pl.: 461. Mai 1791 [*Stercul.*].
Typus: *P. suberifolium* (L.) Willd. (Sp. Pl. 3: 728. 1800) (*Pentapetes suberifolia* L.).

Ptilochaeta Turcz. in Bull. Soc. Imp. Naturalistes Moscou 16: 52. 1843 (prim.) [*Malpigh.*].
Typus: *P. bahiensis* Turcz.

(H) *Ptilochaeta* Nees in Martius, Fl. Bras. 2(1): 147. 1 Apr 1842 [*Monocot.: Cyper.*].
Typus: *P. diodon* Nees

Pupalia Juss. in Ann. Mus. Natl. Hist. Nat. 2: 132. 1803 [*Amaranth.*].
Typus: *P. lappacea* (L.) Juss. (*Achyranthes lappacea* L.).

(≡) *Pupal* Adans., Fam. Pl. 2: 268, 596. Jul-Aug 1763.

Pycnanthemum Michx., Fl. Bor.-Amer. 2: 7. 19 Mar 1803 [*Lab.*].
Typus: *P. incanum* (L.) Michx. (*Clinopodium incanum* L.) (typ. cons.).

(=) *Furera* Adans., Fam. Pl. 2: 193, 560. Jul-Aug 1763.
Typus: *Satureja virginiana* L.

Pyrenacantha Wight in Bot. Misc. 2: 107. Nov-Dec 1830 [*Icacin.*].
Typus: *P. volubilis* Wight

Pyrrhopappus DC., Prodr. 7: 144. Apr (sero) 1838 [*Comp.*].
Typus: *P. carolinianus* (Walter) DC. (*Leontodon carolinianus* Walter) (typ. cons.).

Quinchamalium Molina, Sag. Stor. Nat. Chili: 151, 350. 12-13 Oct 1782 [*Santal.*].
Typus: *Q. chilense* Molina

Rafinesquia Nutt. in Trans. Amer. Philos. Soc., ser. 2, 7: 429. 2 Apr 1841 [*Comp.*].
Typus: *R. californica* Nutt.

(H) *Rafinesquia* Raf., Fl. Tellur. 2: 96. Jan-Mar 1837 [*Dicot.: Legum.*].
≡ *Hosackia* Benth. ex Lindl. 1829.

Ramonda Rich. in Persoon, Syn. Pl. 1: 216. 1 Apr - 15 Jun 1805 [*Gesner.*].
Typus: *R. pyrenaica* Pers., nom. illeg. (*Verbascum myconi* L., *R. myconi* (L.) Rchb.).

(H) *Ramondia* Mirb. in Bull. Sci. Soc. Philom. Paris 2: 179. 20 Jan - 21 Feb 1801 [*Pteridoph.*].
Typus: *R. flexuosa* (L.) Mirb. (*Ophioglossum flexuosum* L.).

Rapistrum Crantz, Class. Crucif. Emend.: 105. Jan-Aug 1769 [*Cruc.*]. Typus: *R. hispanicum* (L.) Crantz (*Myagrum hispanicum* L.) (typ. cons.).

(H) *Rapistrum* Scop., Meth. Pl.: 13. 25 Mar 1754 [*Dicot.: Cruc.*].
     ≡ *Neslia* Desv. 1814 (nom. cons.).

Rechsteineria Regel in Flora 31: 247. 21 Apr 1848 [*Gesner.*]. Typus: *R. allagophylla* (Mart.) Regel (*Gesneria allagophylla* Mart.).

(≡) *Alagophyla* Raf., Fl. Tellur. 2: 33. Jan-Mar 1837.

(=) *Megapleilis* Raf., Fl. Tellur. 2: 57. Jan-Mar 1837.
     Typus: *M. tuberosa* Raf., nom. illeg. (*Gesneria bulbosa* Ker Gawl.).

(=) *Styrosinia* Raf., Fl. Tellur. 2: 95. Jan-Mar 1837.
     Typus: *S. coccinea* Raf., nom. illeg. (*Gesneria aggregata* Ker Gawl.).

(=) *Tulisma* Raf., Fl. Tellur. 2: 98. Jan-Mar 1837.
     Typus: *T. verticillata* (Hook.) Raf. (*Gesneria verticillata* Hook.).

Rehmannia Libosch. ex Fisch. & C. A. Mey., Index Sem. Hort. Petrop. 1: 36. Jan 1835 [*Scrophular.*]. Typus: *R. sinensis* (Buc'hoz) Libosch. ex Fisch. & C. A. Mey. (*Sparmannia sinensis* Buc'hoz) (etiam vide *Sparrmannia* [*Dicot.*]).

Relhania L'Hér., Sert. Angl.: 22. Jan (prim.) 1789 [*Comp.*]. Typus: *R. paleacea* (L.) L'Hér. (*Leysera paleacea* L.).

(=) *Osmites* L., Sp. Pl., ed. 2: 1285. Jul-Aug 1763.
     Typus (vide Bremer in Taxon 28: 412. 1979): *O. bellidiastrum* L., nom. illeg. (*Anthemis bellidiastrum* L., nom. illeg., *Anthemis fruticosa* L.).

Retama Raf., Sylva Tellur.: 22. Oct-Dec 1838 [*Legum.*]. Typus: *R. monosperma* (L.) Boiss. (Voy. Bot. Espagne 2: 144. 10 Feb 1840) (*Spartium monospermum* L.).

(≡) *Lygos* Adans., Fam. Pl. 2: 321, 573. Jul-Aug 1763.

Rhagadiolus Vaill., Königl. Akad. Wiss. Paris Phys. Abh. 5: 737. 1754. [*Comp.*].
Typus: *R. edulis* Gaertn. (Fruct. Sem. Pl. 2: 354. Sep-Dec 1791) (*Lapsana rhagadiolus* L.).

Rhaphiolepis Lindl. in Bot. Reg.: ad t. 468. 1 Jul 1820 (*'Raphiolepis'*) (orth. cons.) [*Ros.*].
Typus: *R. indica* (L.) Lindl. (*Crataegus indica* L.).

(=) *Opa* Lour., Fl. Cochinch.: 304, 308. Sep 1790.
Typus (vide McVaugh in Taxon 5: 144. 1956): *O. metrosideros* Lour.

Rhipsalis Gaertn., Fruct. Sem. Pl. 1: 137. Dec 1788 [*Cact.*].
Typus: *R. cassutha* Gaertn.

(=) *Hariota* Adans., Fam. Pl. 2: 243, 520. Jul-Aug 1763.
Typus: non designatus.

Rhodopis Urb., Symb. Antill. 2: 304. 20 Oct 1900 [*Legum.*].
Typus: *R. planisiliqua* (L.) Urb. (*Erythrina planisiliqua* L.).

(H) *Rhodopsis* Lilja, Fl. Sv. Odl. Vext., Suppl. 1: 42. 1840 [*Dicot.: Portulac.*].
≡ *Tegneria* Lilja 1839.

Rhodothamnus Rchb. in Mössler, Handb. Gewächsk., ed. 2, 1: 667, 688. Jul-Dec 1827 [*Eric.*].
Typus: *R. chamaecistus* (L.) Rchb. (*Rhododendron chamaecistus* L.).

Rhynchanthera DC., Prodr. 3: 106. Mar (med.) 1828 [*Melastomat.*].
Typus: *R. grandiflora* (Aubl.) DC. (*Melastoma grandiflorum* Aubl.) (typ. cons.).

(H) *Rynchanthera* Blume, Tab. Pl. Jav. Orchid.: ad t. 78. 1-7 Dec 1825 [*Monocot.: Orchid.*].
Typus: *R. paniculata* Blume

Rhynchocorys Griseb., Spicil. Fl. Rumel. 2: 12. Jul 1844 [*Scrophular.*].
Typus: *R. elephas* (L.) Griseb. (*Rhinanthus elephas* L.).

(≡) *Elephas* Mill., Gard. Dict. Abr., ed. 4: [461]. 28 Jan 1754.

Rhynchoglossum Blume, Bijdr.: 741. Jul-Dec 1826 (*'Rhinchoglossum'*) (orth. cons.) [*Gesner.*].
Typus: *R. obliquum* Blume

Rhynchosia Lour., Fl. Cochinch.: 425, 460. Sep 1790 [*Legum.*].
Typus: *R. volubilis* Lour.

(=) *Dolicholus* Medik. in Vorles. Chur-pfälz. Phys.-Öcon. Ges. 2: 354. 1787.
Typus: *D. flavus* Medik., nom. illeg. (*Dolichos minimus* L.).

(=) *Cylista* Aiton, Hort. Kew. 3: 36, 512. 7 Aug - 1 Oct 1789.
Typus: *C. villosa* Aiton

Rhytidocaulon P. R. O. Bally in Can-dollea 18: 335. Mar 1963 ('1962') [*Asclepiad.*].
Typus: *R. subscandens* P. R. O. Bal-ly

(H) *Rhytidocaulon* Nyl. ex Elenkin in Izv. Imp. Bot. Sada Petra Velikago 16: 263. 1916 [*Fungi*].
≡ *Chlorea* Nyl. 1855 (nom. rej. sub *Letharia*).

Rhytidophyllum Mart., Nov. Gen. Sp. Pl. 3: 38. Jan-Jun 1829 (*'Rytido-phyllum'*) (orth. cons.) [*Gesner.*].
Typus: *R. tomentosum* (L.) Mart. (*Gesneria tomentosa* L.).

Richea R. Br., Prodr.: 555. 27 Mar 1810 [*Epacrid.*].
Typus: *R. dracophylla* R. Br.

(H) *Richea* Labill., Voy. Rech. Pérouse 1: 186. 22 Feb - 4 Mar 1800 [*Dicot.: Comp.*].
Typus: *R. glauca* Labill.

(=) *Cystanthe* R. Br., Prodr.: 555. 27 Mar 1810.
Typus: *C. sprengelioides* R. Br.

Ricotia L., Sp. Pl., ed. 2: 912. Jul-Aug 1763 [*Cruc.*].
Typus: *R. aegyptiaca* L., nom. illeg. (*Cardamine lunaria* L., *R. lunaria* (L.) DC.) (etiam vide *Scopolia* [*Di-cot.*]).

Rinorea Aubl., Hist. Pl. Guiane: 235. Jun-Dec 1775 [*Viol.*].
Typus: *R. guianensis* Aubl.

(=) *Conohoria* Aubl., Hist. Pl. Guiane 239. Jun-Dec 1775.
Typus: *C. flavescens* Aubl.

Robinsonia DC. in Arch. Bot. (Pa-ris) 2: 333. 21 Oct 1833 [*Comp.*].
Typus: *R. gayana* Decne. (in Ann. Sci. Nat., Bot., ser. 2, 1: 28. Jan 1834) (typ. cons.).

(H) *Robinsonia* Scop., Intr. Hist. Nat.: 218. Jan-Apr 1777 [*Dicot.: Quiin.*].
≡ *Touroulia* Aubl. 1775.

Robynsia Hutch. in Hutchinson & Dalziel, Fl. W. Trop. Afr. 2: 68, 108. Mar 1931 [*Rub.*].
Typus: *R. glabrata* Hutch.

(H) *Robynsia* Drap. in Lemaire, Hort. Universel 2: 127. Aug-Oct 1840 [*Monocot.: Amaryllid.*].
Typus: *R. geminiflora* Drap.

Rochea DC., Pl. Hist. Succ.: ad t. 103. 16 Oct 1802 [*Crassul.*].
Typus: *R. coccinea* (L.) DC. (Pl. Hist. Succ., index) (*Crassula coccinea* L.) (typ. cons.).

(H) *Rochea* Scop., Intr. Hist. Nat.: 296. Jan-Apr 1777 [*Dicot.: Legum.*].
Typus: non designatus.

Rochelia Rchb. in Flora 7: 243. 28 Apr 1824 [*Boragin.*].
Typus: *R. saccharata* Rchb., nom. illeg. (*Lithospermum dispermum* L. f., *R. disperma* (L. f.) Wettst.).

(H) *Rochelia* Roem. & Schult., Syst. Veg. 4: xi, 108. Jan-Jun 1819 [*Dicot.: Boragin.*].
≡ *Lappula* Moench 1794.

Rosa L., Sp. Pl.: 491. 1 Mai 1753 [*Ros.*].
Typus: *R. cinnamomea* L. (typ. cons.: [specimen] Herb. Linnaeus No. 652.8 (LINN)).

Rothia Pers., Syn. Pl. 2: 638, [659]. Sep 1807 [*Legum.*].
Typus: *R. trifoliata* (Roth) Pers. (*Dillwynia trifoliata* Roth) (typ. cons.).

(H) *Rothia* Schreb., Gen. Pl.: 531. Mai 1791 [*Dicot.: Comp.*].
Typus: non designatus.

Rourea Aubl., Hist. Pl. Guiane: 467. Jun-Dec 1775 [*Connar.*].
Typus: *R. frutescens* Aubl.

Rubus L., Sp. Pl.: 492. 1 Mai 1753 [*Ros.*].
Typus: *R. fruticosus* L. (typ. cons.: [specimen] Herb. Linnaeus No. 653.9 (LINN)).

Rulingia R. Br. in Bot. Mag.: ad t. 2191. 1820 [*Stercul.*].
Typus: *R. pannosa* R. Br.

(H) *Ruelingia* Ehrh. in Neues Mag. Aerzte 6: 297. 12 Mai - 7 Sep 1784 [*Dicot.: Portulac.*].
≡ *Anacampseros* L. 1758 (nom. cons.).

Ryania Vahl, Eclog. Amer. 1: 51. 1797 [*Flacourt.*].
Typus: *R. speciosa* Vahl

(=) *Patrisa* Rich. in Actes Soc. Hist. Nat. Paris 1: 110. 1792.
Typus: *P. pyrifera* Rich.

Ryssopterys Blume ex A. Juss. in Delessert, Icon. Sel. Pl. 3: 21. Feb 1838 [*Malpigh.*].
Typus: *R. timoriensis* (DC.) A. Juss. (*Banisteria timoriensis* DC.)

Saccocalyx Coss. & Durieu in Ann. Sci. Nat., Bot., ser. 3, 20: 80. Aug 1853 [*Lab.*].
Typus: *S. satureioides* Coss. & Durieu

Sagotia Baill. in Adansonia 1: 53. 1 Oct 1860 [*Euphorb.*].
Typus: *S. racemosa* Baill.

(H) *Sagotia* Duchass. & Walp. in Linnaea 23: 737. Jan 1851 [*Dicot.: Legum.*].
Typus: *S. triflora* (L.) Duchass. & Walp. (*Hedysarum triflorum* L.).

Salacia L., Mant. Pl.: 159. Oct 1771 [*Hippocrat.*].
Typus: *S. chinensis* L.

(≡) *Courondi* Adans., Fam. Pl. 2: 446, 545. Jul-Aug 1763.
Typus (vide Nicolson & Suresh in Taxon 35: 181. 1986): *Christmannia corondi* Dennst. ex Kostel.

Salix L., Sp. Pl.: 1015. 1 Mai 1753 [*Salic.*].
Typus: *S. alba* L. (typ. cons.: [specimen] Herb. Burser. XXIV: 104, (UPS)).

Salmea DC., Cat. Pl. Horti Monsp.: 140. Feb-Mar 1813 [*Comp.*].
Typus: *S. scandens* (L.) DC. (*Bidens scandens* L.) (typ. cons.).

(H) *Salmia* Cav., Icon. 3: 24. Apr 1795 [*Monocot.: Lil.*].
Typus (vide Rickett & Stafleu in Taxon 9: 156. 1960): *S. spicata* Cav.

Salomonia Lour., Fl. Cochinch.: 1, 14. Sep 1790 [*Polygal.*].
Typus: *S. cantoniensis* Lour.

(H) *Salomonia* Heist. ex Fabr., Enum.: 20. 1759 [*Monocot.: Lil.*].
≡ *Polygonatum* Mill. 1754.

Samadera Gaertn., Fruct. Sem. Pl. 2: 352. Apr-Mai 1791 [*Simaroub.*].
Typus: *S. indica* Gaertn.

(≡) *Locandi* Adans., Fam. Pl. 2: 449, 571. Jul-Aug 1763.
Typus: *Niota pentapetala* Poir. (in Lamarck, Encycl. 4: 490. 1 Nov 1798).

Samyda Jacq., Enum. Syst. Pl.: 4, 21. Aug-Sep 1760 [*Flacourt.*]. Typus: *S. dodecandra* Jacq. (typ. cons.).

(H) *Samyda* L., Sp. Pl.: 443. 1 Mai 1753 [*Dicot.: Mel.*]. Typus: *S. guidonia* L.

Santaloides G. Schellenb., Beitr. Anat. Syst. Connar.: 38. 1910 [*Connar.*]. Typus: *S. minor* (Gaertn.) G. Schellenb. (*Aegiceras minus* Gaertn.) (typ. cons.).

(H) *Santalodes* Kuntze, Revis. Gen. Pl. 1: 155. 5 Nov 1891 [*Dicot.: Connar.*]. Typus: *S. hermanniana* Kuntze, nom. illeg. (*Connarus santaloides* Vahl).

(≡) *Kalawael* Adans., Fam. Pl. 2: 344, 530. Jul-Aug 1763.

Sapindus L., Sp. Pl.: 367. 1 Mai 1753 (gend. masc. cons.) [*Sapind.*]. Typus: *S. saponaria* L.

Sapium Jacq., Enum. Syst. Pl.: 9, 31. Aug-Sep 1760 [*Euphorb.*]. Typus: *S. aucuparium* Jacq., nom. illeg. (*Hippomane glandulosa* L., *S. glandulosum* (L.) Morong).

(H) *Sapium* P. Browne, Civ. Nat. Hist. Jamaica: 338. 10 Mar 1756 [*Dicot.: Euphorb.*]. Typus (vide Kruijt & Zijlstra in Taxon 38: 321. 1989): *Excoecaria glandulosa* Sw.

Sargentia S. Watson in Proc. Amer. Acad. Arts 25: 144. 25 Sep 1890 [*Rut.*]. Typus: *S. greggii* S. Watson

(H) *Sargentia* H. Wendl. & Drude ex Salomon, Palmen: 160. Sep-Oct 1887 [*Monocot.: Palm.*]. ≡ *Pseudophoenix* H. Wendl. ex Sarg. 1886.

Sarmienta Ruiz & Pav., Fl. Peruv. Prodr.: 4. Oct (prim.) 1794 [*Gesner.*]. Typus: *S. repens* Ruiz & Pav. (Fl. Peruv. 1: 8. 1798 (med.)), nom. illeg. (*Urceolaria scandens* J. D. Brandis, *S. scandens* (J. D. Brandis) Pers.).

(≡) *Urceolaria* Molina ex J. D. Brandis in Molina, Naturgesch. Chili: 133. 1786.

Sarothamnus Wimm., Fl. Schles.: 278. Feb-Jul 1832 [*Legum.*]. Typus: *S. vulgaris* Wimm., nom. illeg. (*Spartium scoparium* L., *Sarothamnus scoparius* (L.) W. D. J. Koch).

(≡) *Cytisogenista* Duhamel, Traité Arbr. Arbust. 1: 203. 1755.

Saurauia Willd. in Ges. Naturf. Freunde Berlin Neue Schriften 3: 407. 1801 (post 21 Apr) (*'Saurauja'*) (orth. cons.) [*Dillen.*].
Typus: *S. excelsa* Willd.

Saussurea DC. in Ann. Mus. Natl. Hist. Nat. 16: 156, 198. Jul-Dec 1810 [*Comp.*].
Typus: *S. alpina* (L.) DC. (*Serratula alpina* L.) (typ. cons.).

(H) *Saussuria* Moench, Methodus: 388. 4 Mai 1794 [*Dicot.: Lab.*].
Typus: *S. pinnatifida* Moench, nom. illeg. (*Nepeta multifida* L.).

Scaevola L., Mant. Pl.: 145. Oct 1771 [*Gooden.*].
Typus: *S. lobelia* L. (Syst. Veg., ed. 13: 178. Apr-Jun 1774), nom. illeg. (*Lobelia plumieri* L., *S. plumieri* (L.) Vahl).

Scaligeria DC., Coll. Mém. 5: 70. 12 Sep 1829 [*Umbell.*].
Typus: *S. microcarpa* DC.

(H) *Scaligera* Adans., Fam. Pl. 2: 323, 601. Jul-Aug 1763 [*Dicot.: Legum.*].
≡ *Aspalathus* L. 1753.

Schaueria Nees, Del. Sem. Hort. Vratisl. 1838: [3]. 1838 [*Acanth.*].
Typus: *S. calycotricha* (Link & Otto) Nees (*Justicia calycotricha* Link & Otto).

(H) *Schauera* Nees in Lindley, Intr. Nat. Syst. Bot., ed. 2: 202. Jul 1836 [*Dicot.: Laur.*].
≡ *Endlicheria* Nees 1833 (nom. cons.).

Schefflera J. R. Forst. & G. Forst., Char. Gen. Pl.: 23. 29 Nov 1775 [*Aral.*].
Typus: *S. digitata* J. R. Forst. & G. Forst.

(=) *Sciodaphyllum* P. Browne, Civ. Nat. Hist. Jamaica: 190. 10 Mar 1756.
Typus: *S. brownii* Spreng.

Schisandra Michx., Fl. Bor.-Amer. 2: 218. 19 Mar 1803 [*Magnol.*].
Typus: *S. coccinea* Michx.

(=) *Stellandria* Brickell in Med. Repos. 6: 327. Feb-Mar 1803.
Typus: *S. glabra* Brickell

Schizocalyx Wedd. in Ann. Sci. Nat., Bot., ser. 4, 1: 73. Feb 1854 [*Rub.*].
Typus: *S. bracteosus* Wedd.

(H) *Schizocalyx* Scheele in Flora 26: 568. 14 Sep 1843 [*Dicot.: Lab.*].
Typus: non designatus.

Schkuhria Roth, Catal. Bot. 1: 116. Jan-Feb 1797 [*Comp.*].
Typus: *S. abrotanoides* Roth

(H)   *Sckuhria* Moench, Methodus: 566. 4 Mai 1794 [*Dicot.: Comp.*].
Typus: *S. dichotoma* Moench, nom. illeg. (*Siegesbeckia flosculosa* L'Hér.).

Schlechtendalia Less. in Linnaea 5: 242. Apr 1830 [*Comp.*].
Typus: *S. luzulifolia* Less.

(H)   *Schlechtendalia* Willd., Sp. Pl. 3: 1486, 2125. Apr-Dec 1803 [*Dicot.: Comp.*].
Typus: *S. glandulosa* (Cav.) Willd. (*Willdenowa glandulosa* Cav.).

Schleichera Willd., Sp. Pl. 4: 892, 1096. Apr 1806 [*Sapind.*].
Typus: *S. trijuga* Willd.

(=)   *Cussambium* Lam., Encycl. 2: 230. 16 Oct 1786.
Typus: *C. spinosum* Buch.-Ham. (in Mem. Wern. Nat. Hist. Soc. 5: 357. 1826).

Schotia Jacq. in Collect. Bot. Spectantia (Vienna) 1: 93. Jan-Sep 1787 [*Legum.*].
Typus: *S. speciosa* Jacq., nom. illeg. (*Guajacum afrum* L., *S. afra* (L.) Thunb.).

(≡)   *Theodora* Medik., Theodora: 16. 1786.

Schouwia DC. in Mém. Mus. Hist. Nat. 7: 244. 20 Apr 1821 [*Cruc.*].
Typus: *S. arabica* DC. (Syst. Nat. 2: 644. Mai (sero) 1821.), nom. illeg. (*Subularia purpurea* Forssk., *Schouwia purpurea* (Forssk.) Schweinf.).

Schradera Vahl, Eclog. Amer. 1: 35. 1797 [*Rub.*].
Typus: *S. capitata* Vahl, nom. illeg. (*Fuchsia involucrata* Sw., *S. involucrata* (Sw.) K. Schum.).

(H)   *Schraderia* Heist. ex Medik., Philos. Bot. 2: 40. Mai 1791 [*Dicot.: Lab.*].
≡ *Arischrada* Pobed. 1972.

Schrankia Willd., Sp. Pl. 4: 888, 1041. Apr 1806 [*Legum.*].
Typus: *S. aculeata* Willd., nom. illeg. (*Mimosa quadrivalvis* L., *S. quadrivalvis* (L.) Merr.).

(H)   *Schranckia* J. F. Gmel., Syst. Nat. 2: 312, 515. Sep (sero) - Nov 1791 [*Dicot.: Celastr.*].
Typus: *S. quinquefaria* J. F. Gmel.

Schrebera Roxb., Pl. Coromandel 2: 1. Mai 1799 [*Ol.*].
Typus: *S. swietenioides* Roxb.

(H)   *Schrebera* L., Sp. Pl., ed. 2, 2: 1662. Jul-Aug 1763 [*Dicot.: Convolvul.*]. Typus: *S. schinoides* L., nom. illeg. (*Schinus myricoides* L.).

Schubertia Mart., Nov. Gen. Sp. Pl. 1: 55. 1 Oct 1824 [*Asclepiad.*].
Typus: *S. multiflora* Mart. (typ. cons.).

(H)   *Schubertia* Mirb. in Nouv. Bull. Sci. Soc. Philom. Paris 3: 123. Aug 1812 [*Gymnosp.: Pin.*].
≡ *Taxodium* Rich. 1810.

Schultesia Mart., Nov. Gen. Sp. Pl. 2: 103. Jan-Jul 1827 [*Gentian.*].
Typus: *S. crenuliflora* Mart. (typ. cons.).

(H)   *Schultesia* Spreng., Pl. Min. Cogn. Pug. 2: 17. 1815 [*Monocot.: Gram.*].

≡ *Eustachys* Desv. 1810.

Schulzia Spreng. in Neue Schriften Naturf. Ges. Halle 2(1): 30. 1813 [*Umbell.*].
Typus: *S. crinita* (Pall.) Spreng. (*Sison crinitum* Pall.).

(H)   *Shultzia* Raf. in Med. Repos., ser. 2, 5: 356. Feb-Apr 1808 [*Dicot.: Gentian.*].
Typus: *S. obolarioides* Raf.

Scleropyrum Arn. in Mag. Zool. Bot. 2: 549. 1838 [*Santal.*].
Typus: *S. wallichianum* (Wight & Arn.) Arn. (*Sphaerocarya wallichiana* Wight & Arn.).

(=)   *Heydia* Dennst. ex Kostel., Allg. Med.-Pharm. Fl. 5: 2005. Jan-Sep 1836.
Typus: *H. horrida* Dennst. ex Kostel.

Scolopia Schreb., Gen. Pl.: 335. Apr 1789 [*Flacourt.*].
Typus: *S. pusilla* (Gaertn.) Willd. (Sp. Pl. 2: 981. Dec 1799) (*Limonia pusilla* Gaertn.).

(=)   *Aembilla* Adans., Fam. Pl. 2: 448, 513. Jul-Aug 1763.
Typus: non designatus.

Scopolia Jacq., Observ. Bot. 1: 32. 1764 (*'Scopola'*) (orth. cons.) [*Solan.*].
Typus: *S. carniolica* Jacq.

(H)   *Scopolia* Adans., Fam. Pl. 2: 419, 603. Jul-Aug 1763 [*Dicot.: Cruc.*].
≡ *Ricotia* L., Jul-Aug 1763 (nom. cons.).

Scutia (Comm. ex DC.) Brongn., Mém. Fam. Rhamnées: 55. Jul 1826 (*Ceanothus* sect. *Scutia* Comm. ex DC., Prodr. 2: 29. Nov (med.) 1825) [*Rhamn.*].
Typus: *Ceanothus circumcissus* (L. f.) Gaertn. (*Rhamnus circumscissus* L. f., *S. circumcissa* (L. f.) W. Theob.).

(=)   *Adolia* Lam., Encycl. 1: 44. 2 Dec 1783.
Typus (vide Escalante in Bol. Soc. Argent. Bot. 1: 215. 1946): *A. alba* Lam.

Sechium P. Browne, Civ. Nat. Hist. Jamaica: 355. 10 Mar 1756 [*Cucurbit.*].
Typus: *S. edule* (Jacq.) Sw. (Fl. Ind. Occid.: 1150. Oct 1800) (*Sicyos edulis* Jacq.) (typ. cons.).

Securidaca L., Syst. Nat., ed. 10: 1151, 1155. 7 Jun 1759 [*Polygal.*].
Typus: *S. volubilis* L. 1759, non L. 1753 [= *S. diversifolia* (L.) Blake (*Polygala diversifolia* L.)].

(H)   *Securidaca* L., Sp. Pl.: 707. 1 Mai 1753 [*Dicot.: Legum.*].
Typus: *S. volubilis* L.

Securigera DC. in Lamarck & Candolle, Fl. Franç., ed. 3, 4: 609. 17 Sep 1805 [*Legum.*].
Typus: *S. coronilla* DC., nom. illeg. (*Coronilla securidaca* L., *S. securidaca* (L.) Degen & Dörfl.).

(≡)   *Bonaveria* Scop., Intr. Hist. Nat.: 310. Jan-Apr 1777.

Securinega Comm. ex Juss., Gen. Pl.: 388. 4 Aug 1789 [*Euphorb.*].
Typus: *S. durissima* J. F. Gmel. (Syst. Nat. 2: 1008. Apr (sero) - Oct 1792) (typ. cons.).

Seemannia Regel in Gartenflora 4: 183. 1855 [*Gesner.*].
Typus: *S. ternifolia* Regel

Selinum L., Sp. Pl., ed. 2: 350. Sep 1762 [*Umbell.*].
Typus: *S. carvifolia* (L.) L. (*Seseli carvifolia* L.) (typ. cons.).

(H)   *Selinum* L., Sp. Pl.: 244. 1 Mai 1753 [*Dicot.: Umbell.*].
Typus (vide Hitchcock in Sprague & al., Nom. Prop. Brit. Bot.: 139. 1929): *S. sylvestre* L.

Selloa Kunth in Humboldt & al., Nov. Gen. Sp. 4, ed. f°: 208. 26 Oct 1818 [*Comp.*].
Typus: *S. plantaginea* Kunth

(H)   *Selloa* Spreng., Novi Provent: 36. Dec 1818 [*Dicot.: Comp.*].
Typus: *S. glutinosa* Spreng.

Seringia J. Gay in Mém. Mus. Hist. Nat. 7: 442. 1821 [*Stercul.*].
Typus: *S. platyphylla* J. Gay, nom. illeg. (*Lasiopetalum arborescens* Aiton, *S. arborescens* (Aiton) Druce).

(H)   *Seringia* Spreng., Anleit. Kenntn. Gew., ed. 2, 2: 694. 31 Mar 1818 [*Dicot.: Celastr.*].
≡ *Ptelidium* Thouars 1804.

Sesbania Scop., Intr. Hist. Nat.: 308. Jan-Apr 1777 [*Legum.*].
Typus: *S. sesban* (L.) Merr. (in Philipp. J. Sci. 7 (Bot.): 235. 1912 (*Aeschynomene sesban* L.).

(≡) *Sesban* Adans., Fam. Pl. 2: 327, 604. Jul-Aug 1763.
(=) *Agati* Adans., Fam. Pl. 2: 326, 513. Jul-Aug 1763.
Typus: *Robinia grandiflora* L.

Seymeria Pursh, Fl. Amer. Sept. 2: 736. Dec (sero) 1813-Jan 1814 [*Scrophular.*].
Typus: *S. tenuifolia* Pursh, nom. illeg. (*Afzelia cassioides* J. F. Gmel., *S. cassioides* (J. F. Gmel.) Blake) (typ. cons.) (etiam vide *Afzelia* [*Dicot.*]).

Shepherdia Nutt., Gen. N. Amer. Pl. 2: 240. 14 Jul 1818 [*Elaeagn.*].
Typus: *S. canadensis* (L.) Nutt. (*Hippophaë canadensis* L.) (typ. cons.).

Shortia Torr. & A. Gray in Amer. J. Sci. Arts 42: 48. 1842 [*Diapens.*].
Typus: *S. galacifolia* Torr. & A. Gray

(H) *Shortia* Raf., Autik. Bot.: 16. 1840 [*Dicot.: Cruc.*].
Typus: *S. dentata* Raf. (*Arabis dentata* Torr. & A. Gray 1838, non Clairv. 1811).

Shuteria Wight & Arn., Prodr. Fl. Ind. Orient.: 207. Oct (prim.) 1834 [*Legum.*].
Typus: *S. vestita* Wight & Arn. (typ. cons.).

(H) *Shutereia* Choisy, Convolv. Orient.: 103. Aug 1834 [*Dicot.: Convolvul.*].
Typus: *S. bicolor* Choisy (*Convolvulus bicolor* Vahl 1794, non Desr. 1792).

Siebera J. Gay in Mém. Soc. Hist. Nat. Paris 3: 344. 1827 [*Comp.*].
Typus: *S. pungens* (Lam.) DC. (Prodr. 6: 531. Jan (prim.) 1838) (*Xeranthemum pungens* Lam.).

(H) *Sieberia* Spreng., Anleit. Kenntn. Gew., ed. 2, 2: 282. 20 Apr 1817 [*Monocot.: Orchid.*].
Typus: non designatus.

Silene L., Sp. Pl.: 416. 1 Mai 1753 [*Caryophyll.*].
Typus: *S. anglica* L.

(=) *Lychnis* L., Sp. Pl.: 436. 1 Mai 1753. Typus (vide Britton & Brown, Ill. Fl. N. U.S., ed. 2, 2: 68. 1913): *L. chalcedonica* L.

Silybum Vaill., Königl. Akad. Wiss. Paris Phys. Abh. 5: 173, 605. 1754 [*Comp.*].
Typus: *S. marianum* (L.) Gaertn. (Fruct. Sem. Pl. 2: 378. Sep-Dec 1791) (*Carduus marianus* L.) (typ. cons.).

Simarouba Aubl., Hist. Pl. Guiane: 859. Jun-Dec 1775 [*Simaroub.*].
Typus: *S. amara* Aubl.

(H)   *Simaruba* Boehm. in Ludwig, Def. Gen. Pl., ed. 3: 513. 1760 [*Dicot.: Burser.*].
≡ *Bursera* Jacq. ex L. 1762 (nom. cons.).

Sinocarum H. Wolff ex R. H. Shan & F. T. Pu in Acta Phytotax. Sin. 18: 374. 1980. [*Umbell.*].
Typus: *S. coloratum* (Diels) H. Wolff ex R. H. Shan & F. T. Pu (in Shan & Sheh, Fl. Reipubl. Pop. Sin. 55(2): 33. Aug 1985) (*Carum coloratum* Diels).

(=)   *Dactylaea* H. Wolff in Repert. Spec. Nov. Regni Veg. 27: 304. 20 Feb 1930.
Typus: *D. wolffiana* Fedde ex H. Wolff

Siphonychia Torr. & A. Gray, Fl. N. Amer. 1: 173. Jul 1838 [*Caryophyll.*].
Typus: *S. americana* (Nutt.) Torr. & A. Gray (*Herniaria americana* Nutt.).

Skimmia Thunb., Nov. Gen. Pl.: 57. 18 Jun 1783 [*Rut.*].
Typus: *S. japonica* Thunb.

Smelowskia C. A. Mey. ex Ledebour, Icon. Pl. 2: 17. 1830 [*Cruc.*].
Typus: *S. cinerea* Ledeb., nom. illeg. (*Sisymbrium album* Pall., *Smelowskia alba* (Pall.) B. Fedtsch.)

(=)   *Rzedowskia* Cham. & Schltdl. in Linnaea 1: 32. 1826.
Typus: *R. sophiifolia* Cham. & Schltdl.

Smithia Aiton, Hort. Kew. 3: 496. 7 Aug - 1 Oct 1789 [*Legum.*].
Typus: *S. sensitiva* Aiton

(H)   *Smithia* Scop., Intr. Hist. Nat.: 322. Jan-Apr 1777 [*Dicot.: Guttif.*].
≡ *Quapoya* Aubl. 1775.

(≡)   *Damapana* Adans., Fam. Pl. 2: 323, 548. Jul-Aug 1763.

Soaresia Sch. Bip. in Jahresber. Pol-
lichia 20-21: 376. Jul-Dec 1863
[*Comp.*].
Typus: *S. velutina* Sch. Bip.

(H) *Soaresia* Allemão in Trab. Soc. Ve-
llosiana Rio de Janeiro 1851: 72.
1851 [*Dicot.: Mor.*].
Typus: *S. nitida* Allemão

Solandra Sw. in Kongl. Vetensk. Acad. Nya Handl. 8: 300. 1787 [*Solan.*].
Typus: *S. grandiflora* Sw.

(H) *Solandra* L., Syst. Nat. Ed. 10, 2:
1269. 7 Jun 1759 [*Dicot.: Umbell.*].
Typus: *S. capensis* L.

Sommerfeltia Less., Syn. Gen. Com-
pos.: 189. Jul-Aug 1832 [*Comp.*].
Typus: *S. spinulosa* (Spreng.) Less.
(*Conyza spinulosa* Spreng.).

(H) *Sommerfeltia* Flörke ex Sommerf. in
Kongel. Norske Videnskabersselsk.
Skr. 19de Aarhundr. 2(2): 60. 1827
[*Fungi*].
Typus: *S. arctica* Flörke

Sonerila Roxb., Fl. Ind. 1: 180. Jan-
Jun 1820 [*Melastomat.*].
Typus: *S. maculata* Roxb.

Sonneratia L. f., Suppl. Pl.: 38, 252.
Apr 1782 [*Sonnerat.*].
Typus: *S. acida* L. f.

(=) *Blatti* Adans., Fam. Pl. 2: 88, 526.
Jul-Aug 1763.
Typus: non designatus.

Sorbaria (Ser.) A. Braun in Ascherson, Fl. Brandenburg 1: 177. Jan
1860 (*Spiraea* sect. *Sorbaria* Ser. in
Candolle, Prodr. 2: 545. Nov (med.)
1825) [*Ros.*].
Typus: *Spiraea sorbifolia* L. (*Sorbaria sorbifolia* (L.) A. Braun).

(≡) *Schizonotus* Lindl., Intr. Nat. Syst.,
Bot.: 81. Sep 1830.

Sorocephalus R. Br. in Trans. Linn.
Soc. London 10: 139. Feb 1810
[*Prot.*].
Typus: *S. imbricatus* (Thunb.) R.
Br. (*Protea imbricata* Thunb.) (typ.
cons.).

(=) *Soranthe* Salisb. ex Knight, Cult.
Prot.: 71. Dec 1809.
Typus: non designatus.

Sparrmannia L. f., Suppl. Pl.: 41
(*'Sparmannia'*), 265, [468]. Apr 1782
(orth. cons.) [*Til.*].
Typus: *S. africana* L. f.

(H) *Sparmannia* Buc'hoz, Pl. Nouv. Dé-
couv.: 3. 1779 [*Dicot.: Scrophular.*].
≡ *Rehmannia* Libosch. ex Fisch. &
C. A. Mey. 1835 (nom. cons.).

Spathelia L., Sp. Pl., ed. 2: 386. Sep
1762 [*Rut.*].
Typus: *S. simplex* L.

Spergularia (Pers.) J. Presl & C. Presl, Fl. Cech.: 94. 1819 (*Arenaria* subg. *Spergularia* Pers., Syn. Pl. 1: 504. 1 Apr - 15 Jun 1805) [*Caryophyll.*].
Typus: *Arenaria rubra* L. (*S. rubra* (L.) J. Presl & C. Presl).

(≡)   *Tissa* Adans., Fam. Pl. 2: 507, 611. Jul-Aug 1763 (typ. des.: Swart in Regnum Veg. 102: 1764. 1979).
(=)   *Buda* Adans., Fam. Pl. 2: 507, 528. Jul-Aug 1763.
     Typus: non designatus

Sphacele Benth. in Edwards's Bot. Reg.: ad t. 1289. 1 Dec 1829 [*Lab.*].
Typus: *S. lindleyi* Benth., nom. illeg. (*Stachys salviae* Lindl., *Sphacele salviae* (Lindl.) Briq.) (typ. cons.).

(=)   *Alguelaguen* Adans., Fam. Pl. 2: 505, 515. Jul-Aug 1763.
     Typus: non designatus.
(=)   *Phytoxis* Molina, Sag. Stor. Nat. Chili, ed. 2: 145. 1810.
     Typus: *P. sideritifolia* Molina

Sphenoclea Gaertn., Fruct. Sem. Pl. 1: 113. Dec 1788 [*Campanul.*].
Typus: *S. zeylanica* Gaertn.

Spirolobium Baill. in Bull. Mens. Soc. Linn. Paris: 773. 1889 [*Apocyn.*].
Typus: *S. cambodianum* Baill.

Stachyanthus Engl. in Engler & Prantl, Nat. Pflanzenfam., Nachtr. 2-4, 1: 227. Oct 1897 [*Icacin.*].
Typus: *S. zenkeri* Engl.

(H)   *Stachyanthus* DC., Prodr. 5: 84. 1-10 Oct 1836 [*Dicot.: Comp.*].
     ≡ *Argyrovernonia* MacLeish 1984.

Stachytarpheta Vahl, Enum. Pl. 1: 205. Jul-Dec 1804 [*Verben.*].
Typus: *S. jamaicensis* (L.) Vahl (*Verbena jamaicensis* L.) (typ. cons.).

(≡)   *Valerianoides* Medik., Philos. Bot. 1: 177. Apr 1789.
(=)   *Vermicularia* Moench, Suppl. Meth.: 150. 2 Mai 1802.
     Typus: non designatus.

Stapelia L., Sp. Pl.: 217, 580. 1 Mai 1753 [*Asclepiad.*].
Typus: *S. hirsuta* L. (typ. cons.).

Stemodia L., Syst. Nat., ed. 10: 1091, 1118, 1374. 7 Jun 1759 [*Scrophular.*].
Typus: *S. maritima* L.

(≡)   *Stemodiacra* P. Browne, Civ. Nat. Hist. Jamaica: 261. 10 Mar 1756.

Stenandrium Nees in Lindley, Intr. Nat. Syst. Bot., ed. 2: 444. Jul 1836 [*Acanth.*].
Typus: *S. mandioccanum* Nees

(=)   *Gerardia* L., Sp. Pl.: 610. 1 Mai 1753.
Typus: *G. tuberosa* L.

Stenocarpus R. Br. in Trans. Linn. Soc. London 10: 201. Feb 1810 [*Prot.*].
Typus: *S. forsteri* R. Br., nom. illeg. (*Embothrium umbelliferum* J. R. Forst. & G. Forst., *S. umbelliferus* (J. R. Forst. & G. Forst.) Druce) (typ. cons.).

(≡)   *Cybele* Salisb. ex Knight, Cult. Prot.: 123. Dec 1809.

Stenocereus (A. Berger) Riccob. in Boll. Reale Orto Bot. Palermo 8: 253. Oct-Dec 1909. (*Cereus* subg. *Stenocereus* A. Berger in Rep. (Annual) Missouri Bot. Gard. 16: 66. 31 Mai 1905). [*Cact.*]
Typus: *S. stellatus* (Pfeiff.) Riccob. (*Cereus stellatus* Pfeiff.).

(=)   *Rathbunia* Britton & Rose in Contr. U.S. Natl. Herb. 12: 414. 21 Jul 1909.
Typus: *R. sonorensis* (Runge) Britton & Rose (*Cereus sonorensis* Runge).

Stenogyne Benth. in Edwards's Bot. Reg.: ad t. 1292. 1 Jan 1830 [*Lab.*].
Typus: *S. rugosa* Benth. (typ. cons.).

Stephanomeria Nutt. in Trans. Amer. Philos. Soc., ser. 2, 7: 427. 2 Apr 1841 [*Comp.*].
Typus: *S. minor* (Hook.) Nutt. (*Lygodesmia minor* Hook.) (typ. cons.).

(=)   *Ptiloria* Raf. in Atlantic J. 1: 145. 1832 (sero).
Typus: non designatus.

Steriphoma Spreng., Syst. Veg. 4(2): 130, 139. Jan-Jun 1827 [*Cappar.*].
Typus: *S. cleomoides* Spreng., nom. illeg. (*Capparis paradoxa* Jacq., *S. paradoxum* (Jacq.) Endl.).

(=)   *Hermupoa* Loefl., Iter Hispan.: 307. Dec 1758.
Typus: *H. loeflingiana* DC. (Prodr. 1: 254. Jan (med.) 1824).

Stifftia J. C. Mikan, Del. Fl. Faun. Bras.: ad t. 1. 1820 (sero) [*Comp.*].
Typus: *S. chrysantha* J. C. Mikan

Strophostyles Elliott, Sketch Bot. S.-Carolina 2: 229. 1823 [*Legum.*].
Typus: *S. angulosa* (Willd.) Elliott (*Glycine angulosa* Willd.).

(=)   *Phasellus* Medik. in Vorles. Churpfälz. Phys.-Öcon. Ges. 2: 352. 1787.
Typus: *P. roseus* Medik., nom. illeg. (*Phaseolus farinosus* L.).

Struthanthus Mart. in Flora 13: 102. 21 Feb 1830 [*Loranth.*].
Typus: *S. syringifolius* (Mart.) Mart. (*Loranthus syringifolius* Mart.) (typ. cons.).

(=)   *Spirostylis* C. Presl in Schult. & Schult. f., Syst. Veg. 7: xvii. 1829.
Typus: *S. haenkei* C. Presl ex Schult. & Schult. f. (Syst. Veg. 7: 163. 1829).

Struthiola L., Syst. Nat., ed. 12, 2: 108, 127; Mant. Pl.: 4, 41. 15-31 Oct 1767 [*Thymel.*].
Typus: *S. virgata* L. (typ. cons.).

(=)   *Belvala* Adans., Fam. Pl. 2: 285, 525. Jul-Aug 1763.
Typus: *Passerina dodecandra* L.

Stylidium Sw. ex Willd., Sp. Pl. 4: 7, 146. 1805 [*Stylid.*].
Typus: *S. graminifolium* Sw. (typ. cons.).

(H)   *Stylidium* Lour., Fl. Cochinch.: 219, 220. Sep 1790 [*Dicot.: Alang.*].
Typus: *S. chinense* Lour.

Suaeda Forssk. ex J. F. Gmel., Onomat. Bot. Compl. 8: 797. 1776 [*Chenopod.*].
Typus: *S. vera* Forssk. ex J. F. Gmel.

Suksdorfia A. Gray in Proc. Amer. Acad. Arts 15: 41. 1 Oct 1879 [*Saxifrag.*].
Typus: *S. violacea* A. Gray

(=)   *Hemieva* Raf., Fl. Tellur. 2: 70. Jan-Mar 1837.
Typus (vide Rydberg in N. Amer. Fl. 22: 121. 1905): *H. ranunculifolia* (Hook.) Raf. (*Saxifraga ranunculifolia* Hook.).

Sutherlandia R. Br. in Aiton, Hort. Kew., ed. 2, 4: 327. Dec 1812 [*Legum.*].
Typus: *S. frutescens* (L.) R. Br. (*Colutea frutescens* L.).

(H)   *Sutherlandia* J. F. Gmel., Syst. Nat. 2: 998, 1027. Apr (sero) - Oct 1792 [*Dicot.: Stercul.*].
≡ *Heritiera* Aiton 1789.

Swartzia Schreb., Gen. Pl.: 518. Mai 1791 [*Legum.*].
Typus: *S. alata* Willd. (Sp. Pl. 2: 1220. Dec 1799).

(=)   *Possira* Aubl., Hist. Pl. Guiane: 934. Jun-Dec 1775.
Typus: *P. arborescens* Aubl.

(=)   *Tounatea* Aubl., Hist. Pl. Guiane: 549. Jun-Dec 1775.
Typus: *T. guianensis* Aubl.

Sweetia Spreng., Syst. Veg. 2: 171, 213. Jan-Mai 1825 [*Legum.*].
Typus: *S. fruticosa* Spreng.

Synandrodaphne Gilg in Bot. Jahrb. Syst. 53: 362. 19 Oct 1915 [*Thymel.*].
Typus: *S. paradoxa* Gilg

(H) *Synandrodaphne* Meisn. in Candolle, Prodr. 15(1): 176. Mai (prim.) 1864 [*Dicot.: Laur.*].
≡ *Rhodostemonodaphne* Rohwer & Kubitzki 1985.

Synedrella Gaertn., Fruct. Sem. Pl. 2: 456. Sep-Dec 1791 [*Comp.*].
Typus: *S. nodiflora* (L.) Gaertn. (*Verbesina nodiflora* L.).

(≡) *Ucacou* Adans., Fam. Pl. 2: 131, 615. Jul-Aug 1763.

Syzygium P. Browne ex Gaertn., Fruct. Sem. Pl. 1: 166. Dec 1788 [*Myrt.*].
Typus: *S. caryophyllaeum* Gaertn. (typ. cons.).

(H) *Suzygium* P. Browne, Civ. Nat. Hist. Jamaica: 240. 10 Mar 1756. [*Dicot.: Myrt.*].
Typus: *Myrtus zuzygium* L.

(=) *Caryophyllus* L., Sp. Pl.: 515. 1 Mai 1753.
Typus: *C. aromaticus* L.

(=) *Jambosa* Adans., Fam. Pl. 2: 88, 564. Jul-Aug 1763 (nom. cons.).

Talinum Adans., Fam. Pl. 2: 245, 609. Jul-Aug 1763 [*Portulac.*].
Typus: *T. triangulare* (Jacq.) Willd. (Sp. Pl. 2: 862. Dec 1799) (*Portulaca triangularis* Jacq.) (typ. cons.).

Tammsia H. Karst., Fl. Columb. 1: 179. 29 Nov 1861 [*Rub.*].
Typus: *T. anomala* H. Karst.

(≡) *Wiasemskya* Klotzsch in Bot. Zeitung (Berlin) 5: 594. 20 Aug 1847 (typ. des.: Andersson & Rova in Taxon 43. 667. 1994).

Tapinanthus (Blume) Rchb., Deut. Bot. Herb.-Buch [1]: 73. Jul 1841 (*Loranthus* sect. *Tapinanthus* Blume, Fl. Javae (Loranth.): 15. 16 Aug 1830) [*Loranth.*].
Typus: *Loranthus sessilifolius* P. Beauv. (*T. sessilifolius* (P. Beauv.) Tiegh.).

(H) *Tapeinanthus* Herb., Amaryllidaceae: 59, 73, 190, 414. Apr (sero) 1837 [*Monocot.: Amaryllid.*].
Typus: *T. humilis* (Cav.) Herb. (*Pancratium humile* Cav.).

402

Taraxacum F. H. Wigg., Prim. Fl. Holsat.: 56. 29 Mar 1780 [*Comp.*].
Typus: *T. officinale* W. W. Weber ex F. H. Wigg. (*Leontodon taraxacum* L.) (typ. cons.).

(H) *Taraxacum* Zinn, Cat. Pl. Hort. Gott.: 425. 20 Apr - 21 Mai 1757 [*Dicot.: Comp.*].
≡ *Leontodon* L. 1753 (nom. cons.).

Tauschia Schltdl. in Linnaea 9: 607. 1835 (post Feb) [*Umbell.*].
Typus: *T. nudicaulis* Schltdl.

(H) *Tauschia* Preissler in Flora 11: 44. 21 Jan 1828 [*Dicot.: Eric.*].
Typus: *T. hederifolia* Preissler

Teclea Delile in Ann. Sci. Nat., Bot., ser. 2, 20: 90. Aug 1843 [*Rut.*].
Typus: *T. nobilis* Delile

(=) *Aspidostigma* Hochst in Schimper, Iter Abyss., Sect. 2: No. 1293. 1842-1843 [in sched.].
Typus: *A. acuminatum* Hochst.

Tectona L. f., Suppl. Pl.: 20, 151. Apr 1782 [*Verben.*].
Typus: *T. grandis* L. f.

(≡) *Theka* Adans., Fam. Pl. 2: 445, 610. Jul-Aug 1763.

Telopea R. Br. in Trans. Linn. Soc. London 10: 197. Feb 1810 [*Prot.*].
Typus: *T. speciosissima* (Sm.) R. Br. (*Embothrium speciosissimum* Sm.) (typ. cons.).

(≡) *Hylogyne* Salisb. ex Knight, Cult. Prot.: 126. Dec 1809.

Tephrosia Pers., Syn. Pl. 2: 328. Sep 1807 [*Legum.*].
Typus: *T. villosa* (L.) Pers. (*Cracca villosa* L.) (etiam vide *Cracca* [*Dicot.*]).

(=) *Erebinthus* Mitch., Diss. Princ. Bot.: 32. 1769.
Typus (vide Wood in Rhodora 51: 292. 1948): *Tephrosia spicata* (Walter) Torr. & A. Gray (*Galega spicata* Walter).

(=) *Needhamia* Scop., Intr. Hist. Nat.: 310. Jan-Apr 1777.
Typus: *Vicia littoralis* Jacq.

(=) *Reineria* Moench, Suppl. Meth.: 44. 2 Mai 1802.
Typus: *R. reflexa* Moench

Terminalia L., Syst. Nat., ed. 12, 2: 665, 674 ('638'); Mant. Pl.: 21, 128. 15-31 Oct 1767 [*Combret.*].
Typus: *T. catappa* L.

(=) *Bucida* L., Syst. Nat., ed. 10: 1025. Mai-Jun 1759.
Typus: *B. buceras* L.

(≡) *Adamaram* Adans., Fam. Pl. 2: (23), 445, 513. Jul-Aug 1763 (typ. des.: Exell in Index Nom. Gen.: No. 6700. 1958).

Ternstroemia Mutis ex L. f., Suppl. Pl.: 39, 264. Apr 1782 [*The.*]. Typus: *T. meridionalis* Mutis ex L. f.

(=) *Mokof* Adans., Fam. Pl. 2: 501, 578. Jul-Aug 1763. Typus: non designatus.

(=) *Taonabo* Aubl., Hist. Pl. Guiane: 569. Jun-Dec 1775. Typus: *T. dentata* Aubl.

Tetraena Maxim., Enum. Pl. Mongolia 1: 129. 1889 [*Zygophyll.*]. Typus: *T. mongolica* Maxim.

(=) *Petrusia* Baill. in Bull. Mens. Soc. Linn. Paris 35: 274. 1881. Typus: *P. madagascariensis* Baill.

Tetragonolobus Scop., Fl. Carn., ed. 2, 2: 87, 507. Jan-Aug 1772 [*Legum.*]. Typus: *T. scandalida* Scop., nom. illeg. (*Lotus siliquosus* L., *T. siliquosus* (L.) Roth).

(≡) *Scandalida* Adans., Fam. Pl. 2: 326, 602. Jul-Aug 1763.

Tetralix Griseb., Cat. Pl. Cub.: 8. Mai-Aug 1866 [*Flacourt.*]. Typus: *T. brachypetalus* Griseb.

(H) *Tetralix* Zinn, Cat. Pl. Hort. Gott.: 202. 20 Apr - 21 Mai 1757 [*Dicot.: Eric.*]. Typus: *Erica herbacea* L. (nom. rej.).

Tetramerium Nees in Bentham, Bot. Voy. Sulphur: 147. 8 Mai 1846 [*Acanth.*]. Typus: *T. polystachyum* Nees

(H) *Tetramerium* C. F. Gaertn., Suppl. Carp.: 90. Mai 1806 [*Dicot.: Rub.*]. Typus: *T. odoratissimum* C. F. Gaertn., nom. illeg. (*Ixora americana* L.).

(=) *Henrya* Nees in Bentham, Bot. Voy. Sulphur: t. 49. 14 Apr 1845. Typus: *H. insularis* Nees

Tetranema Benth. in Edwards's Bot. Reg. 29: ad t. 52. 1 Oct 1843 [*Scrophular.*]. Typus: *T. mexicanum* Benth.

Tetrapterys Cav., Diss. 9: 433. Jan-Feb 1790 *('Tetrapteris')* (orth. cons.) [*Malpigh.*]. Typus: *T. inaequalis* Cav.

Thalictrum L., Sp. Pl.: 545. 1 Mai
1753 [*Ranuncul.*].
Typus: *T. foetidum* L. (typ. cons.:
[specimen] Italy, Valle d'Aosta, Cle-
menceau near Lignan, WSW of the
village, rocks in gorge, c. 1650 m, 23
Apr 1994, *R. Hand 933* (B: isotypi:
BM, C, K)).

Thamnea Sol. ex Brongn. in Ann.
Sci. Nat. (Paris) 8: 386. Aug 1826
[*Brun.*].
Typus: *T. uniflora* Sol. ex Brongn.

(H) *Thamnia* P. Browne, Civ. Nat. Hist.
Jamaica: 245. 10 Mar 1756 [*Dicot.:*
Flacourt.].
Typus: *Laetia thamnia* L. (Pl. Ja-
maic. Pug.: 31. 28 Nov 1759).

Thespesia Sol. ex Corrêa in Ann.
Mus. Natl. Hist. Nat. 9: 290. 1807
[*Malv.*].
Typus: *T. populnea* (L.) Sol. ex Cor-
rêa (*Hibiscus populneus* L.).

(≡) *Bupariti* Duhamel, Semis Plantat.
Arbr., Add.: 5. 1760.

Thevetia L., Opera Var.: 212. 1758
[*Apocyn.*].
Typus: *T. ahouai* (L.) A. DC. (Prodr.
8: 345. Mar (med.) 1844) (*Cerbera*
ahouai L.) (typ. cons.).

(≡) *Ahouai* Mill., Gard. Dict. Abr., ed.
4: [42]. 28 Jan 1754.

Thorelia Gagnep. in Notul. Syst.
(Paris) 4: 18. 28 Nov 1920 [*Comp.*].
Typus: *T. montana* Gagnep.

(H) *Thorelia* Hance in J. Bot. 15: 268.
Sep 1877 [*Dicot.: Myrt.*].
Typus: *T. deglupta* Hance

Thouinia Poit. in Ann. Mus. Natl.
Hist. Nat. 3: 70. 1804 [*Sapind.*].
Typus: *T. simplicifolia* Poit. (typ.
cons.).

(H) *Thouinia* Thunb. ex L. f., Suppl. Pl.:
9, 89. Apr 1782 [*Dicot.: Ol.*].
Typus: *T. nutans* L. f.

Thryallis Mart., Nov. Gen. Sp. Pl. 3:
77. Jan-Jun 1829 [*Malpigh.*].
Typus: *T. longifolia* Mart.

(H) *Thryallis* L., Sp. Pl., ed. 2, 1: 554.
Sep 1762 [*Dicot.: Malpigh.*].
Typus: *T. brasiliensis* L.

Thryptomene Endl. in Ann. Wiener
Mus. Naturgesch. 2: 192. 1839 [*Myrt.*].
Typus: *T. australis* Endl.

(=) *Gomphotis* Raf., Sylva Tellur.: 103.
Oct-Dec 1838.
Typus: *G. saxicola* (A. Cunn. ex
Hook.) Raf. (*Baeckea saxicola* A.
Cunn. ex Hook.).

Thunbergia Retz. in Physiogr. Sälsk. Handl. 1(3): 163. 1780 [*Acanth.*]. Typus: *T. capensis* Retz.

(H) *Thunbergia* Montin in Kongl. Vetensk. Acad. Handl. 34: 288. 1773 [*Dicot.: Rub.*].
≡ *Pleimeris* Raf. 1838.

Thymelaea Mill., Gard. Dict. Abr., ed. 4: [1381]. 28 Jan 1754 [*Thymel.*]. Typus: *T. sanamunda* All. (Fl. Pedem. 1: 132. Apr-Jul 1785) (*Daphne thymelaea* L.) (typ. cons.).

Thymopsis Benth. in Bentham & Hooker, Gen. Pl. 2: 201, 407. 7-9 Apr 1873 [*Comp.*]. Typus: *T. wrightii* Benth., nom. illeg. (*Tetranthus thymoides* Griseb., *Thymopsis thymoides* (Griseb.) Urb.).

(H) *Thymopsis* Jaub. & Spach, Ill. Pl. Orient. 1: 72. Oct 1842 [*Dicot.: Guttif.*]. Typus: *T. aspera* Jaub. & Spach

Tiliacora Colebr. in Trans. Linn. Soc. London 13: 53, 67. 23 Mai - 21 Jun 1821 [*Menisperm.*]. Typus: *T. racemosa* Colebr.

(=) *Braunea* Willd., Sp. Pl. 4: 638, 797. Apr 1806. Typus: *B. menispermoides* Willd.

Timonius DC., Prodr. 4: 461. Sep (sero) 1830 [*Rub.*]. Typus: *T. rumphii* DC., nom. illeg. (*Erithalis timon* Spreng., *T. timon* (Spreng.) Merr.) (typ. cons.).

(=) *Porocarpus* Gaertn., Fruct. Sem. Pl. 2: 473. Sep-Dec 1791. Typus: *P. helminthotheca* Gaertn.

(=) *Polyphragmon* Desf. in Mém. Mus. Hist. Nat. 6: 5. 1820. Typus: *P. sericeum* Desf.

(=) *Helospora* Jack in Trans. Linn. Soc. London 14: 127. 28 Mai - 12 Jun 1823. Typus: *H. flavescens* Jack

(=) *Burneya* Cham. & Schltdl. in Linnaea 4: 188. Apr 1829. Typus: *B. forsteri* Cham. & Schltdl., nom. illeg. (*Erithalis polygama* G. Forst.).

Tina Schult. in Roem. & Schult., 5: XXXII, 414. Dec 1819 [*Sapind.*]. Typus: *T. thouarsiana* (Camb.) Capuron in Mém. Mus. Natl. Hist. Nat., Sér. B, Bot., ser. 2, 19: 163, 171, 1969 (*Cupania thouarsiana* Camb.) (typ. cons.).

Tinospora Miers in Ann. Mag. Nat. Hist., ser. 2, 7: 35, 38. Jan 1851 [*Menisperm.*].
Typus: *T. cordifolia* (Willd.) Miers ex Hook. f. & Thomson (Fl. Ind. (2): 184. 1-19 Jul 1855) (*Menispermum cordifolium* Willd.).

(=) *Campylus* Lour., Fl. Cochinch.: 94, 113. Sep 1790.
Typus: *C. sinensis* Lour.

Tithymalus Gaertn., Fruct. Sem. Pl. 2: 115. Sep (sero) - Nov 1790 [*Euphorb.*].
Typus: *T. peplus* (L.) Gaertn. (*Euphorbia peplus* L.).

(H) *Tithymalus* Mill., Gard. Dict. Abr., ed. 4: [1391]. 28 Jan 1754 [*Dicot.: Euphorb.*].
≡ *Pedilanthus* Poit. 1812 (nom. cons.).

Tittmannia Brongn. in Ann. Sci. Nat. (Paris) 8: 385. Aug 1826 [*Brun.*].
Typus: *T. lateriflora* Brongn.

(H) *Tittmannia* Rchb., Iconogr. Bot. Exot. 1: 26. Jan-Jun 1824 [*Dicot.: Scrophular.*].
Typus: *T. viscosa* (Hornem.) Rchb. (*Gratiola viscosa* Hornem.).

Toddalia Juss., Gen. Pl.: 371. 4 Aug 1789 [*Rut.*].
Typus: *T. asiatica* (L.) Lam. (Tabl. Encycl. 2: 116. 6 Nov 1797) (*Paullinia asiatica* L.) (typ. cons.: [specimen] Herb. Hermann 3: 45, No. 143 (BM)).

Tolmiea Torr. & A. Gray, Fl. N. Amer. 1: 582. Jun 1840 [*Saxifrag.*].
Typus: *T. menziesii* (Pursh) Torr. & A. Gray (*Tiarella menziesii* Pursh).

(H) *Tolmiea* Hook., Fl. Bor.-Amer. 2: 44. 1834 (sero) [*Dicot.: Eric.*].
Typus: *T. occidentalis* Hook.

(≡) *Leptaxis* Raf., Fl. Tellur. 2: 75. Jan-Mar 1837.

Tontelea Miers in Trans. Linn. Soc. London 28: 331 (*'Tontelia'*), 382. 17 May - 8 Jun 1872 [*Hippocrat.*].
Typus: *T. attenuata* Miers (typ. cons.).

(H) *Tontelea* Aubl., Hist. Pl. Guiane 1: 31. Jun-Dec 1775 [*Dicot.: Celastr.*].
Typus: *T. scandens* Aubl.

Tournefortia L., Sp. Pl.: 140. 1 Mai 1753 [*Boragin.*].
Typus: *T. hirsutissima* L. (typ. cons.: [icon in] Plumier, Pl. Amer. t. 229. 1760).

Tourrettia Foug. in Mém. Acad. Sci. (Paris) 1784: 205. 1787 *('Tour-retia')* (orth. cons.) [*Bignon.*].
Typus: *T. lappacea* (L'Hér.) Willd. (Sp. Pl. 3: 263. 1800) (*Dombeya lappacea* L'Hér.) (etiam vide *Dombeya* [*Dicot.*]).

Tovaria Ruiz & Pav., Fl. Peruv. Prodr.: 49. Oct (prim.) 1794 [*Tovar.*].
Typus: *T. pendula* Ruiz & Pav. (Fl. Peruv. 3: 73. Aug 1802).

(H) *Tovara* Adans., Fam. Pl. 2: 276, 612. Jul-Aug 1763 [*Dicot.: Polygon.*].
Typus: *T. virginiana* (L.) Raf. (Fl. Tellur. 3: 12. Nov-Dec 1837) (*Polygonum virginianum* L.).

Trachelospermum Lem. in Jard. Fleur. 1: ad t. 61. 1851 [*Apocyn.*].
Typus: *T. jasminoides* (Lindl.) Lem. (*Rhyncospermum jasminoides* Lindl.) (typ. cons.).

Trachyspermum Link, Enum. Hort. Berol. Alt. 1: 267. Mar-Jun 1821 [*Umbell.*].
Typus: *T. copticum* (L.) Link (*Ammi copticum* L.).

($\equiv$) *Ammios* Moench, Methodus: 99. 4 Mai 1794.

Trichilia P. Browne, Civ. Nat. Hist. Jamaica: 278. 10 Mar 1756 [*Mel.*].
Typus: *T. hirta* L. (Syst. Nat., ed. 10: 1020. 7 Jun 1759) (typ. cons.).

Trichocalyx Balf. f. in Proc. Roy. Soc. Edinburgh 12: 87. 1883 [*Acanth.*].
Typus: *T. obovatus* Balf. f. (typ. cons.).

Trichodesma R. Br., Prodr.: 496. 27 Mar 1810 [*Boragin.*].
Typus: *T. zeylanicum* (Burm. f.) R. Br. (*Borago zeylanica* Burm. f.) (typ. cons.).

(=) *Borraginoides* Boehm. in Ludwig, Def. Gen. Pl., ed. 3: 18. 1760.
Typus: *Borago indica* L.

Trichostachys Hook. f. in Bentham & Hooker, Gen. Pl. 2: 24, 128. 7-9 Apr 1873 [*Rub.*].
Typus: *T. longifolia* Hiern (in Oliver, Fl. Trop. Afr. 3: 227. Oct 1877).

(H) *Trichostachys* Welw., Syn. Madeir. Drog. Med.: 19. 1862 [*Dicot.: Prot.*].
Typus: *T. speciosa* Welw.

Trigoniastrum Miq., Fl. Ned. Ind.,
Eerste Bijv.: 394. Dec 1861 [*Trigon.*].
Typus: *T. hypoleucum* Miq.

Trigonostemon Blume, Bijdr.: 600.   (=)   *Enchidium* Jack in Malayan Misc.
24 Jan 1826 *('Trigostemon')* (orth.   2(7): 89. 1822.
cons.) [*Euphorb.*].   Typus: *E. verticillatum* Jack
Typus: *T. serratus* Blume

Triguera Cav., Diss. 2, App.: [1].   (H)   *Triguera* Cav., Diss. 1: 41. 15 Apr
Jan-Apr 1786 [*Solan.*].   1785 [*Dicot.: Bombac.*].
Typus: *T. ambrosiaca* Cav. (typ.   Typus: *T. acerifolia* Cav.
cons.).

Trimenia Seem., Fl. Vit.: 425. Feb   (=)   *Piptocalyx* Oliv. ex Benth., Fl. Aus-
1873 [*Monim.*].   tral. 5: 292. Aug-Oct 1870.
Typus: *T. weinmanniifolia* Seem.   Typus: *P. moorei* Oliv. ex Benth.

Trinia Hoffm., Gen. Pl. Umbell.:
xxix, 92. 1814 [*Umbell.*].
Typus: *T. glaberrima* Hoffm., nom.
illeg. (*Seseli pumilum* L., *T. pumila*
(L.) Rchb.) (typ. cons.).

Triopterys L., Sp. Pl.: 428. 1 Mai
1753 *('Triopteris')* (orth. cons.) [*Malpigh.*].
Typus: *T. jamaicensis* L.

Triplochiton K. Schum. in Bot. Jahrb. (H)   *Triplochiton* Alef. in Oesterr. Bot.
Syst. 28: 330. 22 Mai 1900 [*Triplo-   Z. 13: 13. Jan 1863 [*Dicot.: Bom-
chiton.*].   bac./Malv.*].
Typus: *T. scleroxylon* K. Schum.   Typus: non designatus.

Tripteris Less. in Linnaea 6: 95. 1831
(post Mar) [*Comp.*].
Typus: *T. arborescens* (Jacq.) Less.
(*Calendula arborescens* Jacq.) (typ.
cons.).

Trophis P. Browne, Civ. Nat. Hist.   (=)   *Bucephalon* L., Sp. Pl.: 1190. 1 Mai
Jamaica: 357. 10 Mar 1756 [*Mor.*].   1753.
Typus: *T. americana* L. (Syst. Nat.,   Typus: *B. racemosum* L.
ed. 10: 1289. 7 Jun 1759).

Tuberaria (Dunal) Spach in Ann. Sci. Nat., Bot., ser. 2, 6: 364. Dec 1836 (*Helianthemum* sect. *Tuberaria* Dunal in Candolle, Prodr. 1: 270. Jan (med.) 1824) [*Cist.*].
Typus: *T. guttata* (L.) Fourr. (in Ann. Soc. Linn. Lyon, ser. 2, 16: 340. 1868) (*Cistus guttatus* L.).

(=)   *Xolantha* Raf., Caratt. Nuov. Gen.: 73, 74, t. 18, f. 1. Apr-Dec 1810.
Typus: *X. racemosa* Raf.

Turpinia Vent. in Mém. Cl. Sci. Math. Inst. Natl. France 1807(1): 3. Jul 1807 [*Staphyl.*].
Typus: *T. paniculata* Vent.

(H)   *Turpinia* Bonpl. in Humboldt & Bonpland, Pl. Aequinoct. 1: 113. Apr 1807 [*Dicot.: Comp.*].
Typus: *T. laurifolia* Bonpl.

(=)   *Triceros* Lour., Fl. Cochinch.: 100, 184. Sep 1790.
Typus: *T. cochinchinensis* Lour.

Umbellularia (Nees) Nutt., N. Amer. Sylva 1: 87. Jul-Dec 1842 (*Oreodaphne* subg. *Umbellularia* Nees, Syst. Laur.: 381, 462. 30 Oct - 5 Nov 1836) [*Laur.*].
Typus: *Oreodaphne californica* (Hook. & Arn.) Nees (*Tetranthera californica* Hook. & Arn., *U. californica* (Hook. & Arn.) Nutt.).

(≡)   *Sciadiodaphne* Rchb., Deut. Bot. Herb.-Buch [1]: 70; [2]: 118. Jul 1841.

Uncaria Schreb., Gen. Pl.: 125. Apr 1789 [*Rub.*].
Typus: *U. guianensis* (Aubl.) J. F. Gmel. (Syst. Nat. 2: 370. Sep (sero) - Nov 1791) (*Ourouparia guianensis* Aubl.).

(≡)   *Ourouparia* Aubl., Hist. Pl. Guiane: 177. Jun-Dec 1775.

Urbania Phil., Verz. Antofagasta Pfl.: 60. Sep-Oct 1891 [*Verben.*].
Typus: *U. pappigera* Phil. (typ. cons.).

(H)   *Urbania* Vatke in Oesterr. Bot. Z. 25: 10. Jan 1875 [*Dicot.: Scrophular.*].
Typus: *U. lyperiifolia* Vatke

Urceola Roxb. in Asiat. Res. 5: 169. 1799 [*Apocyn.*].
Typus: *U. elastica* Roxb.

(H)   *Urceola* Vand., Fl. Lusit. Bras. Spec.: 8. 1788 [*Spermatoph.*].
Typus: non designatus.

Urnularia Stapf in Hooker's Icon. Pl.: ad t. 2711. Sep 1901 [*Apocyn.*]. Typus: *U. beccariana* (Kuntze) Stapf (*Ancylocladus beccarianus* Kuntze).

(H)   *Urnularia* P. Karst. in Not. Sällsk. Fauna Fl. Fenn. Förh. 8: 209. 1866 [*Fungi*]. Typus: *U. boreella* P. Karst.

Ursinia Gaertn., Fruct. Sem. Pl. 2: 462. Sep-Dec 1791 [*Comp.*]. Typus: *U. paradoxa* (L.) Gaertn. (*Arctotis paradoxa* L.).

Vahlia Thunb., Nov. Gen. Pl.: 36. 10 Jul 1782 [*Saxifrag.*]. Typus: *V. capensis* (L. f.) Thunb. (*Russelia capensis* L. f.).

(=)   *Bistella* Adans., Fam. Pl. 2: 226, 525. Jul-Aug 1763. Typus: *B. geminiflora* Delile (in Cailliaud, Voy. Méroé 2: 97. Jan-Jun 1826).

Vaupelia Brand in Repert. Spec. Nov. Regni Veg. 13: 82. 30 Jan 1914 [*Boragin.*]. Typus: *V. barbata* (Vaupel) Brand (*Trichodesma barbatum* Vaupel).

(H)   *Vaupellia* Griseb., Fl. Brit. W. I.: 460. Mai 1862 [*Dicot.: Gesner.*]. Typus: *V. calycina* (Sw.) Griseb. (*Gesneria calycina* Sw.).

Verbesina L., Sp. Pl.: 901. 1 Mai 1753 [*Comp.*]. Typus: *V. alata* L. (typ. cons.).

Vernonia Schreb., Gen. Pl.: 541. Mai 1791 [*Comp.*]. Typus: *V. noveboracensis* (L.) Willd. (Sp. Pl. 3: 1632. Apr-Dec 1803) (*Serratula noveboracensis* L.) (typ. cons.).

Verticordia DC., Prodr. 3: 208. Mar (med.) 1828 [*Myrt.*]. Typus: *V. fontanesii* DC., nom. illeg. (*Chamelaucium plumosum* Desf., *V. plumosa* (Desf.) Druce) (typ. cons.).

Vigna Savi in Nuov. Giorn. Lett. 8: 113. 1824 [*Legum.*].
Typus: *V. villosa* Savi, nom. illeg. (*Dolichos luteolus* Jacq., *V. luteola* (Jacq.) Benth.).

(=) *Voandzeia* Thouars, Gen. Nov. Madagasc.: 23. 17 Nov 1806.
Typus: *V. subterranea* (L. f.) DC. (Prodr. 2: 474. Nov (med.) 1825) (*Glycine subterranea* L. f.).

(=) *Candelium* Medik., Vorles. Churpfälz. Phys.-Ökon. Ges. 2: 352. 1787. Typus (vide Pasquet in Taxon 51: 819. 2002): *C. nigrum* Medik. [= *V. radiata* (L.) R. Wilczek (*Phaseolus radiata* L.)].

Villanova Lag., Gen. Sp. Pl.: 31. Jun-Dec 1816 [*Comp.*].
Typus: *V. alternifolia* Lag. (typ. cons.).

(H) *Villanova* Ortega, Nov. Pl. Descr. Dec.: 47. 1797 [*Dicot.: Comp.*].
Typus: *V. bipinnatifida* Ortega

(=) *Unxia* L. f., Suppl. Pl.: 56, 368. Apr 1782.
Typus: *U. camphorata* L. f.

Villaria Rolfe in J. Linn. Soc., Bot. 21: 311. 12 Dec 1884 [*Rub.*].
Typus: *V. philippinensis* Rolfe

(H) *Vilaria* Guett. in Mém. Minéral. Dauphiné 1: clxx. 1779 [*Dicot.: Comp.*].
Typus: *V. subacaulis* Guett.

Villarsia Vent., Choix Pl.: ad t. 9. 1803 [*Gentian.*].
Typus: *V. ovata* (L. f.) Vent. (*Menyanthes ovata* L. f.) (typ. cons.).

(H) *Villarsia* J. F. Gmel., Syst. Nat. 2: 306, 447. Sep (sero) - Nov 1791 [*Dicot.: Gentian.*].
Typus: *V. aquatica* J. F. Gmel.

Virgilia Poir. in Lamarck, Encycl. 8: 677. 22 Aug 1808 [*Legum.*].
Typus: *V. capensis* (L.) Poir. (*Sophora capensis* L.) (typ. cons.).

(H) *Virgilia* L'Hér., Virgilia: ad t. [1]. Jan-Jun 1788 [*Dicot.: Comp.*].
Typus: *V. helioides* L'Hér.

Viscaria Bernh., Syst. Verz. 261. 1800.
Typus: *V. vulgaris* Bernh. (≡ *Lychnis viscaria* L.)

(≡) *Steris* Adans., Fam. Pl. 255, 607. 1763 (typ. des.: Šourková in Novit. Bot. Inst. Horto Bot. Univ. Carl. Prag 1973-1975: 25-28. 1976).

Vismia Vand., Fl. Lusit. Bras. Spec.: 51. 1788 [*Guttif.*].
Typus: *V. cayennensis* (Jacq.) Pers. (Syn. Pl. 2: 86. Nov 1806) (*Hypericum cayennense* Jacq.) (typ. cons.).

(=) *Caopia* Adans., Fam. Pl. 2: 448. Jul-Aug 1763.
Typus: non designatus.

Vochysia Aubl., Hist. Pl. Guiane: 18.
Jun-Dec 1775 *('Vochy')* (orth. cons.)
[*Vochys.*].
Typus: *V. guianensis* Aubl.

Wahlenbergia Schrad. ex Roth, Nov.　　(=)　*Cervicina* Delile, Descr. Egypte,
Pl. Sp.: 399. Apr 1821 [*Campanul.*].　　　Hist. Nat. 2: 150. 1813 (sero) - 1814
Typus: *W. elongata* (Willd.) Schrad.　　　(prim.).
ex Roth (*Campanula elongata* Willd.)　　Typus: *C. campanuloides* Delile
[= *W. capensis* (L.) A. DC. (*Cam-*
panula capensis L.)].

Wallenia Sw., Prodr.: 2, 31. 20 Jun -
29 Jul 1788 [*Myrsin.*].
Typus: *W. laurifolia* Sw.

Walpersia Harv. in Harvey & Son-　　(H)　*Walpersia* Reissek ex Endl., Gen. Pl.:
der, Fl. Cap. 2: 26. 16-31 Oct 1862　　　1100. Apr 1840 [*Dicot.: Rhamn.*].
[*Legum.*].　　　　　　　　　　　　　≡ *Trichocephalus* Brongn. 1826.
Typus: *W. burtonioides* Harv.

Warburgia Engl., Pflanzenw. Ost-　　(=)　*Chibaca* Bertol. in Mem. Reale Ac-
Afrikas C: 276. Jul 1895 [*Canell.*].　　　cad. Sci. Ist. Bologna 4: 545. 1853.
Typus: *W. stuhlmannii* Engl.　　　　　Typus: *C. salutaris* Bertol.

Wedelia Jacq., Enum. Syst. Pl.: 8,　　(H)　*Wedelia* Loefl., Iter Hispan.: 180.
28. Aug-Sep 1760 [*Comp.*].　　　　　Dec 1758 [*Dicot.: Nyctagin.*].
Typus: *W. fruticosa* Jacq.　　　　　　≡ *Allionia* L. 1759 (nom. cons.).

Weihea Spreng., Syst. Veg. 2: 559,　　(≡)　*Richaeia* Thouars, Gen. Nov. Mada-
594. Jan-Mai 1825 [*Rhizophor.*].　　　gasc.: 25. 17 Nov 1806.
Typus: *W. madagascarensis* Spreng.

Weinmannia L., Syst. Nat., ed. 10:　　(≡)　*Windmannia* P. Browne, Civ. Nat.
997, 1005, 1367. 7 Jun 1759 [*Cu-*　　　Hist. Jamaica: 212. 10 Mar 1756.
non.].
Typus: *W. pinnata* L.

Wendlandia Bartl. ex DC., Prodr. 4:　　(H)　*Wendlandia* Willd., Sp. Pl. 2: 6, 275.
411. Sep (sero) 1830 [*Rub.*].　　　　　Mar 1799 [*Dicot.: Menisperm.*].
Typus: *W. paniculata* (Roxb.) DC.　　　≡ *Androphylax* J. C. Wendl. 1798
(*Rondeletia paniculata* Roxb.) (typ.　　(nom. rej. sub *Cocculus*).
cons.).

Wendtia Meyen, Reise 1: 307. 23-　　(H)　*Wendia* Hoffm., Gen. Pl. Umbell.:
31 Mai 1834 [*Geran.*].　　　　　　　136. 1814 [*Dicot.: Umbell.*].
Typus: *W. gracilis* Meyen　　　　　Typus: *W. chorodanum* Hoffm.

Wiborgia Thunb., Nov. Gen. Pl.: 137. 3 Jun 1800 [*Legum.*]. Typus: *W. obcordata* (Berg.) Thunb. (*Crotalaria obcordata* Berg.) (typ. cons.).

(H) *Viborgia* Moench, Methodus: 132. 4 Mai 1794 [*Dicot.: Legum.*]. Typus: non designatus.

Wigandia Kunth in Humboldt & al., Nov. Gen. Sp. 3, ed. 4°: 126; ed. f°: 98. 8 Feb 1819 [*Hydrophyll.*]. Typus: *W. caracasana* Kunth. (typ. cons.).

Wikstroemia Endl., Prodr. Fl. Norfolk.: 47. 1833 (post 12 Mai) ('*Wickstroemia*') (orth. cons.) [*Thymel.*]. Typus: *W. australis* Endl.

(H) *Wikstroemia* Schrad. in Gött. Gel. Anz. 1821: 710. 5 Mai 1821 [*Dicot.: The.*]. Typus: *W. fruticosa* Schrad.

(=) *Capura* L., Mant. Pl.: 149, 225. Oct 1771. Typus: *C. purpurata* L.

Willughbeia Roxb., Pl. Coromandel 3: 77. 18 Feb 1820 [*Apocyn.*]. Typus: *W. edulis* Roxb.

(H) *Willughbeja* Scop. ex Schreb., Gen. Pl.: 162. Apr 1789 [*Dicot.: Apocyn.*]. ≡ *Ambelania* Aubl. 1775.

Wisteria Nutt., Gen. N. Amer. Pl. 2: 115. 14 Jul 1818 [*Legum.*]. Typus: *W. speciosa* Nutt., nom. illeg. (*Glycine frutescens* L., *W. frutescens* (L.) Poir.).

(≡) *Phaseoloides* Duhamel, Traité Arbr. Arbust. 2: 115. 1755.

(=) *Diplonyx* Raf., Fl. Ludov.: 101. Oct-Dec (prim.) 1817. Typus: *D. elegans* Raf.

Withania Pauquy, Belladone: 14. Apr 1825 [*Solan.*]. Typus: *W. frutescens* (L.) Pauquy (*Atropa frutescens* L.) (typ. cons.).

Xanthophyllum Roxb., Pl. Coromandel 3: 81. 18 Feb 1820 [*Polygal.*]. Typus: *X. flavescens* Roxb. (typ. cons.).

(=) *Pelaë* Adans., Fam. Pl. 2: 448, 589. Jul-Aug 1763. Typus: non designatus.

(=) *Eystathes* Lour., Fl. Cochinch.: 223, 234. Sep 1790. Typus: *E. sylvestris* Lour.

Xanthostemon F. Muell. in Hooker's J. Bot. Kew Gard. Misc. 9: 17. Jan 1857 [*Myrt.*]. Typus: *X. paradoxus* F. Muell.

(=) *Nani* Adans., Fam. Pl. 2: 88, 581. Jul-Aug 1763. Typus: *Metrosideros vera* Lindl.

Xerocarpa H. J. Lam, Verben. Malay. Archip.: 98. 7 Apr 1919 [*Verben.*].
Typus: *X. avicenniifoliola* H. J. Lam

(H) *Xerocarpa* (G. Don) Spach, Hist. Nat. Vég. 9: 583. 15 Aug 1840 (*Scaevola* sect. *Xerocarpa* G. Don, Gen. Hist. 3: 728. 8-15 Nov. 1834) [*Dicot.: Gooden.*].
Typus: non designatus.

Xylopia L., Syst. Nat., ed. 10: 1241, 1250, 1378. 7 Jun 1759 [*Annon.*].
Typus: *X. muricata* L. (typ. cons.).

(≡) *Xylopicrum* P. Browne, Civ. Nat. Hist. Jamaica: 250. 10 Mar 1756.

Xylosma G. Forst., Fl. Ins. Austr.: 72. Oct-Nov 1786 [*Flacourt.*].
Typus: *X. orbiculata* (J. R. Forst. & G. Forst.) G. Forst. (*Myroxylon orbiculatum* J. R. Forst. & G. Forst.) (etiam vide *Myroxylon* [*Dicot.*]).

Zaluzianskya F. W. Schmidt, Neue Selt. Pfl.: 11. 1793 (ante 17 Jun) [*Scrophular.*].
Typus: *Z. villosa* F. W. Schmidt

(H) *Zaluzianskia* Neck. in Hist. & Commentat. Acad. Elect. Sci. Theod.-Palat. 3: 303. 1775 [*Pteridoph.*].
≡ *Marsilea* L. 1753 (nom. cons.).

Zelkova Spach in Ann. Sci. Nat., Bot., ser. 2, 15: 356. 1 Jun 1841 [*Ulm.*].
Typus: *Z. crenata* Spach, nom. illeg. (*Rhamnus carpinifolius* Pall., *Z. carpinifolia* (Pall.) K. Koch).

Zinnia L., Syst. Nat., ed. 10: 1189, 1221, 1377. 7 Jun 1759 [*Comp.*].
Typus: *Z. peruviana* (L.) L. (*Chrysogonum peruvianum* L.).

(≡) *Crassina* Scepin, Acid. Veg.: 42. 19 Mai 1758.
(=) *Lepia* Hill, Exot. Bot.: 29. Feb-Sep 1759.
Typus: non designatus.

Zollingeria Kurz in J. Asiat. Soc. Bengal, Pt. 2, Nat. Hist. 41: 303. 1872 [*Sapind.*].
Typus: *Z. macrocarpa* Kurz

(H) *Zollingeria* Sch. Bip. in Flora 37: 274. 14 Mai 1854 [*Dicot.: Comp.*].
Typus: *Z. scandens* Sch. Bip.

Zuccagnia Cav., Icon. 5: 2. Jun-Sep 1799 [*Legum.*].
Typus: *Z. punctata* Cav.

(H) *Zuccangnia* Thunb., Nov. Gen. Pl.: 127. 17 Dec 1798 [*Monocot.: Lil.*].
Typus: *Z. viridis* (L.) Thunb. (*Hyacinthus viridis* L.).

Zuccarinia Blume, Bijdr.: 1006. Oct 1826 - Mar 1827 [*Rub.*].
Typus: *Z. macrophylla* Blume

(H)   *Zuccarinia* Maerkl. in Ann. Wetterauischen Ges. Gesammte Naturk. 2: 252. 28 Apr 1811 [*Spermatoph.*].
Typus: *Z. verbenacea* Maerkl.

F. FOSSIL PLANTS (EXCL. DIATOMS)

Araucarites C. Presl in Sternberg, Versuch Fl. Vorwelt 2: 203. 1 Sep – 7 Oct 1838.
Typus: *Araucarites goeppertii* C. Presl

(H)   *Araucarites* Endl., Gen. 263. Oct 1837 [*Foss.*].
Typus: non designatus

Asterophyllites Brongn., Prodr. Hist. Vég. Foss.: 159. Dec 1828.
Typus: *A. equisetiformis* (Sternb.) Brongn. (*Bornia equisetiformis* Sternb.).

(H)   *Asterophyllites* Brongn. in Mém. Mus. Hist. Nat. 8: 210. Mai 1822 [*Foss.*].
Typus: *A. radiatus* Brongn.

(≡)   *Bornia* Sternb., Vers. Fl. Vorwelt 4: xxviii. 1825.

(=)   *Bechera* Sternb., Vers. Fl. Vorwelt 4: xxx. 1825.
Typus: *B. ceratophylloides* Sternb.

(=)   *Brukmannia* Sternb., Vers. Fl. Vorwelt 4: xxix. 1825.
Typus: *B. tenuifolia* (Sternb.) Sternb. (*Schlotheimia tenuifolia* Sternb.).

Baiera Braun in Beitr. Petrefacten-Kunde 6: 20. 1843.
Typus: *B. muensteriana* (C. Presl) Heer (in Mém. Acad. Imp. Sci. Saint Pétersbourg, ser. 7, 22: 52. 1876) (*Spaerococcites muensterianus* C. Presl) (typ. cons.).

(H)   *Bajera* Sternb., Vers. Fl. Vorwelt 4: xxviii. 1825 [*Foss.*].
Typus: *B. scanica* Sternb.

Calamites Brongn., Prodr. Hist. Vég. Foss.: 121. Dec 1828.
Typus: *C. suckowii* Brongn. (typ. cons.).

(H)   *Calamitis* Sternb., Vers. Fl. Vorwelt 1(1): 22. 31 Dec 1820 [*Foss.*].
Typus: *C. pseudobambusia* Sternb.

Cardiocarpus Brongn., Rech. Graines Foss. Silic.: 20. 1 Dec 1880.
Typus: *C. drupaceus* Brongn.

(H)   *Cardiocarpus* Reinw. in Syll. Pl. Nov. 2: 14. 1825 [*Dicot.: Simaroub.*].
Typus: *C. amarus* Reinw.

Classopollis Pflug in Palaeontographica, Abt. B, Paläophytol. 95: 91. 1953 [Fossil pollen morphogenus]. Typus *C. classoides* Pflug.

(=) *Corollina* Malyavk. in Trudy Vse-soyuzn. Neftian. Nauchno-Issl. Inst. Geol.-Razvedochn., ser. 2, 33: 120. 1949 (post 19 Jul) [Fossil pollen morphogenus]. Typus: *C. compacta* Malyavk.

(=) *Circulina* Malyavk. in Trudy Vse-soyuzn. Neftian. Nauchno-Issl. Inst. Geol.-Razvedochn., ser. 2, 33: 120. 1949 (post 19 Jul) [Fossil pollen morphogenus]. Typus (vide R. Potonié in Beih. Geol. Jahrb. 72: 147. 1966): *C. funifera* Malyavk.

Cordaianthus Grand'Eury in Mém. Acad. Roy. Sci. Inst. France 24: 227. 1877. Typus: *C. gemmifer* Grand'Eury

(=) *Botryoconus* Göpp. in Palaeontographica 12: 152. 1864. Typus: *B. goldenbergii* Göpp.

Cordaites Unger, Gen. Sp. Pl. Foss.: 277. 17-20 Apr 1850 [*Cordaitales*]. Typus: *C. borassifolius* (Sternb.) Unger (*Flabellaria borassifolia* Sternb.).

(≡) *Neozamia* Pomel in Bull. Soc. Géol. France, ser. 2, 3: 655. 1846 (post 15 Jun).

Cupressinoxylon Göpp. in Natuurk. Verh. Hollandische Maatsch. Wetensch. Haarlem, ser. 2, 6: 196. 1850 (med.). Typus: *Cupressinoxylon gothanii* Kräusel in Jahrb. Preuss. Geol. Landesant. 39: 436. 1920 (typ. cons).

(=) *Retinodendron* Zenker, Beitr. Naturgesch. Urwelt 2. 1833. Typus: *R. pityoides* Zenker

Cyclopteris Brongn. in Prodr. Hist. Vég. Foss.: 51. Dec 1828. Typus: *C. orbicularis* Brongn. (Hist. Vég. Foss. 1: 220. 6 Jun 1831).

(H) *Cyclopteris* Schrad. ex Gray, Nat. Arr. Brit. Pl. 2: 9. 1 Nov 1821 [*Pteridoph.*]. ≡ *Cystopteris* Bernh. 1805 (nom. cons.).

Diphyes Cookson in Proc. Roy. Soc. Victoria 78: 85. 1965. Typus: *D. colligerum* (Deflandre & Cookson) Cookson (*Hystrichosphaeridium colligerum* Deflandre & Cookson).

(H) *Diphyes* Blume, Bijdr.: 310. 20 Sep-7 Dec 1825 [*Monocot.: Orchid.*]. Typus: non designatus.

Dolerotheca T. Halle in Kongl. Svenska Vetenskapsakad. Handl., ser. 3, 12(6): 42. Jul-Dec 1933.
Typus: *D. fertilis* (Renault) T. Halle (*Dolerophyllum fertile* Renault).

(=) *Discostachys* Grand'Eury, Géol. Paléontol. Bassin Houillier Gard: t. 8, f. 2. 1890.
Typus: *D. cebennensis* Grand'Eury

Doliostrobus Marion in Ann. Sci. Géol. 20(3): 2. 1888.
Typus: *D. sternbergii* Marion (1888) nom. illeg. (*D. taxiformis* var. *sternbergii* Mai & Walther).

(H) *Doliostrobus* Marion in C. R. Acad. Sci. Paris 99: 823. 1884 [Foss.].
Typus (vide Andrews in Bull. U.S. Geol. Surb. 1013: 150. 1955): *D. sternbergii* (Corda) Marion (*Araucaria sternbergii* Corda)

Glossopteris Brongn., Prodr. Hist. Vég. Foss.: 54. Dec 1828.
Typus: *G. browniana* Brongn. (Hist. Vég. Foss. 1: 222. 6 Jun 1831).

(H) *Glossopteris* Raf., Anal. Nat.: 205. Apr-Jul 1815 [*Pteridoph.*].
≡ *Phyllitis* Hill 1757.

Lycopodites Lindl. & Hutton, Foss. Fl. Gr. Brit. 1: 171. Jan-Apr 1833.
Typus: *L. falcatus* Lindl. & Hutton (typ. cons.).

(H) *Lycopodites* Brongn. in Mém. Mus. Hist. Nat. 8: 231. Mai 1822 [Foss.].
Typus: *L. taxiformis* Brongn.

Megalopteris (J. W. Dawson) E. B. Andrews in Rep. Geol. Surv. Ohio 2(2): 415. 1875 (*Neuropteris* subg. *Megalopteris* J. W. Dawson, Fossil Pl. Canada: 51. 1871).
Typus: *M. dawsonii* (Hartt) E. B. Andrews (*Neuropteris dawsonii* Hartt).

(=) *Cannophyllites* Brongn., Prodr. Hist. Vég. Foss.: 130. Dec 1828.
Typus: *C. virletii* Brongn.

Neuropteris (Brongn.) Sternb., Vers. Fl. Vorwelt 4: 16. Sep 1825 (*Filicites* [unranked] *Neuropteris* ('*Nevropteris*') Brongn. in Mém. Mus. Hist. Nat. 8: 233. Mai 1822) (orth. cons.).
Typus: *N. heterophylla* (Brongn.) Sternb. (*Filicites heterophyllus* Brongn.) (typ. cons.: [specimen (Mus. Natl. Hist. Nat. Paris, palaeobot. coll. No. 448) illustrated in] Brongniart, Hist. Vég. Foss. 1: t. 71. 6 Jun 1831).

Odontopteris (Brongn.) Sternb., Vers. Fl. Vorwelt 4: xxi. 1825 (*Filicites* sect. *Odontopteris* Brongn. in Mém. Mus. Hist. Nat. 8: 234. Mai 1822). Typus: *Filicites brardii* Brongn. (*O. brardii* (Brongn.) Sternb.).

(H) *Odontopteris* Bernh. in J. Bot. (Schrader) 1800(2): 7, 106. Oct-Dec 1801 [*Pteridoph.*].
≡ *Lygodium* Bernh. 1801.

Pitys Witham in Int. Struct. Foss. Veg. 1833. (*'Pitus'*) (orth. cons.). Typus: *P. antique* Witham

Protopodocarpoxylon Eckhold in Palaeontographica Abt. B, Paläophytol. 89: 144. 1949. Typus: *P. befordense* (Stopes) Kräusel (*Podocarpoxylon bedfordense* Stopes) (typ. cons.).

Sigillaria Brongn. in Mém. Mus. Hist. Nat. 8: 209, 222. May 1822. Typus: *S. scutellata* Brongn.

(H) *Sigillaria* Raf. in Amer. Mag. & Crit. Rev. 4: 192. Jan. 1819 [*Monocot.: Lil.*].
≡ *Smilacina* Desf. 1807 (nom. cons.).
(=) *Rhytidolepis* Sternb., Vers. Fl. Vorwelt 1(2): 32 Jan-Aug 1821. Typus: *R. ocellata* Sternb.

Sphenophyllum Brongn., Prodr. Hist. Vég. Foss.: 68. Dec 1828. Typus: *S. emarginatum* (Brongn.) Brongn. (*Sphenophyllites emarginatus* Brongn.).

(≡) *Sphenophyllites* Brongn. in Mém. Mus. Hist. Nat. 8: 209, 234. Mai 1822.
(=) *Rotularia* Sternb., Vers. Fl. Vorwelt 1(2): 33. Jan-Aug 1821. Typus: *R. marsiliifolia* Sternb.

APPENDIX IV

NOMINA SPECIFICA CONSERVANDA ET REJICIENDA

In the following list the nomina conservanda have been inserted in the left column, in **bold-face italics**. They are arranged in alphabetical sequence within the major groups. Rejected synonyms (nomina rejicienda) are listed in the right column.

Names listed in this Appendix fall under the special provisions of Art. 14.4. Species names with a type conserved under Art. 14.3 are listed in full in App. IIIA, with a cross-reference in the present Appendix.

Neither a rejected name, nor any combination based on a rejected name, may be used for a taxon that includes the type of the corresponding conserved name (Art. 14.7; see also Art. 14 Note 2). Combinations based on a conserved name are therefore, in effect, similarly conserved. When such a later combination is in current use, it is cross-referenced (by "vide" = see) to its conserved basionym.

typ. cons.   typus conservandus, type to be conserved (Art. 14.9; see also Art. 14.3 and 10.4); as by Art. 14.8, listed types of conserved names may not be changed even if they are not explicitly designated as typ. cons.

(H)   homonym (Art. 14.10; see also Art. 53), only the earliest being listed.

(≡)   nomenclatural synonym (i.e., homotypic synonym, based on the same nomenclatural type as the conserved name), usually only the earliest legitimate one being listed (Art. 14.4).

(=)   taxonomic synonym (i.e., heterotypic synonym, based on a type different from that of the conserved name), to be rejected only in favour of the conserved name (Art. 14.6 and 14.7).

Some names listed as conserved have no corresponding nomina rejicienda because they were conserved solely to maintain a particular type.

A. ALGAE

Achnanthes quadricauda Turpin in Mém. Mus. Hist. Nat. 16: 311. 1828 [*Chloroph.*].
Typus: [specimen from strain] Hungary, Lake Belsö-tó, *Hegewald 1971/ 256* (Kernforschungsanlage Jülich, Germany) (typ. cons.).

Antithamnion antillanum Børgesen in Dansk Bot. Ark. 3: 226. 17 Oct 1917 [*Rhodoph.*].
Typus: Virgin Islands, St. Thomas, 12 Dec 1895, *Børgesen 56* (C).

(=)   *Callithamnion lherminieri* P. Crouan & H. Crouan ex Mazé & Schramm, Essai Alg. Guadeloupe, ed. 2: 144. 1878.
Typus (vide Athanasiadis in Bot. Mar. 28: 460. 1985): Guadeloupe, Vieux-Fort, Petite-Fontaine, 29 Feb 1870, *Mazé 1259* (PC).

Coleochaete orbicularis Pringsh. in Jahrb. Wiss. Bot. 2: 35. 1860 [*Chloroph.*].
Typus: [icon in] Pringsheim in Jahrb. Wiss. Bot. 2: t. III, f. 6. 1860.

(=)   *Phyllactidium pulchellum* Kütz., Phycol. General.: 295. 14-16 Sep 1843.
Typus: Germany, Nordhausen, *Kützing* (L).

Coleochaete soluta (Bréb.) Pringsh. in Jahrb. Wiss. Bot. 2: 34. 1860 (C. *scutata* var. *soluta* Bréb. in Ann. Sci. Nat. Bot., ser. 3, 1: 29. 1844) [*Chloroph.*].
Typus: France, Calvados, Falaise, *Brébisson* (L).

(≡)   *Coleochaete prostrata* Kütz., Tab. Phycol. 4: 20. 1854.

Craticula ambigua (Ehrenb.) D. G. Mann, vide *Navicula ambigua*
Craticula cuspidata (Kütz.) D. G. Mann, vide *Frustulia cuspidata*

Cyclotella comta Kütz., Sp. Alg.: 20. 23-24 Jul 1849 [*Bacillarioph.*].
Typus: Germany, Hochsimmer am Rhein, Grunow 1298 (W) (typ. cons.).

421

Cyclotella kurdica Håk. in Diatom
Res. 8: 315. Nov 1993 [*Bacillari-
oph.*].
Typus: Turkey, Araxes area, Hassan
Kaleh, *Ehrenberg 1 bv* (BHUPM)
(typ. cons.).

Frustulia cuspidata Kütz. in Linnaea 8: 549. 1834 [*Bacillarioph.*].
Typus: England, Sussex, Lewes, Sep
1850, *Smith 5b* (BM No. 23449) (typ.
cons.).

Glenodinium elpatiewskyi (Lemmerm.) Schiller, vide *Peridinium elpatiewskyi*

Gomphonema vibrio Ehrenb., Verbr.
Mikrosk. Lebens Amerika: 128. Mai-
Jun 1843 [*Bacillarioph.*].
Typus: Seychelles, Mahe, *Van Heurck*
in Types Syn. Diatom. Belgique No.
213 (BORD) (typ. cons.).

Navicula ambigua Ehrenb., Verbr.
Mikrosk. Lebens Amerika: 129. Mai-
Jun 1843 [*Bacillarioph.*].
Typus: s. loc., Dec 1853, *Smith 25b*
(BM No. 23489) (typ. cons.).

Navicula angulata E. J. Quekett [*Bacillarioph.*], vide sub *Pleurosigma* (p. 164).
Navicula cuspidata (Kütz.) Kütz., vide *Frustulia cuspidata*
Peridiniopsis elpatiewskyi (Lemmerm.) Bourrelly, vide *Peridinium elpatiewskyi*

Peridinium elpatiewskyi Lemmerm.
in Krypt.-Fl. Brandenburg 3: 670.
15 Jun 1910 [*Dinoph.*].
Typus: Germany, Plußsee, 2 Aug
1976, *Meyer 244* (B No. A-36959;
isotypus: Max-Planck-Institut für
Limnologie, Plön, Germany) (typ.
cons.).

Pinnularia gastrum Ehrenb. in Abh. Königl. Akad. Wiss. Berlin, Phys. Kl. 1841: 384, 421. 1843 [*Bacillarioph.*]
Typus: Ireland, Lough Mourne, 1870, *Donkin* (BM) (typ. cons.)

Placoneis gastrum (Ehrenb.) Mereschk., vide *Pinnularia gastrum*

Pleurosigma angulatum (E: J. Quekett) W. Sm., vide *Navicula angulata*

Porphyra purpurea (Roth) C. Agardh, vide *Ulva purpurea*

Scenedesmus quadricauda (Turpin) Bréb., vide *Achnanthes quadricauda*

Synedra nitzschioides Grun. in Verh. K.K. Zool.-Bot. Ges. Wien 12: 403. 1862 (*'nitschioides'*) (orth. cons.) [*Bacillarioph.*].
Typus: Mexico, Baja California, c. 1860, comm. *Weisse* (W Coll. Grunow No. W 2948d) (typ. cons.).

Thalassionema nitzschioides (Grun.) Mereschk., vide *Synedra nitzschioides*

Ulva purpurea Roth, Catal. Bot. 1: 209. Jan-Feb 1797 [*Rhodoph.*].
Typus: Germany, Helgoland, 17 Oct 1996, *Wagner* (BM No. 54930) (typ. cons.).

(=) *Ulva purpureoviolacea* Roth, Tent. Fl. Germ. 1: 524. Feb-Apr 1788. Typus: destr.

Utriculidium durvillei Skottsb. in Wiss. Ergebn. Schwed. Südpolar-Exped. 1901-1903, 4(6): 36. 1907 [*Phaeoph.*].
Typus: Falkland Islands, Stanley Harbour, 18 Aug 1902, *Skottsberg 565* (S) (typ. cons.).

B. FUNGI

Agaricus lycoperdoides Bull., Herb. France 4 (37-48): t. 166. 1784 (*'lycoperdonoides'*) (orth. cons.) (*Asterophora lycoperdoides* (Bull.) Ditmar). Typus: [icon in] Bulliard, Herb. France t. 166. Epitypus (vide Redhead & Seifert in Taxon 50: 279, 2001): on decaying *Russula*, Börje parish, Uppland, Sweden, 24 Oct 1949, *A. Melderis* (DAOM 65245).

(≡) *Asterophora agaricoides* Fr., Symb. Gasteromyc. 1: 3. 1817 : Fr., Elench. Fung. 1: 19. 1828.

(H) *Asterophora lycoperdoides* Fr., Symb. Gasteromyc. 1: 8. 1817 : Fr., Syst. Mycol. 3: 205. 1829. Lectotypus (vide Redhead & Seifert in Taxon 50: 279. 2001): [icon in] Sowerby, Col. Fig. Engl. Fung., t. 383. 1803.

Armillaria matsutake S. Ito & S. Imai in Bot. Mag. (Tokyo) 39: 327. 1925. Typus: [icon in] Kawamura, Ill. Jap. Fungi, t. VIII, fig. 11-12 (as *Cortinellus edodes*). 1913.

(=) *Armillaria nauseosa* A. Blytt in Videnskabs-Selskabets Skr. I, 6: 22. May 1905. Holotypus: [Norway], Akershus, Oslo, Sognsvand, 4 Sep 1887, *A. N. Blytt* (O).

Aspergillus nidulans (Eidam) G. Winter, vide *Sterigmatocystis nidulans*

Aspergillus niger Tiegh. in Ann. Sci. Nat., Bot., ser. 5, 8: 240. Oct 1867. Typus: [specimen] (CBS No. 554.65; isotypus: IMI No. 50566) (typ. cons.).

(=) *Ustilago phoenicis* Corda, Icon. Fung. 4: 9. 1840 (*Aspergillus phoenicis* (Corda) Thom). Typus: non designatus.

(=) *Ustilago ficuum* Reichardt in Verh. K.K. Zool.-Bot. Ges. Wien 17, Abh.: 335. 1867 (*Aspergillus ficuum* (Reichardt) Thom & Church). Typus: [specimen] (IMI No. 91881).

Aspicilia calcarea (L.) Körb., vide *Lichen calcareus*

Baeomyces bacillaris Ach., Methodus: 329. 1803. Typus: England, Durham, Cleveland, Ayton Moor, *W. Mudd* in Mudd, Monogr. Brit. Cladon. [exs.] No 70 (BM; isotypi: FH, UPS) (typ. cons.).

424

Baeomyces byssoides (L.) Pers., vide *Lichen byssoides*

Biatora vernalis (L.) Fr., vide *Lichen vernalis*

Bryoria chalybeiformis (L.) Brodo & D. Hawksw., vide *Lichen chalybeiformis*

Cantharellus lutescens Fr. : Fr., Syst. Mycol. 1: 320. 1821.
Typus: Sweden, Uppland, Silva Nosten, close to Läbyvad (near Uppsala), among mosses in boggy coniferous wood, 16 Sep 1932, *Lundell & Nannfeldt* in Fungi Exs. Suec. No 42 (UPS F-10762; isotypi: BPI, C, K, LE, PC, PRM, S, W) (typ. cons.).

Cantharellus tubaeformis Fr. : Fr., Syst. Mycol. 1: 319. 1821.
Typus: Sweden, Uppland, Silva Nosten, close to Läbyvad (near Uppsala), among mosses in boggy coniferous wood, 16 Sep 1932, *Lundell & Nannfeldt* in Fungi Exs. Suec. No 43 (UPS F-10763; isotypi: BPI, C, K, LE, PC, PRM, S, W) (typ. cons.).

Cenomyce coniocraea Flörke, Deutsche Lich. 7: 14. 1821.
Typus: Sweden, Närke, Svennevad, Korsmon, 1950, *G. Kjellmert* in Magnusson, Lich. Sel. Scand. Exs. No 388 (UPS; isotypi: B, H, US) (typ. cons.).

Cenomyce polydactyla Flörke, Deutsche Lich. 10: 13. 1821.
Typus: [Germany, Mecklenburg-Vorpommern], Rostock, *H. G. Flörke* in Flörke, Deutsche Lich. No 195A (UPS).

(=) *Lichen ventricosus* Huds., Fl. Angl.: 458. 1762.
Lectotypus (vide Ahti & DePriest in Taxon 54: 184. 2005): [icon in] Dillenius, Hist. Musc.: t. 15, f. 17B. 1742. Epitypus (vide Ahti and DePriest in Taxon 54: 184. 2005): Herb. Dillenius No. 94.17 (OXF).

(=) *Lichen difformis* Huds., Fl. Angl.: 458. 1762.
Lectotypus (vide Ahti & DePriest in Taxon 54: 185. 2005): [icon in] Dillenius, Hist. Musc.: t.15, f. 18. 1742. Epitypus (vide Ahti & Depriest in Taxon 54: 185. 2005): Herb. Dillenius No. 94.17B (OXF).

(=) *Cenomyce conglomerata* Dufour, Rév. Clad.: 25. Mai 1821.
Lectotypus (vide Ahti & DePriest in Taxon 54: 185. 2005): France, *J.-M. Dufour* (PC-Lenormand).

Cenomyce stellaris Opiz, Böh. Phan. Crypt. Gew.: 141. 1823 (*Lichen rangiferinus* var. *alpestris* L., Sp. Pl.: 1153. 1 Mai 1753).
Typus: Herb. Dillenius No. 107.29E, right-hand side specimen (OXF) (typ. cons.).

Chaetosphaeria myriocarpa (Fr. : Fr.) Booth, vide *Sphaeria myriocarpa*

Cladonia digitata (L.) Hoffm., vide *Lichen digitatus*

Cladonia macilenta Hoffm., Deutschl. Fl. 2: 126. 1796.
Typus: Germany, Niedersachsen, Oldenburg, Litteler Fuhrenkamp, 1919, *H. Sandstede* in Sandstede, Cladon. Exs. No. 477 (UPS; isotypi: FH, H, TNS (typ. cons.).

Cladonia ochrochlora Flörke, De Cladon.: 75. Jul 1828.
Typus: Germany, Niedersachsen, Oldenburg, Oldenburger Sand, 1918, *H. Sandstede* in Sandstede, Cladon. Exs. No. 241 (UPS; isotypi: FH, H, MIN, TUR-V No. 19413, US-Evans) (typ. cons.).

(=)   *Cenomyce carneopallida* (Flörke) Sommerf. Suppl. Fl. Lapp.: 129. 1826 (*Capitularia pyxidata* var. *carneopallida* Flörke, Beitr. Naturk. 2: 281. Sep 1810).
Lectotypus (vide Ahti, Fl. Neotrop. Mon. 78: 127. 2000): Germany, Harrz, *H. G. Flörke 17* (H-ACH No. 1706A).

Cladonia rangiformis Hoffm., Deutschl. Fl. 2: 114. 1796.
Typus: [Germany, Niedersachsen], "Auf begrastem Heideboden bei Wenden, Hannover, 1921 Okt., leg. Sandstede", Sandstede, Cladoniae exsiccatae No. 803 (H; isotypus: UPS) (typ. cons.).

Cladonia transcendens (Vain.) Vain. in Hue in Nouv. Arch. Hist. Mus. Nat., sér. 3, 10: 262. 1898 (*Cladonia corallifera* var. *transcendens* Vain. in Acta Soc. Fauna Fl. Fenn. 4: 179. Dec 1887).
Typus: Canada, British Columbia, Queen Charlotte Islands, Graham Island, McClinton Bay, 1967, *I. M. Brodo 13003* (CANL; isotypus: H) (typ. cons.).

Cladonia uncialis (L.) F. H. Wigg., vide *Lichen uncialis*
Collema cristatum (L.) F. H. Wigg., vide *Lichen cristatus*

Collema phyllocarpum Pers. in Gaudichaud, Voy. Uranie, Bot.: 204. 1827.
Typus: Brazil, Rio Grande do Sul, Serra dos Vallos per Cruz Alta, in arbore solitaria ripae rivuli, 21 Apr. 1893, *G. A. Malme 1265* (S) (typ. cons.).

Cryptococcus gattii (Vanbreus. & Takashio) Kwon-Chung & Boekhout in Taxon 51: 806. 2002. (*C. neoformans* var. *gattii* Vanbreus. & Takashio in Ann. Soc. Belg. Méd. Trop. 50: 701. 1970).
Typus: [lyophilized culture] from spinal fluid of *Homo sapiens* L.; Zaïre (Mycological Department, Institut de Médecine Tropicale, Antwerp No RV 20186, now at BCCM/IHEM).

(=) *Cryptococcus hondurianus* Castell. in Med. Press Circ. 136: 440. 1933. Neotypus (vide Kwon-Chung & al. in Taxon 51: 805. 2002): [lyophilized culture] from skin of a patient with blastomycosis, *A. Castellani* (ATCC 14248; iso-neotypus: CBS No 883).

(=) *Cryptococcus bacillisporus* Kwon-Chung & J. E. Bennett in Int. J. Syst. Bacteriol. 28: 618. 1978. Holotypus: [lyophilized culture] ex cerebral spinal fluid, from a patient from the San Fernando Valley, California, U.S.A. in the Veteran's Administration hospital, Los Angeles, isolated by M. Huppert prior to 1971 (ATCC No 32608; isotypi: CBS No 6955, NIH No 191).

Cryptococcus neoformans (San Felice) Vuill., vide *Saccharomyces neoformans*

Fusarium sambucinum Fuckel in Jahrb. Nassauischen Vereins Naturk. 23-24: 167. 1870.
Typus: Fuckel, Fungi Rhen. No. 211 (G).

(=) *Fusarium roseum* Link in Ges. Naturf. Freunde Berlin Mag. Neuesten Entdeck. Gesammten Naturk. 3: 10. 1809.
Typus: "*Fusarium roseum* Link" [manu Link] (B).

(=) *Fusarium sulphureum* Schltdl., Fl. Berol. 2: 139. 1 Jun - 15 Aug 1824 : Fr., Syst. Mycol. 3: 471. 1832.
Typus: Germany, Berlin, "in tuberis vetustis *Solani tuberosi*", 1820 (HAL).

(=) *Fusarium maydis* Kalchbr. in Math. Term. Közlem. 3: 285. 1865.
Typus: non designatus.

(=) *Fusisporium ricini* Berenger in Mem. Accad. Agric. Verona 44: 257. 1865.
Typus: non designatus.

(=) *Fusarium subcarneum* P. Crouan, Fl. Finistère: 14. 1867.
Typus: non designatus.

Gyalecta suaveolens Fr., Syst. Orb. Veg. 1: 285. Dec 1825.
Typus: *"Aspicilia chrysophana"*, Sudeten, *Körber 12* ex Typenherb. Körber (L) (typ. cons.).

Helminthosporium avenae Eidam in Lanwirth (Breslau) 27: 509. 1891. Typus: [Italy, Tuscany], *"Helminthosporium teres* f. *avenae sativae*, Dintorni di Pavia"* summer 1889, *Briosi & Cavara*, Fungi Parass. No. 80 on *Avena sativa* leaves (DAOM) (typ. cons.).

(=) *Helminthosporium avenaceum* M. A. Curtis ex Cooke in Grevillea 17: 67. 8 Mar 1889.
Typus: [U.S.A., N. Carol.], "Hillsborough", *Curtis 6515* (FH).

Ionaspis suaveolens (Fr.) Th. Fr., vide *Gyalecta suaveolens*

Lecanora subimmergens Vain. in Bot. Mag. (Tokyo) 35: 51 (1921).
Typus: Japan, Prov. Kozuke, 19.2.1918, *A. Yasuda 355* (TUR-V 6093; isotypus: BM).

(=) *Lecanora argillaceofusca* Müll. Arg. in Nuov. Giorn. Bot. Ital. 21: 358. 1889.
Typus: Brazil, Prov. Rio de Janeiro, 1889. *A. Glaziou* (G).

Lecidea euphorea (Flörke) Nyl. in Mém. Soc. Sci. Nat. Cherbourg 5: 126. 24-30 Mar 1858 (*Lecidea sabuletorum* var. *euphorea* Flörke in Ges. Naturf. Freunde Berlin Mag. Neuesten Entdeck. Gesammten Naturk. 3: 311. 1808).
Typus: [Austria, Salzburg], "Lungau, Weg von Mauterndorf auf den Moserkopf, knapp N von Stampfl", 21 Sep 1985, *Wittmann* (SZU No. 4161) (typ. cons.).

Lecidea pulveracea (Schaer.) Th. Fr., Lichenogr. Scand.: 549. 1874 (*Lecidea enteroleuca* var. *pulveracea* Schaer., Enum. Crit. Lich. Eur.: 128. Aug-Sep 1850).
Typus: *"Lecidea elaeochroma* var. *pulveracea*(Flk.)"*, Flotow, Lich. Exs. No 102A (REG).

(≡) *Lepra cyanescens* Rabenh., Deutschl. Krypt.-Fl. 2(1): 3. Mar 1845.

429

Lichen byssoides L., Syst. Nat., ed.
12, 2: 709; Mant. Pl.: 133. 15-31
Oct 1767.
Typus: Herb. Linnaeus No. 1273.2
(LINN) (typ. cons.).

Lichen calcareus L., Sp. Pl.: 1140.
1 Mai 1753.
Typus: Sweden, Gotland, Visby, 26
Jun 1918, *Malme* in Lich. Suec. Exs.
No. 772 (UPS) (typ. cons.).

Lichen calicaris L., Sp. Pl.: 1146.
1 Mai 1753.
Typus: [specimen] Herb. Dillenius,
t. 23, No. 62B (OXF) (typ. cons.).

Lichen chalybeiformis L., Sp. Pl.:
1155. 1 Mai 1753.
Typus: Herb. Linnaeus No. 1273.291
(LINN) (typ. cons.).

Lichen cristatus L., Sp. Pl.: 1152.
1 Mai 1753.
Typus: Italy, Trentino, Cortina d'Am-
pezzo, Pocol, 1948, *Degelius* (UPS)
(typ. cons.).

Lichen cylindricus L., Sp. Pl.: 1144.
1 Mai 1753.
Typus: Sweden, "ad flumen Kama-
jock prope Qvickjock Lapponiae
Lulensis", 1871, *Hellbom & Hellbom* (UPS).(typ. cons.).

Lichen deustus L., Sp. Pl.: 1150.
1 Mai 1753.
Typus: Sweden, Närke, Örebro, *Hellbom* in Rabenhorst, Exs. No. 812
(UPS) (typ. cons.).

Lichen digitatus L., Sp. Pl.: 1152.
1 Mai 1753.
Typus: Sweden, Ostrogothia, *Stenhammar* in Lich. Suec. Exs. No. 195
(UPS) (typ. cons.).

Lichen hirtus L., Sp. Pl.: 1155.
1 Mai 1753.
Typus: Sweden, Fries, Lich. Suec.
Exs. No. 150 (UPS) (typ. cons.).

Lichen juniperinus L., Sp. Pl.:
1147. 1 Mai 1753.
Typus: Sweden, Härjedalen, Storsjö,
Flatruet W of Falkvålen, 2 Aug
1991, *Mattsson 2340* (LD; isotypi:
H, HMAS, LE, LINN, M, O, TNS,
US) (typ. cons.).

Lichen leptaleus Ach., Lichenogr.
Suec. Prodr.: 108. Mai 1799.
Typus: Scotland, Perthshire, Killin,
Crombie in Lich. Brit. Exs. No. 151
(UPS) (typ. cons.).

Lichen olivaceus L., Sp. Pl.: 1143.
1 Mai 1753.
Typus: Sweden, Härjedalen, Fjell-
näs, *Vrang* in Crypt. Exs. Mus. Hist.
Nat. Vindob. No. 3063 (UPS) (typ.
cons.).

Lichen pallescens L., Sp. Pl.: 1142.
1 Mai 1753.
Typus: Sweden, Härjedalen, Ram-
undberget, NE of Kvarbäckstjärn, 27
Jun 1973, *Santesson 24384* (UPS)
(typ. cons.).

Lichen proboscideus L., Sp. Pl.:
1150. 1 Mai 1753.
Typus: Sweden, Uppland, Boo, Värm-
dö, Skepparholmen 1906, *Malme* in
Lich. Suec. Exs. No. 56 (UPS) (typ.
cons.).

Lichen tartareus L., Sp. Pl.: 1141.
1 Mai 1753.
Typus: Herb. Linnaeus No. 1273.31
(LINN) (typ. cons.).

Lichen tenellus Scop., Fl. Carniol.,
ed. 2, 2: 394. Jan-Aug 1772.
Typus: Czech Republic, Moravia,
Mor. Herálec, 7 Aug 1942, *Nád-vorník* in Physc. Exs. No. 8 (UPS)
(typ. cons.).

Lichen uncialis L., Sp. Pl.: 1151.
1 Mai 1753.
Typus: Sweden, Dalarna, Stora Kop-parberg, Rotneby, *Stenhammar* in
Lich. Suec. Exs., ed. 2, No. 210
(UPS) (typ. cons.).

Lichen vernalis L., Syst. Nat., ed.
12, 3: 234. Dec 1768.
Typus: Sweden, Fries, Lich. Suec.
Exs. No. 224 (UPS) (typ. cons.).

Ochrolechia pallescens (L.) A. Massal., vide *Lichen pallescens*

Ochrolechia tartarea (L.) A. Massal., vide *Lichen tartareus*

Parmelia olivacea (L.) Ach., vide *Lichen olivaceus*

Penicillium chrysogenum Thom in
U.S.D.A. Bur. Anim. Industr. Bull.
118: 58. 1910.
Typus: [specimen] (IMI No. 24314)
(typ. cons.).

(=) *Penicillium griseoroseum* Dierckx
in Ann. Soc. Sci. Bruxelles 25: 86.
1901.
Neotypus (vide Pitt, Genus Penicillium: 249. 1980): [specimen] (IMI No. 92220).

(=) *Penicillium citreoroseum* Dierckx in
Ann. Soc. Sci. Bruxelles 25: 86.
1901.
Neotypus (vide Pitt, Genus Penicillium: 250. 1980): [specimen] (NRRL No. 889).

(=) *Penicillium brunneorubrum* Dierckx
in Ann. Soc. Sci. Bruxelles 25: 88.
1901.
Neotypus (vide Pitt, Genus Penicillium: 250. 1980): [specimen] (IMI No. 92198).

Peronospora lunariae Gäum. in Beih. Bot. Centralbl. 35(1): 526. 1918. Typus: Switzerland, Canton Bern, Taubenloch Pass near Biel, on *Lunaria rediviva* L., 17 Jul 1915, *E. Gäumann* (BERN).

(=) *Peronospora senecionis* Fuckel in Jahrb. Nassauischen Vereins Naturk. 23-24: 69. 1870. Typus: Germany: Oberbayern, near Hohenschwangau, summer [pre 1870] on "*Senecio cordatus* Koch" [*Lunaria rediviva*] [? L. Fuckel] (G).

Phaffia rhodozyma M. W. Mill. & al. in Int. J. Syst. Bacteriol. 26: 287-288. 1976. Typus: [lyophilized culture ex] CBS 5905.

(=) *Rhodomyces dendrorhous* F. Ludw. in Centralbl. Bakteriol. Parasitenk. 10: 13. 1891. Typus: non designatus.

Phoma betae A. B. Frank in Z. Vereins Rübenzuckerindustr. Deutsch. Reiches 42: 904, t. 20. Dec 1892. Typus: [Canada, B.C.], "*Phoma betae* on *Beta vulgaris* (sugar beet), N. Saanich", Sep 1938, *Jones* (DAOM No. 118567) (typ. cons.).

(=) *Phyllosticta tabifica* Prill. in Bull. Soc. Mycol. France 7: 19, t.3, f. 1a-c. 1891. Lectotypus (vide Shoemaker & Redhead in Taxon 48: 381. 1999): [icon in] Bull. Soc. Mycol. France 7: t. 3, f. 1a-c. 1891.

Physcia adscendens H. Olivier, Fl. Lich. Orne 1: 79. Mar-Apr 1882. Typus: Sweden, *Acharius* (H-ACH No. 1428) (typ. cons.).

Physcia leptalea (Ach.) DC., vide *Lichen leptaleus*

Physcia tenella (Scop.) DC., vide *Lichen tenellus*

Pleospora tritici-repentis Died. in Centralbl. Bakteriol., 2. Abth., 11: 56. Sep 1903. Typus: "auf den überwinterten Blättern von *Triticum repens*", 7 May 1901, *Diedicke* (JE).

(=) *Sphaeria sarcocystis* Berk. & M. A. Curtis in Grevillea 4: 152. Jun 1876. Typus: [U.S.A.], "Carolina", [*Curtis*] *6358 in N. Amer. Fungi No. 961*, "on wheat" (BM).

Podospora fimiseda (Ces. & De Not.) Niessl., vide *Sordaria fimiseda*

Pyrenophora avenae S. Ito & Kurib.
in Proc. Imp. Acad. Japan 6: 354. Oct
1930.
Typus: "On *Avena sativa* L. and *Avena fatua* L. half rotten straw, grains
or stubble" (SAP).

(=) *Pyrenophora chaetomioides* Speg. in
Anales Mus. Nac. Buenos Aires 6:
285. Apr 1899.
Lectotypus (vide Shoemaker & Redhead in Taxon 48: 383. 1999): [Argentina], "s[obre] graminea, La Plata", Aug 1888, *Spegazzini* (LPS No.
2114; isolectotypus: DAOM No.
70588b).

Ramalina calicaris (L.) Fr., vide *Lichen calicaris*

Rhizoctonia solani J. G. Kühn,
Krankh. Kulturgew.: 224. 1858.
Typus: [dried culture] CBS 239.95
(typ. cons.).

(=) *Rhizoctonia napae* Westend. & Wallays, Herb. Crypt.: No 225. 1846
(*'napaeae'*).
Typus: *Westendorp & Wallays* Herb.
Crypt. No 225 (BR).

Rinodina cacuminum (Th. Fr.) Malme in Bot. Not. 1896: 176. 18896 (*R.
sophodes* f. *cacuminum* Th. Fr., Lichenogr. Scand.: 201. 1871).
Typus: "*Rinodina sophodes* β *milvina* f. *cacuminum* Th. Fr., Norge,
Dovre, Harbakken", 1863, *T. M.
Fries* (UPS).

(H) *Rinodina cacuminum* (A. Massal.)
Anzi, Lich. Rar. Ven. 2: No. 48.
1863 (*Diploica cacuminum* A. Massal., Symm. lich. nov.: 52. 1855).
Lectotypus (vide Laundon in Lichenologist 24: 345. 1992): "*Diploicia
cacuminum*", Italy, *Anzi*, Lich. Rar.
Ven., No. 48 (UPS).

Saccharomyces neoformans San Felice, vide sub *Cryptococcus* (p. 200).
Sordaria fimiseda Ces. & De Not., vide sub *Podospora* (p. 216).

Sphaeria myriocarpa Fr. in Kongl.
Vetensk. Acad. Handl. 1817 (pars
post.): 267. 1817 (post 1 Jul) : Fr.,
Syst. Mycol. 2: 459. 1823.
Typus: Netherlands, near Baarn,
Landgoed Pijnenburg, 19 Aug 1972,
Gams ([dried culture ex] CBS No.
264.76) (typ. cons.).

Sterigmatocystis nidulans Eidam in
Beitr. Biol. Pflanzen 3: 392. 1883.
Typus: [specimen] (IMI No. 86806,
stat. anamorph.) (typ. cons.).

Umbilicaria cylindrica (L.) Delise ex Duby, vide *Lichen cylindricus*
Umbilicaria deusta (L.) Baumg., vide *Lichen deustus*
Umbilicaria proboscidea (L.) Schrad., vide *Lichen proboscideus*
Usnea hirta (L.) F. H. Wigg., vide *Lichen hirtus*

Ustilago hordei (Pers. : Pers.) La-gerh. in Mitt. Bad. Bot. Vereins 59: 70. 1889 (*Uredo segetum* var. *hordei* Pers., Tent. Disp. Meth. Fung.: 57. 14 Oct - 31 Dec 1797 : Pers., Syn. Meth. Fung.: 224. 31 Dec 1801). Typus: *"Uredo Ustilago* var. *Hordei",* ex herb. Persoon, specimen med. (in hordeo) (L No. 910264-12).

(H) *Ustilago hordei* Bref. in Nachr. Klub Landw. Berlin 221: 1593. 1888. Typus: non designatus.

Ustilago scitaminea Syd. in Ann. Mycol. 22: 281. 1924. Typus: India, Bengal, Bhagalpur, [on *Saccharum officinarum* L.] 26.VIII.1907, *E. J. Butler* in Syd., Ust. 384 (as *Ustilago sacchari*) (Herb. Ustilaginales Vánky).

(=) *Ustilago amadelpha* Syd. & al. in Ann. Mycol. 10: 249. 1912. Holotypus: India, Bengal Muzaffar-pur District, Awapur, in paniculis et apice culmorum Andropogonii spec. [prob. misident.] 15.4.1911, *E. J. Butler 1425* (HCLO).

Vulpicida juniperina (L.) Mattsson & M. J. Lai, vide *Lichen juniperinus*

C. BRYOPHYTA

Anthoceros agrestis Paton, J. Bryol. 10: 257. 1979 [*Hepat.*]. Typus: England, arable field near Oxton, NE of Nottingham, 22 Oct 1965, *J. A. Paton 3004* (BM).

(≡) *Anthoceros nagasakiensis* Steph., Sp. Hepat. 5: 1005. 1916. Typus: Japan, Nagasaki, *Faurie 308* (G).

Arctoa hyperborea (Dicks.) Bruch & Schimp., vide *Bryum hyperboreum*
Brachythecium plumosum (Hedw.) Schimp., vide *Hypnum plumosum*
Brachythecium salebrosum (Hoffm. ex F. Weber & D. Mohr) Schimp., vide *Hypnum salebrosum*

Bryum hyperboreum Dicks., Fasc.
Pl. Crypt. Brit. 4: [29]. 4 Oct 1801
[*Musci*].
Typus: [Norway] "unterhalb Kongs-
vold Dovrefjeld", 23 Jul 1843, *Schim-
per* (BM, herb. Schimper) (typ. cons.).

Calliergon megalophyllum Mikut.,   (≡)   *Hypnum moldavicum* Velen., Rozpr.
Bryoth. Balt. No. 141. 28 Sep 1908   České Akad. Císaře Františka Josefa
[*Musci*].   Vědy, Tř. 2, Vědy Math. Přír.
Typus: [Latvia] "... Mündung eines   12(11): 19. 1903.
kleinen Baches in den Babit-See, am   Typus: [Czech Republic] "Štěcho-
SW-Ufer, ¼ km W von Gesinde   vice, pravý břeh", Apr 1896, ex herb.
Perkoni", 11 Jun [i.e. 24 Jun] 1906,   Velenovský (PRC).
Mikutowicz, Bryoth. Balt. No. 141
(S; isotypi: S, H).

Calypogeia fissa (L.) Raddi, vide *Mnium fissum*

Cnestrum schisti (F. Weber & D. Mohr) I. Hagen, vide *Grimmia schisti*

Grimmia crinita Brid., Muscol. Re-   (≡)   *Dicranum phascoideum* P. Beauv.,
cent. Suppl. 1: 95. 20 Apr 1806   Prodr. Aethéogam.: 54. 10 Jan 1805
[*Musci*].   ('*phascoidum*').
Typus: [Switzerland] Avril 02. *Ro-   Lectotypus (vide J. M. Muñoz in
ger* (B, herb. Bridel No. 119).   Taxon 49: 289. 2000): [France]
                                             "Herbier Palisot de Beauvois. Type
                                             de *Dicranum ? phascoideum*. Sur
                                             les murs du bois de Boulogne.
                                             Echantillon original provenant de
                                             l'herbier Desportes." (G; isolecto-
                                             typus: PC-P.Beauv.).

Grimmia schisti F. Weber & D.
Mohr, Index Mus. Pl. Crypt.: fol. 2,
recto. Aug-Dec 1803 [*Musci*].
Typus: "*Weissia schisti*, e Lappo-
nia", 1802, *Wahlenberg* per Weber
& Mohr 1804 (B, herb. Bridel No.
277) (typ. cons.).

Gymnomitrion concinnatum (Lightf.) Corda, vide *Jungermannia concinnata*

Hypnum plumosum Hedw., Sp.
Musc. Frond.: 257. 1 Jan 1801 [*Musci*].
Typus: "Österreich, Ost-Steiermark:
Waldbach am Nordfuss des Ringkogels bei Hartberg", c. 600 m, Feb
1843, *Baumgartner* in Crypt. Exs.
Mus. Hist. Nat. Vindob. No. 3733
(S) (typ. cons.).

Hypnum salebrosum Hoffm. ex F.
Weber & D. Mohr, Index Mus. Pl.
Crypt.: fol. 2, verso. Aug-Dec 1803
[*Musci*].
Typus: Austria, "bei Kremsmünster
in Oberösterreich", 2 Nov 1859,
Bötsch in Rabenhorst, Bryoth. Eur.
No. 350, ex herb. Milde (S) (typ.
cons.).

Jungermannia concinnata Lightf.,
Fl. Scot. 2: 786. 1777 [*Hepat.*].
Typus: Scotland, Perthshire, Ben
Lawers, Creag Loistge east cliffs,
ridge at end of low cliffs, on thin
schistose soil over rocks with *Salix
herbacea*, 1030 m, 24 Sep 2003, *D.
G. Long 32138* (E; isotypi: BM, F,
G, H, JE, NY, S) (typ. cons.).

Jungermannia exsecta Schmidel,
Icon. Pl., ed. 2: 241, t. 62. 1796 [*Hepat.*].
Typus: S. loc. (?C. Europe), *Schrader*, Syst. Samml. Crypt. Gew. No.
98 (M; isotypi: G, KIEL, L) (typ.
cons.).

Jungermannia palmata Hedw., Theoria Generat. 87. 1784 [*Hepat.*].
Typus: [icon in] Hedwig l.c. t. XVIII ("XV") f. 93-95. 1784. Epitypus: Germany, Bavaria, "Bernau am Chiemsee; an modernden Baumstümpfen", Oct 1903, *H. Paul* (Schiffner, Hepat. Europ. exs. 1272-a] (JE; dupl.: E, G, H, S, U, W).

(=) *Riccia fruticulosa* O. F. Müll., Fl. Danica 5(15): 5. t. 898, f. 3. 1782. Lectotypus (vide Grolle & So in Bot. J. Linn. Soc. 142: 231. 2003): [icon in] O. F. Müller, Fl. Danica 5(15): 5. t. 898, f. 3. 1782. Epitypus (vide Grolle & So l.c.): Norway, Sogne og Fjordane, Flora District, Endestadbogen, found on a decaying log in a moist spruce forest, 25 Jul 2002, *L. Söderström 2002/164* (C; dupl. BM, JE, TRH).

Mnium fissum L., Sp. Pl.: 1114. 1 Mai 1753 [*Hepat.*].
Typus: [icon in] Micheli, Nov. Pl. Gen.: t. 5, f. 14. 1729 (typ. cons.).

Tritomaria exsecta (Schmidel) Loeske, vide *Jungermannia exsecta*

Tortula solmsii Limpr. in Rabenh. Krypt.-Fl., ed. 2, 4(1): 660. Dec 1888 [*Musci*].
Typus: [Portugal] "San Bartholomeo dos Messines in via [...] ad rupes arenarias", 1866, *Solms-Laubach* (BM: "Flora Lusitanica Algarve"; isotypi: BM "Hampe Herbarium", JE, M).

(≡) *Tortula limbata* Lindb., Öfvers. Förh. Kongl. Svenska Vetensk.-Akad. 21: 238. 21 Aug 1864. Lectotypus (vide M. J. Cano in Taxon 53: 198. 2004): [Italy] "Italia orientalis, terra de Otranto ad muros" *Rabenhorst* (RO).

D. PTERIDOPHYTA

Blechnum vestitum (Blume) Kuhn in Ann. Mus. Bot. Lugduno-Batavum 4: 284. 1869 (*Lomaria vestita* Blume, Enum. Pl. Javae 2: 203. 1828).
Typus:"Crescit in montibus altissimis Javae" *Blume* (L barcode L0051111).

(H) *Blechnum vestitum* T. Moore, Index Fil.: 207. 1857. Typus: non designatus.

Davallia repens (L. f.) Kuhn, Filices Africanae 27. 1868. (*Adiantum repens* L. f., Suppl. Pl. 446. 1782).
Typus: Ile de France [Mauritius], "*Sonnerat* par *Thouin* (*Commerson*) 74" (P; isotypus: L).

(H)   *Davallia repens* (Bory) Desv., Mém. Soc. Linn. Paris 6(3): 314. 1827. (*Dicksonia repens* Bory, Voy. Îles Afrique. 2: 323. 1804).
Typus: Bourbon [Réunion], *Bory s.n.*, (P; isotypi: B, BM)

Nephrolepis cordifolia (L.) C. Presl, vide *Polypodium cordifolium*

Polypodium cordifolium L., Sp. Pl.: 1089. 1 Mai 1753.
Typus: Hispaniola, Dominican Republic, prov. de Azua, San José de Ocoa, slope of Loma del Rancho, 23 Feb 1929, *Ekman H 11627* (K; isotypi: B, LD, S, UPS) (typ. cons.).

E. SPERMATOPHYTA

Acalypha virginica L., Sp. Pl.: 1003. 1 Mai 1753 [*Dicot.: Euphorb.*].
Typus: *Clayton 201* (BM) (typ. cons.).

Achyranthes aspera L. [*Dicot.: Amaranth.*], vide sub *Achyranthes* (p. 287).

Adonis annua L. [*Dicot.: Ranunc.*], vide sub *Adonis* (p. 288).

Aechmea distichantha Lem., Jard. Fleur. 3: ad t. 269. 15 Feb 1853 [*Monocot.: Bromel.*].
Typus: [icon in] Lemaire, Jard. Fleur.: t. 269. 1853.

(≡)   *Billbergia distichostachya* Lem. in Jard. Fleur. 2: 96. Sep 1851 (*'distichostachia'*) (neotyp. des.: Grant in Taxon 45: 547. 1996).

(=)   *Billbergia polystachya* Lindl. & Paxton, Paxt. Fl. Gard. 3: ad t. 80. Mai 1852.
Typus: [icon in] Lindley & Paxton, Paxt. Fl. Gard.: t. 80. Mai 1852.

Aegilops truncialis L. [*Monocot.: Gram.*], vide sub *Aegilops* (p. 251).

Agrostis canina L. [*Monocot.: Gram.*], vide sub *Agrostis* (p. 252).

Aira praecox L. [*Monocot.: Gram.*], vide sub *Aira* (p. 252).

Allasia payos Lour., Fl. Cochinch.: 85. Sep 1790 [*Dicot.: Verben.*].
Typus: Tanzania, Tanga, Jan 1893, *Volkens 1* (BM) (typ. cons.).

Allium ampeloprasum L., Sp. Pl.:
294. 1 Mai 1753 [*Monocot.: Lil.*].
Typus: Holms Isles, *Newton* in Peti-
ver, Hort. Sicc. Angl. (BM-SL 152:
153).

(=) *Allium porrum* L., Sp. Pl.: 295. 1
Mai 1753.
Lectotypus (vide Wilde-Duyfjes in
Taxon 22: 59, 77. 1973): [icon] *"Por-
rum"* in Dodoens, Stirp. Hist. Pempt.,
ed. 2: 688. 1616.

Allium christophii Trautv., Trudy
Imp. S-Petersburgsk. Bot. Sada 9(1):
268. 1884 *('cristophi')* [*Monocot.:
Lil.*].
Typus: "Achalteke?", *Cristoph 7511*,
Smirnow a 1883 (LE) (typ. cons.).

(=) *Allium bodeanum* Regel, Trudy Imp.
S.-Petersbursk. Bot. Sada 3(2): 238.
Jan-Mai 1875.
Typus: [Iran] "Persia". *Bode* (LE).

Allium nigrum L., Sp. Pl., ed. 2, 1:
430. Sep 1762. [*Monocot.: Lil.*].
Typus: "Cyprus, 1-2 km N.E. of
Syso, 10 km S.E. of Polis (Akmas)"
450 m, 14 Apr 1979, *Edmondson &
McClintock, E.2822* (K; isotypus: E)
(typ. cons.).

(=) *Allium magicum* L., Sp. Pl.: 296,
errata. 1 Mai 1753.
Lectotypus (vide Seisums in Taxon
47: 712. 1998): *"Moly latifolium lili-
florum* Bauh.*, Moly Theophrasti* Clus.
Basiliae in horto Heinzmanni" herb.
Burser III: 106 (UPS).

Amaryllis belladonna L. [*Monocot.: Amaryllid.*], vide sub *Amaryllis* (p. 252).

Ammannia octandra L. f., Suppl.
Pl.: 127. Apr 1782 [*Dicot.: Lythr.*].
Typus: India, Madepala, *Koenig* (BM)
(typ. cons.).

Andropogon bicornis L., Sp. Pl.:
1046. 1 Mai 1753 [*Monocot.: Gram.*].
Typus: [Puerto Rico] Mayagüez, between Monte Mesa and the sea, 27
Oct 1913, *A. Chase* in Amer. Gr.
Nat. Herb. No. *247.* (MO) (typ.
cons.).

Andropogon distachyos L. [*Monocot.: Gram.*], vide sub *Andropogon* (p. 253).

Anemone coronaria L. [*Dicot.: Ranunc.*], vide sub *Anemone* (p. 293).

Anemone narcissiflora L., Sp. Pl.:
542. 1 Mai 1753 *('uarcissifolia')*
(orth. cons.) [*Dicot.: Ranuncul.*].
Typus: Herb. Linnaeus No. 710.31
(LINN).

Astragalus garbancillo Cav., Icon. 1: 59, t. 85. Dec 1791 [*Dicot.: Legum.*].
Typus: [icon] *"Astragalus garvanzillo"* in Cavanilles, Icon. t. 85. Dec 1791.

(=) *Astragalus unifultus* L'Her., Stirp. Nov.: 168. Sep 1791.
Typus: non designatus.

Atalantia monophylla (L.) DC., vide *Limonia monophylla*

Atriplex hortensis L. [*Dicot.: Ranunc.*], vide sub *Atriplex* (p. 298).

Bactris gasipaes Kunth in Humb. & al., Nov. Gen. Sp. 1, ed 4°: 302; ed. f°: 242. 1816 [*Monocot.: Palm.*].
Typus: Colombia, Ibagué, *Bonpland* (P).

(=) *Martinezia ciliata* Ruiz & Pav., Syst. Veg. Fl. Peruv. Chil.: 275. Dec 1798 (*Bactris ciliata* (Ruiz & Pav.) Mart.).
Typus: Peru, Pozuzo, *Pavón* (MA).

Baltimora recta L. [*Dicot.: Comp.*], vide sub *Baltimora* (p. 299).

Bejaria aestuans Mutis [*Dicot.: Eric.*], vide sub *Bejaria* (p. 300).

Berlinia polyphylla Harms in Engler & Diels, Veg. Erde 9(3,1): 472. Feb-Sep 1915 [*Dicot.: Legum.*]
Typus: "Gabon: Plantation de Ninghé-Ninghé sur la Bokoué, près du poste de Kango sur le Komo", 6 Oct 1912, *Fleury in Chevalier 26690* (P) (typ. cons.).

Biscutella didyma L. [*Dicot.: Cruc.*]., vide sub *Biscutella* (p. 302).

Bobartia indica L. [*Monocot.: Irid.*], vide sub *Bobartia* (p. 256).

Boerhavia diffusa L., Sp. Pl.: 3. 1 Mai 1753 [*Dicot.: Nyctagin.*].
Typus: Virgin Isl., St. Croix, Teague Bay, West Indes Laboratory, 30 Mai 1977, *Fosberg 56776* (BM) (typ. cons.).

Bombax ceiba L. [*Dicot.: Bombac.*], vide sub *Bombax* (p. 303).

Bouteloua gracilis (Kunth) Lag. ex Griffiths, Contr. U.S. Natl. Herb. 14: 375. 1912 (*Chondrosum gracile* Kunth in Humb. & al., Nov. Gen. Sp. 1, ed. 4°: 176; ed. f°: 142. Mai 1816) [*Monocot.: Gram.*].
Typus: Mexico: "crescit in crepidinibus et devexis montis porphyritici La Buffa de Gaunaxuato Mexicanorum" *Humboldt & Bonpland* (B-W No. 1628; isotypi: ?P, US).

(H) *Bouteloua gracilis* Vasey in Watson & Rothrock, Rep. U.S. Geogr. Surv. 6: 287. 1878.
Typus: U.S.A., Arizona, Riley's Well, Wheeler's Exped., 1874, *J. Rothrock 701* (US).

Bromus secalinus L. [*Monocot.: Gram.*], vide sub *Bromus* (p. 257).

Bromus sterilis L., Sp. Pl.: 77. 1 Mai 1753 [*Monocot.: Gram.*].
Typus: England, Surrey, Tothill, near Hedley, 15 Jun 1932, *Hubbard 9045* in Gram. Brit. Exs. Herb. Kew. Distrib. No. 69 (E; isotypus: K) (typ. cons.).

Bulbostylis lanata (Kunth) Lindm. in Bih. Kongl. Svenska-Acad. Handl. 26(3, 9): 18. 11 Apr 1900 (*Isolepis lanata* Kunth in Humboldt & al., Nov. Gen. Sp. 1, ed. 4°: 220; ed. f°: 177. Mai 1816). [*Monocot.: Cyper.*].
Typus: Venezuela, "propre Atures (Missionibus Orinocensium)", *Humboldt & Bonpland* (P).

(H) *Bulbostylis lanata* DC., Prodr. 7: 268. Apr 1838 [*Dicot.: Comp.*]
Typus: Mexico, "Leon à l'ouest de Gaunasuato", *Mendez* (G-DC).

Bulbostylis pauciflora (Liebm.) C. B. Clarke in Bull. Misc. Inform. Kew, add. ser. 8: 26. Jun 1908 (*Oncostylis pauciflora* Liebm. in Kongel. Dansk Vidensk. Selsk. Naturvidensk. Math. Afn., ser. 2, 2: 241. 1850) [*Monocot.: Cyper.*].
Typus: Ste Croix, 1848 *Oersted* (C; isotypus: NY).

(H) *Bulbostylis pauciflora* (Kunth) DC., Prodr. 5: 139. 1-10 Oct 1836 (*Eupatorium pauciflorum* Kunth in Humboldt & al, Nov. Gen. Sp. 4: ed. f°: 94. 26 Oct 1818) [*Dicot.: Comp.*].
Typus: "Nova Granata?", *Humboldt & Bonpland* (P).

Cactus cruciformis Vell., Fl. Flumin.: 207. 7 Sep - 28 Nov 1829 [*Dicot.: Cact.*].
Typus: [icon in] Vellozo, Fl. Flumin. Icon. 5: t. 29. 1831 (typ. cons.).

(≡)   *Cereus squamulosus* Salm-Dyck ex DC., Prodr. 3: 125. Mar (med.) 1828 (neotyp. des.: Taylor in Taxon 43: 469. 1994).

(≡)   *Cereus tenuispinus* Haw. in Philos. Mag. 1: 125. 1827.
Neotypus (vide Taylor in Taxon 43: 469. 1994): [icon in] Loddiges, Bot. Cab.: t. 1887. 1833.

(≡)   *Cereus myosurus* Salm-Dyck ex DC., Prodr. 3: 469. Mar (med.) 1828.
Neotypus (vide Taylor in Taxon 43: 469. 1994): [unpubl. icon dated 20 Jan 1829, by Bond] "*Cereus myosurus*, received in 1827 from Mr Hitchin of Norwich" (K).

(≡)   *Cereus tenuis* DC., Prodr. 3 469. Mar (med.) 1828 (neotyp. des.: Taylor in Taxon 43: 469. 1994).
≡ *Cereus myosurus* Salm-Dyck ex DC.

Caesalpinia sinensis (Hemsl.) Vidal in Bull. Mus. Natl. Hist. Nat. Paris ser. 3, 395 (Bot. 27): 90. 1976 (*Mezoneuron sinense* Hemsl. in J. Linn. Soc. 23: 204. 1887) [*Dicot.: Legum.*].
Typus: China, Hubei (Hupeh), Yichang (Ichang), *Henry 3113* (K).

(H)   *Caesalpinia chinensis* Roxb., Fl. Ind., ed. 1832, 2: 361. 1832.
Typus: non designatus.

Canna jaegeriana Urban in Repert. Spec. Nov. Regni Veg. 15: 102. 1917 [*Monocot.: Cann.*].
Typus: Haiti, "in locis humidis Montis Nigri, le grand fond et Sourçailla", alt. 300-1200 m., *Jaeger 165* (K; isotypi: LE MO).

(≡)   *Canna leucocarpa* Bouché, Linnaea 18: 493. Jun 1845.
Neotypus (vide Maas van der Kamer in Taxon 53: 835. 2004) [icon in] Engler, Pflanzenr. IV.47 (Heft 56): f. 11E. 1912.

443

Canna tuerckheimii Kraenzl. in Engler, Pflanzenr. IV.47(Heft 56): 70. 22 Oct 1912 [*Monocot.: Cann.*].
Typus: Guatamala, Depto. Alta Verapaz: near Cubilguïtz, 350 m, Dec 1901, *von Tuerckheim II 513* (US; isotypi: G, GH, K).

(=) *Canna latifolia* Mill., Gard. Dict., ed. 8: *Canna* No. 2. 16 Apr 1768.
Neotypus (vide Tanaka in Makinoa, ser. 2, 1: 48. 2001): [icon in] Roscoe, Monandr. Pl. Scitam.: t. 4. 1825.

(=) *Canna gigantea* F. Delaroche in Redouté, Liliac. 6: t. 331. 1811.
Lectotypus (vide Maas van der Kamer in Taxon 53: 833. 2004): [icon in] Redouté, Liliac. t. 133. 1811.

(=) *Canna neglecta* Weinm. in Flora 3: 607. 1820.
Neotypus (vide Maas van der Kamer in Taxon 53: 833. 2004): "*Canna gigantea* Desfont. = *Canna neglecta* Weinm. Bot. Zeit." (LE).

(=) *Canna violacea* Bouché in Linnaea 12: 146. Mar-Jul 1838.
Neotypus (vide Maas van der Kamer in Taxon 53: 833. 2004): [icon in] Kraenzl. in Engler, Pflanzenr. IV.47(Heft 56): f. 12. 1912.

Capparis cartilaginea Decne in Ann. Sci. Nat. Bot., sér 2, 3: 273. 1835 [*Dicot.: Cappar.*].
Typus: Egypt, Sinai Desert, Jun 1832, *Thomson* (K).

(=) *Capparis inermis* Forssk., Fl. Aegypt.-Arab. Ind.: CXIII & 100. 1775.
Neotypus (vide Rivera & al., Taxon 52: 357. 2003): S. Yemen, Jebel Shamsan, Tower of Silence & vicinity, 150 m, 7 Jun 1987, *L. Boulos & al. 15560* (K).

Carduus thomsonii Hook. f., Fl. Brit. India 3: 361. Mar 1881 *('thomsoni')* [*Dicot.: Comp.*].
Typus: [India/Pakistan] "Ladak", alt. 12-13,000 ft. 20 Sep 1848, *Bové* (P, "type ... R.S. Rhagavan 11 Aug 1981"; isotypus: K).

(≡) *Carduus ladak* C. B. Clarke, Compos. Ind.: 215. Sep 1876.

Carex filicina Nees in Wight, Contr. Bot. India: 123. Dec 1834 [*Monocot.: Cyper.*].
Typus: "Carex filicina α N[ees ab] E[senbeck]", peninsula Indiae orientalis, Herb. Wight No. 1916.a/127 ex herb. Arnott (E).

(=) *Cyperus caricinus* D. Don, Prodr. Fl. Nepal.: 39. 26 Jan - 1 Feb 1825. Holotypus: "*Cyperus caricinus* D. Don", Nepal, from D. Don (K).

Carex lachenalii Schuhr, Beschr. Riedgräs.: 51. Apr-Mai 1801 [*Monocot.: Cyper.*].
Typus: "*Carex lagenalii* n. 79", ex herb. Schkuhr (HAL).

(=) *Carex tripartita* All., Fl. Pedem. 2: 265. Apr-Jul 1785.
Lectotypus (vide Turland in Taxon 46: 341. 1997): [icon in] Allioni, Fl. Pedem.: t. 92, f. 5. 1785.

Cassia ligustrina L., Sp. Pl.: 378. 1 Mai 1753 [*Dicot.: Legum.*].
Typus: [icon in] Dillenius, Hort. Eltham.: t. 269, f. 338. 1732, excl. legum. (typ. cons.).

Cassine barbara L., Mant. Pl.: 220. Oct 1771 [*Dicot.: Celastr.*].
Typus: S. Africa, Western Cape, Rhenosterkop, *Schlechter 10574* (PRE) (typ. cons.).

Cassine peragua L. [*Dicot.: Celastr.*], vide sub *Cassine* (p. 312).

Centaurea pumilio L., Cent. Pl. 1: 30. 19 Feb 1755 *('pumilis')* (orth. cons.) [*Dicot.: Comp.*]
Typus: Egypt, Alexandria, in arenosis maritimis ad Sidi-Gaber, 8 Apr 1908, *Bornmüller 10781* (B) (typ. cons.).

Cereus jamacaru DC., Prodr. 3: 467. Mar 1828 [*Dicot.: Cact.*].
Typus: Brazil, Bahia, Mun. Curaçá, north of Barro Vermelho towards Curaçá, caatinga, 395 m, 7 Jan 1991, *Taylor & al. 1369* (CEPC; isotypi: K, ZSS, HRCB) (typ. cons.).

Chelone hirsuta L., Sp. Pl.: 611.
1 Mai 1753 [*Dicot.: Scrophular.*].
Typus: U.S.A., Maryland, Montgomery County, along River Road
0.9 miles west of the junction of
West Willard Road, 9 Jun 1995, *Reveal 7413* (BM), (typ. cons.).

Chondrilla nudicaulis L., Mant. Pl.:
278. Oct 1771 [*Dicot.: Comp.*].
Typus: Egypt, "Kairo, in palmetis
ad El Marg", 27 Apr 1908, *Bornmüller 10830* (JE; isotypi: G, LD,
LE) (typ. cons.).

Chrysanthemum coronarium L.,
Sp. Pl.: 890. 1 Mai 1753 [*Dicot.:
Comp.*].
Typus: Greece, Kriti (Crete) Nomos
Irakliou, Eparhia Kenourgiou, 500
m E of Gangales, E side of road to
Vali (35°03'39"N, 25°00'57"E), 250
m, large field with *Hordeum* crop,
13 Apr 2003, *Kyriakopoulos & Turland sub Turland 1166* (UPA; isotypi: B, BM, MO) (typ. cons.).

Cinchona dichotoma Ruiz & Pav.,
Fl. Peruv. 2: 53. Sep 1799 [*Dicot.:
Rub.*].
Typus: Peru, Huánuco, Maquizapa
on road to Mozón, 780 m, 20 Feb
1966, *Schunke Vigo 1100* (NY; isotypi: F, MO, VEN) (typ. cons.).

Clausena pentaphylla DC., Prodr. 1:
538. Jan (med.) 1824 [*Dicot.: Rut.*].
Typus: India, Uttar Pradesh, "Cawnpore" [Kanpur], *Roxburgh 2484*
(BM) (typ. cons.).

Commelina benghalensis L., Sp. Pl.:
41. 1 Mai 1753 [*Monocot.: Commelin.*].
Typus: Herb. Linnaeus No. 65.16
(LINN) (typ. cons.).

446

Cordia myxa L. [*Dicot.: Borag.*]., vide sub *Cordia* (p. 321).

Coronilla valentina L. [*Dicot.: Legum.*], vide sub *Coronilla* (p. 321).

Corydalis solida (L.) Clairv., vide *Fumaria bulbosa*

Cotyledon orbiculata L. [*Dicot.: Crassul.*]., vide sub *Cotyledon* (p. 322).

Crotalaria lotifolia L. [*Dicot.: Legum.*]., vide sub *Crotalaria* (p. 323).

Cuscuta capitata Roxb., Fl. Ind. 1.: 468. 1820 [*Dicot.: Convolvul.*].
Typus: [India] "Hab. Himal. Bor. Occ. Regio temp. Kunawur" 1800-2400 m, on *Thymus*, *Thomson* (MO; isotypus: K) (typ. cons.).

Cyperus sanguinolentus Vahl, En-um. Pl. 2: 351. Oct-Dec 1805 [*Mo-nocot.: Cyper.*].
Typus: India, Uttar Pradesh, N.W. Himalaya, Distr. Tehri-Garhwál, Oct 1894, *Gamble 15117* (L No. 951.65-240) (typ. cons.).

Dioscorea sativa L. [*Monocot.: Dioscor.*]., vide sub *Dioscorea* (p. 263).

Eremurus spectabilis M. Bieb., Fl. Taur.-Caucas. 3: 269. 1819 [*Mono-cot. Lil.*].
Typus: *Erymnurus* [sic!] *caucasicus*. "In montibus haud procul Sabli pro-mont. Caucasici ad viam publicam quae ... ad pagum Alexandria, 1813" (LE) (typ. cons.).

Erica calycina L., Sp. Pl. ed. 2: 507. 1-20 Sep 1762 [*Dicot.: Eric.*].
Typus: South Africa: Cape Town, 3318CD, Tafelberg, plateau near re-servoir, 28 Nov 1897, *W. Froem-bling 325* (NBG) (typ. cons.).

Erica carnea L., Sp. Pl.: 355. 1 Mai 1753 [*Dicot.: Eric.*].
Typus: [icon] "*Erica Coris folio* IX", Clusius, Rar. Pl. Hist.: 44. 1601.

(=)   *Erica herbacea* L., Sp. Pl.: 352. 1 Mai 1753.
Lectotypus (vide Brickell & McClin-tock in Taxon 36: 480. 1987): [icon] "*Erica Coris folio* VII", Clusius, Rar. Pl. Hist.: 44. 1601.

Erica corifolia L., Sp. Pl.: 355. 1
Sep 1762 [*Dicot.: Eric.*].
Typus: South Africa, Cape District,
Bothasig (Bosmansdam) 3318DC,
27 Oct 1965, *E. Esterhuysen 31332*
(NBG; isotypus: BOL) (typ. cons.).

Erica imbricata L., Sp. Pl. ed. 2:
503. 1-20 Sep 1762 [*Dicot.: Eric.*].
Typus: South Africa, Cape Penisula,
3418AB, Tokai Flats, 1 Oct 1916,
W. Foley 10 (NBG; isotypus: BM)
(typ. cons.).

Erica vagans L., Mant. Pl.: 230. Oct
1771 [*Dicot.: Eric.*].
Typus: U.K., Cornwall, Goonhilly,
12 Jul 1932, *Turrill* (K) (typ. cons.).

Eriocaulon lineare Small, Fl. S.E.
U.S.: ed. 2: 236. 1328. 22 Jul 1903
[*Monocot.: Eriocaul.*].
Typus: USA: Florida, Leon County,
c. 4 miles S of Tallahassee, 15 Aug
1962, *R. Kral & R. K. Godfrey* (NY;
isotypi: DS, F, FSU, GA, GH, ILL,
LAF, LL, MO, MSC, NCU, NCSC,
NO, P, PAC, PH, POM, RSA, SMS,
SMU, TEX, UC, US, VDB) (typ.
cons.).

Eryngium bourgatii Gouan, Ill. Observ. Bot.: 7, t. 3. 1773 [*Dicot.: Umbell.*]
Typus: [icon. in] Gouan, Ill. Observ.
Bot., t. 3. 1773.

(=) *Eryngium pallescens* Mill., Gard.
Dict., ed. 8: *Eryngium* No. 5. 16 Apr
1768 *('pallescente')*.
Lectotypus (vide Feliner in Taxon
50: 585. 2001): "*Eryngium pallescens* Miller. Type specimen. J. Bot.
1913. 132." / "Hort. Habitat in Hispania, Jacquin." *Miller* (excl. right-
hand spec.) (BM).

Eugenia nitida Vell., Fl. Flumin.:
208. 7 Sep - 28 Nov 1829 [*Dicot.:
Myrt.*].
Typus: [Brazil, Rio de Janeiro] *Mikan
1047* (W; isotypi: F, K) (typ. cons.).

Euphorbia falcata L., Sp. Pl.: 456. 1 Mai 1753 [*Dicot.: Euphorb*].
Typus: *Arduino* in Herb. Linnaeus No. 630.26, right-hand plant (LINN) (typ. cons.).

Eyrea rubelliflora F. Muell. in Linnaea 25: 403. 1853 [*Dicot.: Comp..*]
Typus: [W. Australia] Nickol River between Karratha and Roebourne, NW Costal Hwy, 20°48'S, 116°55'E, *Hunger & Killian 3726* (MEL; isotypi: B, BM, CANB, K, Perth) (typ. cons.).

Festuca elmeri Scribn. & Merr., Bull. Torrey Bot. Club 29: 468. 1902 [*Monocot.: Gram.*].
Typus: U.S.A., California, Santa Clara Co., Stanford University, Apr. 1900, *A. D. E. Elmer* 2101 (US).

(=) *Festuca howellii* Hack. ex Beal, Grass. N. Amer. 2: 591. 1887.
Typus: U.S.A., Oregon, Josephine Co., Deer Creek Mts. 5 Jul 1887, *T. Howell 248* (MSC).

Ficus americana Aubl., Hist.. Pl. Guinane 2: 952. Jun-Dec 1775 [*Dicot.: Mor.*].
Typus: [icon in] Plumier, Pl. Amer. (ed. Burman): t. 132, fig. 2. 1757.

(≡) *Ficus perforata* L., Pl. Surin.: 17. 23 Jun 1775.

Ficus crassipes F. M. Bailey, Rep. Pl. Prelim. Gen. Rep. Bot. Meston's Exped. Bellenden-Ker Range: 2. 1889. [*Dicot.: Mor.*].
Typus: Australia, Queensland, Harvey's Creek, *Russell River, Bailey & Meston [s.n.]* (BRI).

(H) *Ficus crassipes* (Heer) Heer, Fl. Foss. Arct. 6(2): 70. 1882. (*Proteoides crassipes* Heer, Fl. Foss. Arct. 3: 110. 1874 [Foss.].
Typus: non designatus.

Ficus maxima Mill., Gard. Dict., ed. 8: *Ficus* No. 6. 16 Apr 1762 (*'maxima'*) [*Dicot.: Mor.*].
Typus: Brazil, Amazonas: mun. Humaitá, nr. Tres Casas, 14 Sep-11 Oct 1934, *Krukoff 6413* (NY; isotypi: A, F, G, K, MO, U (typ. cons.).

Ficus tiliifolia Baker in J., Linn. Soc., Bot. 21: 443. 1885. [*Dicot.: Mor.*]. Typus: Central Madagascar, *Baron 3285* (P).

(H) *Ficus tiliifolia* (A. Braun) Heer, Fl. Tert. Helv. 2: 68. 1856 (*'tiliaefolia'*) (*Cordia tiliifolia* A. Braun, Jahrb. Mineral. Geognosie Geol. Petrefaktenk.: 170. 1845 (*'Cordia? tiliaefolia'*)) [Foss.]. Typus: non designatus.

Ficus tremula Warb. in Bot. Jahrb. Syst. 20: 171. 1894. [*Dicot.: Mor.*]. Typus: Tanzania, "Sansibarküste: Bagamoyo", *Stuhlmann 274* (B).

(H) *Ficus tremula* Heer in Abh. Schweiz. Paläontol. Ges. 1: 11. 1874 [Foss.]. Typus: non designatus.

Fraxinus angustifolia Vahl, Enum. Pl. 1: 52. Jul-Dec 1804 [*Dicot.: Ol.*]. Typus: Spain, *Schousboe* (C).

(=) *Fraxinus rotundifolia* Mill., Gard. Dict., ed. 8: *Fraxinus* No. 2. 16 Apr 1768.
Lectotypus (vide Green in Kew Bull. 40: 131. 1985): [icon] *"Fraxinus rotundiore folio"*, Bauhin, Hist. Pl. Univ. 1(2): 177. 1650.

Fumaria bulbosa L., nom. utique rej. [*Dicot.: Papaver.*], vide sub *Corydalis* (p. 322).

Galanthus elwesii Hook. f. in Bot. Mag.: ad t. 6166. 1 Mai 1875 [*Monocot.: Amaryllid.*]. Typus: Turkey, Adana Prov., north part of Giaour dağ, 600 m, 18 Mai 1879, *Danford* (K) (typ. cons.).

Gilia splendens Douglas ex H. Mason & A. D. Grant, Madroño 9: 212. 1948 [*Dicot.: Polemon.*] Typus: [U.S.A.] California, Monterey Co., Santa Lucia Mts., Tassajara Hot Springs, alt. 1530 ft., 26 Apr 1933, *Roxana S. Ferris 8317* (UC No. 524203; isotypi: DS, GH), (typ. cons.).

Glycine umbellata Muhl. ex Willd., Sp. Pl. 3: 1058. 1-10 Nov 1802 [*Dicot.: Leg.*].
Typus: U.S.A., S. Carolina, Georgetown County, E. side of Waccamaw River, 7 mi. NE of Georgetown, 12 Sep 1996, *B. Seckinger 406* (USCH: isotypi: MEXU, MONT), (typ. cons.).

Gomphrena ficoidea L., Sp. Pl.: 225. 1 Mai 1753 [*Dicot.: Amaranth.*].
Typus: Herb. Linnaeus No. 290.23 (LINN) (typ. cons.).

Grewia mollis Juss. in Ann. Mus. Natl. Hist. Nat. 4: 91. 1804 [*Dicot.: Til.*].
Typus: Nigeria, Nupe, *Barter 1097* (K) (typ. cons.).

Hedysarum cornutum L., Sp. Pl., ed. 2: 1060. Jul 1763 [*Dicot.: Legum.*].
Typus: *Gérard 18* in Herb. Linnaeus No. 921.71 (LINN).

 (≡) *Hedysarum spinosum* L., Syst. Nat., ed. 10: 1171. Mai-Jun 1759.

Hedyotis fruticosa L. [*Dicot.: Rub.*], vide sub *Hedyotis* (p. 343).

Hibiscus sabdariffa L., Sp. Pl.: 695. 1 Mai 1753 [*Dicot.: Malv.*].
Typus: Herb. Clifford: p. 350, *Hibiscus 6* (BM) (typ. cons.).

Hieracium gronovii L., Sp. Pl.: 802. 1 Mai 1753 [*Dicot.: Comp.*].
Typus: U.S.A., Virginia, *Clayton 447* (BM) (typ. cons.).

Hypericum ×desetangsii Lamotte in Bull. Soc. Bot. France 21: 121. 1874 [*Dicot.: Gutt.*].
Typus: France, Aisne, Villers-Cotterets, 17 Aug 1879, *Bonnet & Delacour* in exsicc. Soc. Dauphinoise 1880, No. 2409 (P) (typ. cons.).

Inga marginata Willd., Sp. Pl. 4: 1015. Apr 1806. [*Dicot.: Legum.*]. Typus: Venezuela, Caracas, *Bredemeyer 5* (B-W No. 19031) (typ. cons.).

Inula verbascifolia (Willd.) Hausskn. in Mitth. Thüring. Bot. Vereins, ser. 2, 7: 32. 28 Mai 1895 (*Conyza verbascifolia* Willd., Sp. Pl. 3(3): 1924. 19 Mar 1803) [*Dicot.: Comp.*]. Typus: [Croatia] Flora Hungarica Exsiccata [Cent. III, Angiosp. 166] No. 256: Comit. Lika-Krbava, Montes Velebit. ... 2 Jul 1911, *F. Filaszky & J. B. Kümmerle* [*s.n.*], sub "*Inula candida* (L.) Cass." (B; isotypi: BP, BR, C, G, GB, GZU, K, M, MO, S, SOM, US, W, Z, ZT) (typ. cons.).

(H) *Inula verbascifolia* Poir., Encycl. Méth. Bot. Suppl. 3: 154. 1813. Typus: "... originaire du Caucase ... cultive au Jardin des Plantes de Paris (P-herb. Desfon.)" [deest].

Ipomoea discolor (Kunth) G. Don, Gen. Hist. 4: 270. 1838 (*Convolvulus discolor* Kunth in Humboldt & al., Nov. Gen. Sp.: t. 212. Sep 1818 [*Dicot.: Convolv.*]. Typus: Venezuela, Amazonas, banks of the Orinoco River, near Carichana, *Humboldt & Bonpland 1045* (P).

(H) *Ipomoea discolor* Jacq., Pl. Hort. Schoenbr. 3: 6. 1798. Typus: [icon in] Jacquin, Pl. Hort. Schoenbr.: t. 261. 1798.

Jatropha peragua L. [*Dicot.: Euphorb.*], vide sub *Jatropha* (p. 350).

Joosia dichotoma (Ruiz & Pav.) H. Karst., vide *Cinchona dichotoma*

Juncus arcuatus Wahlenb., Fl. Lapp.: 87, t. 4. 1812 [*Monocot.: Junc.*]. Typus: Sweden, Piteå Lappmark på Örfjället, 6 Sep 1807, *Wahlenberg* (UPS) (typ. cons.).

Juncus debilis A Gray, Man. Bot.: 506. 1848 [*Monocot.: Junc.*]. Typus: U.S.A., Kentucky, Crab Orchard, 1 Aug 1835, *Short* (P).

(=) *Juncus radicans* Schlechtend., Linnaea 18: 442. Jun 1845. Lectotypus (vide Balslev in Fl. Neotrop. Monogr, 68: 118. 1996): Mexico, Veracruz, in rivulis Jalapa, *Schiede* (HAL).

Lantana camara L. [*Dicot.: Solan.*], vide sub *Lantana* (p. 352).

Launaea nudicaulis (L.) Murray, vide *Chondrilla nudicaulis*

Lepismium cruciforme (Vell.) Miq., vide *Cactus cruciformis*

Limonia monophylla L. [*Dicot.: Rut.*], vide sub *Atalantia* (p. 298).

Ludwigia repens J. R. Forst., Fl. Amer. Sept.: 6. 1771 [*Dicot.: Onagr.*]. Typus: [U.S.A.], Florida, near Jacksonville, Duval Co., 19 Mai 1894, *A. H. Curtiss 4836* (MO; isotypi: FLAS, GA, GH, MSC, NA, NY, P, W) (typ. cons.).

Luzula nodulosa E. Mey., Linnaea 22: 410. 1849 [*Monocot.: Junc.*]. Typus: [Greece, Pylos, Arcadia], *Chaubert* (P).

(≡) *Juncus graecus* Chaub. & Bory, Nouv. Fl. Pélop.: 23, t. 12, fig. 1. 1838.

Lycopersicon esculentum Mill., Gard. Dict., ed. 8: *Lycopersicon* No. 1. 16 Apr 1768 (*Solanum lycopersicum* L., Sp. Pl.: 185. 1 Mai 1753) [*Dicot.: Solan.*]. Typus: Herb. Linnaeus No. 248.16 (LINN).

(≡) *Lycopersicon lycopersicum* (L.) H. Karsten, Deut. Fl.: 966. Mai 1882 (*'Lycopersicum lycopersicum'*).

Maerua crassifolia Forssk., Fl. Aegypt.-Arab.: 104. 1 Oct 1775 [*Dicot.: Cappar.*]. Typus: Yemen, between Watadah and Sirwah, *Wood 3153* (BM) (typ. cons.).

Magnolia kobus DC., Syst. Nat. 1: 456. 1-15 Nov 1817 [*Dicot.: Magnol.*].
Typus: [icon in] Kaempfer, Icon. Sel. Pl.: t. 42. 1791 (typ. cons.).

Matricaria recutita L. [*Dicot.: Comp.*], vide sub *Matricaria* (p. 362).

Melica transsilvanica Schur, Enum. Pl. Transsilv. 764. Apr-Jun 1866 [*Monocot.: Gram.*].
Typus: [Germany] (Siebenburgen): Auf Wiesen um Hermannstadt in Grasgärten ("*Melica caespitosa* Schur"), 6 Aug 1850, *Schur* (WU).

(=) *Melica caricina* d'Urv., in Mém., Soc. Linn. Paris: 1: 263. 1822.
Typus: [Ukraine, Crimea] "In pascuis Tauris frequens, ad littore Bosphori Cimmeriani [Kerch Strait]. 1820", *Dumont d'Urville* (P).

Melochia corchorifolia L. [*Dicot.: Stercul.*], vide sub *Melochia* (p. 362).

Mespilus cotoneaster L., Sp. Pl.: 479. 1 Mai 1753 [*Dicot.: Ros.*].
Typus: Sweden, Uppland, Norby lund, at the Linnaean path, 15 Jul 2002, *S. Ryman 9126* (UPS) (typ. cons.).

Mimosa pigra L., Cent. Pl. I: 13. 19 Feb 1755 [*Dicot.: Legum.*].
Typus: Mozambique, Gaza District, between Chibuto and Canicado by R. Limpopo, *Barbosa & Lemos 7999* (K; isotypi: COI, LISC, LMJ) (typ. cons.).

Mitrephora maingayi Hook. f. & Thomson, Fl. Brit. India 1: 77. 1872 [*Dicot.: Annon.*].
Typus: *Maingay* (K; isotypi: BM, GH, L).

(=) *Mitrephora teysmannii* Scheff., Natuurk. Tijdschr. Ned.-Indië 31: 12. 1868 (*'teysmanni'*).
Typus: Cult. Hort. Bogor. [orig. coll. in Bangka], *R. H. C. C. Scheffer* (BO).

Myrtus fragrans Sw., Prodr.: 79. 20 Jun-29 Jul 1788 [*Dicot.: Myrt.*]. Typus: Jamaica, *Swartz* (BM) (typ. cons.).

Nicotiana plumbaginifolia Viv., Elench. Pl.: t. 1. 1802 [*Dicot.: Solan.*] Typus: [icon in] Vivani, Elench. Pl.: t. 1. 1802.

(=)   *Nicotiana pusilla* L., Syst. Nat., ed. 10: 933. Mai-Jun 1759. Lectoypus (vide Knapp & Clarkson in Taxon 53: 844. 2004): [icon in] Miller, Fig. Pl. Gard. Dict.: t. 185, f. 2. 1757.

(=)   *Nicotiana humilis* Mill., Gard. Dict., ed. 8: *Nicotiana* No. 10. 16 Apr 1768. Lectotypus (vide Knapp & Clarkson in Taxon 53: 844. 2004): "herb. Miller E Vera Cruz Houston" (BM No. 776388).

(=)   *Nicotiana tenella* Cav., Descr. Pl. 1: 105. 1801. Lectotypus: (vide Knapp & Clarkson in Taxon 53: 844. 2004): "Ex Acapulco, Née dedit." (MA No. 307526).

Onobrychis cornuta (L.) Desv., vide *Hedysarum cornutum*

Ononis spinosa L., Sp. Pl.: 716. 1 Mai 1753 [*Dicot.: Legum.*]. Typus: Herb. Burser XXI: 79 (UPS) (typ. cons.).

Ophrys barbata Walter, Fl. Carol.: 221. Apr-Jun 1788 [*Monocot.: Orchid.*]. Typus: United States, Florida, Baker County, Osceola National Forest, 21 Mar 1991 *Orzell & Bridges 16163* (TEX; isotypi: FLAS, FTG, NY, US, USF) (typ. cons.).

Ophrys speculum Link in J. Bot.
(Schrader) 1799(2): 324. Apr 1800
[*Monocot.: Orchid.*].
Typus: "*Ophrys Speculum* nb." [Portugal, near Setubal], Link (B-W No.
16940-1, sub *Ophrys scolopax*) (typ.
cons.).

Orchis majalis Rchb., Iconogr. Bot.,
Pl. Crit. 6: 7, t. DLXV. 1828 [*Monocot.: Orchid.*]
Typus: [icon] "*Orchis latifolia*" in,
Sturm, Deutschl. Fl. Abt. 1, 2(7): t.
XX(1). 1799.

(=) *Orchis vestita* Lag. & Rodr. in Anales Ci. Nat. 16: 142. 1803.
Typus: "Tanger", s. ann. et coll.
(non loc.).

(=) *Orchis sesquipedalis* Willd., Sp. Pl.
4: 30. 1805.
Typus: "Lusitania", s. ann., *Linck*
(B, Herb. Willd. No. 16838).

(=) *Orchis elata* Poir., Voy. Barbarie 2:
248. 1789.
Typus: "Numidia", 1785 or 1786,
Poiret (W, Herb. Rchb. f. No.
12846).

Oreodoxa regia Kunth in Humboldt
& al., Nov. Gen. Sp. 1, ed. 4°: 305;
ed. f°: 244. Aug (sero) 1816 [*Monocot.: Palm.*].
Typus: Cuba, Havana, *Bonpland
1276* (P-Bonpl.).

(=) *Palma elata* W. Bartram, Travels
Carolina: iv, 115-116. 1791.
Neotypus (vide Zona in Taxon 43:
662. 1994): U.S.A., Florida, Collier
County, *Zona & Bigelow 406*
(FLAS).

Osteospermum spinosum L. [*Dicot.: Comp.*], vide sub *Osteospermum* (p. 372).

Paspalum dasypleurum Kunze ex
E. Desv. in Gay, Fl. Chil. 6: 242.
1854 [*Monocot.: Gram.*]
Typus: Chile, VIII Región: Antuco,
Poeppig 19(39) (M).

(=) *Paspalum cumingii* Nees ex Steud.,
Syn. Pl. Glumac. 1: 23. 1853.
Typus: Chile, V Región: Valparaíso,
Cuming 756 (P).

(=) *Paspalum pachyrrhizum* Steud., Syn.
Pl. Glumac. 1: 23. 1853.
Lectotypus (vide Zuloaga & Morrone Taxon 49: 561. 2000): Chile, X
Región: Valdivia, *Lechler 310* (P).

(=) *Paspalum paradisiacum* Steud., Syn.
Pl. Glumac. 1: 23. 1853.
Typus: Chile, V Región, Valparaíso,
Bertero 1222 (P).

Paullinia asiatica L. [*Dicot.: Rut.*], vide sub *Toddalia* (p. 407).

Passiflora incarnata L. [*Dicot.: Passiflor.*], vide sub *Passiflora* (p. 374).

Patosia clandestina (Phil.) Buchenau., vide *Rostkovia clandestina* Phil.

Penstemon hirsutus (L.) Willd., vide *Chelone hirsuta*

Peperomia nitida Dahlst., Kongl. Svenska Veterskapsakad Handl. 33(2): 92. 1900 [*Dicot. Piper.*].
Typus: Brazil, São Paulo, Campinas supra saxa humo tecta umbrosa. *Mosen 3986* ['*3985*'] (S).

(H) *Peperomia nitida* Sessé & Moc., Fl. Mex. ed. 1: 11. 1893.
Typus: non designatus.

Persicaria maculosa Gray, Nat. Arr. Brit. Pl. 2: 269. 1 Nov 1821 (*Polygonum persicaria* L. Sp. Pl.: 361. 1 Mai 1753) [*Dicot.: Polygon.*].
Typus: Herb. Burser IV: 101, left-hand plant (UPS).

(≡) *Persicaria mitis* Delarbre, Fl. Auvergne, ed. 2: 518. Aug 1800.

Persicaria mitis (Schrank) Assenov in Jordanov, Fl. Nar. Rep. Bălg. 3: 239. 19 Dec 1966 (*Polygonum mite* Schrank, Baier. Fl.: 668. Jun-Dec 1789) [*Dicot. Polygon.*].
Typus: [Germany] "Bayerischer Wald, Wiesentener Forst nördlich Wiesent an der Strasse nach Frauenzell, feuchter Waldwegrand in Fichtenforst", 430-445 m, 1 Sep 1993, *Schuwerk 93/475* (M) (typ. cons.).

(H) *Persicaria mitis* Delarbre, Fl. Auvergne, ed. 2: 518. Aug 1800.
≡ *Persicaria maculosa* Gray (nom. cons.).

Peucedanum capense (Thunb.) Sond. in Harvey & Sonder, Fl. Cap. 2: 554. 15-31 Oct 1862 (*Laserpitium capense* Thunb., Prodr. Pl. Cap.: 50. 1794) [*Dicot.: Umbell.*].
Typus: S. Africa, "e Cap. b. Spei", *Thunberg* (UPS, Herb. Thunberg No. 6950).

(H) *Peucedanum capense* (Eckl. & Zeyh.) D. Dietr., Syn. Pl. 2: 967. 1 Dec 1840 (*Oreoselinum capense* Eckl. & Zeyh., Enum. Pl. Afric. Austral.: 350. Apr 1837).
Typus: S. Africa, Cape, Stellenbosch, Klapmuts, *Ecklon & Zeyher 2239* (S).

Peucedanum nodosum L., Sp. Pl.: 246. 1 Mai 1753 [*Dicot.: Umbell.*].
Typus: [Greece, Crete] "in pratis supra Kastelli Pedhiada" 2 Jul 1899, *Baldacci 317* (BM) (typ. cons.).

Phalaenopsis sumatrana Korthals & Rchb. f. in Hamburger Garten-Blumenzeitung, 16: 115. Mar 1860 [*Monocot.: Orchid.*].
Typus: [icon] *"Phalaenopsis zebrina"* in Ann. Hort. Bot. 4: 145. 1860 (typ. cons.).

Phaseolus helvolus L., Sp. Pl.: 724. 1 Mai 1753 *('helvulus')* (orth. cons.). [*Dicot.: Legum.*].
Typus: U.S.A., S. Carolina, George-town Co., edge of marsh on ocean side of Beach Rd., about 1 mile N of tis terminus on South Island, 22 Aug 1991, *J. B. Nelson 11,140* (USCH; isotypus: MEXU) (typ. cons.).

Physocarpus opulifolius (L.) Maxim. in Trudy Imp. S.-Petersburgsk. Bot. Sada 6: 220. Jul-Aug 1879 (*Spiraea opulifolia* L., Sp. Pl.: 489. 1 Mai 1753) [*Dicot.: Ros.*].
Typus: U.S.A., Virginia, Herb. Linnaeus No. 651.12 (LINN).

(H)   *Physocarpus opulifolius* Raf., New Gl. 3: 73. Jan-Mar 1838 *('Physocarpa opulifolia')*.
Typus: non designatus.

Pinus maximinoi H. E. Moore in Baileya 14: 8. 15 Jan 1966 [*Gymnosp.: Pin.*].
Typus: Guatamala, *Hartweg 620* (K; isotypi: B, BM, CGE, FI, MO, NY, P, W).

(=)   *Pinus escandoniana* Roezl, Cat. Grain. Conif. Mexic.: 24. Jun 1857
Typus (vide Farjon in Neotrop. Monogr. 75: 144. 1997): Mexico, Mt. Tzompoli, *Roezl* (FI).

(=)   *Pinus hoseriana* Roezl, Cat. Grain. Conif. Mexic.: 24. Jun 1857
Typus (vide Farjon in Neotrop. Monogr. 75: 144. 1997): Mexico, Mt. Tzompoli, *Roezl* (FI).

(=)   *Pinus tzompoliana* Roezl, Cat. Grain. Conif. Mexic.: 24. Jun 1857.
Typus (vide Farjon in Neotrop. Monogr. 75: 144. 1997): Mexico, Mt. Tzompoli, *Roezl* (FI).

Poa pratensis L., Sp. Pl.: 67. 1 Mai 1753 [*Monocot.: Gram.*].
Typus: Rossia, Prov. Sanct-Petersburg, 5 km australi-occidentem versus a st. viae ferr., Mga., pratulum ad ripam dextra fl,. Mga. 26 Jun 1997, *Tzevlev N-257* (BM) (typ. cons.).

Podocarpus affinis Seem., Fl. Vit.: 266. 28 Feb 1868 [*Gymnosp.: Podocarp.*].
Typus: Fiji, Viti Levu: Namosi, Voma Peak, *Seemann 574* (K; isotypi: BM, E, GH).

(H)   *Podocarpus affinis* A. Massal., Syll. Pl. Foss.: 41. 1859 [Foss.].
Typus: non designatus.

Polygonum barbatum L., Sp. Pl.: 362. 1 Mai 1753 [*Dicot.: Polygon.*].
Typus: China, Canton, Sep 1885, *T. Sampson 541* (BM) (typ. cons.).

Potamogeton maackianus A. Benn. in J. Bot. 42: 74. Mar 1904 [*Monocot.: Potamogeton.*].
Typus: "Ussurien, am Flusse Sungatchi", *Maack* (LE).

(≡)   *Potamogeton serrulatus* Regel & Maack in Mém., Acad. Imp. Saint Petersbourg, sér. 7, 4(4): 139. Dec 1861.

Potamogeton schweinfurthii A. Benn. in Dyer, Fl. Trop. Afr. 8: 220. 1901 [*Monocot.: Potamogeton.*].
Typus: "Plantae Abyssinicae, Im Zana [sic! = Lake T'ana] (im offenem Wasser, nah am Ufer) bei Angasha, 9 Nov [18]63", *Schimper-1359* (K) (typ. cons.).

Potentilla nivea L., Sp. Pl.: 499. 1 Mai 1753 [*Dicot.: Ros.*].
Typus: Sweden, Torne Lappmark, Abisko area, Latnjajaure, *Eriksen 620* (GB; isotypi: ALA, BM, C, H, LE, G, S, UPS) (typ. cons.).

Pycreus sanguinolentus (Vahl) Nees, vide *Cyperus sanguinolentus*

Pyracantha coccinea M. Roem., Fam. Nat. Syn. Monogr. 3: 219. Apr. 1847 [*Dicot.: Ros.*].
Typus: [Austria] "In Comitatu Tyroliensi ad rivum in sabulo", Herb. Burser XIV: 8 (UPS).

(=) *Mespilus pauciflora* Poir. in Lamarck, Encycl., 4: 441. 1 Nov 1798.
Lectotypus (vide Muñoz & Aedo in Taxon 47: 171. 1998): "De Vidi vers Lausanne", *Reynier* (P-LAM).

Quercus pubescens Willd., Berlin. Baumz.: 279. 1796 [*Dicot.: Fag.*].
Typus: France, Vaucluse, Mont Ventoux, 19 Sep 1877, *Reverchon* (K) (typ. cons.).

(=) *Quercus humilis* Mill., Gard. Dict., ed. 8: *Quercus* No. 4. 16 Apr 1768.
Lectotypus (vide Franco & López in Anales Jard. Bot. Madrid 44: 556. 1987): *"Quercus robur* V *humilis"*, *Miller* (BM).

Ranunculus lobatus Jacquem. ex Cambess. in Jacquemont, Voy. Inde 4(Bot.): 4, t. 1B. 1841 [*Dicot.: Ranuncul.*].
Typus: *"Ranunculus lobatus"* [icon in] Jacquemont, Voy. Inde 4(Bot.): t. 1B. 1841.

(H) *Ranunculus lobatus* Moench, Methodus: 214. 1794.
Lectotypus (vide Lourteig in Darwiniana 9: 487. 1951): Herb. Linnaeus No. 715.66 (LINN).

Ribes leptostachyum Benth., Pl. Hartw.: 186. 1845 [*Dicot.: Grossular.*].
Typus: Chile, Coquimbo, Los Patos, *R. A. Philippi* (SGO; isotypus: K).

(H) *Ribes leptostachyum* Decne. in Jacquemont, Voy. Inde 4(Bot.): 65, t. 76. 1841.
Typus: India, "In Kanawer inter Soungnum et in montibus Lapann," alt. 3500 m, *Jacquemont* (P).

Rosa cinnamomea L. [*Dicot.: Ros.*], vide sub *Rosa* (p. 389).

Rostkovia clandestina Phil. in Linnaea 29: 76. 1857 [*Monocot.: Junc.*].
Typus: Colombia, Cauca, "Popayán, in ascensu ad Páramo de Guanacas, alt. 10,000 ad 10,500 ped." *Hartweg 1027* (K; isotypi: G, K, P, W).

(=) *Rostkovia brevifolia* Phil. in Linnaea 29: 76. 1857.
Lectotypus (vide Balslev in Fl. Neotrop. Mongr. 64: 46. 1996): Chile, Laguna de Malvarco, Jan 1856, *Germain* (SGO No. 63044).

Roystonea regia (Kunth) O. F. Cook, vide *Oreodoxa regia*

Rubus fruticosus L. [*Dicot.: Ros.*], vide sub *Rubus* (p. 389).

Rudbeckia purpurea L., Sp. Pl.: 907. 1 Mai 1753 [*Dicot.: Comp.*].
Typus: [U.S.A., Arkansas] *"Echinacea serotina* Arkansa", *Nuttall* (BM No. 541360) (typ. cons.).

Rumex acetosa L., Sp. Pl.: 337. 1 Mai 1753 [*Dicot.: Polygon.*].
Typus: Sweden, Södermanland, Salem, Wiksberg, *Jonsell 7110* (UPS; isotypi: BM, C, H, K, LD, O, S) (typ. cons.).

Rumex alpinus L., Sp. Pl.: 334. 1 Mai 1753 [*Dicot.: Polygon.*].
Typus: Herb. Linn. 464.35 (LINN) (typ. cons.).

Salix alba L. [*Dicot.: Salic.*], vide sub *Salix* (p. 390).

Salsola rosacea L., Sp. Pl.: 222. 1 Mai 1753 [*Dicot.: Chenopod.*].
Typus: [Kazakhstan] Taldy-Kurgan distr., NE corner of Katutau, upper part of gorge Kybe Sang c. 60 km W of Dzharkent (Panfilov) at road to Sary-Ozek, c. 1200 m, 9 Sep 1992, (LE; isotypi: AA, B, C, K, KAS) (typ. cons.).

Sarracenia purpurea L., Sp. Pl.: 510. 1 Mai 1753 [*Dicot.: Sarracen.*].
Typus: Canada, Quebec, Amos, East Abitibi County: about 4 miles west around Lake Beauchamp, 7 Jan 1952, *Baldwin & Breitung 2910* (K; isotypi: CAN, GH, MT, O, WLU) (typ. cons.).

Saussurea piptanthera Edgew. in Trans. Linn. Soc. London 20: 76. 29 Aug 1846 [*Dicot.: Comp.*].
Typus: Kashmir, 1833 *Royle* (G.-DC).

(≡) *Saussurea indica* Sch. Bip. in Linnaea 19(3): 331. Jul 1846.
Typus: non designatus.

Schedonorus arundinaceus (Schreb.) Dumort., Obser. Gramin. Belg.: 106. 1824 (*Festuca arundinacea* Schreb., Spic. Fl. Lips.: 57. 1771) [*Monocot.: Gram.*].
Typus: [icon in] Scheuzer, Agrostographia, t. 5, f. 18. 1719.

(H) *Schedonorus arundinaceus* Roem. & Schult., Syst. Veg. 2: 700. 1817 (*'Schenodorus'*).
Typus: Sweden, Norshom, Ostrogothia, *Liljeblad* (UPS).

Scleria flagellum-nigrorum Berg. in Kongl. Vetensh. Acad. Handl. 26: 142, t. 4-5. 1765 [*Monocot.: Cyper.*]. Typus: Surinam, *Rolander* (SBT) (typ. cons.).

Scleria pauciflora Muhl. ex Willd., Sp. Pl. 4: 318. 1805 [*Monocot.: Cyper.*]. Typus: [U.S.A.], "in Pennsylvania". *Muhlenberg* (B-W No. 17333) (typ. cons.).

Scrophularia auriculata L., Sp. Pl.: 620. 1 Mai 1753 [*Dicot.: Scrophular.*]. Typus: "Habitat in Hispania" Herb. Linnaeus No. 773.4 (LINN) (typ. cons.).

Senna ligustrina (L.) H. S. Irwin & Barneby, vide *Cassia ligustrina*

Sesamum indicum L., Sp. Pl.: 634. 1 Mai 1753 [*Dicot.: Pedal.*]. Typus: Herb. Linnaeus No. 802.3 (LINN).

(=) *Sesamum orientale* L., Sp. Pl.: 634. 1 Mai 1753. Lectotypus (vide Bruce in Turrill & Milne-Redhead, Fl. Trop. E. Afr. Pedal.: 19. 1953): Herb. Clifford: 318, *Sesamum* No. 1 (BM).

Silene apetala Willd., Sp. Pl. 2: 703. Mar 1799 [*Dicot.: Caryophyll.*]. Typus: Spain, Cádiz, "sec. aggeres viae ferreae inter Algeciras et S. Roque", 19 Apr 1895, *Porta & Rigo 78* (B) (typ. cons.).

462

Silene gallica L., Sp. Pl.: 417. 1 Mai 1753 [*Dicot.: Caryophyll.*].
Typus: Herb. Linnaeus No. 583.11 (LINN).

(=) *Silene anglica* L., Sp. Pl.: 416. 1 Mai 1753.
Lectotypus (vide Talavera & Muñoz in Anales Jard. Bot. Madrid 45: 408. 1989): Herb. Linnaeus No. 583.1 (LINN).

(=) *Silene lusitanica* L., Sp. Pl.: 416. 1 Mai 1753.
Lectotypus (vide Talavera & Muñoz in Anales Jard. Bot. Madrid 45: 408. 1989): Herb. Linnaeus No. 583.6 (LINN).

(=) *Silene quinquevulnera* L., Sp. Pl.: 416. 1 Mai 1753.
Lectotypus (vide Talavera & Muñoz in Anales Jard. Bot. Madrid 45: 409. 1989): "in Lusatia, Italia, Gallia", Herb. Burser XI: 72 (UPS).

Silene sieberi Fenzl, Pug. Pl. Nov. Syr.: 8. Mai-Jun 1842 [*Dicot.: Caryophyll.*].
Typus: Creta, "*Silene caesia* Sbr." [1817, *Sieber*] (G-BOIS) (typ. cons.).

Spartium capense L., Pl. Rar. Afr.: 14. 20 Dec 1760 [*Dicot.: Legumin.*].
Typus: [S. Africa, Western Cape Prov., Cape Penins.], "Cape of Good Hope Nature Reserve, c. 0.9 km. from entrance gate on gravel slope along roadside near resting place", 100 m., 10 Dec 1995, *Campbell & Van Wyk 151* (NBG: isotypi: K, MO, PRE) (typ. cons.).

Stipa robusta (Vasey) Scribn. in U.S.D.A. Bull. (1895-1901) 5: 23. 19 Feb 1897 (*Stipa viridula* var. *robusta* Vasey in Contr. U.S. Natl. Herb. 1: 56. 13 Jun 1890) [*Monocot.: Gram.*].
Typus: U.S.A., New Mexico, 1881, *Vasey* (US No. 993051) (typ. cons.).

Strophostyles helvola (L.) Elliott, vide *Phaseolus helvolus*

Tetracera volubilis L., Sp. Pl.: 533.
1 Mai 1753 [*Dicot.: Dillen.*].
Typus: Mexico, Veracruz, Zacupan,
Sulphur Spring, Dec 1906, *Purpus
2206* (F; isotypus: US No. 840326)
(typ. cons.).

Thalictrum foetidum L. [*Dicot.: Ranuncul.*], vide sub *Thalictrum* (p. 405).

Thyrsostachys siamensis Gamble in   (≡)   *Bambusa regia* Munro in Trans.
Ann. Roy. Bot. Gard. (Calcutta) 7:   Linn. Soc. London 26: 116. 5 Mar -
59. 1896 [*Monocot.: Gram.*].   11 Apr 1868.
Typus: Thailand, *Kurz* (K).

Tilia platyphyllos Scop., Fl. Carniol.,
ed. 2, 1: 373. 1772 [*Dicot.: Til.*].
Typus: cultivated in England, Cam-
bridge; seedling from N. Slovenia,
Bohinjska Bela, 2 Aug 1995, *Pigott
95-97* (BM; isotypi: B, K) (typ. cons.).

Toddalia asiatica (L.) Lam., vide *Paullinia asiatica*
Tournefortia hirsutissima L. [*Dicot.: Boragin.*], vide sub *Tournefortia* (p. 407).

Triticum aestivum L., Sp. Pl.: 85. 1   (=)   *Triticum hybernum* L., Sp. Pl.: 86.
Mai 1753 [*Monocot.: Gram.*].   1 Mai 1753.
Typus: Herb. Clifford: 24, *Triticum*   Lectotypus (vide Hanelt & al. in
No. 3. (BM).   Taxon 32: 492. 1983): Herb. Clif-
                                             ford: 24, *Triticum* No. 2. (BM).

Uvularia perfoliata L., Sp. Pl.: 304.
1 Mai 1753 [*Monocot.: Lil.*].
Typus: Virginia, *Clayton 258* (BM)
(typ. cons.).

Vellozia candida J. C. Mikan, Del.   (=)   *Vellozia tertia* Spreng., Neue Entd.
Fl. Faun. Bras. 2: ad t. [7]. 1822   2: 108. Jan 1821 [*Velloz.*].
[*Monocot.: Velloz.*].   Neotypus (vide Mello-Silva & Nic
Typus: [icon in] Mikan, Del. Fl. Faun.   Lughadha in Taxon 48: 581. 1999):
Bras. 2: t. [7]. 1822 (typ. cons.).   [icon in] Mikan, Del. Fl. Faun. Bras.
                                             2: t. [7]. 1822.

Veronica cinerea Boiss. & Balansa, Diagn. Pl. Orient., sér. 2, 6: 131. Jul-Dec 1859 [*Dicot.: Scrophular.*]. Typus: Turkey, Kayseri Prov., "regioni subalpina montis Karamas dagh ... propre Caesaream" [Kayseri], 3 Jul 1856, *Balansa 1025* (W).

(H)   *Veronica cinerea* Raf., New. Fl.: 4: 39. 1838. Typus: non designatus.

Vitex payos (Lour.) Merr., vide *Allasia payos.*

Wormia suffruticosa Griff. ex Hook. f. & Thomson in Hooker, Fl. Brit. India 1: 35. Mai 1872. [*Dicot.: Dillen.*]. Typus: Malacca, 1845, Griffith 55 (K).

(=)   *Wormia subsessilis* Miq., Fl. Ned. Ind., Eerste Bijv. 3: 619. Dec 1861. Typus: Banka, *Teysmann 3203 HB* (U).

F. FOSSIL PLANTS (EXCL. DIATOMS)

Filicites heterophyllus Brongn., vide sub *Neuropteris* (p. 418).

Neuropteris heterophylla (Brongn.) Sternb., vide *Filicites heterophyllus*

APPENDIX V

NOMINA UTIQUE REJICIENDA

The names printed in ***bold-face italics***, and all combinations based on these names, are ruled as rejected under Art. 56, and none is to be used. The rejected names are arranged in alphabetical sequence, irrespective of rank, within each major group. Cross-references to the rejected basionyms are provided under some combinations based on them.

The rejected names are neither illegitimate nor do they cease to be validly published (Art. 6). Later homonyms of a rejected name (Art. 53), and names illegitimate because of inclusion of the type of a subsequently rejected name (Art. 52), are not to be used unless they are conserved.

A. ALGAE

Fucus verrucosus Huds., Fl. Angl.: 470. Jan-Jun 1762.
Lectotypus (vide Irvine & Steentoft in Taxon 44: 223. 1995): England, Dillenius Herb. no. 50 (OXF).

Gracilaria verrucosa (Huds.) Papenf., vide *Fucus verrucosus*

Scytosiphon simplicissimus (Clemente) Cremades, vide *Ulva simplicissima*

Ulva simplicissima Clemente, Ensayo Var. Vid: 320. 1807.
Lectotypus (vide Cremades in Anales Jard. Bot. Madrid 51: 29. 1993): Spain, Cádiz, Puerto de Santa Maria, *Clemente 216* (MA).

B. FUNGI

Alphitomorpha communis Wallr. in Verh. Ges. Naturf. Freunde Berlin 1: 31. 1819 : Fries, Syst. Mycol. 3(1): 239. 1829.
Typus: non designatus.

Arthonia lurida Ach., Lichenogr. Universalis: 143. Apr-Mai 1810.
Holotypus: Helvetia, *Schleicher* (UPS).

Botrydina Bréb. in Mém. Soc. Acad. Agric. Industr. Instruct. Arrondissement Falaise 1839: 36. 1839.
≡ *Phytoconis* Bory (per typ. des.) (nom. utique rej.).

Byssus lacteus L., Sp. Pl.: 1169. 1 Mai 1753.
Holotypus: [icon in] Dillenius, Hist. Musc.: t. 1, f. 3. 1742.

Calicium lichenoides (L.) Schumach., vide *Mucor lichenoides*

Cladonia sylvatica (L.) Hoffm., vide *Lichen rangiferinus* var. *sylvaticus*

Collema proboscidale Mont. in Ann. Sci. Nat., Bot., ser. 2, XVIII: 20.1842.
Lectotypus (vide Jørgensen in Taxon 51: 567. 2002): [India], Nilgherrie (G ex herb. Montagne).

Coriscium Vain. in Acta Soc. Fauna Fl. Fenn. 7(2): 188. 1-22 Nov 1890.
Typus: *C. viride* (Ach.) Vain. (*Endocarpon viride* Ach.).

Dactylium Nees, Syst. Pilze: 58. Mar-Mai 1816 : Fr., Syst. Mycol. 3: 382, 412. 1832.
Typus: *D. candidum* Nees : Fr. (nom. utique rej.).

Dactylium candidum Nees, Syst. Pilze: 58. Mar-Mai 1816 : Fr., Syst. Mycol. 3: 412. 1832.
Typus: non designatus.

Erysiphe communis (Wallr. : Fr.) Schltdl., vide *Alphitomorpha communis*

Helotium Tode, Fungi Mecklenb. Sel. 1: 22. 1790 : Fr., Syst. Mycol. 3, index: 94. 1832.
Typus (vide Donk in Beih. Nova Hedwigia 5: 123. 1962): *H. glabrum* Tode : Fr.

Hueëlla Zahlbr. in Engler & Prantl, Nat. Pflanzenfam., ed. 2, 8: 180. 1926.
Typus: *H. fauriei* (Hue) Zahlbr. (*Pannaria fauriei* Hue) [= *Fuscopannaria leucophaea* (Hue) P. M. Jørg.].

Homodium pernigratum Nyl. in Bol. Soc. Broter. 3: 131. 1885.
Lectotypus (vide Jørgensen & Paz-Bermúdez in Taxon 53: 557. 2004): [Angola], Huilla [Huíla], 1883, *F. N. [F. Newton]* (PO No 5011L).

Lecanora anomala var. **tenebricosa** Ach., Lichenogr. Universalis: 382. Apr-Mai 1810.
Lectotypus (vide Printzen in Biblioth. Lichenol. 60: 228. 1995): Helvetia, [*Schleicher*] *346* (H-ACH No. 323).

Lecanora subfusca (L.) Ach., vide *Lichen subfuscus*

Lecanora tenebricosa (Ach.) Röhl., vide *L. anomala* var. *tenebricosa*

Lecidea atroalba var. **concreta** Ach. in Kongl.Vetensk. Acad. Nya Handl. 29: 233. 1808.
Lectotypus (vide Fryday in Lichenologist 34: 454. 2002): ad saxa in montosis Sveciae (UPS-ACH).

Lecidea epixanthoidiza Nyl. in Flora 58: 10. 1 Jan 1875.
Lectotypus (vide Tønsberg in Sommerfeltia 14: 169. 1992): Finland, Tavastia australis, Padasjoki, Nyystölä, 1872, *Lang 206* (H).

Lecidea flavocoerulescens Hornem. in Oeder, Fl. Dan. 8(24): 5, t. 1431, f. 1. 1810.
Lectotypus (vide Degelius in Acta Horti Gothob. 12: 114. 1937): Norway, Upper Telemark, *Smith* (UPS).

Lecidea synothea Ach. in Kongl. Vetensk. Acad. Nya Handl. 29: 236. 1808.
Lectotypus (vide Cannon & Hawksworth in Taxon 32: 479. 1983): Suecia, 1807, *Swartz* (BM).

Lichen atro-albus L., Sp. Pl.: 1141. 1 Mai 1753.
Typus: non designatus.

Lichen atro-virens L., Sp. Pl.: 1141. 1 Mai 1753.
Typus: non designatus.

Lichen coeruleonigricans Lightf., Fl. Scot. 2: 805. 20-23 Sep 1777.
Lectotypus (vide Timdal in Opera Bot. 110: 121. 1991): [icon in] Dillenius, Hist. Musc.: t. 82, f. 2. 1742.

Lichen cornucopioides L., Sp. Pl.: 1151. 1 Mai 1753.
Lectotypus (vide Jørgensen & al. in Bot. J. Linn. Soc. 115: 297, 374. 1994): Herb. Linnaeus No. 1273.217 p.p., young podetium (LINN).

Lichen daedaleus Sm. in Smith & Sowerby, Engl. Bot. 30: 2129. 1810.
Holotypus: "Scotland, Mr Menzies" (LINN-Sm. No. 1694.83).

Lichen fagineus L., Sp. Pl.: 1141. 1 Mai 1753.
Typus: non designatus.

Lichen fahlunensis L., Sp. Pl.: 1143. 1 Mai 1753.
Lectotypus (vide Jørgensen & al. in Bot. J. Linn. Soc. 115: 308, 375. 1994): Herb. Linnaeus No. 1273.70 (LINN).

Lichen hispidus Schreb., Spic. Fl. Lips.: 126. 9 Jul-25 Oct 1771.
Lectotypus (vide Laundon in Taxon 44: 246. 1995): [icon in] Vaillant, Bot. Paris.: t. 20, f. 5. 1727.

Lichen jubatus L., Sp. Pl.: 1155. 1 Mai 1753.
Lectotypus (vide Hawksworth in Taxon 19: 238. 1970): Herb. Linnaeus No. 1273.281 (LINN).

Lichen monocarpus Ach., Lichenogr. suec. prodr.: 196. 1799.
Lectotypus (vide Stenroos in Acta Bot. Fenn. 150: 180. 1994): [South Africa, Cape of Good Hope] "E cap. B. Spei", *C.P. Thunberg* (UPS-Thunberg No. 26451).

Lichen plicatus L., Sp. Pl.: 1150. 1 Mai 1753.
Lectotypus (vide Jørgensen & al. in Bot. J. Linn. Soc. 115: 379. 1994): [icon in] Dillenius, Hist. Musc.: t. 11, f. 1. 1742.

Lichen rangiferinus var. *sylvaticus* L., Sp. Pl.: 1153. 1 Mai 1753.
Lectotypus (vide Jørgensen & al. in Bot. J. Linn. Soc. 115: 349, 380. 1994): [icon in] Dillenius, Hist. Musc.: t. 16, f. 30. 1742.

Lichen semipinnatus J. F. Gmel., Syst. Nat., ed. 13bis, 2: 1372. Apr-Oct 1792 (typ. des.: Laundon in Taxon 44: 246. 1995).
≡ *Lichen hispidus* Schreb. (nom. utique rej.).

Lichen subfuscus L., Sp. Pl.: 1142. 1 Mai 1753.
Lectotypus (vide Brodo & Vitikainen in Mycotaxon 21: 294. 1984): [icon in] Dillenius, Hist. Musc.: t. 18, f. 16B. 1742.

Lycoperdon aurantium L., Sp. Pl.: 1053. 1 Mai 1753.
Lectotypus (vide Demoulin in Bull. Jard. Bot. Belg. 37: 297. 1967): [icon in] Vaillant, Bot. Paris.: t. 16, f. 9-10. 1727.

Monacrosporium candidum (Nees : Fr.) X. Z. Liu & K. Q. Zhang, vide *Dactylium candidum*

Mucor fulvus L., Sp. Pl.: 1185. 1 Mai 1753.
Typus: non designatus.

Mucor lichenoides L., Sp. Pl.: 1185. 1 Mai 1753.
Lectotypus (vide Jørgensen & al. in Bot. J. Linn. Soc. 115: 369, 383. 1994): [icon in] Dillenius, Hist. Musc.: t. 14, f. 3 (infer.). 1742.

Mucor sphaerocephalus L., Sp. Pl.: 1185. 1 Mai 1753.
Typus: non designatus.

Pachybasium niveum O. Rostr. in Dansk Bot. Ark. 2(5): 41. Aug 1916.
Neotypus (vide Bissett in Canad. J. Bot. 61: 1313. 1983): [isolate from soil from] Canada, Alberta, Kananaskis, Mt Allen, 7 Mai 1969, *Bissett* (DAOM No. 167322).

Parmelia olivaria (Ach.) Hue, vide *Parmelia perlata* var. *olivaria*

Parmelia perlata var. *olivaria* Ach., Methodus: 217. 1803.
Lectotypus (vide Hale & Fletcher in Bryologist 93: 28. 1990): [Spain, on *Olea*, 1791-93, leg. *P. K. A. Schousboë*] (H-ACH No. 1327C).

Pertusaria faginea (L.) Tuck., vide *Lichen fagineus*

Pertusaria lactea (L.) F. Arnold, vide *Byssus lacteus*

Peziza [unranked] *Phialea* Pers., Mycol. Eur. 1: 276. 1 Jan-14 Apr 1822 (*Peziza* ser. *Phialea* (Pers. : Fr.) Fr., Syst. Mycol. 2: 116. 1822).
Typus: *Peziza phiala* Vahl : Fr.

Phacidium musae Lév. in Ann. Sci. Nat., Bot., ser. 3, 5: 303. 1846.
Holotypus: *Bonpland* (PC, Herb. Amér. Equat.).

Phialea (Pers. : Fr.) Gillet, vide *Peziza* [unranked] *Phialea*

Phytoconis Bory, Mém. Conferva Byssus: 52. 1797.
Typus: *P. botryoides* (L.) Bory (*Byssus botryoides* L.).

Porpidia flavocoerulescens (Hornem.) Hertel & A. J. Schwab, vide *Lecidea flavocoerulescens*

Reticularia segetum Bull., Herb. France: t. 472, f. 2. 1790.
Typus: non designatus.

Rhizocarpon concretum (Ach.) Elenkin, vide *Lecidea atroalba* var. *concreta*

Scleroderma aurantium (L. : Pers.) Pers., vide *Lycoperdon aurantium*

Stilbum cinnabarinum Mont. in Ann. Sci. Nat., Bot., ser. 2, 8: 360. 1837.
Holotypus: Cuba, *Ramon* ex Herb. Montagne (PC).

Tolypocladium niveum (O. Rostr.) Bissett, vide *Pachybasium niveum*

Umbilicaria exasperata Hoffm., Descr. Pl. Cl. Crypt. 1: 7. 1789.
Typus: non designatus.

Uredo segetum (Bull. : Pers.) Pers., vide *Reticularia segetum*

Usnea plicata (L.) F. H. Wigg., vide *Lichen plicatus*

Ustilago segetum (Bull. : Pers.) Roussel, vide *Reticularia segetum*

C. BRYOPHYTA

Calypogeia trichomanis (L.) Corda, vide *Mnium trichomanis*

Campylium polymorphum (Hedw.) Pilous, vide *Hypnum polymorphum*

Cephaloziella byssacea (Roth) Warnst., vide *Jungermannia byssacea*

Dawsonia longifolia (Bruch & Schimp.) Zanten, vide *Polytrichum longifolium*

Grimmia alpicola Sw. ex Hedw., Sp. Musc. Frond. 77. 1 Jan 1801 [*Musci*]. Holotypus: Sweden, *Swartz* ex Herb. Hedwig-Schwägrichen (G).

Hypnum polymorphum Hedw., Sp. Musc. Frond.: 259. 1 Jan 1801 [*Musci*].
Lectotypus (vide Hedenäs in Taxon 45: 689. 1996): Austria, Carinthia, *Wulfen* ex Herb. Hedwig-Schwägrichen, specimen marked "d" (G).

Jungermannia alpestris Schleich. ex F. Weber, Hist. Musc. Hepat. Prodr.: 80. 1815 [*Hepat.*].
Holotypus: Switzerland, *Schleicher* in Pl. Crypt. Helv. Cent. 2 No. 59 ex Herb. Weber (S).

Jungermannia byssacea Roth, Tent. Fl. Germ. 3: 387. Jan-Apr 1800 [*Hepat.*].
Lectotypus (vide Schuster & Damsholt in Meddel. Grønland 189: 305. 1974): *Lindenberg* in Hep. No. 3440 p.p., female juvenile plants in a mixture (W).

Jungermannia globulifera Pollich, Hist. Pl. Palat. 3: 182. no. 1061. 1777 [*Hepat.*].
Typus: non designatus.

Jungermannia lanceolata L., Sp. Pl.: 1131. 1 Mai 1753 [*Hepat.*].
Lectotypus (vide Grolle in Taxon 15: 189. 1966): [icon in] Dillenius, Hist. Musc.: t. 70, f. 10A. 1742.

Lophozia alpestris (Schleich. ex F. Weber) A. Evans, vide *Jungermannia alpestris*

Mnium trichomanis L., Sp. Pl.: 1114. 1 Mai 1753 [*Hepat.*].
Lectotypus (vide Isoviita in Acta Bot. Fenn. 89: 15. 1970): [icon in] Dillenius, Hist. Musc.: t. 31, f. 5. 1742.

Polytrichum longifolium Bruch & Schimp. in Bruch & al., Bryol. Europ. [4: 256]. Jan-Oct 1844 [*Musci*].
Lectotypus (vide Zanten in Lindbergia 4: 133. 1977): Australia, "Neuholland", *Hügel* (W).

Riccia minima L., Sp. Pl. 1139. 1 Mai 1753 [*Hepat.*].
Typus: non designatus.

Schistidium alpicola (Sw. ex Hedw.) Limpr., vide *Grimmia alpicola*

D. PTERIDOPHYTA

Adiantum dissimile Schrad. in Gött. Gel. Anz. 87: 872. 29 Mai 1824.
Holotypus: Brazil, "Bahia", *Prinz von Wied-Neuwied* (BR).

Asplenium ramosum L., Sp. Pl.: 1082. 1 Mai 1753.
Lectotypus (vide Jermy & Jarvis in Bot. J. Linn. Soc. 109: 321. 1992): Herb. Burser XX: 16 (UPS).

Cheilanthes dalhousiae Hook., Sp. Fil. 2: 80. 1852.
Lectotypus (vide Fraser-Jenkins in Pakistan System. 5: 89. 1991): India: Simla, *Lady Dalhousie*, (K, neg. no. 7107, frond no. 4),

Nephrolepis auriculata (L.) Trimen, vide *Polypodium auriculatum*

Polypodium auriculatum L., Sp. Pl.: 1088. 1 Mai 1753.
Lectotypus (vide Trimen in J. Linn. Soc., Bot. 24: 152. 1887): Ceylon, Herb. Hermann 1: 39, No. 383 (BM).

E. SPERMATOPHYTA

Acantholimon echinus (L.) Bunge, vide *Statice echinus*

Actaea spicata var. *alba* L., Sp. Pl.: 504. 1 Mai 1753 [*Dicot.: Ranuncul.*].
Typus: non designatus.

Adansonia gibbosa (A. Cunn.) Guymer ex D. Baum, vide *Capparis gibbosa*

Alyxia glaucescens Wall. in Roxb., Fl. Ind. 2: 542. Mar-Jun 1824 [*Dicot.: Apocyn.*].
Holotypus: Penang, *Porter* (K-W No. 1609).

Anthospermum ciliare L., Sp. Pl., ed. 2: 1512. Jul-Aug 1763 [*Dicot.: Rub.*].
Lectotypus (vide Brummitt in Taxon 36: 73-74. 1987): Herb. Linnaeus No. 1233.4 (LINN).

Apocynum vincifolium Burm. f., Fl. Indica: 71. 1768 *('vincaefolium')* [*Dicot.: Apocyn.*].
Holotypus: *"Asclepias javanica – foliis ovatis petiolatis florum pedunculus umbellatus"*, Herb. Burman (G!).

Aristolochia longa L., Sp. Pl.: 962. 1 Mai 1753 [*Dicot.: Aristoloch.*].
Lectotypus (vide Nardi in Taxon 32: 654. 1983): Herb. Linnaeus No. 1071.10 (LINN).

Armeria juniperifolia (Vahl) Hoffmanns. & Link, vide *Statice juniperifolia*

Astragalus capitatus L., Sp. Pl.: 755. 1 Mai 1753 [*Dicot.: Legum.*].
Typus: non designatus.

Atriplex hastata L., Sp. Pl.: 1053. 1 Mai 1753 [*Dicot.: Chenopod.*].
Lectotypus (vide Rauschert in Feddes Repert. 85: 643. 1974): Herb. Linnaeus No. 1221.17 (LINN).

Betula alba L., Sp. Pl.: 982. 1 Mai 1753 [*Dicot.: Betul.*].
Lectotypus (vide Holub in Folia Geobot. Phytotax. 24: 410. 1989): Herb. Linnaeus No. 1109.1 (LINN).

Biarum orientale (L.) Druce, vide *Calla orientalis*

Bromus purgans L., Sp. Pl.: 76. 1 Mai 1753 [*Monocot.: Gram.*].
Lectotypus (vide Hitchcock in Contr. U.S. Natl. Herb. 12: 122. 1908): *Kalm* in Herb. Linnaeus No. 93.11 (LINN).

Buchnera euphrasioides Vahl, Symb. Bot. 3: 81. 1794 [*Dicot.: Scrophular.*].
Lectotypus (vide Hepper in Taxon 35: 390. 1986): Ghana, *König* ex Herb. Vahl (C).

Cacalia L., Sp. Pl.: 834. 1 Mai 1753 [*Dicot.: Comp.*].
Typus (vide Rydberg in Bull. Torrey Bot. Club 51: 370. 1924): *C. alpina* L.

Calla orientalis L., Sp. Pl., ed. 2: 1373. Jul-Aug 1763 [*Monocot.: Ar.*]
Holotypus: Herb. Rauwolff No. 100 (L).

Capparis gibbosa A. Cunn. in J. Bot (Hooker) 4: 261. Oct 1841 [*Dicot.: Bombac.*].
Typus: [Western Australia] Careening Cove, *Cunningham 308* (BM).

471

Carex bipartita All., Fl. Pedem. 2: 265. Apr-Jul 1785 [*Monocot.: Cyper.*]. Lectotypus (vide Turland in Taxon 46: 343. 1997): [Italy?] *Bellardi* (TO, Herb. Allioni).

Carex pedata L., Sp. Pl. ed. 2: 1384. Jul-Aug 1763 [*Monocot.: Cyper.*]. Typus: non designatus.

Carex uliginosa L., Sp. Pl.: 973. 1 Mai 1753 [*Monocot.: Cyper.*]. Typus: non designatus.

Cassia biflora L., Sp. Pl.: 378. 1 Mai 1753 [*Dicot.: Legum.*]. Neotypus (vide Wit in Webbia 11: 238. 1956): Jamayca, *Browne* in Herb. Linnaeus No. 528.21 (LINN).

Cassia chamaecrista L., Sp. Pl.: 379. 1 Mai 1753 [*Dicot.: Legum*]. Lectotypus (vide Irwin & Barneby in Mem. New York Bot. Gard. 35: 802. 1982): Herb. Linnaeus No. 528.30 (LINN).

Cassia galegifolia L., Syst. Nat. ed. 10: 1017. 7 Jun 1759 [*Dicot.: Legum.*]. Lectotypus (vide Irwin & Barneby in Mem. New York. Bot. Gard. 35: 56. 1982): [icon in] Plumier, Pl. Amer.: t. 78, f. 1. 1756.

Cassia tenuissima L., Sp. Pl.: 378. 1 Mai 1753 [*Dicot.: Legum.*]. Neotypus (vide Reveal in Taxon 41: 132. 1992): near Kingston, Jamayca, *Houstoun* (BM).

Cenchrus carolinianus Walter, Fl. Carol.: 79. Apr-Jun 1788 [*Monocot.: Gramin.*]. Neotypus (vide Reveal in Taxon 39: 354. 1990): U.S.A., S. Carolina, Beaufort County, St. Helena Island, ... 12 Sep 1982, *Boufford & al. 23096* (BM).

Centaurium minus Moench, Methodus: 349. 4 Mai 1794 [*Dicot.: Gentian.*]. Lectotypus (vide Melderis in Bot. J. Linn. Soc. 65: 229. 1972): "*Gentiana Centaurium minus* C.B.", Herb. Clifford: 278 (BM).

Cerastium viscosum L., Sp. Pl.: 437. 1 Mai 1753 [*Dicot.: Caryophyll.*]. Typus: non designatus.

Cerastium vulgatum L., Fl. Suec., ed. 2: 158. Oct 1755 [*Dicot.: Caryophyll.*]. Typus: non designatus.

Ceratoschoenus macrophyllus Tuck. in Amer. J. Sci. Arts 6(17): 232. Sep 1848 [*Monocot.: Cyper.*]. Typus: non designatus.

Chamaedorea donnell-smithii Dammer in Gard. Chron., ser. 3, 38: 43. 15 Jul 1905 [*Monocot.: Palm.*]. Holotypus: Honduras, Cortés, "Río Chamelecon", 16 Dec 1888, *Thiem 5537* ex Herb. J. Donnell Smith (US No. 932194).

Chrysodendron tinctorium Terán & Berland. in Berlandier, Mem. Comis. Límites: 7. 1832 (*'tinctoria'*) [*Dicot.: Berberid.*]. Lectotypus (vide Marroquín in Cuad. Inst. Invest. Ci. Univ. Autón. Nuevo León 15: 12. 1972): Mexico, Tamaulipas, "in montibus prope San Carlos" [c. 100 km N. of Ciudad Victoria], Nov 1831, *Berlandier 927* (GH).

Citta nigricans Lour., Fl. Cochinch.: 456. Sep 1790 [*Dicot.: Legum.*]. Lectotypus (vide Wilmott-Dear in Taxon 40: 517. 1991): Vietnam, *Loureiro* (BM).

Clausena san-ki (Perr.) Molino, vide *Illicium san-ki*

Cleistanthus orygalis (Blanco) Merr., vide *Gluta orygalis*

Crataegus oxyacantha L., Sp. Pl.: 477. 1 Mai 1753 [*Dicot.: Ros.*].
Lectotypus (vide Dandy in Bot. Soc. Exch. Club Brit. Isles 12: 867. 1946): Herb. Linnaeus No. 643.12 (LINN).

Croton racemosus Burm. f., Fl. Ind. 206 ('306'), t. 62, fig. 2. 1768 [*Dicot.: Euphorb.*].
Lectotypus (vide Esser in Taxon 50: 1211. 2002: [Sri Lanka] *Anonymous* (G, Herb. Houttuyn).

Cyclamen europaeum L., Sp. Pl.: 145. 1 Mai 753 [*Dicot.: Primul.*].
Lectotypus (vide S. Cafferty & C. Grey-Wilson in Taxon 47: 479. 1998): Herb. Burser XVII: 89 (UPS).

Cymbidium longifolium D. Don, Prodr. Fl. Nepal.: 36. 26 Jan - 1 Feb 1825 [*Monocot.: Orchid.*].
Holotypus: Nepal, "Gosaingsthan" [Gossainkunde], 1819, *Wallich* (BM).

Cypripedium hirsutum Mill., Gard. Dict., ed. 8: *Cypripedium* No. 3. 16 Apr 1768 [*Monocot.: Orchid.*].
Typus: non designatus.

Datisca hirta L., Sp. Pl.: 1037. 1 Mai 1753 [*Dicot.: Anacard.*].
Lectotypus (vide Britton in Bull. Torrey Bot. Club 18: 269. 1891): Philadelphia, *Kalm* in Herb. Linnaeus No. 1196.5 (LINN).

Dianthus arboreus L., Sp. Pl.: 413. 1 Mai 1753 [*Dicot.: Caryophyll.*].
Lectotypus: (vide Greuter in Candollea 20: 186. 1966): [icon.] *"Betonica coronaria arborea Cretica"* in Bauhin & al., Hist. Pl. 3: 328. 1651.

Dipleina umbellata Raf., Autik. Bot. 1: 54. 1840 [*Dicot.: Ranuncul.*].
Typus: Siberia: *"Dipleina umbellata* Raf.", s. coll. (DWC).

Dolichos pubescens L., Sp. Pl., ed. 2: 1021. Jul-Aug 1763 [*Dicot.: Legum.*].
Lectotypus (vide Verdcourt in Taxon 45: 329. 1996): Herb. Linnaeus No. 900.18 (LINN).

Drosera longifolia L., Sp. Pl.: 252. 1 Mai 1753 [*Dicot.: Droser.*].
Typus: non designatus.

Dryobalanops sumatrensis (J. F. Gmel.) Kosterm., vide *Laurus sumatrensis*

Echinocactus smithii Muehlenpf, in Allg. Gartenzeitung 14: 360. 1846 [*Dicot.: Cact.*].
Neotypus (vide Mosco & Zanovello in Taxon 48: 177. 1999): Mexico, San Luis Potosí: Loma Bonita, Matehuala. *Sánchez-Mejorada 2105* (MEXU).

Echites trichotoma Desf., Tabl. École Bot., ed. 3: 398. 15-21 Mar 1829 [*Dicot.: Apocyn.*].
Typus: [Cult. Paris] *Desfontaines* (FI-W).

Echium lycopsis L., Fl. Angl.: 12. 3 Apr 1754 [*Dicot.: Boragin.*].
Lectotypus (vide Stearn in Ray Soc. Publ. 148, Introd.: 65. 1973): [icon] *"Echii altera species"*, Dodoens, Stirp. Hist. Pempt.: 620. 1583.

Edgeworthia tomentosa (Thunb.) Nakai, vide *Magnolia tomentosa*

Epilobium alpinum L., Sp. Pl.: 348. 1 Mai 1753 [*Dicot.: Onagr.*].
Lectotypus (vide Marshall in J. Bot. 45: 367. 1907): Herb. Linnaeus No. 486.8 (LINN).

Epilobium junceum Spreng. in Biehler, Pl. Nov. Herb. Spreng.: 17. 30 Mai 1807 [*Dicot.: Onagr.*].
Lectotypus (vide Garnock-Jones in Taxon 32: 656. 1983): New Zealand, *Forster* (K).

Eriophorum polystachion L., Sp. Pl.: 52. 1 Mai 1753 [*Monocot.: Cyper.*].
Lectotypus (vide Novoselova in Bot. Žurn. 79(11): 86. 1994): Herb. Linnaeus No. 72.2, planta media (LINN).

Euphorbia pilulifera L., Sp. Pl.: 454. 1 Mai 1753 [*Monocot.: Gram.*].
Lectotypus (vide N.E. Brown & al. in Oliver, Fl. Trop. Afr. 6(1): 497-498. 1911): Herb. Linnaeus No. 630.8 (LINN).

Excoecaria integrifolia Roxb., Fl. Ind., ed. 1832, 3: 757. Oct-Dec 1832 [*Dicot.: Euphorb.*].
Lectotypus (vide Schot in Taxon 45: 553. 1996): s. loc., *Roxburgh* (BR).

Festuca elatior L., Sp. Pl.: 75. 1 Mai 1753 [*Monocot.: Gram.*].
Lectotypus (vide Linder in Bothalia 16: 59. 1986): Herb. Linnaeus No. 92.17 (LINN).

Festuca vizzavonae Ronniger, Verh. Zool.-Bot. Ges. Wien 68: 226. 1918 [*Monocot.: Gram.*].
Lectotypus (vide Foggio & Signorini in Parlatorea 2: 130. 1997): ... Locus: Korsika, Col. de Vizzavona, 1200 m, 25 Jun 1914, legit.: *Ronniger* (W 22344 Herb. Ronniger).

Ficus caribaea Jacq., Observ. Bot. 2: 30. 1767 [*Dicot.: Mor.*].
Neotypus (vide C.C. Berg in Taxon 52: 368. 2003): Puerto Rico, Adjuntas road, 5 miles from Ponce, 24 Dec 1902, *Heller 6330* (L no. 908118-847).

Ficus ciliolosa Link, Enum. Hort. Berol. Alt. 2: 450. Jan-Jun 1822 [*Dicot.: Mor.*].
Typus: "Hort. Bot. Berol." (B No. 10 000 2774).

Ficus novae-walliae Dum. Cours., Bot. Cult. 3: 681. 1-4 Jul 1802 (*'novae Walliae'*) [*Dicot.: Mor.*].
Typus: deest.

Fraxinus tetragona Cels ex Dum. Cours., Bot. Cult. 1: 712. 1-4 Jul 1802 [*Dicot.: Ole.*].
Typus: non designatus.

Fumaria bulbosa L., Sp. Pl.: 699. 1 Mai 1753 [*Dicot.: Papaver.*].
Typus: Herb. Linnaeus No. 881.5 (LINN) (typ. cons. sub *Corydalis*, nom. cons.).

Galanthus reflexus Herb. in Edwards's Bot. Reg. 31: Misc. 35 (No. 44). 1845 [*Monocot.: Amaryllid.*].
Typus: non designatus.

Gaura mollis James, Long Exped. (Philad. ed.) 2: 77. Jan 1823 [*Dicot.: Onagr.*].
Typus: U.S.A., SW branches of the Arkansa[s] [Colorado, Las Animas Co., Brachicha Canyon] *E. James 7* (NY).

Geranium rapulum A. St.-Hil. & Naudin., Ann. Sci. Nat., Bot., sér. 2, 18: 25. 1842 [*Dicot.: Geran.*].
Lectotypus: (vide Aedo in Taxon 53: 1074. 2005: Brazil, Rio Grande do Sul, 1833, *Gaudichaud 1204* (P, also K).

Gilia grinnellii Brand in Engler, Pflanzenr. IV. 250 (Heft 27): 101. 19 Feb 1907 [*Dicot.: Polemon.*].
Typus: non designatus.

Gluta orygalis Blanco, Fl. Filip., ed. 2: 451. 1845 [*Dicot.: Euphorb.*].
Typus: non designatus.

Glycicarpus edulis Dalzell in J. Asiat. Soc. Bombay 3: 70. t. 5-6. Jan 1849 [*Dicot.: Anacard.*]
Lectotypus (vide Gandhi & al. in Taxon 50: 583. 2001): [icon in] J. Asiat. Soc. Bombay 3: t. 6. Jan 1849.

Heliconia humilis (Aubl.) Jacq., vide *Musa humilis*

Holcus saccharatus L., Sp. Pl.: 1047. 1 Mai 1753 [*Monocot.: Gram.*].
Typus: non designatus.

Hypericum chinense Osbeck, Dagb. Ostind. Resa: 244. 1757. [*Dicot.: Guttif.*].
Holotypus: China, Kwangtung [Guangdong], Danish Island, 24 Oct 1751, *Osbeck* (S).

Hypericum quadrangulum L., Sp. Pl.: 785. 1 Mai 1753 [*Dicot.: Guttif.*].
Lectotypus (vide Robson in Taxon 39: 135. 1990): Herb. Clifford.: 380, *Hypericum* No. 5 (BM).

Hypoëstes verticillaris (L. f.) Sol. ex Roem. & Schult., vide *Justicia verticillaris*

Illicium san-ki Perr., Cat. Pl. Intr. Colon.: 33. Mai-Dec 1824 [*Dicot.: Rut.*].
Neotypus (vide Molino in Bull. Mus. Natl. Hist. Nat., B, Adansonia 16(1): 134. 1994): Philippines, Luzon, Benguet subprov., Mai 1914, *Merrill 1791* (PNH; isoneotypi: BM, BO, P, PE, SING).

Inga juglandifolia Willd., Sp. Pl. 4: 1018. Apr 1806 [*Dicot.: Legum.*].
Holotypus: Venezuela, Caracas, *Bredemeyer 3* (B-W No. 10040).

Inga pisana G. Don, Gen. Hist. 2: 388. Oct 1832 [*Dicot.: Legum.*].
Holotypus: *"Mimosa pisana vulgo Pisana"*, Peru, Chinchao, 1795, *Ruiz & Pavón* (BM; isotypi: B, MA).

Juncus cymosus Lam., Encycl. 3: 267. 1789 [*Monocot.: Junc.*].
Neotypus (vide Phipps in Taxon 48: 831. 1999): [icon in] Mill., Fig. Pl. Gard. Dict.: t. 179. 1757.

Justicia verticillaris L. f., Suppl. Pl.: 85. Apr 1782 [*Dicot.: Acanth.*].
Lectotypus (vide Brummitt & al. in Taxon 32: 658. 1983): Herb. Thunberg No. 427 (UPS).

Lachnaea conglomerata L., Sp. Pl.: 560. 1 Mai 1753 [*Dicot.: Rhamn.*].
Typus: non designatus.

Lactuca flava Forssk., Fl. Aegypt.-Arab.: 143. 1 Oct 1775 [*Dicot.: Comp.*].
Typus deest.

Lassonia heptapeta Buc'hoz, Pl. Nouv. Découv.: 21. t. 19, f. 1. 1779-1784. [*Dicot.: Magnol.*].
Typus: [icon in] Buc'hoz, Pl. Nouv. Découv.: t. 19, f. 1. 1779-1784.

Lassonia quinquepeta Buc'hoz, Pl. Nouv. Découv.: 21. t. 19, f. 2. 1779-1784. [*Dicot.: Magnol.*].
Typus: [icon in] Buc'hoz, Pl. Nouv. Découv.: t. 19, f. 2. 1779-1784.

Laurus sumatrensis J. F. Gmel., Syst. Nat. 2: 650. Sep (sero) - Nov 1791 [*Dicot.: Laur.*].
Holotypus: [icon in] Verh. Holl. Maatsch. Weetensch. Haarlem 21: 271, t. B. 1784.

Lavandula spica L., Sp. Pl.: 572. 1 Mai 1753 [*Dicot.: Labiat.*].
Lectotypus (vide López González in Jarvis et. al. in Reg. Veg. 127: 60. 1993): Herb. Burser XII: 64 (UPS).

Lotus glaber Mill., Gard. Dict., ed. 8: *Lotus* No. 3. 16 Apr 1768. [*Dicot.: Legum.*].
Typus: non designatus.

Lupinus hirsutus L., Sp. Pl.: 721. 1 Mai 1753 [*Dicot.: Legum.*].
Lectotypus (vide Lee & Gladstone in Taxon 28: 619-620. 1979): "*Lupinus hirsutus* L. 1015 – Roy. prodr. 367", ex Herb. van Royen (L No. 908.119-125).

Luzula capillaris Steud., Syn. Pl. Cyper.: 293. 1855 [*Monocot.: Junc.*].
Typus: [U.S.A.] Amer. sept. unita, 1826, *Leman* [in herb. Lenormand] (P).

Luzula hyperborea R. Br., Chloris Melvilliana [Parry's 1st Voyage App.]: 283. 1823 [*Monocot.: Junc.*].
Lectotypus (vide Kirschner & Kaplan, Taxon 50: 1195. 2001): [U.S.A.] "49 *Luzula hyperborea*, Melville Island, *Cptn Parry* s. d." (BM – upper left hand specimen).

Luzula interrupta Desv., J. Bot. 162, t. 6, fig. 3. 1808 [*Monocot.: Junc.*].
Typus: "in America calidiore", *s. coll.* (P).

Magnolia tomentosa Thunb. in Trans. Linn. Soc. London 2: 336. 1794 [*Dicot.: Thymel.*].
Lectotypus (vide Rehder & Wilson in Sargent, Pl. Wilson. 1: 400, 408. 1913): Herb. Thunberg No. 12886 (UPS).

Mahonia tinctoria (Terán & Berland.) I. M. Johnst., vide *Chrysodendron tinctorium*

Malva rotundifolia L., Sp. Pl.: 688. 1 Mai 1753 [*Dicot.: Malv.*].
Lectotypus (Riedl in Rechinger, Fl. Iran. 120: 26. 1976): Herb. Linnaeus No. 870.18 (LINN).

Melianthus minor L., Sp. Pl.: 639. 1 Mai 1753 [*Dicot.: Melianth.*].
Lectotypus (vide Wijnands in Taxon 34: 314. 1985): [icon in] Commelijn, Hort. Med. Amstel. Pl.: t. 4. 1706.

Mespilus cordata Mill., Fig. Pl. Gard. Dict.: 116, t. 179. Aug 1757 [*Dicot.: Ros.*].
Lectotypus (vide Phipps in Taxon 48: 831. 1999): [icon in] Mill., Fig. Pl. Gard. Dict.: t. 179. 1757.

Mimosa cinerea L., Sp. Pl.: 517 (no. 10). 1 Mai 1753 [*Dicot.: Leg.*].
Lectotypus (vide Rico in Taxon 46: 476. 1997): [icon.] "*Acacia* Maderaspat. *spinosa Intsiae accedens, cortice cinereo, ramulis communi pedicula binis*" in Plukenet, Phytographia: t. 2, f. 1. 1691.

Mucuna nigricans (Lour.) Steud., vide *Citta nigricans*

Musa humilis Aubl., Hist. Pl. Guiane: 931. Jun-Dec 1775 [*Monocot.: Mus.*].
Lectotypus (vide Andersson in Taxon 33: 524. 1984): French Guiana, *Aublet* (BM).

Nardus gangitis L., Sp. Pl.: 55. 1 Mai 1753 [*Monocot.: Gram.*].
Typus: non designatus.

Nepeta hirsuta L., Sp. Pl.: 571. 1 Mai 1753 [*Dicot.: Labiat.*].
Typus: non designatus.

Nymphaea pentapetala Walter, Fl. Carol.: 155. Apr-Jun 1788. [*Dicot.: Nymph.*].
Neotypus (vide Wiersema & Reveal in Taxon 40: 509. 1991): U.S.A., S. Carolina, Charleston County, Mayrant Backwater, ... 5 Jun 1943, *Hunt & Martin 2056* (CLEMS).

Nymphaea reniformis Walter, Fl. Carol.: 155. Apr-Jun 1788. [*Dicot.: Nymph.*].
Neotypus (vide Wiersema & Reveal in Taxon 40: 512. 1991): U.S.A., S. Carolina, Berkeley County, Santee Canal, ... 12 Jul 1939, *Godfrey & Tryon 471* (DUKE).

Ocimum vaalae Forssk., Fl. Aegypt-Arab.: 111. 1775 [*Dicot.: Labiat.*].
Typus: non designatus.

Orchis latifolia L., Sp. Pl.: 941. 1 Mai 1753 [*Monocot.: Orchid.*].
Typus: non designatus.

Orobanche laevis L., Sp. Pl.: 632. 1 Mai 1753 [*Dicot.: Orobanch.*].
Typus: non designatus.

Orobanche major L., Sp. Pl.: 632. 1 Mai 1753 [*Dicot.: Orobanch.*].
Lectotypus (vide Turland & Rumsey in Taxon 46: 787. 1997): Herb. Clifford: 321, *Orobanche* No. 1, fol. A (BM).

Panicum divergens Kunth in Humb. & al., Nov. Gen. Sp. 1, ed 4°: 102. 1816 [*Monocot.: Gram.*].
Typus: Ecuador: Pichincha: Salgolquí, *Humboldt & Bonpland* (P; isotype frag. US-80645).

Papilionopsis Steenis, Nova Guinea, Bot. 3: 17. 1960 [*Dicot.: Legum.*].
Typus: *P. stylidioides* Steenis

Pepo indicus Burm., Auctuarium (Index Univ.): [6]. after 20 Jun 1755 [*Dicot.: Cucurbit.*].
Lectotypus (vide Merrill, Interpr. Herb. Amboin.: 494. 1917): [icon] *"Pepo indicus"* in Rumph., Herb. Amboina 5: t. 145. 1747.

Pentocnide Raf., Fl. Tellur. 3: 48. Nov-Dec 1837 [*Dicot.: Urtic.*].
Typus: *P. glomerata* Raf.

Phaca trifoliata L., Mant. Pl.: 270, Oct 1771 [*Dicot.: Legum.*].
Typus: non designatus.

Phaulopsis longifolia Sims in Bot. Mag.: ad t. 2433. 1 Oct 1823 [*Dicot.: Acanth.*].
Lectotypus (vide Manktelow in Taxon 44: 641. 1995): [icon in] Bot. Mag.: t. 2433. 1823.

Phyllanthus cyclanthera Baill. in Adansonia 1: 31. 1860 [*Dicot.: Euphorb.*].
Typus: Mexico, *Sessé & Mocino* (G).

Phyllanthus hamrur Forssk., Fl. Aegypt.-Arab.: 159. 1 Oct 1775 [*Dicot.: Euphorb.*].
Typus: non designatus.

Phyteuma pauciflorum L., Sp. Pl.: 170. 1 Mai 1753 [*Dicot.: Campanul.*].
Lectotypus (vide Cafferty & Sales in Taxon 48: 601. 1999): [icon] *"Rapunculus alpinus parvus comosus"* in Bauhin & al., Hist. Pl. 2: 811. 1651.

Poa malabarica L., Sp. Pl.: 69. 1 Mai 1753 [*Monocot.: Gram.*].
Lectotypus (vide Merrill in Bull. Torrey Bot. Club 60: 633-635. 1933): [icon in] Rheede, Hort. Malab. 12: t. 45. 1693.

Polypodiopsis Carrière, Traité Gén. Conif., ed. 2, 710. 15 Jan 1867 [*Gymnosp.: Podocarp.*[?]].
Typus: *P. muelleri* Carrière (nom. utique rej.).

Polypodiopsis muelleri Carrière, Traité Gén. Conif., ed. 2, 710. 15 Jan 1867 (*'muellerii'*) [*Gymnosp.: Podocarp.*[?]].
Typus: New Caledonia (spec. ignota).

Potamogeton oblongifolius J. R. Forst., Fl. Amer. Sept.: 7. 1771 [*Monocot.: Potamogeton.*]
Lectotypus (vide Reveal & al. in Taxon 51: 865. 2003): [U.S.A.,] Virginia, near a bridge at Lancaster, Lancaster Co., *Mitchell* (BM).

Potamogeton rotundifolius J. R. Forst., Fl. Amer. Sept.: 7. 1771 [*Monocot.: Potamogeton.*]
Holotypus: [U.S.A.,] Virginia, at Bullock's Bridge near Hanover, Hanover Co., *Clayton* (deest).

Pterocephalus papposus (L.) Coult., vide *Scabiosa papposa*

Quercus aegilops L., Sp. Pl.: 996. 1 Mai 1753 [*Dicot.: Fag.*].
Lectotypus (Menitsky in Novosti Sist. Vysš. Rast. 9: 126. 1972): [icon] *"Cerri glans Aegilops aspris"* in Bauhin & al., Hist. Pl. 1(2): 77. 1650.

Quercus esculus L., Sp. Pl.: 996. 1 Mai 1753 [*Dicot.: Fag.*].
Lectotypus (Govaerts in Taxon 44: 631. 1995): Herb. Linnaeus No. 1128.28 (LINN).

Quercus palensis Palassou, Essai Monts Pyr.: 317. 1784 [*Dicot.: Fag.*].
Typus: non designatus.

Rhus hirta (L.) Sudw., vide *Datisca hirta*

Rhus sinuata Thunb., Prodr. Pl. Cap. 1: 52. 1794 [*Dicot.: Anacard.*].
Holotypus: Herb. Thunberg No. 7368 (UPS).

Rosa eglanteria L., Sp. Pl.: 491. 1 Mai 1753 [*Dicot.: Ros.*].
Lectotypus (vide Heath in Calyx 1: 153. 1992): [icon] *"Rosa Eglanteria"* in Tabernaemontanus, Eicon. Pl.: 1087. 1590.

Rotala decussata DC., Prodr. 3: 76. Mar 1828 [*Dicot.: Lythr.*].
Holotypus: Australia, Queensland, Endeavour River, *Brown* (G-DC).

Salix fluviatilis Nutt., N. Amer. Sylva 1: 73. Jul-Dec 1842 [*Dicot.: Salic.*].
Neotypus (vide Dorn in Taxon 47: 459. 1998): [United States, Oregon] "Yamhill County: Willamette River at Wheatland Ferry", 3 Jun 1996, *Dorn 6899* (RM; isoneotyp.: NY).

Savia clusiifolia Griseb. in Nachr. Königl. Ges. Wiss. Georg-Augusts-Univ. 6: 164. 1865 [*Dicot.: Euphorb.*].
Lectotypus (vide Hoffmann in Taxon 43: 465. 1994): Cuba, "circa Punta Brava propre Matanzas", *Rugel 321* (GOET).

Scabiosa papposa L., Sp. Pl.: 101. 1 Mai 1753 [*Dicot.: Dipsac.*].
Lectotypus (vide Meikle in Taxon 31: 542. 1982): "14. *Scabiosa*" ex Herb. van Royen (L No. 902.125-731).

Scirpus miliaceus L., Syst. Nat., ed. 10: 868. 7 Jun 1759 [*Monocot.: Cyper.*].
Lectotypus (vide Blake in J. Arnold Arbor. 35: 216-219. 1954): Herb. Linnaeus No. 71.40 (LINN).

Scrophularia aquatica L., Sp. Pl.: 620. 1 Mai 1753 [*Dicot.: Scrophular.*]. Lectotypus (Dandy in Watsonia 7: 164. 1969): Herb. Linnaeus No. 773.3 (LINN).

Secale creticum L., Sp. Pl.: 84. 1 Mai 1753 [*Monocot.: Gram.*]. Typus: non designatus.

Senna galegifolia (L.) Barneby & Lourteig, vide *Cassia galegifolia*

Silene polyphylla L., Sp. Pl.: 420. 1 Mai 1753 [*Dicot.: Caryophyll.*]. Typus: non designatus.

Silene rubella L., Sp. Pl.: 419. 1 Mai 1753 [*Dicot.: Caryophyll.*]. Lectotypus (vide Oxelman & Lidén in Taxon 36: 477. 1987): Herb. Linnaeus No. 583.43 (LINN).

Smilax humilis Mill., Gard. Dict., ed. 8: unpaged. 1768 [*Monocot.: Smilac.*]. Lectotypus (vide Reveal in Taxon 49: 297. 2000): [icon. in] Catesby, Nat. Hist. Carolina 1: t. 47. 1730.

Solanum indicum L., Sp. Pl.: 187. 1 Mai 1753 [*Dicot.: Solan.*]. Lectotypus (vide Hepper in Bot. J. Linn. Soc. 76: 288. 1978): Herb. Hermann 3: 16, No. 94 (BM).

Solanum quercifolium L., Sp. Pl.: 185. 1 Mai 1753 [*Dicot.: Solan.*]. Lectotypus (vide Knapp & Jarvis in Bot. J. Linn. Soc. 104: 355. 1990): Herb. Linnaeus No. 248.8 (LINN).

Solanum sodomeum L., Sp. Pl.: 187. 1 Mai 1753 [*Dicot.: Solan.*]. Lectotypus (vide Hepper in Bot. J. Linn. Soc. 76: 290. 1978): Herb. Hermann 3: 30, No. 95 (BM).

Solanum verbascifolium L., Sp. Pl.: 184. 1 Mai 1753 [*Dicot.: Solan.*]. Lectotypus (vide Roe in Taxon 17: 177. 1968): Herb. Linnaeus No. 248.1 (LINN).

Statice echinus L., Sp. Pl.: 276. 1 Mai 1753 [*Dicot.: Plumbagin.*]. Lectotypus (vide Meyer in Haussknechtia 3: 7. 1987): [icon] *"Limonium cespitosum, foliis aculeatis"*, Buxbaum, Pl. Min. Cogn. Cent. 2: t. 10. 1728.

Statice juniperifolia Vahl, Symb. Bot.: 25. Aug-Oct. 1790 [*Dicot.: Plumbagin.*]. Typus: non designatus.

Stipa columbiana Macoun, Cat. Canad. Pl. 4-5: 101. 1888 [*Monocot.: Gram.*]. Lectotypus (vide Hitchcock in Contr. U.S. Natl. Herb. 24: 253. 1925): British Columbia, Yale, 17 Mai 1875, *Macoun* (CAN No. 9899).

Striga euphrasioides (Vahl) Benth., vide *Buchnera euphrasioides*

Strychnos colubrina L., Sp. Pl.: 189. 1 Mai 1753 [*Dicot.: Logan.*]. Lectotypus (vide Bisset & al., Lloydia 36: 183. 1973): [icon.] *"Modiracanirum"* in Rheede, Hort. Malab. 8: 47, t. 24. 1688.

Syringa buxifolia Nakai in Bot. Mag. (Tokyo) 32: 131. Jun 1918 [*Dicot.: Ol.*]. Holotypus: China, in hortis Lan-chau, Kansu, *Umemura* (TI)

Tabernaemontana echinata Aubl., Hist. Pl. Guiane: 263. 1775 [*Dicot.: Apocyn.*]. Lectotypus (vide Allgore in Mem. Mus. Nation. Hist. Nat. Nouv. ser., B (Paris) 30: 134. 1985): [icon] Aubl., Hist. Pl. Guiane: t. 103. 1775 (excl. fruit).

Tilia officinarum Crantz, Stirp. Austr. Fasc. 2: 61. 1763 [*Dicot.: Til.*].
Holotypus: Herb. Crantz No. 801 (BP).

Tontelea aubletiana Miers, Trans. Linn. Soc. London 28: 383. 1872 [*Dicot.: Celastr.*].
Typus: Guiana, s. loc., *Aublet* (BM, excl. fl. shoots in upper half of the sheet).

Tontelea scandens Aubl., Hist. Pl. Guiane: 31. 1775 [*Dicot.: Celastr.*].
Typus: Guiana, s. loc., *Aublet* (BM, excl. sterile shoots on lower part of the sheet).

Toxicodendron crenatum Mill., Gard. Dict., ed. 8, *Toxicodendron* No. 5. 16 Apr 1768 [*Dicot.: Anacard.*].
Neotypus (vide Reveal in Taxon 40: 334. 1991): 1769, [*Miller*] ex Chelsea Physic Garden No. 2392 (BM).

Toxicodendron volubile Mill., Gard. Dict., ed. 8, *Toxicodendron* No. 6. 16 Apr 1768 (*'volubilis'*) [*Dicot.: Anacard.*]
Neotypus (vide Reveal in Taxon 40: 334. 1991): "unattributed sheet" [sic!] (BM).

Trichostema spirale Lour., Fl. Cochinch. 2: 451. Sep 1790 (*'spiralis'*) [*Dicot.: Labiat.*].
Typus: non designatus.

Trifolium agrarium L., Sp. Pl.: 772. 1 Mai 1753 [*Dicot.: Legum.*].
Typus: non designatus.

Trifolium filiforme L., Sp. Pl.: 773. 1 Mai 1753 [*Dicot.: Legum.*].
Typus: non designatus.

Trifolium procumbens L., Sp. Pl.: 772. 1 Mai 1753 [*Dicot.: Legum.*].
Typus: non designatus.

Trigonella hamosa L., Syst. Nat., ed. 10: 1180. 7 Jun 1759 [*Dicot.: Legum.*].
Lectotypus (vide Lassen in Taxon 36: 478. 1987): Herb. Linnaeus No. 932.5 (LINN).

Veronica decussata Moench, Verz. Ausländ. Bäume: 137. 1785 [*Dicot.: Plantagin.*].
Typus: non designatus.

Villamillia Ruiz & Pav., Fl. Peruv.: t. 402. 1830-1833 [*Dicot.: Phytolacc.*].
Typus: *V. tinctoria* Ruiz & Pav. (nom. utique rej.).

Villamillia tinctoria Ruiz & Pav., Fl. Peruv.: t. 402. 1830-1833 [*Dicot.: Phytolacc.*].
Typus: [icon in] Ruiz & Pavon, Fl. Peruv.: t. 402.

APPENDIX VI

OPERA UTIQUE OPPRESSA

Publications are listed alphabetically by authors. Numbers of the relevant entries in TL-2 (Stafleu & Cowan, *Taxonomic literature* 1-7; in Regnum Veg. 94, 98, 105, 110, 112, 115, 116. 1976-1988) are added parenthetically in **bold-face-type** when available. Names appearing in the listed publications in any of the ranks specified in square brackets at the end of each entry are not accepted as validly published under the present *Code* (Art. 32.7).

Agosti, J. 1770. *De re botanica tractatus.* Belluno. (TL-2 No. **66**.) [Genera.]

Buc'hoz, P. J. 1762-1770. *Traité historique des plantes qui croissent dans la Lorraine et les Trois Evêchés.* 10 vol. Nancy & Paris. (TL-2 No. **872**.) [Species and infraspecific taxa.]

Buc'hoz, P. J. 1764. *Tournefortius Lotharingiae, ou catalogue des plantes qui croissent dans la Lorraine et les Trois Evêchés.* Paris & Nancy. (TL-2 No. **873**.) [Species and infraspecific taxa.]

Buc'hoz, P. J. 1770. *Dictionnaire raisonné universel des plantes, arbres et arbustes de France.* Vol. 1, 2, and 3 (pp. 1-528) [but not vol. 3 (pp. 529-643), nor vol. 4]. Paris. (TL-2 No. **874**.) [Species and infraspecific taxa.]

Donati, V. 1753. *Auszug seiner Naturgeschichte des adriatischen Meers.* Halle. (TL-2 sub No. **1500**.) [All ranks.]

Ehrhart, J. B. 1753-1762. *Oeconomische Pflanzenhistorie.* 12 vol. Ulm & Memmingen. (TL-2 No. **1647**.) [Genera.]

Ehrhart, J. F. 1780-1785. *Phytophylacium ehrhartianum.* 10 decades. Hannover. (vide TL-2, 1: 731.) [Genera.]

Ehrhart, J. F. 1789. Index phytophylacii ehrhartiani. Pp. 145-150 *in:* Ehr, J. F.: *Beiträge zur Botanik,* 4. Hannover & Osnabrück. (TL-2 sub No. **1645**.) [Genera.]

Feuillée, L. 1756-1757, 1766. *Beschreibung zur Arzeney dienlicher Pflanzen, welche in den Reichen des mittägigen America in Peru und Chily vorzüglich im Gebrauch sind.* 2 vol. [and re-issue]. Nürnberg. (TL-2 sub No. **1767**.) [Genera.]

Gandoger, M. 1883-1891. *Flora Europae terrarumque adjacentium.* 27 vol. Paris, London & Berlin. (TL-2 No. **1942**.) [Species.]

Garsault, F. A. P. de, 1764. *Les figures de plantes et animaux d'usage en médecine, décrites dans la matière médicale de Mr. Geoffroy.* 5 vol. Paris. (TL-2 No. **1959**.) [Species and infraspecific taxa.]

Garsault, F. A. P. de, 1764-1767. *Description, vertus et usages de sept cent dix-neuf plantes, tant étrangères que de nos climats.* 5 vol. Paris. (TL-2 No. **1961**.) [Genera, species and infraspecific taxa.]

Garsault, F. A. P. de, 1765. *Explication abrégée de sept cent dix-neuf plantes, tant étrangères que de nos climats.* Paris. (TL-2 No. **1960**.) [Species and infraspecific taxa.]

Gilibert, J. E. 1782. *Flora lituanica inchoata.* 2 vol. Grodno. (TL-2 No. **2012**.) [Species and infraspecific taxa.]

Gilibert, J. E. 1782. *Exercitium botanicum, in schola vilnensi peractum, seu enumeratio methodica plantarum tam indigenarum quam exoticarum quas proprio marte determinaverunt alumni in campis vilniensibus aut in horto botanico universitatis.* Wilnius. (TL-2 No. **2013**.) [Species and infraspecific taxa.]

Gilibert, J. E. 1785-1787. *Caroli Linnaei botanicorum principis systema plantarum Europae.* 7 vol. Vienne. (TL-2 No. **2014**.) [Species and infraspecific taxa.]

Gilibert, J. E. 1792. *Exercitia phytologica, quibus omnes plantae Europae, quas vivas invenit in variis herbationibus, seu in Lithuania, Gallia, Alpibus, analysi nova proponuntur.* 2 vol. Lyon. (TL-2 No. **2015**.) [Species and infraspecific taxa.]

Gleditsch, J. G. 1753. Observation sur la pneumonanthe, nouveau genre de plante, dont le caractère diffère essentiellement de celui de la gentiane. *Hist. Acad. Roy. Sci. (Berlin)* 1751: 158-166. [All ranks.]

Guettard, J. E. 1755. Cinquième [Sixième] mémoire sur les glandes des plantes, et le quatrième [cinquième] sur l'usage que l'on peut faire de ces parties dans l'établissement des genres des plantes. *Hist. Acad. Roy. Sci. Mém. Math. Phys.* (Paris, 4°) 1749: 322-377, 392-443. (TL-2 No. **2208**.) [All ranks.]

Haller, A. von, 1753. *Enumeratio plantarum horti regii et agri gottingensis.* Göttingen. (TL-2 No. **2309**.) [All ranks.]

Heister, L. 1753. *Descriptio novi generis plantae rarissimae et speciosissimae africanae.* Braunschweig. (TL-2 No. **2592**.) [All ranks.]

Hill, J. 1753. [Entries on natural history.] *In:* Scott, G. L. (ed.), *A supplement of Mr. Chambers's Cyclopaedia: or a universal dictionary of the arts and sciences.* 2 vol. London. [All ranks.]

Hill, J. 1753-1754. [Entries on natural history.] *In:* Society of Gentlemen (ed.), *A new and complete dictionary of arts and sciences.* 4 vol. London. [All ranks.]

Hill, J. 1754, etc. *The useful family herbal.* [Including subsequent re-issues and editions]. London. (TL-2 No. **2768**.) [Genera, species and infraspecific taxa.]

Hill, J. 1756-1757 *The British herbal.* London. (TL-2 No. 2769.) [Species and infraspecific taxa.]

Motyka, J. 1995-1996. *Porosty* (Lichenes). *Rodzina* Lecanoraceae. 4 vol. Lublin. [Genera and species.]

Necker, N. J. de, 1790-1791, 1808. *Elementa botanica.* [All issues and editions]. Neuwied, Paris, Strasbourg, Mainz. (TL-2 No. **6670**.) [Genera.]

Rumphius, G. E. 1755. *Herbarii amboinensis auctuarium.* Amsterdam. (TL-2 No. **9785**.) [Genera; this does not affect species names published by Burman in the "Index", pp. 75-94, in that work.]

Secretan, L. 1833. *Mycographie suisse, ou description des champignons qui croissent en Suisse.* 3 vol. Genève. (TL-2 No. **11595**.) [Species and infraspecific taxa.]

Trew, C. J. [and others] [1747-]1753-1773. *Herbarium blackwellianum emendatum et auctum.* 6 vol. Nürnberg. (TL-2 No. **546**.) [Genera.]

APPENDIX VII

GLOSSARY OF TERMS USED AND DEFINED IN THIS CODE

The particular usage of a few other words, not defined in the Code, is also indicated; these are italicized in the list below and are accompanied by editorial explanation of their use.

admixture. [Not defined] – something mixed in, especially a minor ingredient, used of components of a gathering that represent a taxon or taxa other than that intended by the collector, and which do not preclude the gathering, or part thereof, being a type specimen, the admixture being disregarded (Art. 8.2).

alternative family names. The eight family names, regularly formed in accordance with Art. 18.1, allowed as alternatives (Art. 18.6) to the family names of long usage treated as validly published under Art. 18.5.

alternative names. Two or more different names proposed simultaneously for the same taxon by the same author (Art. 34.2).

analysis. See *illustration with analysis*.

anamorph. A mitotic asexual morph in pleomorphic fungi (Art. 59.1).

ascription. The direct association of the name of a person or persons with a new name or description or diagnosis of a taxon (Art. 46.3).

automatic typification. (1) Typification of a nomenclaturally superfluous and illegitimate name by the type of the name which ought to have been adopted under the rules (Art.7.5). (2) Typification of the name of a taxon above the rank of genus by the type of the generic name on which it is based (Art.10.6 and 10.7).

autonym. A generic name or specific epithet repeated without an author citation as the final epithet in the name of a subdivision of a genus or of an infraspecific taxon that includes the type of the adopted, legitimate name of the genus or species, respectively (Art. 22.1 and 26.1).

available. [Not defined] – applied to an epithet in a legitimate (Art. 11.5 and 15.5) or illegitimate (Art. 58.1) name, the type of which falls within the circumscription of the taxon under consideration and where the use of the epithet would not be contrary to the rules (see also ***available name***)].

available name. A name published under the *International Code of Zoological Nomenclature* with a status equivalent to that of a validly published name under the *International Code of Botanical Nomenclature* (Art. 45.4 footnote).

avowed substitute (replacement name, nomen novum). A name proposed as a substitute for a previously published name (Art. 7.3 and 33.4).

basionym. A previously published legitimate name-bringing or epithet-bringing synonym from which a new name is formed for a taxon of different rank or position (Art. 33.4, 49.1 and 52.3).

binary combination. A generic name combined with a specific epithet to form a specific name (Art. 23.1).

binary designation. [Not defined] – an apparent binary combination that has not been validly published (Art. 46 Note 2; see also Art. 6.3).

combinatio nova (comb. nov.). See ***new combination***.

combination. A name of a taxon below the rank of genus, consisting of the name of a genus combined with one or two epithets (Art. 6.7).

compound. A name or epithet which combines elements derived from two or more Greek or Latin words, a regular compound being one in which a noun or adjective in a non-final position appears as a modified stem (Rec. 60G.1) (see also ***pseudocompound***).

confusingly similar names. Orthographically similar names of genera or epithets of names of subdivisions of genera, of species, or of infraspecific taxa likely to be confused (Art. 53.3 and 53.4).

conserved name (nomen conservandum). (1) A name of a family, genus, or species ruled as legitimate and with precedence over other specified names even though it may have been illegitimate when published or lack priority (Art. 14.1–14.7). (2) A name for which its type, orthography, or gender has been fixed by the conservation process (Art. 14.1, 14.9–14.11).

correct name. The name of a taxon with a particular circumscription, position, and rank that must be adopted in accordance with the rules (Art. 6.6, 11.1, 11.3, and 11.4).

cultivar. A special category of plants used in agriculture, forestry, and horticulture defined and regulated in the *International Code of Nomenclature for Cultivated Plants* (Art. 28 Notes 2, 4, and 5).

date of name. The date of valid publication of a name (Art. 45.1).

descriptio generico-specifica. A single description simultaneously validating the names of a genus and its single species (Art. 42.1).

description. [Not defined] – a written statement of a feature or features of a taxon required for valid publication of its name (cf. Art. 32.1 (d) and 32.3).

descriptive name. A name of a taxon above the rank of family not based on a generic name (Art. 16.1).

designation. [Not defined] – the term used for what appears to be a name but that has not been validly published (Art. 23.6, 46 Note 2; see also Art. 6.3).

diagnosis. A statement of that which in the opinion of its author distinguishes the taxon from other taxa (Art. 32.2).

duplicate. Part of a single gathering of a single species or infraspecific taxon made by the same collector(s) at one time (Art. 8.3 footnote).

effective publication. Publication in accordance with Art. 29-31 (Art. 6.1).

epithet. The final word in a binary combination and the word following the connecting term denoting rank in other combinations (Art. 6.7, 11.4, 21.1, 23.1, and 24.1).

epitype. A specimen or illustration selected to serve as an interpretative type when the holotype, lectotype, or previously designated neotype, or all original material associated with a validly published name cannot be identified for the purpose of precise application of the name of a taxon (Art.9.7).

exsiccata. [Not defined] – Latin adjective used as noun, nominative plural "exsiccatae", refers to a set of dried specimens, usually numbered and with printed labels, distributed by sale, gift, or exchange (cf. Art. 30.4 and 30 Note 1)].

ex-type (ex typo) [also ex-holotype (ex holotypo), ex-isotype (ex isotypo)]. A living isolate obtained from the type of a name when this is a culture permanently preserved in a metabolically inactive state (Rec. 8B.2).

final epithet. The last epithet in sequence in any particular combination, whether in the rank of a subdivision of a genus, or of a species, or of an infraspecific taxon (Art. 11.4, footnote).

forma specialis. See *special form*.

form-taxon. In pleomorphic fungi, a taxon typified by an anamorph (Art. 59.3).

fossil taxon. A taxon the name of which is based on a fossil type (Pre. 7 footnote and Art. 13.3).

gathering. [Not defined] – something brought together, used for a collection of one or more specimens made at the same place and time (Art. 8.2).

heterotypic synonym (taxonomic synonym). A synonym based on a type different from that of the accepted name (Art. 14.4); termed a "subjective synonym" in the *International Code of Zoological Nomenclature* and the *Bacteriological Code* (Art. 14.4 footnote).

holomorph. A pleomorphic fungal species in all its morphs (Art. 59.1).

holotype. The one specimen or illustration used by the author or designated by the author as the nomenclatural type (Art. 9.1).

homonym. A name spelled exactly like another name published for a taxon of the same rank based on a different type (Art. 53.1). Note. Names of subdivisions of genera or infraspecific taxa with the same epithet even if of different rank are treated as homonyms disregarding the connecting term (Art. 53.4).

homotypic synonym (nomenclatural synonym). A synonym based on the same type as that of another name in the same rank (Art. 14.4); termed an "objective synonym" in the *International Code of Zoological Nomenclature* and the *Bacteriological Code* (Art. 14.4 footnote).

hybrid formula. An expression consisting of the names of the parent taxa of a hybrid with a multiplication sign placed between them (Art. H.2.1).

illegitimate name. A validly published name that is not in accordance with one or more rules (Art. 6.4), principally those on superfluity (Art. 52) and homonymy (Art. 53 and 54).

illustration with analysis. An illustration with a figure or group of figures, in vascular plants commonly separate from the main illustration, showing details aiding identification (Art. 42.4).

improper Latin termination. A termination of a name or epithet not agreeing with the termination mandated by the Code (Art. 16.3, 18.4, 19.6, and 32.7).

indelible autograph. Handwritten material reproduced by some mechanical or graphic process (such as lithography, offset, or metallic etching) (Art. 30.2).

indirect reference. A clear (if cryptic) indication, by an author citation or in some other way, that a previously and effectively published description or diagnosis applies (Art. 32.6).

informal usage. Usage of rank-denoting terms at more than one non-successive position in the taxonomic sequence. Note: names involved in such usage are validly published but unranked (Art. 33.11).

isonym. The same name based on the same type, published independently at different times by different authors. Note: only the earliest isonym has nomenclatural status (Art. 6 Note 1).

isosyntype. A duplicate of a syntype (Art. 9.10).

isotype. A duplicate specimen of the holotype (Art. 9.3).

later homonym. A homonym published later than another (Art. 53.1).

lectotype. A specimen or illustration designated from the original material as the nomenclatural type if no holotype was indicated at the time of publication, or if it is missing, or if it is found to belong to more than one taxon (Art. 9.2).

legitimate name. A validly published name that is in accordance with all rules (Art. 6.5).

misplaced term. A rank-denoting term used contrary to the relative order specified in the Code (Art. 18.2, 19.2, 33.9, and 33 Note 3).

monotypic genus. A genus for which a single binomial is validly published (Art. 42.2).

morphotaxon. A fossil taxon which, for nomenclatural purposes, comprises only one part, life-history stage, or preservational state represented by the corresponding nomenclatural type (Art. 1.2).

name. A name that has been validly published, whether it is legitimate or illegitimate (Art. 6.3).

neotype. A specimen or illustration selected to serve as nomenclatural type if no original material is extant or as long as it is missing (Art. 9.6).

new combination. A combination formed from a previously published legitimate name and employing the same final epithet (or employing the name itself if formed from a generic name) (Art. 7.4).

new name. A newly published name. Note: this name may be the name of a new taxon, a new combination, a name at a new rank (status novus), or an avowed substitute (nomen novum) for an existing name (Art. 7.3, 7.4, 9 Note 1, Rec. 45A.1).

nomen conservandum (nom. cons.). See *conserved name*.

nomen novum (nom. nov.). See *avowed substitute*.

nomen nudum (nom. nud.). A name of a new taxon published without a description or diagnosis or reference to a description or diagnosis (Rec. 50B.1).

nomen rejiciendum (nom. rej.). A name rejected in favour of a name conserved under Art. 14 or a name ruled as rejected under Art. 56 (see also "*rejected name*") (App. II, III, IV, and V).

nomen utique rejiciendum. A name ruled as rejected under Art. 56. Note: it and all combinations based on it are not to be used (see App. V).

nomenclatural novelties. New names and descriptions or diagnoses of new taxa (Rec. 30A).

nomenclatural synonym. See *homotypic synonym*.

nomenclatural type. The element to which the name of a taxon is permanently attached (Art. 7.2).

non-fossil taxon. A taxon the name of which is based on a non-fossil type (Pre. 7 footnote and Art. 13.3).

nothogenus. A hybrid genus (Art. 3.2).

nothomorph. A rank-denoting term formerly used for a subordinate taxon within a nothospecies. Names published as nothomorphs are now treated as names of varieties (Art. H.12.2 and footnote).

nothospecies. A hybrid species (Art. 3.2).

nothotaxon. A hybrid taxon (Art. 3.2 and H.3.1).

objective synonym. See *homotypic synonym*.

opera utique oppressa. Works, ruled as suppressed, in which names in specified ranks are not validly published (Art. 32.9 and App. VI).

original material. Specimens and illustrations indicated in the protologue of a name (see Art. 9 Note 2 for details).

original spelling. The spelling employed when a name was validly published (Art. 60.2).

orthographic variants. Various spelling, compounding, and inflectional forms of a name or its epithet, only one nomenclatural type being involved (Art. 61.2).

page reference. Citation of the page or pages on which the basionym or replaced synonym was validly published (Art. 33 Note 1).

paratype. A specimen cited in the protologue that is neither the holotype nor an isotype, nor one of the syntypes if two or more specimens were simultaneously designated as types (Art. 9.5).

plant. Any organism traditionally studied by botanists (Pre. 1 footnote and Pre.7).

position. [Not defined] – used to denote the placement of a taxon relative to other taxa in a classification, regardless of rank (Prin. IV, Art. 6.6 and 11.1).

priority. A right to precedence established by the date of valid publication of a legitimate name (Art. 11) or of an illegitimate earlier homonym (Art. 45.3), or by the date of designation of a type (Art. 7.10, 7.11).

protologue. Everything associated with a name at its valid publication, i.e. description or diagnosis, illustrations, references, synonymy, geographical data, citation of specimens, discussion, and comments (Rec. 8A footnote).

provisional name. A name proposed in anticipation of the future acceptance of the taxon concerned, or of a particular circumscription, position, or rank of the taxon (Art. 34.1).

pseudocompound. A name or epithet which combines elements derived from two or more Greek or Latin words and in which a noun or adjective in a non-final position appears as a word with a case ending, not as a modified stem (Rec.60G.1(b)) (see also *compound*).

rank. [Not defined] – used for the relative position of a taxon in the taxonomic hierarchy (Art.2.1).

rejected name. A name the use of which is prohibited, either by formal action under Art. 14 or 56 overriding other provisions of the Code (see ***nomen rejiciendum*** and ***nomen utique rejiciendum***) or because it was nomenclaturally superfluous when published (Art. 52) or a later homonym (Art. 53 and 54).

replaced synonym. The name replaced by an avowed substitute (nomen novum, replacement name) (Art. 33.4).

replacement name. See *avowed substitute*.

sanctioned name. The name of a fungus treated as if conserved against earlier homonyms and competing synonyms, through acceptance in one of two sanctioning works (Art. 15).

special form (forma specialis). A taxon of parasites, especially fungi, characterized from a physiological standpoint but scarcely or not at all from a morphological standpoint, the nomenclature of which is not governed by this Code (Art. 4 Note 4).

specimen. A gathering, or part of a gathering, of a single species or infraspecific taxon made at one time, disregarding admixtures (Art. 8.2).

status. (1) Nomenclatural standing with regard to effective publication, valid publication, legitimacy, and correctness (Art. 6 and 12.1). (2) Rank of a taxon within the taxonomic hierarchy (see *status novus*) (Art. 7.4).

status novus (stat. nov.). Assignment of a taxon to a different rank within the taxonomic hierarchy, e.g. when an infraspecific taxon is raised to the rank of species or the inverse change occurs (Art. 7.4, Rec. 21B.4 and 24B.2).

subdivision of a family. Any taxon of a rank between family and genus (Art. 4 Note 1).

subdivision of a genus. Any taxon of a rank between genus and species (Art. 4 Note 1).

subjective synonym. See *heterotypic synonym*.

superfluous name. A name applied to a taxon circumscribed by the author to definitely include the type of a name which ought to have been adopted, or of which the epithet ought to have been adopted under the rules (Art. 52.1).

synonym. A name considered to apply to the same taxon as the accepted name (Art. 7.2).

syntype. Any specimen cited in the protologue when no holotype was designated, or any of two or more specimens simultaneously designated as types (Art. 9.4).

tautonym. A binary combination in which the specific epithet exactly repeats the generic name (Art. 23.4).

taxon (taxa). A taxonomic group at any rank (Art. 1.1).

taxonomic synonym. See *heterotypic synonym*.

teleomorph. Meiotic sexual morph in pleomorphic fungi (Art. 59.1).

type. See *nomenclatural type*.

validate. [Not defined] – to make valid, used in the context of valid publication of a name, either with reference to an existing designation (e.g. Art. 42 Ex. 1), or in describing the method by which this is effected (e.g. Rec. 32A.1).

validly published name. A name effectively published and in accordance with Art. 32-45 or H.9 (Art. 6.2).

voted example. An Example mandated by a Congress to be inserted in the Code in order to legislate nomenclatural practice when the corresponding Article is open to divergent interpretation or does not adequately cover the matter (as contrasted with other Examples provided by the Editorial Committee) (Art. 7 Ex. 10 footnote).

INDEX TO APPENDICES II-V

As in the St Louis Code, references are to the printed page, not to the header or entry number as in earlier editions and the index covers all appendices of nomina conservanda and rejicienda.

Rejected names, and their equally rejected synonyms, are printed in *italics.* Nomina utique rejicienda (App. V), and combinations based on them, are in addition denoted by an asterisk (*). All other names, including those that are conserved, appear in roman type.

Page numbers of main entries of conserved names and outright rejected names are in **bold-face type.**

Rejected homonyms (or parahomonyms) of conserved names are not indexed separately. Orthographical variants are not indexed, nor are the names that appear in the type paragraphs.

513

515

INDEX OF SCIENTIFIC NAMES

This index includes the names appearing in the text of the *Code* and in Appendix I; for names in Appendices II-V, see the preceding index. As with the Subject Index (p. 544) the references are not to pages but to the Articles, Recommendations, etc. of the *Code*.

543

SUBJECT INDEX

The references in this index are not to pages but to the Articles, Recommendations, etc. of the *Code,* as follows: Div. = Division; Pre. = Preamble; Prin. = Principles; arabic numerals = Articles or, when followed by a letter, Recommendations; Ex. = Examples; N. = Notes; fn. = footnotes; H. = App. I (hybrids); App. = other Appendices.

For ease of reference, a few sub-indices have been included under the following headings: Abbreviations, Definitions, Epithets, Publications, Transcriptions (and related subjects), and Word elements.

Scientific (Latin) names appearing in the body of the *Code* plus Appendix I are not included in this Subject index, but in the preceding "Index of scientific names"; those in Appendices II-V appear in an index directly following the appendices.

557

Subject Index